MW00387350

FORWARD/COMMENTARY

The **OPFOR Worldwide Equipment Guide (WEG)** is published in three volumes, (Ground; Airspace & Air Defense Systems; and Naval & Littoral Systems) the WEG is the approved document for OPFOR equipment data used in U.S. Army training. The equipment portrayed in the WEG represents military systems, variants, and upgrades that US forces may encounter now and in the foreseeable future. The authors continually analyze real world developments, capabilities, and trends to guarantee the OPFOR remains relevant.

This book is Volume 2. In order to keep under the 630-page limit from the printer, pages 612 to 724 of Volume 1 were moved to the front of Volume 2. Volume 2 contains selected weapons systems and equipment that are included in the categories of Air and Air Defense Systems, Tier Tables – Fixed Wing, Rotary Wing, UAVs, Air Defense, Aviation Countermeasures, Upgrades, Emerging Technology, Unconventional and SPF Aerial Systems
Theatre Missiles, and Air Defense Systems.

The importance of these manuals cannot be overstated due to the proliferation of weapons through sales and resale, wartime capture, and licensed or unlicensed production of major end items, distinctions between equipment as friendly or OPFOR have blurred. Sales of upgrade equipment and kits for application to weapon systems have further blurred distinctions between old or obsolete systems and modern systems. The OPFOR Worldwide Equipment Guide describes base models listed in the FMs or upgrades of those base models, which reflect current capabilities. Many of the less common variants and upgrades are also addressed.

Why buy a book you can download for free? We print this so you don't have to.

Some are available only in electronic media. Some online docs are missing pages or barely legible.

We at 4th Watch Publishing are former government employees, so we know how government employees actually use the standards. When a new standard is released, an engineer prints it out, punches holes and puts it in a 3-ring binder. While this is not a big deal for a 5 or 10-page document, many NIST documents are over 100 pages and printing a large document is a time consuming effort. So, an engineer that's paid $75 an hour is spending hours simply printing out the tools needed to do the job. That's time that could be better spent doing engineering. We publish these documents so engineers can focus on what they were hired to do – engineering. It's much more cost-effective to just order the latest version from Amazon.com. SDVOSB

If there is a standard you would like published, let us know. Our web site is www.usgovpub.com

Many of our titles are available as ePubs for Kindle, iPad, Nook, remarkable, BOOX, and Sony ereaders. Please visit our web site to see our recommendations.

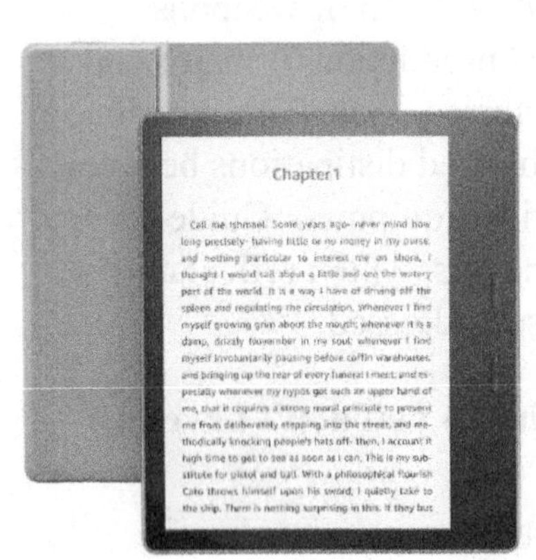

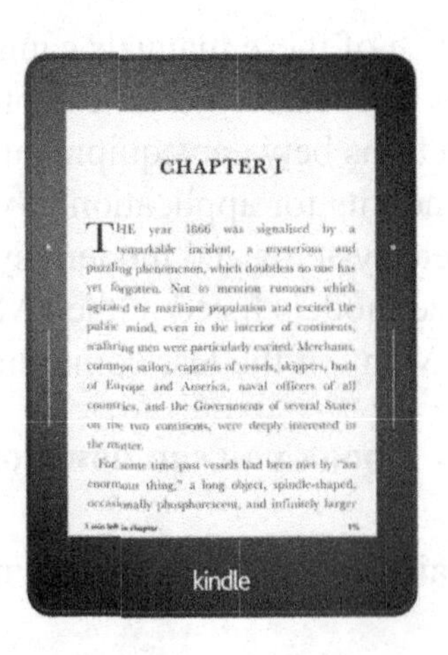

Why buy an eBook when you can access data on a website for free? HYPERLINKS

Yes, many books are available as a PDF, but not all PDFs are bookmarked? Do you really want to search a 6,500-page PDF document manually? Load our copy onto your Kindle, PC, iPad, Android Tablet, Nook, or iPhone (download the FREE kindle App from the APP Store) and you have an easily searchable copy. Most devices will allow you to easily navigate an ePub to any Chapter. Note that there is a distinction between a Table of Contents and "Page Navigation". Page Navigation refers to a different sort of Table of Contents. Not one appearing as a page in the book, but one that shows up on the device itself when the reader accesses the navigation feature. Readers can click on a navigation link to jump to a Chapter or Subchapter. Once there, most devices allow you to "pinch and zoom" in or out to easily read the text. (Unfortunately, downloading the free sample file at Amazon.com does not include this feature. You have to buy a copy to get that functionality, but as inexpensive as eBooks are, it's worth it.) Kindle allows you to do word search and Page Flip (temporary place holder takes you back when you want to go back and check something). Visit **www.usgovpub.com** to learn more.

Pages 612 to 724 of Volume 1 (Ground Systems) are inserted before Volume 2

Volume 2 starts after page 110

INFOWAR

INFOWAR is defined as specifically planned and integrated actions taken to achieve an information advantage at critical points and times. The goal is to influence an enemy's decision-making through his collected and available information, information systems, and information-based processes, while retaining the ability to employ friendly information, information-based processes, and systems to control the use of the electromagnetic spectrum at critical locations and times in the battle space or to attack the enemy.

All INFOWAR elements are mutually supporting. INFOWAR occurs through the combinations of seven elements:

- Electronic warfare (EW).
- Deception.
- Physical destruction.
- Protection and security measures.
- Perception management.
- Information attack (IA).
- Computer warfare.

EW capabilities allow foreign forces to exploit, deceive, degrade, disrupt, damage, or destroy sensors, processors, communications, and command and control (C^2) nodes. Information supremacy, delay, and denial, or distortions of the adversary's use of the electromagnetic spectrum and information infrastructure are the objectives. Electronic warfare (EW) is a perfect example of the integrated nature of foreign forces INFOWAR elements. The EW section in this chapter provides basic characteristics of selected systems either in use or readily available to the foreign forces.

Computer warfare includes capabilities that allow the foreign forces to conduct network warfare (NETWAR) to attack and exploit information systems by attacking key information technology systems within cyberspace, and to conduct network operations (NETOPS) to establish and protect C4ISR networks and information. In NETWAR the foreign forces obtains access through social engineering processes such as phishing schemes but can also employ complex technologies to intercept communications through man in the middle (MITM) attacks and hack into a system remotely. Once access is obtained, foreign forces will attempt to degrade the system or to exploit the system to collect intelligence. Some forms of attack can be launched unexpectedly through a data driven attack known as a "drive-by- download" that is embedded into a website or uploaded to a commonly used system, the victim in these cases releases malware by attempting to access what is believed to be a legitimate site or program. There are various methods used to accomplish this for the purpose of releasing viruses and other malware designed to give the attacker control over the victim's computer. The methods used to gain access involve highly developed social engineering techniques and or network attacks such as a spoofed email account or a false website.

Because these types of attacks are launched by the end user they are a concern since the attack is able to legitimately pass through various electronic defense systems such as firewalls. If access is obtained an electronic beachhead is established on the infected computer exposing the rest of the network to further attacks such as distributed denial of service attacks that overwhelm internet servers. Attacks can include vandalizing or sabotaging a website, downloading sensitive information or degrading a key war fighting function through the release of malware.

Cyber espionage describes those INFOWAR actions that involve collecting sensitive and proprietary information such as plans, capabilities or personal data. The threat will exploit poor information technology security that exposes sensitive data to risk of exploitation or manipulation.

Information attack is a type of action that focuses on the intentional disruption or distortion of information in a manner that supports accomplishment of the mission. Unlike computer warfare attacks that target the information systems, information attacks target the information itself. Attacks on the commercial Internet by civilian hackers have demonstrated the vulnerability of cyber and information systems to innovative and flexible penetration, disruption, or distortion techniques.

Computer warfare consists of attacks that focus specifically on the computer systems, networks, and/or nodes. This includes a wide variety of activities, ranging from unauthorized access (hacking) of information systems for intelligence-collection purposes, to the insertion of malicious software (viruses, worms, logic bombs, or Trojan horses). Such attacks concentrate on the denial, disruption, or manipulation of the infrastructure's integrity. Terrorist organizations use a variety of encryption techniques such as embedding communications into innocuous computer applications in order to transmit data in a surreptitious manner. Other methods are the use of code words to conceal the meaning of topics and swapping Subscriber Identity Module (SIM) cards or cell phones to prevent electronic surveillance systems from identifying the user of a particular phone.

Evolving mobile technology has increased the portability of battlefield automated systems improving the integration and capabilities of many military functions. The proliferation of these capabilities increases the military's dependence on mobile devices and the networks that support them. Referred to as the "edge of the network," mobile systems provide improved situational awareness. However, the complexity associated with these enhancements in functionality, has introduced additional vulnerabilities. Vulnerabilities associated with wireless networking, and the need to support an expanding list of military applications, make mobile technology an important, feasible and valuable target for INFOWAR operations. Mobile devices share many of the vulnerabilities of personal computers. However, the attributes that make mobile phones easy to carry, use, and modify as well as comparatively low security standards open them to a range of attacks.

This selection of systems is not intended to be complete; rather, it is representative of the types and capabilities that are currently fielded or available. Later WEG updates will include equipment for other elements of INFOWAR operations. For more information on the INFOWAR tactics techniques and procedures see chapter 7 of TC 7-100.2 OPFOR Tactics.

Table 5: Tactical Electronic Warfare Systems

System	Country of Origin	Description	Frequency Range (MHz)	Vs. GPS?	Vs. FH?	Range (km)	Power Output
MEERKAT-S	UK	ESM/ELINT	2-40,000	No	Yes	500km	Passive
WEASEL 2000	UK	ESM/ELINT	.5-10,000	-	Yes	500km	Passive
EULe	Germany	ESM/ELINT	.9-3,000	-	Yes	450km	Passive
MCS90 TAMARA	Czech Rep	ESM/ELINT	820-3,000	-	No	450km	Passive
R-703 /709	Russia	ESM/ELINT	1.5-2,000	Yes	n/a	Unk	Unk
CICADA-C	Germany	Mounted ESM/ECM HF/VHF/UHF	.525-3,000	Yes	Yes	100km	10kw
TRC274	France	Mounted ESM/ECM HF/VHF/UHF	20-3,000	Yes	Yes	150km	4kW 1.2kW on the move
GSY1800	S. Africa	Mounted ESM/ECM HF/VHF/UHF	1-3000	Yes	Yes	100km	(ECM: 1kW)
PELENA-6	Russia	Mounted ECM HF/VHF/UHF	20-1,000	No	No	60 km	60W
R-330 ZH	Russia	Mounted ECM HF/VHF	100-2000	Yes	No	60km	1kW
CICADA-R	Germany	Mounted ESM/ECM HF/VHF/UHF	6,000-18,000	No	Yes	100km	1kW
LIMAN P2	Ukraine	Mounted ECM VHF/UHF	225-1,215*	Yes	Yes	100km	Unk
R-934B	Russia	Mounted ECM VHF/UHF	100-400	No	No	50km	500W
BOQ-X300	Sweden	Mounted ECM S/C/X/Ku/K	2-40,000	Yes	n/a	Unk	Unk
CBJ-40 BOME	France	Mounted ECM S/C/X/Ku	2-20,000	Yes	n/a	Unk	Unk
PELENA-1	Russia	Mounted ECM S/C	1,000-4,000	Yes	n/a	250 km	Unk
SPN-2/4	Russia	Mounted ECM X/K	6,000-17,500	No	No	130km	(ECM: 1kW)

SGS2000	Germany	Mounted ESM/ECM HF/VHF/UHF	1.5-1000	No**	Yes	100km	(ECM: 1kW)
System	**Country of Origin**	**Description**	**Frequency Range (MHz)**	**Vs. GPS?**	**Vs. FH?**	**Range (km)**	**Power Output**
JN-1102	China	Mounted ECM VHF (UAV Mounted)	20-500	No	Unk	Unk	Unk
BARRAGE	France	Mounted ECM VHF (UAV Mounted)	1-3000	Yes	No	Unk	Unk
AJ-045A	Bulgaria	Mounted ECM VHF (UAV Mounted)	20-100	No	No	10km	Unk
HUMMEL	Germany	Mounted ESM/ECM VHF	20-80	No	Yes	100km	(ECM: 1kW)
STORM-H	France	Manpack ECM HF/VHF/UHF/ SHF	20-470	No	No	5km	1kW (est.)
EL/K 7029/A/B	Israel	Mounted ESM /VHF /UHF/ESM	116-400	-	Yes	100km	Passive
ORION	Russia	ELINT	200-18,000	-	No	400km	Passive
AVTOBAZA-M	Russia	ELINT	200-18,000	-	No	400km	Passive

*Liman does not DF in the 960 to 1,215 MHz range

**SGS 2000 frequency range can be extended up to 3000MHz

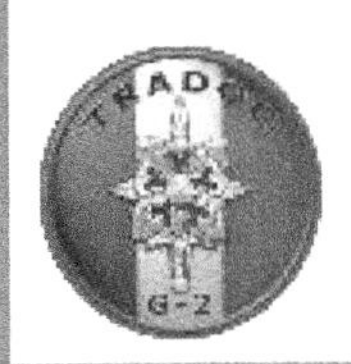

Russian Airborne EA Radar Jammer 1L245

SYSTEM	SPECIFICATIONS	PERFORMANCE	SPECIFICATIONS
Alternative Designations:	Vstrecha, Ground-Based Weapons Control Radar Suppression System	Fording Depths (m):	Amphibious
Date of Introduction:	1979	Communications:	Combat Net Radios Sets
Proliferation:	At least 2 countries	Protection:	Against 5.56 ball 5.56, all-around
		Transmitter:	
Description:		Frequency Range (GHz):	8 - 18
Crew:	3	Output Power (W):	1,000
Platform (chassis):	MT-LBu	Simple pulse signals received and analyzed 0.25 – 10 khz (µs):	1 - 5
Combat Weight (mt):	15.7	Pulse linear-frequency modulated signals 1 – 20 MHz (µs):	1 - 25
Chassis Length Overall (m):	7.26	Polarization of signals received:	vertical and horizontal
		Vertical:	YES
		Horizontal:	YES

Height Overall (m):	3	Sector, deg:	Combat operation (azimuth x elevation) 120 x 15 ECM with respect to the main radar 2 x 2 directional lobe of the reconnaissance strike system
Width Overall (m):	2.85	Radar detection and suppression range:	
		Reconnaissance strike system (km):	80 - 200
Automotive Performance:		Tactical aircraft (km):	30 - 100
Engine Type:	YaMZ-238, 240hp diesel	Polarization of signals transmitted:	chaotic
Cruising Range (km):	500 km	Types of jamming:	
Speed (km/h):		Noise:	Yes
Max Road:	60	Spot Programmed:	Yes
Max Off-Road:	26	Range Programmed:	Yes
Cross-Country:	INA	Interval Programmed:	Yes
Max Swim:	4.5		

NOTES

THE 1L245 IS MAINLY USED TO DEFEAT AIRBORNE RADAR SYSTEMS AND JSTARS. HOWEVER, IT HAS THE CAPABILITY TO AND JAM GROUND SYSTEMS AS WELL

Russian Airborne EA Jammer Infauna

SYSTEM	SPECIFICATIONS	PERFORMANCE	SPECIFICATIONS
Alternative Designations:	None	Engine Type:	INA
Date of Introduction:	2012	Cruising Range (km):	600
Proliferation:	Russian Airborne Forces (VDV)	Speed (km/h):	
Description:	Modified BTR 80 chassis identified as the K1Sh1.	Max Road:	85
Height (m):	2.21	Max Off-Road:	60
Weight (kg):	12,000	Average Cross-Country:	40
Length (m):	7.65	Max Swim:	10
Crew:	2	Fording Depths (m):	Amphibious
Troop Capacity:	8	EW/JAMMING SYSTEM:	
Combat Weight (mt):	14	Frequency Jamming Range (MHz):	20 - 2020
Chassis Length Overall (m):	7.55	Output Power (W):	60
Height Overall (m):	2.41	Voltage Supply (V):	12.6
Width Overall (m):	2.95	Weight of System (Separated from vehicle) (kg):	14
Ground Pressure (kg/cm2):	INA	Deployment:	Protects combat vehicles and troops against radio controlled mines and explosives (IEDs).
Drive Formula:	8 x 8	APS Capability	Softkill aerosol interference against high-precision weapons with laser and video-management systems.

NOTES

FIELDED TO: EW BATTALIONS IN AIRBORNE FORCES OF THE RUSSIAN ARMY. AS PART OF THE AIRBORNE FORCES, THIS SYSTEM WILL BE USED TO ACCOMPANY INITIAL ENTRY TROOPS AND ASSIST WITH SECURING KEY TERRAIN TO ENABLE FOLLOW-ON FORCES (MOTORIZED, MECHANIZED, AND ARMOR) TO DEPLOY INTO COMBAT AREAS.

Russia Mobile EW System LEER-2

SYSTEM	SPECIFICATIONS	PERFORMANCE	SPECIFICATIONS
Alternative Designations:	Tigr-M REI PP	Crew:	6
Date of Introduction:	2012	Engine MZ-5347-10 multi-fuel diesel engine (hp):	215
Proliferation:	Russia	Max Speed (kph):	125
Jamming Range (MHz):	20-1000	Protection:	7.62
Automotive Range (km):	400		

NOTES

E TIGR-M MKTK REI PP MOBILE TECHNICAL CONTROL, ELECTRONIC EMULATION AND ELECTRONIC COUNTERMEASURES SYSTEM IS DESIGNED FOR DEVELOPING RADIO EMITTERS, JAMMING AND SUPPRESSING RADIO-ELECTRONIC MEANS INCLUDING CELLULAR PHONE SYSTEMS.

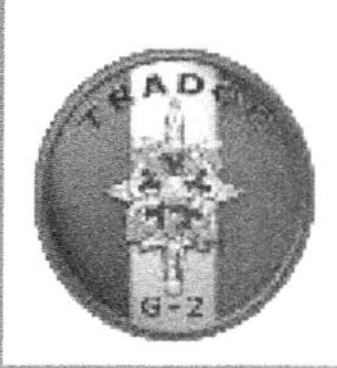

Russian GNSS Jamming Transmitter Aviaconversia

SYSTEM	SPECIFICATIONS	PERFORMANCE	SPECIFICATIONS
Alternative Designations:	GPS / GLONASS Jammer	Fixed:	Yes
Date of Introduction:	1990s	Manportable:	Yes
Proliferation:	Iran, Iraq, North Korea	Vehicle:	Yes
Description:	Aviaconversia portable GPS and GLONASS jammer	Airborne:	Yes
SPECIFICATIONS		Antenna and Transmitter:	
Jamming Range (km):	150 - 200	Antenna Type:	Omni directional or directional yagi
Power Supply DC (V):	15	Transmitter:	Continuous-wave with cable link to antenna
Configuration Weight (kg):	8 – 12	Signal:	Coherent pulse Doppler, J-band
Frequency range:		Power (kw):	
Civilian Channel (MHz):	1,577	Consumed (W):	25
Military Channel (MHz):	1,230	Emitted (W):	4
Target Systems:		Width of radiation pattern lobe (°):	360
Glonass:	Yes	Operation:	Designed to affect C2, maneuver and fire support activities. May affect high precision munitions.
GPS:	Yes		
Mount:			

NOTES

CONTINUOUSLY JAMS GPS SIGNALS. 6 MHZ FREQUENCY DEVIATION TO COUNTER EP MEASURES. ADJUSTABLE MODULATION FROM 44 TO 270 HZ TO COUNTER VARIOUS GNSS SYSTEMS.

Belarusian GNSS Distributed Jamming Complex Optima-3

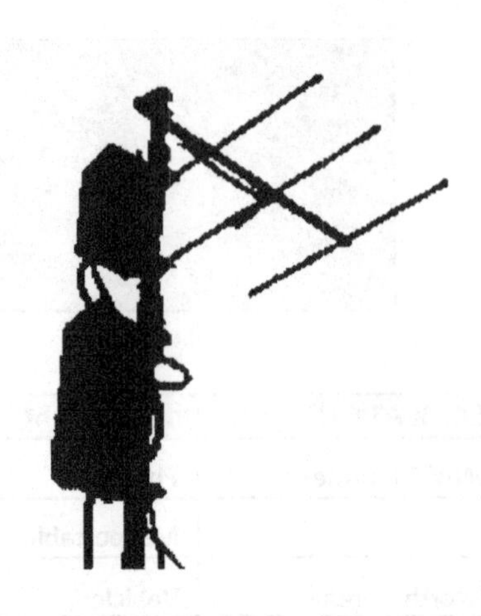

SYSTEM	SPECIFICATIONS	PERFORMANCE	SPECIFICATIONS
Alternative Designations:	GPS / GLONASS Jammer	Width of radiation pattern lobe (°):	
Date of Introduction:	2003	Horizontal (°):	60 - -10
Proliferation:	Fielded in one country and marketed for sale in at least 2 others	Vertical (°):	40 - -10
Jammer Range (km):	100	Components:	The main components are the operator's automated workstation and a control unit for each transmitter. .
Power Supply AC/DC:	220/24	Number of Transmitters:	9
Weight Transmitter (kg):	10	Control System:	ASU-PP automated control system. System status and situational awareness is monitored by the operator at the remote workstation
Frequency range:		Command Links:	
Civilian (MHz):	1,575.42	Command VHF:	Yes
Military (MHz):	1,227.6	Command GSM:	Yes
Antenna and Transmitter:		Control Link Range (km):	30-50
Mount:		Response time:	
Fixed:	Yes	Command VHF (seconds):	5
Vehicle:	Yes	Command GSM (seconds):	≤ 60

Antenna Type:	Omni directional	Operation:	Designed to affect C2, maneuver and fire support activities. May affect high precision munitions.
Transmitter:	Continuous-wave with cable link to antenna		
Signal:	Complex frequency and interval waveform degrade the GNSS code.		
Output Power (W):	20		

NOTES

OPTIMA-3 GNSS JAMMING COMPLEX IS A NETWORK OF JAMMING TRANSMITTERS THAT CAN BE CONTROLLED BY A CENTRAL COMMAND STATION BY EITHER GSM CELL PHONE OR ULTRA-SHORT WAVE RADIO.

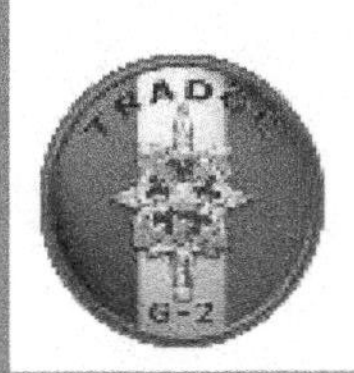

Germany Electronic Support / Attack Jamming Transmitter Cicada-C and R

SYSTEM	SPECIFICATIONS	PERFORMANCE	SPECIFICATIONS
Alternative Designations:	Hummel	FM Morse Code:	YES
Date of Introduction:	1990s	AM Telegraph	YES
Proliferation:	Holland, Spain, Australia	AM Voice:	YES
Description:	Threat multi-range jammer.	AM Morse Code:	YES
Vehicle Range (KM):	150-200	AM Facsimile:	YES
Power Supply (AC/DC):	6 / 15	Fast scan speed (MHz/s):	250 / 1000 (upgraded variant)
Weight (Mt.):	8 – 12	Passive Antennas:	
Frequency Range (MHz):	0.525 - 3,000		dipole or monopole antennas
JAMMING TRANSMITTER:			vertical or horizontal polarized omnidirectional antennas
Transmit modulation:	deception and burn through		antennas for operation on the move
FM Voice:	YES	AMPLIFIER:	
FM Telegraph / Teletype :	YES	Amplifier types:	broadband, solid-state, liquid cooled
FMFacsimile:	YES	Max output power kW into 50 Ω	10
AM Telegraph:	YES	Transmitter:	Continuous-wave with cable link to antenna

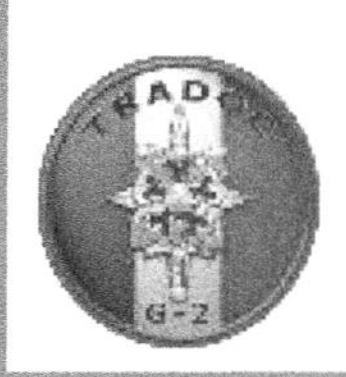

AM Voice:	YES	Amplifier Power Consumption(kW):	25.4
Jamming operation modes:		Width of radiation pattern lobe (°):	360
PRESET FREQ:	YES	OPERATION:	
BROAD JAM:	YES	Frequency Range (MHz):	1.5 - 3,000
ATTACK JAM:	YES	Automatic computer controlled jamming sequences	YES
MANUAL JAM:	YES	High power Amplifier MAX (Kw):	YES
DECEPTION:	YES	"Look Through" Capability ensures jammer is only active when target signal is on the air	YES
Broad jam bandwidth:	up to 240 MHz, in 0.1 MHz steps, simultaneous generation of up to 16 separate broadband segments	Deception jamming	YES
Jamming efficiency:	up to 16 simultaneously active radio nets	Broadband TDM (barrage) jamming against simultaneous spread spectrum transmitters	YES
ANTENNA AND TRANSMITTER:		Programming protected frequencies:	YES
Mount:	Fixed site, vehicle and airborne platforms available.	Local or remote control	YES
Antenna Type:	Omni directional or directional yagi	Environmental conditions:	
RECEIVER:		Operation (°C):	-25 - + 55
Receive modes:		Storage (°C):	-40 - +70
FM Voice:	YES	VARIANTS:	
FM Telegraph / Teletype:	YES	Cicada-R Radar Jammer	
FM Facsimile:	YES	Frequency range (GHz):	6-18

NOTES

CAN BE MOUNTED ON AN ARMORED SIX-WHEEL APC, TRACKED VEHICLE, OR IN A FIXED SITE. ABLE TO REMOTELY OPERATE WITH EITHER RADIO OR WIRED LINKS. ADDITIONAL JAMMING MODES AGAINST NEW THREATS (E.G. FREQUENCY HOPPERS, MOBILE TELEPHONES, SATELLITE NAVIGATION). COMPACT DESIGN FOR HIGHLY MOBILE APPLICATIONS.

Germany Ground Based SIGINT Light Electronic Support System (EULe)

Source: Panzerbaer

SYSTEM	SPECIFICATIONS	PERFORMANCE	SPECIFICATIONS
Alternative Designations:	Owl (English), (TMS) 210	Fast scan speed (GHz/s):	1.3
Date of Introduction:	2000's	Scan Mode:	Frequency, memory and radio frequency panorama
Proliferation:	Germany.	Target Data:	Identifies and locates LPI, LPD emissions Frequency Hop, and Spread Spectrum
Description:	Transportable Monitoring System	Frequency identification accuracy:	± 30MHz
Vehicle Range (km) est.:	150-200	Accuracy of DF, degrees:	2
Engine:	Mercedes Benz 306 hp	ANTENNAS:	
Drive:	4x4	Passive Antennas Frequency Range:	
Length (m):	6.6	ADD 195(MHz):	20-1300
Width (m):	2.39	ADD 170 UHF DF antenna for GSM (MHz):	800-2,000
Crew:	2	HE 500 (MHz):	20-3,000
Weight:	14 t	antennas for operation on the move	
Max Speed (km/h):	80	STORAGE and C2 INTERFACE:	
RECEIVER:		Storage (Tbyte):	1
Designation:	ESMB	Remote Data Rates:	

Power Supply AC/DC:	11/ 32	LAN 500 m, (copper pair) (kbps)	128
Power Supply Weight (kg):	36.8	Serial 300m, (kbps)	20
Operational Frequency Ranges (MHz):	.9 – 3,000	GSM Link, (kbps)	10
DF and Intercept (MHz):	20 - 1300	OPERATION:	
Intercept only (MHz):	≥1300		
Receive modes:		Software Defined (RAMON) Receiver	
FM Voice:	YES	Uses two or more station for DF	
FM Telegraph / Teletype:	YES	Local or remote control	
FM Facsimile:	YES	Operations Time (hours):	12
FM Morse Code:	YES	Environmental conditions:	
AM Telegraph	YES	Operation:	40 °C to +65 °C
AM Voice:	YES	Storage:	-40 °C to +65 °C
AM Morse Code:	YES	VARIANTS:	
AM Facsimile:	YES	ESMD	
GSM:	YES	Receiver Frequency Range (MHz):	.9 – 26,500

NOTES

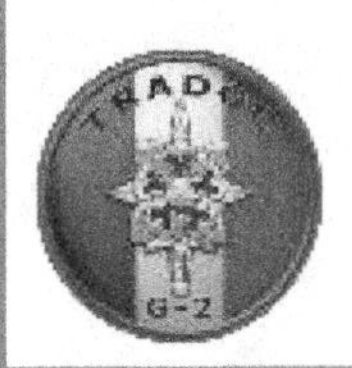

Russian Ground Based ES, ELINT System Avtobaza

SYSTEM	SPECIFICATIONS	PERFORMANCE	SPECIFICATIONS
Alternative Designations:	1L222	Fast scan speed Targets Per Second:	15
Date of Introduction:	1980s	Frequency identification accuracy:	
Proliferation:	At least 4 countries	Accuracy of DF:	
Description:	Passive ELINT signals intercept system designed to intercept and locate pulsed airborne radars including fire control radars, terrain following radars and ground mapping radars as well as weapon (missile) data links.	Azimuth (degrees):	.3 - .5
Vehicle Range:	INA	Elevation (degrees):	3
Engine:	INA	ANTENNAS:	
Drive:	INA	Rotating Parabolic:	Yes
Length:	INA	Azimuth:	360
Width:	INA	Rotation Orbits per minute:	6 - 12
Crew:	4	Local or remote control	Yes
Weight (mt):	13.3	Range of Remote Operations (m):	100
Max Speed:	INA	LAN copper twisted pair:	INA
		Serial 300m (kbps):	INA

RECEIVER:		Radio Link (kbps):	INA
Designation:	INA	OPERATION:	
Power Supply:	6V or 15V DC	Prioritization Targets:	Yes
Operational Frequency Range (GHz):	8 – 17.5	Number of Targets Monitored:	60
DF and Intercept:	Yes	Local or remote control	Yes
Receive modes:		Set Up Time (minutes):	25
SLAR:	Yes	Environmental conditions:	
PGM Targeting Radar:	Yes	Operation Temperature:	-45 - 40
Nap of the Earth (NOE) Radar:	Yes	Humidity (%):	98
Early Warning Radar:	Yes	VARIANTS:	Avtobaza-M
SAT Phones:	Yes	Receiver Range (km):	400
		Frequency range (GHz):	2 - 18

NOTES
MAY HAVE BEEN MODIFIED TO RECEIVE AND LOCATE EMISSIONS ASSOCIATED WITH SATELLITE TELEPHONES. REPORTEDLY OPERATED IN SYRIA IN 2011 -2012.

United Kingdom Ground Based ES, ELINT System Weasel 2000

SYSTEM	SPECIFICATIONS	PERFORMANCE	SPECIFICATIONS
Alternative Designations:	None	Operation:	
Date of Introduction:	2005	Coverage sector deg.:	0 - 180
Proliferation:	At least 2 countries	Travel mode:	
Description:	Passive ELINT System	Fixed:	Yes
Crew:	2	Mobile:	No
Platform:	Mercedes Ax or 4 ton	Remote operation:	
		Number of Sensors:	≥ 1
Combat Weight (Tons):	12.6	Remote Range wired(m):	85
Antenna and Receiver:		Direction Finding::	
Mount:	Motor-driven	Library::	Yes
Antenna Type:	Omni directional	Other links:	
	Rotating Dish	PERFORMANCE	
Azimuth coverage º:	360	Surveillance range (km):	Passive system
Direction Finding assembly:	Dual 8 port switched amplitude comparison system	Variants:	
Frequency Range (GHz):	0.4 - 10		TAC Weasel
Bearing Accuracy (DF) 1 Deg. RMS (GHz):	3 - 10		Weasel II
Polarization Range (GHz):	2 - 18		Weasel III
Receiver sensitivity (dBmi):	-62		
Range (dB):	60		

NOTES
AT THE NETWORK CONTROL CENTER NCC. SENSORS OPERATE AUTOMATICALLY AFTER SETUP. SEARCH RECEIVER AUTOMATICALLY TUNES TO CORRECT FREQUENCY ONCE THE EMISSION IS IDENTIFIED. ANALYSIS AND THREAT DATA CAN BE TRANSMITTED TO REMOTE USERS. THE DATA CAN ALSO BE FUSED WITH OTHER SENSOR SYSTEMS FOR TARGET LOCATION BY AZIMUTH INTERSECTION. WEASEL SYSTEMS CAN BE USED IN CONJUNCTION WITH THE SCORPION JAMMING SYSTEM.

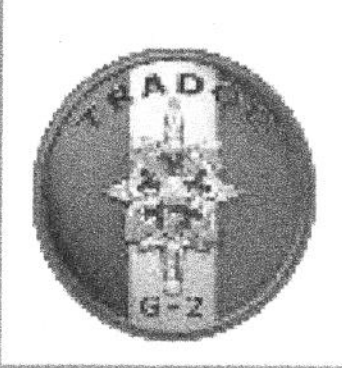

Finland ESM Wideband COMINT Sensor Elektrobit

SYSTEM	SPECIFICATIONS	PERFORMANCE	SPECIFICATIONS
Alternative Designations:	N/A	Antennas for operation on the move:	Yes
		Amplifier types:	broadband, solid-state, liquid cooled
Date of Introduction:	2010s	Output power (kW): into 50 Ω	10
Proliferation:	Finland.	frequency change time at the power output < 100 µs	
Description:	Wideband COMINT Sensor.	Transmitter:	Continuous-wave with cable link to antenna
SPECIFICATIONS:		Amplifier Power (kW):	
Range est. (km):	5 - 10	Consumption:	25
		Emitted:	4
		Width of radiation pattern lobe:	360
Power Supply (AC/DC):	6 / 15	OPERATION:	
Weight (kg):	2	Storage:	High capacity raw data recording of predetermined area
Frequency Range (MHz):	30 - 40		Reconfigurable waveforms broadband wireless signals
RECEIVER:		LPI:	Yes
Receive modes:	VHF/UHF/SHF	Remote Operations:	Yes
Fast scan speed (MHz / s):	250 - 1000 (upgrade)	Environmental conditions:	
Passive Antennas:		VARIANTS:	INA
Dipole or monopole antennas:	Yes	Frequency range (GHz):	INA

Vertical or horizontal polarized omnidirectional antennas:	Yes		

NOTES

China Unmanned Aerial Vehicle EW, ECM ASN-207

SYSTEM	SPECIFICATIONS	PERFORMANCE	SPECIFICATIONS
Alternative Designations:	D-4	Dimensions (m):	
Date of Introduction:	2002	Wing Span:	6
Proliferation:	At least 1 country	Length (fuselage):	3.8
		Height:	1.4 (excluding skids)
Description:		Launch Method:	Solid rocket booster on a zero length launcher.
Engines (HP):	51	Recovery Method:	Parachute (nonsteerable)
HS-700 four-cylinder, two-stroke gasoline air-cooled piston		Landing Method:	2 spring loaded skids
Propulsion:	propeller	Maximum Flights Per Aircraft:	INA
Takeoff:	222	Survivability/Countermeasures:	INA
Fuel and Payload (combined):	50	Pre-programmable waypoints for self-correcting:	Yes
Speed (km/h):		EW/ECM	
Maximum (level):	210	Payload Type:	JN-1102 EW/ECM suite
Cruise:	150	Frequency Range (MHz):	20 -.500
Ceiling (m):		Intercept:	Yes
Maximum:	5,000 - 6,000	VARIANTS:	
Minimum:	100	ASN-206:	
Fuel (liters):	INA	Date of Introduction:	1990's
Endurance (hr.):	8 - 16		

Range (km):			
RPV Mode:	600		
Pre-programmed Mode:	600		

NOTES
THE UAV IS LAUNCHED FROM A ZERO-LENGTH LAUNCHER USING A SOLID ROCKET BOOSTER THAT IS JETTISONED AFTER TAKE-OFF.

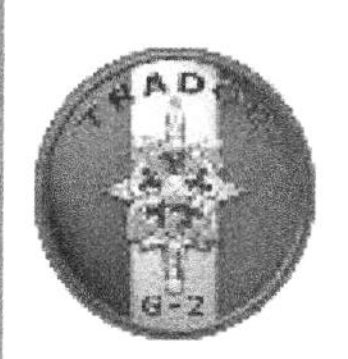

Sweden Airborne ECM/EW Pod, Saab BOQ X-300 (on JAS39/Gripen)

SYSTEM	SPECIFICATIONS	PERFORMANCE	SPECIFICATIONS
Alternative Designations:	None	Intern Range (km):	
Date of Introduction:	1997	Combat Radius:	800
Proliferation:	Sweden (Hungary and South Africa – planned)	Ferry:	3,000
Description:		Takeoff Run/Landing Roll (m):	800/800al:
Crew:	1 (pilot) (JAS 39A/C), 2 pilots (JAS 39B/D)	External:	3,300.
Appearance:		Dimensions (m):	
Wings:	Multi-sparred delta.	Length:	14.1 (A/C), 14.8 (B/D)
Engines:	Turbofan with intake boxes on both	Wingspan (m):	8.4
sides of fuselage.		Height:	4.5
Tail:	Leading edge swept fin with upright inset rudder.	BOQ-X300 ECM/EW POD.	
Engines:	1 x 12,140 lbs thrust Volvo Aero RM12, 18,200 lbs thrust with afterburner	Alternative Designations:	None
Weight (kg):		DATE OF INTRODUCTION:	2012
Takeoff:	12,500 (A/C), 14,000 (B/D)	PROLIFERATION:	Sweden
Empty:	6.500 (A/C), 7,100 (B/D)	COUNTRY OF ORIGIN:	Sweden
Speed (km/h):		FREQ. BANDS:	S/C/X/Ku/K
Maximum (at altitude):	2,150, Mach 1.8+	FREQ. RANGE (MHz):	2-40,000
Max "G" Force (g):	+9/-3 g	RANGE:	INA
Ceiling (m):	16,000	POWER OUTPUT:	INA
Fuel (liters):		TYPE:	Airborne Electronic Countermeasures (ECM), radar jamming system.

NOTES
THE BOQ-X300 HIGH-PERFORMANCE JAMMING POD IS THE LATEST POD BEING DEVELOPED BY SAAB FOR THE GRIPEN FIGHTER. THE POD IS A MODULAR SYSTEM THAT INTEGRATES A SOPHISTICATED JAMMER, SUPPORTED BY A RWR AND ESM SYSTEM. AS AN OPTION, THE POD CAN BE CONFIGURED WITH A DUAL FIBER OPTIC TOWED DECOY TO PROVIDE EFFECTIVE COUNTERMEASURES AGAINST MONOPULSE THREAT. THE BOQ-X300 PROVIDES SELF-PROTECTION FOR HIGH VALUE ASSETS SUCH AS FIGHTER, ATTACK AND RECONNAISSANCE AIRCRAFT. THE POD IS DESIGNED TO SUPPRESS LEGACY THREATS, SURFACE BASED AS WELL AS AIRBORNE. A SECONDARY ROLE FOR THE BOQ-X300 IS TO PROVIDE JAMMING FOR TRAINING OF RADAR OPERATORS IN AIRBORNE AS WELL AS GROUND- OR SEA-BASED ENVIRONMENTS.

Worldwide Equipment Guide

Chapter 10: Countermeasures, Upgrades, and Emerging Technology

Chapter 10: Countermeasures, Upgrades, and Emerging Technology

Chapter 10 includes information on countermeasure techniques, weapon system upgrades, and emerging technology. The section on countermeasures will detail how the OPFOR can employ a variety of countermeasures in order to secure the advantage over the enemy. The section on weapon system upgrades provides an overview of the types of upgrades common OPFOR weapons systems have. The section on emerging technology highlights advancements of weapon system technology for near- and mid-term time periods.

Mrs. Jennifer Dunn

DSN: 552-7962 Commercial (913) 684-7962

E-mail address: jennifer.v.dunn.civ@mail.mil

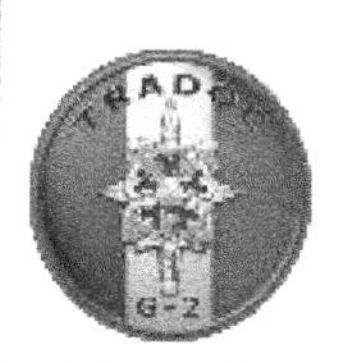

Countermeasures

Countermeasures (CMs) are survivability measures to preserve the integrity of assets and personnel by degrading enemy sensors and weapons effectiveness. These measures often fit within the US Army term CCD (camouflage, concealment and deception) or within the OPFOR term C3D (camouflage, cover, concealment and deception). Decoys used by tactical units within branch operations are designed to aid survivability, and are considered to be countermeasures. Countermeasures can take the form of tactical CMs (or reactive measures), or they can be technical CMs. The variety of tactical CM changes with new unit tactics techniques and procedures (TTP), to adapt to a given situation, within rules of engagement. This document focuses on technical CM. In specialized branches new technical CMs continue to appear.

Modern forces will upgrade systems with selected countermeasures. Many CMs noted are intended to protect combat vehicles from anti-armor sensors and weapons. Although the below CM can be used to counter precision weapons, many were developed for use against conventional weapons. Priorities for countermeasures are dictated by the goals of survival, mission success, and maintaining effectiveness. The first CM priority is to avoid detection until you can control the events. Among goals for using countermeasures, the highest is mission success.

Survival ("Don't Be Killed") is defined holistically, including the following requirements in order of priority: operating system or network survival, vehicle survival, vehicle avoidance of major damage, crew survival, and vehicle avoidance of minor repair. A compatible suite of countermeasures may be limited to a more modest goal, to preserve a measure of effectiveness, even at the risk of system survival. Effectiveness in this context could be defined as - ability to effectively execute the immediate and subsequent missions, until system or subsystem failure interrupts this process. Effectiveness includes: crew effectiveness, mission success, operating system effectiveness, and vehicle/soldier readiness for employment.

Several factors must be considered when selecting countermeasures:

- Countermeasures should be fielded and mounted on systems with a holistic and rational approach to assure survivability. The rational developer will focus his countermeasures with the highest priority given to assure protection against the most likely and most lethal threats. However, with changing threat capabilities over time, and conflicting priorities, the current CM mix may not be successful. Most CM are responses to specific perceived threats, and are limited by cost and weight budget concerns. With the modern reliance on precision weapons, military forces may develop complex and expensive countermeasure "suites" to degrade their effects.
- Some countermeasures can degrade a variety of sensors and weapons capabilities. They can be grouped by threat to be countered, such as artillery or ATGM CMs. Others are more adversary and technology-specific, and may not be fielded until that technology is fielded. Driven by threatening technologies, designers may launch a short-response program to produce or purchase countermeasures for rapid mounting.
- The R&D process has led to the development of counter-countermeasures, intended to negate the effects of CMs. However, at some level, these are also CMs. To avoid confusion on labeling, these will also be called countermeasures.
- When countermeasures are added to a vehicle or within close proximity, they must be mutually compatible and compatible with other subsystems. Thus issues such as electromagnetic interference and self-blinding with smokes must be considered.

- Although a variety of countermeasures are now marketed, many technical and financial factors can negate their advantages. Countermeasure development may be restricted due to resource, technology, and fabrication limitations, which vary by country and time frame. Budget limitations may limit fielding of feasible and valuable CM, or compel selection of less capable countermeasures. For instance, active protection systems can counter some weapons; but they are expensive, hazardous to soldiers, and ineffective against many weapons. Thus they may be unsuitable and unlikely for application to many systems. OPFOR users should consult the POC below for assistance in selecting CMs for a specific system.
- Countermeasures will not replace the need for armor protection and sound tactics.

Lethality Component versus Countermeasure Responses

This table is intended to assist in selection of CM and understanding the categorization for use in upgrade schemes. Many of the more widely-fielded countermeasures are designed to degrade a variety of sensors and munitions, for minimal upgrade cost. Thus, countermeasure types may be repeated under several functions. Because new technologies are emerging rapidly, and systems are finding applications which can place them in several CM types, the placement of CMs can be somewhat arbitrary. Use against artillery vs ATGMs vs ground vehicle weapons will vary. The following list of CM can be used for artillery, air defense, antitank, armor, aircraft, theater missile, and other systems, depending on the platform, gun, sensor, and munition configuration of the system.

Capability to Be Degraded	Type of Countermeasure
Detection and location	Camouflage: nets, paints, fasteners for added natural materials Cover: entrenching blades, hole-blast device, underground facilities Concealment: screens, skirts, thermal engine covers, scrim, other signature reduction Deformers, engine exhaust diversion, other signature alteration measures Aerosols: smoke and flares, water spray systems Decoys, clutter, and acoustic countermeasures Counter-location measures: GPS jammers, laser and radar warning systems
C2/sensor-shooter links	See Information Warfare (IW) Chapter
Platform or weapon	Counterfire: directional warning systems, laser radars, for rapid response Directed energy weapons (DEW), such as high-energy lasers System prioritization for hard-kill, e.g., anti-helicopter mines (See Ch 7)
Weapon sensors and fire control	CCD as noted above. Directed energy weapons, such as low-energy lasers (LEL) Electro-optical countermeasures (EOCMs)
Submunition dispensing/activation	Global positioning system (GPS) jammer Fuze (laser/IR/RF), RF barrage jammers, acoustic jammers
Precision munition and submunition sensors	CCD as noted above. False-target generator (visual, IR, RF/acoustic) Electromagnetic mine countermeasure system, to pre-detonate or confuse Fuze jammers (laser/IR/RF), RF barrage jammers, acoustic jammers
Munition/submunition in-flight, and its effects	Sensors to detect munitions: MMW radars, RF/IR/UV passive sensors Air watch and air defense/NBC warning net, to trigger alarm signal Active protection systems, for munition/submunition hard kill Cover, additional armor to reduce warhead effects
Other system effects	Miscellaneous CM (See below)

Countermeasures against sensors

Type Countermeasure	Countermeasure	Example	Application
Camouflage	Camouflage nets Camouflage paints, IR/radar/and laser-absorptive materials/paints Fasteners, belts for attaching natural materials	Russian MKS and MKT Salisbury screen rubber epoxy Chinese "grass mat" set	Variety of vehicles Variety systems Uniforms and vehicles
Cover	Natural and manmade cover, civilian buildings Entrenching blade to dig in vehicles Hole-blast devices for troop positions, spider holes Underground facilities, bunkers, firing positions	Tree cover, garages, underpasses T-80U tank, BMP-3 IFV, 2S3 arty Hardened artillery sites, bunkers	TELs, vehicles, troops IFVs, tanks, SP arty Infantry, SOF Iraqi and NK sites
Concealment	Screens, overhead cover for infantry (conceal IR/visible signature) Canvas vehicle cover, to conceal weapons Thermal covers, vehicle screens Scrim, side skirts and skirting around turret	Colebrand netting Cover on Chinese Type 90 MRL Kintex thermal blanket over engine French "Ecrim" track cover scrim	Infantry, weapon, sensor Truck-based weapons For combat vehicles Combat vehicles
Deformers/ signature modification	"Wummels" (erectable umbrellas to change/conceal shape/edges) Exhaust deformers (redirect exhaust under/behind vehicle) Engine and running gear signature modification (change sound) IR/radar deformers (in combination with RAM and RAP, etc)	Barracuda RAPCAM/TOPCAM Russian exhaust deflectors Track pads, road wheel/exhaust change Cat-eyes, Luneburg lens	Vehicles, sites, weapons Combat vehicles Tracked, other vehicles Tracked, other vehicles
Aerosols	Visual suppression measures, smokes, WP rounds Multi-spectral smokes for IR and or MMW bands, Flares, chaff, WP, to create false targets, disrupt FLIR Toxic smokes (irritants to disrupt infantry and weapons crews) Water spray systems (to reduce thermal contrast)	Smoke generators, fog oil, S-4, RPO-D ZD-6 Smoke grenades (visual/IR) WP rounds, Galix 6 flare system, Adamsite and CN in smoke mix Add-on kits for vehicles	Blinding, screening Vehicle protection Combat vehicles, arty Smoke generators Recon, C2, AD, arty
Decoys	Clutter (civilian/military vehicles, structures, burning equipment) Low to high-fidelity (multi-spectral) decoys Radar/IR decoy supplements (to add to visual/fabricated decoys) Acoustic countermeasures (to deceive reconnaissance, sensors)	Log site, truck park, tank farm, derricks IMT-72 "dummy tank", Shape Intl Corner reflectors, KFP-1-180 IR heater Acoustic tape/speaker systems	Artillery, combat vehicles TBM, vehicle decoys Vehicle/site radar decoys Vehicles, sites
Counter-location measures	Degrade GPS by jamming to reduce precision location capability Jam radars/IR sensors Laser, IR, and radar warning systems (to trigger move/CM)	Aviaconversia GPS jammer SPN-2 truck-borne jammer set Slovenian LIRD laser warner	Infantry and others tactical/operational area Combat vehicles

Countermeasures Against Weapons And Weapon Sensors

Type Countermeasure	Countermeasure	Example	Application
Added protection (supplements to armor in reaction to specific capability)	Armor supplements (ERA, screens, bar or box armor, sand bags) Armor skirts over road wheels Mine rollers, plows and flails Vehicle belly armor, raised or redesigned belly design, skirt Vertical smoke grenade launchers (to counter PGM top attack)	Barracuda, SNPE ERA KMT-5, KMT-6	
EOCM	Use EOCMs such as IR jammer/IR searchlights to redirect ATGM	KBCM infrared CM system	Combat vehicles
False-target Generators	Acoustic jammers and directed acoustic countermeasure Laser false-target generator (against semi-active laser homing) Electromagnetic mine countermeasure system, counters fuzes	In development, can be improvised In development	To distract acoustic seekers Combat vehicles
Jammers	Altimeter jammer (counters submunition dispersion altimeter) Fuze jammers (to spoof RF proximity fuzes on munitions) Incoherent infrared jamming (to jam IR fuzes on munitions) GPS jammers to confuse navigation and course correction systems	SPR-1 armored ECM vehicle	High priority sites, CPs etc.
Active countermeasures	Active protection systems, for munition hard kill. High energy laser weapons to destroy munitions or sensors Low energy lasers to blind or dazzle. Radio-frequency weapons to burn electronics and detonate munitions Directed MGs	Arena hard-kill system ZM-87 laser weapon VEMASID counter-mine system	Tanks, recon vehicle, IFVs AT, AD systems
Counterfire/ Threat response warners	Directional warning system (locate laser/radar, to direct weapons) Employ sensors (RF/IR/UV- to detect munitions) Acoustic directional systems (to detect munitions) Laser radars (laser scanner to locate optics and direct weapons) Directed energy weapons (against optics) Anti-helicopter mines (against aircraft) Employ air watch/security, AD, NBC, nets to trigger alarm signal Dazzle grenades (temporarily blind personnel)	Pilar acoustic detection system Star-burst grenades	Infantry
Miscellaneous CM	Optical filters to degrade effect of battlefield lasers. Pulse code/thermal CCM beacons on SACLOS ATGMs (to counter EOCM)	HOT-3 ATGM	

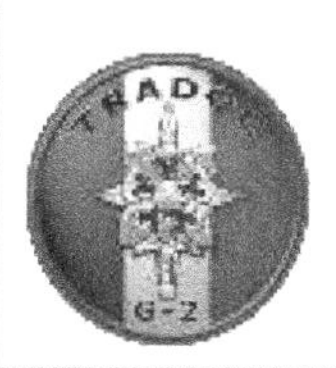

Countermeasures By Functional Area And Type System

Functional Area	System	Type Countermeasure	Countermeasure
Infantry, Special Forces, Reconnaissance, Military Police/Security	Dismounted soldier, Utility vehicle troops	Camouflage Cover Concealment Aerosols CM Operational Technologies	Camouflage nets Fasteners, belts for attaching natural materials Natural and manmade cover, civilian buildings Hole-blast devices for troop positions, spider holes Underground facilities, bunkers, firing positions Screens, overhead cover for infantry (conceal IR/visible signature) Visual suppression measures, smoke grenades, WP rounds Multi-spectral smokes for IR and or MMW bands, Flares, chaff, WP, to create false targets, disrupt FLIR Vertical smoke grenade launchers (to counter PGM top attack) Toxic smokes (irritants to disrupt infantry and weapons crews) Acoustic directed counterfire system Dazzle grenades (temporarily blind personnel)
Mechanized Infantry, Reconnaissance, Military Police/Security, Antitank	Armored personnel carrier Armored scout cars (Less costly LAVs) Light tanks Self-propelled AT Guns (HACVs)	Camouflage Cover Concealment Deformers/signature modification Aerosols Counter-location measures CM Operational Technologies	Camouflage paints, IR/radar/and laser-absorptive materials/paints Fasteners, belts for attaching natural materials Natural and manmade cover, civilian buildings Underground facilities, bunkers, firing positions Armor supplements (stand-off screens, bar armor, sand bags) Thermal covers, vehicle screens Scrim, side skirts and skirting around turret Exhaust deformers (redirect exhaust under/behind vehicle) Engine and running gear signature modification (change sound) IR/radar deformers (in combination with RAM and RAP, etc) Visual suppression measures, smokes, WP rounds Multi-spectral smoke grenades for IR and or MMW bands, Flares, chaff, WP, to create false targets, disrupt FLIR Toxic smokes (irritants to disrupt infantry and weapons crews) Clutter (civilian/military vehicles, structures, burning equipment) Laser, IR, and radar warning systems (to trigger move/CM) Beyond line-of-sight modes Remote-controlled missiles and guns Mine rollers, plows and flails Air watch/security, AD, NBC, nets to trigger alarm signal Optical filters to degrade effect of battlefield lasers. Encoded SACLOS ATGMs (to counter EOCM)

Functional Area	System	Type Countermeasure	Countermeasure
Air Defense, Artillery, Radar units, Theater Missile Units, Aviation, Headquarters,	Command and communications vehicles, Radars, missile launchers, Aircraft (High value targets)	Camouflage Cover Concealment Deformers/signature modification Aerosols Counter-location measures Decoys CM Operational Technologies	Camouflage paints, IR/radar/and laser-absorptive materials/paints Natural and manmade cover, civilian buildings Entrenching blade to dig in vehicles Underground facilities, bunkers, firing positions Canvas vehicle cover, to conceal weapons when not in use Thermal covers, vehicle screens Scrim, side skirts and skirting around turret "Wummels" (erectable umbrellas to change/conceal shape/edges) Exhaust deformers (redirect exhaust under/behind vehicle) Engine and running gear signature modification (change sound) IR/radar deformers (in combination with RAM and RAP, etc) Visual suppression measures, smokes, WP rounds Multi-spectral smoke grenades for IR and or MMW bands, Flares, chaff, WP, to create false targets, disrupt FLIR Degrade GPS by jamming to reduce precision location capability Jam radars/IR sensors Laser, IR, and radar warning systems (to trigger move/CM) Clutter (civilian/military vehicles, structures, burning equipment) Low to high-fidelity (multi-spectral) decoys Radar/IR decoy supplements (to add to visual/fabricated decoys) Acoustic countermeasures (to deceive reconnaissance, sensors) Anti-helicopter mines (against aircraft) Beyond line-of-sight modes Non-ballistic launch modes Anti-radiation missiles Low energy lasers to blind/dazzle optics on designators/aircraft Encoded laser target designators to foil false target generators Radio-frequency weapons - burn electronics/detonate munitions High energy laser weapons to destroy munitions or sensors Laser false-target generator (against semi-active laser homing) Altimeter jammer (counters submunition dispersion altimeter) Fuze jammers (to spoof RF proximity fuzes on munitions) Incoherent infrared jamming (to jam IR fuzes on munitions)

Functional Area	System	Type Countermeasure	Countermeasure
			GPS jammers to confuse navigation and course correction systems Optical filters to degrade effect of battlefield lasers
Information Warfare/ Deception Units	IW vehicles	Camouflage Cover Deformers/signature modification Aerosols Counter-location measures Decoys	Camouflage paints, IR/radar/and laser-absorptive materials/paints Natural and manmade cover, civilian buildings Underground facilities, bunkers, firing positions "Wummels" (erectable umbrellas to change/conceal shape/edges) IR/radar deformers (in combination with RAM and RAP, etc) Visual suppression measures, smokes, WP rounds Multi-spectral smoke grenades for IR and or MMW bands, Flares, chaff, WP, to create false targets, disrupt FLIR Degrade GPS by jamming to reduce precision location capability Jam radars/IR sensors Laser, IR, and radar warning systems (to trigger move/CM) Clutter (civilian/military vehicles, structures, burning equipment) Low to high-fidelity (multi-spectral) decoys Radar/IR decoy supplements (to add to visual/fabricated decoys) Acoustic countermeasures (to deceive reconnaissance, sensors)
All Units	Combat support vehicles (Light strike vehicles, Tactical utility vehicles, Motorcycles, ATVs, Armored CSVs, etc), Trucks	Camouflage Cover Concealment Deformers/signature modification Aerosols Decoys CM Operational Technologies	Camouflage paints, IR/radar/and laser-absorptive materials/paints Fasteners, belts for attaching natural materials Natural and manmade cover, civilian buildings Underground facilities, bunkers, firing positions Armor supplements (ERA, screens, bar or box armor, sand bags) Thermal covers, vehicle screens Engine and running gear signature modification (change sound) IR/radar deformers (in combination with RAM and RAP, etc) Multi-spectral smoke grenades for IR and or MMW bands, Flares, chaff, WP, to create false targets, disrupt FLIR Clutter (civilian/military vehicles, structures, burning equipment) Air watch/security, AD, NBC, nets to trigger alarm signal Acoustic directed counter-fire system

Trophy "Family" HV (Heavy Vehicle) Armor Countermeasure

SYSTEM	SPECIFICATIONS
Alternative Designations:	"Wind Breaker" systems ELM-2133 (WindGuard) Radar
Date of Introduction:	Declared operational by Israeli Defense Force (IDF) in 2009; Started full scale development in 2005.
Proliferation:	Fielded and successfully combat tested in one country (Israel). Tested by US in 2006. In 2013, Canada was conducting tests.
Description:	Active Protection Systems (APS)/Anti-armor countermeasures. System can engage anti-tank guided missiles (ATGM), rocket propelled grenades (RPG) and tank launched high explosive anti-tank projectiles.
Automated Hard-Kill System:	Usually 2 platforms per tank. One mounted on each side of the turret. When traveling the system normally face inwards to the turret. Each platform has Multiple Explosive Formed Penetrators (MEFP) that fire a multiple "spray" of projectiles that engages and neutralizes the warhead (at a certain point on the projectile) before detonation. System can engage on the move or at a standstill. System can engage multiple targets and auto reloads. Minimum collateral damage (estimated at less than 1 percent) to dismounted infantry and non-combatants in area of engagement.
Antenna and Transmitter:	Four (bullet/fragment resistance) flat-panel phased array antennas; pulse Doppler active electronically scanned array (AESA) radar. One (each) radar provides a 90° coverage area. Radars are located on front, rear and sides of turret providing 360⁰ coverage including defense from "top down" threats.
Operation:	ELM-2133 radar actively scans for threats. If a threat is located the radar provides identification, tracks the vector of incoming threat, and delivers the point of origin to the battle management system (BMS). The computer for the BMS determines if the threat will engage the tank and if that is the case, begins the automated countermeasure process. The ballistic automated hard-kill system uncovers and aims launchers. The computer through advanced algorithms and logics continues to track the threat and determines the best intercept point. The countermeasure is then launched and neutralizes the threat. The system takes no action if the threat is going to miss the tank.
Performance:	The Trophy (HV) system is currently fielded as an APS for the Israeli Merkava Mk 4 main battle tank (MBT). Trophy has an extensive history of successful tests. The system has

	been challenged by different anti-tank platforms including: RPG-7, RPG-29, AT-3, among others. The first successful combat engagement took place on March 2011 in Gaza. An Israeli tank was on patrol near the border and an anti-tank weapon was fired at the tank but the Trophy system neutralized the threat. The system had a number of successful engagements (reporting indicates 5 or more) during the conflict between Hamas and Israeli in August 2014.	
Range (m):	10-60	
Power Supply:	UNK	VARIANTS
Weight (kg):	850	Trophy Medium Vehicle (MV)
Frequency Range (radar):	S Band	Trophy Light Vehicle (LV)

Equipment Upgrades

Armed forces worldwide employ a mix of legacy systems and selected modern systems. In the current era characterized by constrained military budgets, the single most significant modernization trend impacting armed forces worldwide is upgrades to legacy systems. Other factors impacting this trend are:

- A need for armed forces to reduce force size, yet maintain overall force readiness for flexibility and adaptability.
- Soaring costs for modern technologies, and major combat systems.
- Personnel shortages and training challenges.
- Availability of a wide variety of upgrade packages and programs for older as well as newer systems.
- New subsystem component technologies (lasers, GPS, imaging sensors, microcircuits, and propellants), which permit application to platforms, weapons, fire control systems, integrated C2, and munitions old and new.
- An explosion of consortia and local upgrade industries, which have expanded worldwide and into countries only recently introduced to capitalism.

The upgrade trend is particularly notable concerning aerial and ground vehicles, weapons, sensors, and support equipment. From prototype, to low-rate initial production (LRIP), to adoption for serial production, minor and major improvements may be incorporated. Few major combat systems retain the original model configuration five or more years after the first run. Often improvements in competing systems will force previously unplanned modifications.

Upgrades enable military forces to employ technological niches to tailor their force against a specific enemy, or to integrate niche upgrades in a comprehensive and well-planned modernization program. Because of the competitive export market and varying requirements from country to country, a vehide may be in production simultaneously in many different configurations, as well as a dozen or more support vehicle variants fulfilling other roles. In light of this trend, OPFOR equipment selected for portrayal in simulations and training should not be limited to the original production model of a system, rather a version of the system that reflects the armed forces strategic and modernization plans and likely constraints that would apply.

The adaptive OPFOR will introduce new combat systems and employ upgrades on existing systems to attain a force structure that supports its plans and doctrine. Because the legacy force mix and equipment were selected in accordance with earlier plans and options, use of upgrades versus costly new acquisitions will always be an attractive option. A key consideration is the planned fielding date. For this document, the most widely portrayed OPFOR time frame is the current Contemporary Operational Environment. Only upgrades currently available (or marketed with production and fielding expected in the near term) are considered in COE Tiers 1-4. Also, system costs and training and fielding constraints must be considered. However, in the Emerging Technology Trends section of this chapter, we anticipate a wide variety of upgrades that could be currently applied to fielded systems.

The selection of equipment upgrades is not a simple matter. Most forces have limited budgets, competing upgrade priorities, and a substantial inventory of outdated equipment. A specific subsystem upgrade (gun, fire control system, etc.) may only slightly improve a generally obsolete system. Another option is an upgrade package, with compatible subsystem upgrades. The surest approach is to refurbish a system

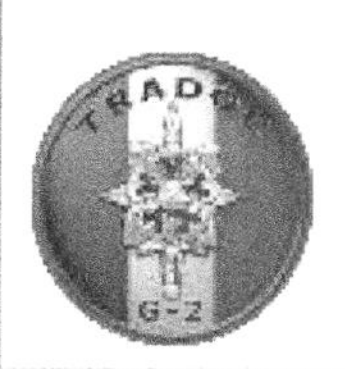

into a new model with all application problems resolved. A critical factor is assurance that the modernized equipment is tested and successful. The best test remains performance in combat.

The following tables describe selected upgrades available for system modernization. The lists are not intended to be comprehensive. Rather, they are intended to highlight major trends in their respective areas. For instance, for armored combat vehicles, the focus is on upgrades in mobility, survivability, and lethality.

The category of survivability upgrades includes countermeasures (CM). The CM upgrades can apply not only to branch-specific systems (tanks, IFV, and artillery), but to general use systems subject to similar threats. An example of this is the proliferation of smoke grenade launchers on artillery and reconnaissance vehicles.

Implementation of all upgrade options for any system is generally not likely. Because of the complexity of major combat systems and need for equipment subsystem integration and maintenance, most force developers will chose a mix of selected upgrades to older systems, as well as limited purchases of new and modern systems. Please note that systems featured in this document may be the original production system or a variant of that system. On data sheets, the variants section describes other systems available for portrayal in training and simulations. Also, equipment upgrade options (such as night sights) and different munitions may be listed, which allow a user to consider superior or inferior variants. Within the document chapters, multiple systems are listed to provide other substitution options. Of course there are thousands of systems and upgrade options worldwide, which could be considered by an adaptive OPFOR.

An OPFOR trainer has the option to portray systems or upgrade packages not included in the OPFOR Worldwide Equipment Guide, to reflect an adaptive thinking OPFOR. In future WEG updates, the authors will expand on the upgrade tables with names and descriptions of upgrade options and specific systems applications which have been noted in the current document. Chapter authors are available to assist users in selecting reasonable upgrade options for system configuration in specific force portrayals.

OPFOR Antitank Weapon Upgrades

GRENADE LAUNCHER	TOWED AT GUN	GROUND ATGM LAUNCHER
Improved AT and dual-purpose rifle grenades permit riflemen to supplement shoulder-mount grenade launchers.	Auxiliary propulsion unit for local movement	Man-portable/ground launch and shoulder launch
Accurate low trajectory longer range grenades for shoulder launchers	Take-apart capability for lighter guns	Take-apart launcher and sub-systems
Parachute-drop overhead camera grenades for shoulder/ground launch	Improved gun and recoil system	Pintle mount/dismount for variety of vehicles/platforms
Tripod, bipod, pintle mounts convert launchers for vehicle/ground use	Ballistic computer/laser rangefinder sights	1st or 2nd generation thermal night sights
Take-apart AT grenade launchers or disposable launch tubes	MMW radar target auto-tracker day/night FCS	Extended range missiles
	Image intensifier/thermal night sights	Soft-launch for use from bunkers and buildings

<table>
<tr><th>GRENADE LAUNCHER</th><th>TOWED AT GUN</th><th>GROUND ATGM LAUNCHER</th></tr>
<tr>
<td>Larger, more lethal disposable AT grenade launchers supplement grenadier reusable launchers at critical times.

Reduced noise, smoke, and flash signature grenades for AT launchers

Improved reusable sights for disposable launchers, including ballistic computer/laser rangefinder sights

Image intensifier/thermal night sights

Counter-charge AT grenades for firing from inside of buildings

Dual-purpose (HE/AT) longer range rounds

HE longer range rounds

Multi-purpose (HE/AT/anti-bunker) rounds

Tandem shaped-charge (HEAT) warhead

Thermobaric Frag-HE warhead

Guided (SAL-H) grenades for shoulder/ground/vehicle launchers

UPGRADE PRIORITY
Computer/LRF FCS
II night sights
Tandem AT grenades, HE/DP grenades, thermobaric grenades</td>
<td>Automated battle management system with graphic flat panel display

Indirect fire rounds/FCS for fire support role

Increased DF range, new tank/AT gun rounds

Improved, heavier, more lethal, and longer range APFSDS-T round

Tandem or triple-charge HEAT round

Improve Frag-HE round and DPICM submunition

Canister/flechette round

New type lethalities (DPICM submunition, etc.)

Gun-launched ATGM (100 mm+), including tandem HEAT

UPGRADE PRIORITY
APU and take-apart for lighter guns
Improved gun and recoil system
Improved sights, 1st gen thermal night sights
Automated battle management system
Improved ammunition, inc ATGM.</td>
<td>Launcher countermeasures (CM), such as reduced noise, smoke, flash

SACLOS Guidance CCM, e.g., pulsed codes

Increased ATGM velocity/reduced flight time

New guidance modes: Semi-active laser beam rider and laser-homing, Fiber-optic guided missile (FOG-M) guidance, Fire and forget imaging infrared seeker, Radar homing, Multi-mode (FOG/IR homing, etc.)

Helicopter stand-off launch using ground guidance

High velocity MANPADS missiles used for AT

Larger warhead/tandem warhead HEAT ATGM

Sensor-fuzed EFP/HEAT top-attack

Thermobaric HE warhead, for new applications

UPGRADE PRIORITY
Take-apart launcher, with pintle mount
Improved 1st gen thermal night sights
SACLOS CCM
Reduced signature
Improved ATGMs (tandem HEAT, etc.)</td>
</tr>
</table>

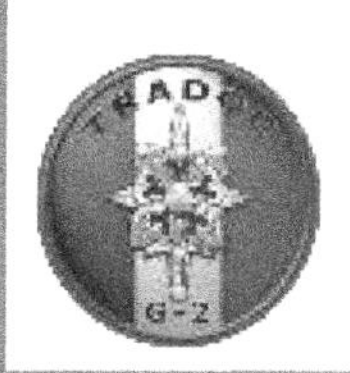

OPFOR Light Armored Vehicle Upgrades

COMBAT SUPPORT VEHICLE	APC/IFV, INFANTRY FIRE SUPPORT VEHICLE	ATGM LAUNCHER VEHICLE
Used/adapted for various roles, e.g., infantry (less than squad), combat support, and support vehicles. Most are light, 4x4 wheeled, van or light utility vehicle; but auxiliary wheel, 6x6 or tracked versions exist. Many of these are being converted to or replaced by mine-resistant vehicles. Included are motorcycles, ATVs, and light strike vehicles (e.g., jeep-type 4x4 vehicles or recreational dune buggies). Gun trucks, riot control vehicles, and amphibious/over-snow all terrain vehicles are used. Add encrypted voice and digital data capability. Graphic display battle management system Central tire inflation system and/or run-flat tires GPS hand-held or bracket mount Ford capability, swim capability desired Hybrid (diesel/electric) drive kits Add-on light armor, mine protection desired CM, such as multi-spectral smoke grenades Laser warning receiver desired 7.62-14.5-mm MG or 20-40-mm automatic grenade launcher main weapon Remote or overhead weapon station (RWS/OWS) Individual weapons, RPG, MANPADS, or ATGM launcher for secondary weapons Day sight and II or thermal night sight UPGRADE PRIORITY	Must be able to carry a squad Higher horsepower diesel engine GPS and inertial land navigation, graphic display battle management system, IFF Swim or deep ford (due to armor increases). Amphibious conversion with compartments for high sea state capability. Add-on armor, ERA, and improved mine protection. Fire and blast suppression CM, e.g., multi-spectral smoke grenades, LWR Firing ports (or forego due to armor increases, use periscopes or side and rear view cameras) IFV/IFSV: 20-100 mm stabilized gun, and 2-man turret. Active protection system (APS) or defensive aids suite (DAS). Upgraded FCS: Cdr's independent viewer, 2-plane stabilized TV sights, 1 - 2 gen FLIR. Upgraded secondary MG or grenade launcher with superior sights (integrated, high-angle, night). Additional remote MGs/AGLs for high-angle fires security. Improved KE, HEAT, Frag-HE rounds, ATGMs APC/IFSV: Includes truck/light vehicle conversions Remote weapon station or 1-man turret with high-angle-of fires 7.62-	Use APC/IFV wheeled/tracked chassis or tank chassis, with mobility and protection upgrades, Side and rear-view cameras Graphic display battle management system CM, e.g., multi-spectral smoke grenades, LWR Active protection system or other DAS. 1-2 man turret, or turretless design. Alternative design: 1-5 pedestal/turret or mast-elevated ATGM launchers on remote or overhead weapon station (RWS/OWS) Autoloader or manual loader under armor Multiple ATGM launch and targeting capability Improved ATGMs, as noted in above table, or RF, laser-beam rider, SAL-H/IIR ATGMs 7.62-12.7-mm MG secondary arms FCS with commander's independent viewer, 2-plane stabilized sights, TV, and target tracking. Use 1st or 2nd gen FLIR Most common ATGM vehicles are combat support vehicles with pintle-mount ATGM launcher (see above table for ground launcher). Recent development: motorcycle with sidecar and pintle-mount ATGM launcher

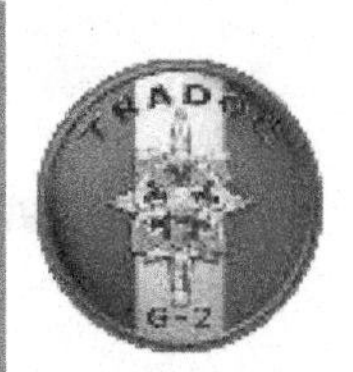

COMBAT SUPPORT VEHICLE	APC/IFV, INFANTRY FIRE SUPPORT VEHICLE	ATGM LAUNCHER VEHICLE
Light armor and smoke grenade launchers Remote MG or auto grenade launcher Day/night (thermal sights), RPG GPS, secure comms	23 mm MG, grenade launcher (some with 20-30-mm auto-cannon and ATGM launcher) UPGRADE PRIORITY Add-on armor, ERA, LWR, new grenades Add auto grenade launcher, upgrade ATGM, and KE round to APFSDS. FCS, stabilized sights, Imp 1st gen FLIR	UPGRADE PRIORITY RWS multiple ATGM launchers (APC/IFV/tank conversion), pintle-mount for light combat support vehicles (motorcycle, ATV, LSV, TUV, truck, etc.) Stabilized sights and 1st gen thermal sights Improved ATGMs

OPFOR Reconnaissance And Assault Vehicle Upgrades

RECONNAISSANCE VEHICLE	HEAVY ARMORED COMBAT VEHICLES	MAIN BATTLE TANK
Light recon vehicle: Combat support vehicle with light armor and TV, thermal sights, Add encrypted voice and digital data capability Combat recon vehicle: See IFV upgrades, e.g.: GPS and inertial land navigation. Digital real-time link to subscriber map overlay display, IFF, force tracker battle management system, Swim capability, winch, central tire inflation for wheeled Upgraded FCS: Cdr's independent viewer, 2-plane stabilized TV camera sights, 1 - 2 gen FLIR Elevated battlefield surveillance radar/TV/FLIR sensor suite with TV, encrypted voice, and digital data transmission capability Launch UAVs and/or robots (unmanned ground vehicles) CM, e.g., multi-spectral smoke grenades, LWR, IR/radar skirts Active protection system (APS) or other defensive aids suite (DAS). 20-100 mm gun with 2-plane stabilization, and 2-man turret. Improved secondary MG or automatic grenade launcher and sights.	Distinction among heavy recon, infantry fire support, assault gun, light tank has blurred APC/IFV chassis with increased armor and higher horsepower diesel engine. GPS and inertial land navigation, graphic display battle management system, IFF Swim or deep ford capability Add-on armor, ERA, improved mine protection, fire and blast suppression. CM, e.g., multi-spectral smoke grenades, LWR Side and rear-view cameras for security Active protection system (APS) or other defensive aides suite (DAS). 76-125 mm tank gun with 2-plane stabilization, Improved MG or auto grenade launcher, sights FCS with commander's independent viewer, 2-plane stabilized sights, TV, and target tracking. Use of 1st or 2nd generation FLIR. Side and rear-view cameras	Higher horsepower diesel engine power packs and add-on reserve fuel tanks GPS and inertial land navigation, graphic display battle management system, IFF Deep ford snorkel capability Welded turret, blow-out panels, ERA, improved mine and turret protection, fire and blast suppression. CM suite, including multi-spectral and vertical smoke grenade mix, LWR, VEESS capability Active protection system (APS) or other defensive aides suite (DAS), self-entrenching blade Side/rear-view security cameras Tank gun with 2-plane stabilization FCS with commander's independent viewer, 2-plane stabilized sights, TV, and target tracking. 2nd or 3rd generation FLIR. Auto-tracker. Hunter-tracker FCS.

RECONNAISSANCE VEHICLE	HEAVY ARMORED COMBAT VEHICLES	MAIN BATTLE TANK
Man-portable SAMs (MANPADS) for self-protection Named and targeted areas of interest link to indirect fire and missile units for real-time targeting. Laser target designator guides munitions. Sensor vehicle: APC/IFV or combat support vehicle and mast-mounted sensor pod: radar, thermal and TV Encrypted voice SATCOM/digital data systems UPGRADE PRIORITY Add higher HP diesel engine Add-on armor, ERA, LWR, new grenades Imp 1st gen FLIR, gunner and commander, Add auto grenade launcher, upgrade ATGM, and KE round to APFSDS. Elevated sensor suite and transmission capability.	Improved KE, electronic fuzed Frag-HE, and tandem HEAT rounds Gun-launched ATGMs (100+ mm) UPGRADE PRIORITY Add higher HP diesel engine Add-on armor, ERA, LWR, new grenades Imp 1st gen FLIR, gunner and commander, Larger stabilized gun, gun-launch ATGM, and KE round to APFSDS.	Heavier and longer range APFSDS-T rounds, electronic fuzed Frag-HE, and tandem HEAT rounds Gun-launched ATGMs and IR homing rounds (100+ mm). Semi-active laser homing munitions permit ATGMs to deliver indirect fire precision strikes. Improved remote-firing MG, high-angle AD sights UPGRADE PRIORITY Add higher HP diesel engine Land navigation and deep ford snorkel Add-on armor, ERA, CM suite Imp stabilization and FCS, 1st gen FLIR, Remote MG, Imp ammo (sabot, Frag-HE, and HEAT) Gun-launch ATGM.

OPFOR Artillery Upgrades

ARTILLERY RSTA/C2 SUPPORT	TOWED AND SELF-PROPELLED CANNON	MULTIPLE ROCKET LAUNCHER
Automated secure digital joint C2 network with SATCOM, linking artillery, air, EW, and reconnaissance units Integrated artillery recon vehicle with sensor mast Reconnaissance strike and fire complexes Forward air controllers linked to artillery units Artillery surveillance vehicles with ground surveillance radars, sensor suite and networked Observation teams with goniometers, thermal sights, digital comms, and laser target designators Artillery links to selected special purpose forces	Conventional munitions, e.g., controlled fragmentation, proximity and multi-option fuzes, special munitions, and propellants (modular propellants) Artillery delivered high precision munitions e.g., SAL-H, sensor-fuzed, course corrected, terminal-homing IR Self-Propelled: Automated fire control with barrel cooling and thermal warning systems Auxiliary power unit Mobility and weight improvements, Muzzle velocity analyzer	Mobility and weight improvements, truck-based launchers which conceal the MRL signature Rapid emplace-displace and response capabilities CM, such as smoke grenade launcher and LWR On-board computer-based fire direction and land navigation systems, which permit autonomous launcher, platoon, and battery operations Tube-launched UAVs linked to the launchers and to the fire control network for real-time acquisition

ARTILLERY RSTA/C2 SUPPORT	TOWED AND SELF-PROPELLED CANNON	MULTIPLE ROCKET LAUNCHER
Acoustic vehicle detection and location Phased array counter-battery radars, networked to automated artillery net, with increased range, lower probability of error, windows-based man-machine interface Target-acquisition UAVs, networked to artillery net Automated battle management equipment use for towed and SP guns, mortars and MRLs Navigation system with GPS/inertial update, linked to automated net UPGRADE PRIORITY Integrated artillery recon vehicle, sensor mast Reconnaissance strike and fire complexes Target-acquisition UAVs, networked Observation teams, radars, acoustic sensors	CM, such as smoke grenade launcher and LWR Upgrade to 52-caliber cannon for longer range Truck-mounted high-mobility systems with long-range cannons Towed: Addition of auxiliary propulsion unit On board technical fire control computer Reduced weight and emplace/displace times Muzzle velocity analyzer Onboard or portable digital linked fire control computer Upgrade to 52-caliber cannon for longer range UPGRADE PRIORITY Mobility and weight improvements On-board navigation and fire direction systems Use of modular propellant Procurement of ADHPM Overall range and accuracy improvements	Improved lethality improved conventional munitions and special purpose (mines, jam, etc,) munitions Extended-range and course-corrected rockets, as well as addition of artillery/cruise missiles Computer-based fire control system for electronically-fuzed rockets Artillery delivered high precision munitions (ADHPM), e.g., sensor fuzed, laser-homing rockets Special munitions, such as FASCAM, chemical warhead, RF jammer rockets Mine clearer and fuel-air explosive rocket MRLs UPGRADE PRIORITY Autonomous/semi-autonomous launcher Countermeasures Improved munitions, e.g., extended range, DPICM and thermobaric ADHPM, e.g., sensor-fuzed munitions and course corrected rounds or rockets

Emerging Technology Trends

In order to provide a realistic OPFOR for use in Army training simulations, we must describe the spectrum of contemporary and legacy OPFOR forces in the current time frame, as well as capabilities in emerging and subsequent operational environments (OEs). This chapter does not predict the future, rather notes emerging adversary capabilities which can affect training.

The OPFOR timeframes for emerging OPFOR are: 2015-2020 (Near Term) and 2021-2028 (Mid-Term). The subsequent time frame is "future" OPFOR time frame. Time lines were determined in part to assist in building OPFOR systems and simulators and for use in Army training simulations. The timeframes are arbitrary and selected for ease in focusing and linking various trends. However, they also generally match force developments for U.S. Army forces, as well as thresholds in emerging and advanced technologies which will pose new challenges to military force planners and developers.

In these time frames, the mix of forces will continue to reflect tiered capabilities. The majority of the force mix, as with all military forces, will use legacy systems. Periods 2015 and after will also see new OPFOR systems and whole new technologies. The most notable difference between the OPFOR force mix and U.S. forces is that the OPFOR will have a broader mix of older systems and a lower proportion of state-of-the-art systems. Rather, OPFOR will rely more on adaptive applications, niche technologies, and selected proven upgrades to counter perceived capabilities of their adversaries. Force developers for OPFOR will retain expensive legacy systems, with affordable upgrades and technology niches. A judicious mix of equipment, strategic advantages, and sound OPFOR principles can enable even lesser (lower-tier) forces to challenge U.S. military force capabilities.

The OPFOR systems must represent reasonable responses to U.S. force developments. A rational thinking OPFOR would study force developments of their adversaries as well as approaches of the best forces worldwide, then exploit and counter them. Thus worldwide and adversary equipment upgrades will trigger OPFOR forces to modify their equipment and tactics to deter, match, overmatch, or counter those changes.

OPFOR Technologies And Emerging Operational Environments

As noted in Chapter 1 on COE OPFOR, the adaptive OPFOR will introduce new combat systems and employ upgrades on existing systems to attain a force structure which supports its plans and doctrine. Because a legacy force mix and equipment were historically selected earlier in accordance with plans and options, upgrades versus costly new acquisitions will always be an attractive option. A key consideration is the planned fielding date. To project OPFOR capabilities in the future, we should look at the technologies in various stages of research and development today, as well as those in the concept stage for applications in the Future OPFOR time frame. Military engineering experience has demonstrated that the process of formulating military requirements, as well as technology, engineering, and budgeting factors can dramatically affect equipment modernization time lines. In addition, scientific discoveries and breakthroughs in the civilian sector have greatly contributed to the so-called "Revolution in Military Affairs", which has increased the capability for battlefield awareness, integration, timeliness, and lethality. The table below shows OPFORs in emerging and Future OEs, and some considerations.

Considerations in Determining Emerging OPFOR Technologies by Time Frame

OPFOR Consideration	Near-Term (2015-2020)	Mid-Term (2021-2025)
Challenging OPFOR	Emerging OPFOR	Objective OPFOR
Technology Source	Current marketed/fielded systems and subsystems	Recent major weapons, upgrade applications
Budget	Constricted but available for niche technologies	Improved, some major system acquisitions
Implications for OPFOR equipment	Many subsystem upgrades, BLOS weapons, remote sensors, countermeasures	More costly subsystems, recent major weapons, competitive in some areas.
Implications for OPFOR tactics and organization, Implications for U.S.	COE tactics with contingency TTP updates. Slight subunit changes add BLOS and AT systems for integrated RISTA and strikes.	Integrated RISTA with remotes. Strikes all levels. Combined arms integrated in small units for increased lethality and autonomy.

The information revolution has also decreased response time in which system developers in the military marketplace can seize a new technology and apply it in new systems or in upgrades to older systems. The following technologies and possible applications of those technologies will influence R&D as well as

fielding decisions for future force modernization and expected OPFOR capabilities to be portrayed in future operating environments.

Technologies And Applications For Use By OPFOR: Near And Mid-Term

TECHNOLOGY CATEGORY	TECHNOLOGY	TECHNOLOGY APPLICATION
Psychological Operations	Mood altering aerosols Reproductive terrorism Non-lethal technologies	Military and civilian targets, for short-term and long-term goals.
Information Operations: Sensors	Higher-resolution multispectral satellite images New sensor frequencies for acquisition New sensor frequencies operational security Use of other light bandwidths (ultraviolet, etc) Passive detection technologies and modes Auto-tracking for sensors and weapons Image processing and display integration Micro-sensors/imaging system miniaturization Unmanned surveillance, target acq/designation Multispectral integrated sensors and Multispectral integrated transmission modes Precision navigation (cm/mm three-dimension) Undersea awareness (sensors, activity) Underground awareness (sensors/mines)	High-intensity use of LITINT (internet, periodicals, forums) Increased use of information from commercial, industrial, scientific and military communities Increased use dual-use technologies
Information Operations: Computers and Comms	Low-Probability-of-Intercept communications New power sources and storage technologies: Micro-power generation Energy cells Advanced Human/Computer Interface Automatic Language Translators	New communities (Blogs, flash mobs, etc, to coordinate and safeguard comms) Secure encryption software New communications tools (internet and subscriber links)
Electronic Attack	Anti-Satellite weapons for RF, EMP, Hard kill Wide area weapons (EMP graphite bombs, etc) EMP Precision (small area) weapons Computer Network Attack Worms, viruses, trojan horses Net-centric warfare Spoofing sensors Spoofing/Intercepting data stream/spyware	Attack electronic grid or nodes at critical times
Chem/Bio/ Radiological Attack	Dirty bombs Genetic/Genomic/DNA tagging to assassinate Genetic/Genomic/DNA targeting for Bio attack Designer Drugs/Organisms/Vectors Biologically based chem (Mycotoxins) Anti-materiel corrosive agents and organisms	Agricultural attack (animal and plant stocks and supplies) Use of tagging to incapacitate political leaders.
Physical Attack	Mini-cruise/ballistic missiles for precision, surgical strikes, and widespread use Atk UAVs (land, sea, undersea-UUV, Micro-aerial vehicles-widespread use Swarming for coordinated attack Notebook command and semi-autonomous links Vehicle launch for NLOS attack/defense Multi-mode guidance: pre-programmed/ guided/homing New types of warheads	

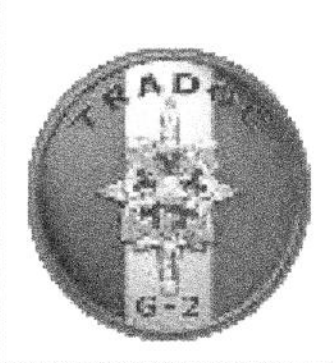

TECHNOLOGY CATEGORY	TECHNOLOGY	TECHNOLOGY APPLICATION
Sustainment, Protection	New battery/power cell technologies Neurological performance enhancers Better lightweight seamless body armor Personal actuators, exoskeletons, anti-RF suits Active armor and active protection systems Countermeasures to defeat rounds and sensors Counter-precision jammers, esp GPS All-spectrum low observable technologies Anti-corrosives Biometric prosthesis and cybernetics Robots assist dismounts, sensors, and logistics Robotic weapon systems	Battlefield fabrication of spare parts Airborne/ship borne refineries Potable water processing systems Transportable power generation systems

OPFOR Capabilities: Near-Term And Mid-Term

The next table provides projected system description and capabilities for analysis of the OPFOR environment facing U.S. forces in subsequent time frames. Data for the first timeframe (2013-2019) reflects generally known systems and subsystems, with their introduction to the emerging OPFOR adversary force. Timelines reflect capability tiers for systems which may be fully fielded (not Interim Operational Capability or First Unit Equipped) in brigade and division unit levels during respective time frames.

The systems projections are not comprehensive, and represent shifting forecasts. They may accordingly shift as we approach the specified time frames. Once we get beyond the turn of the decade, our current view of the future trends becomes less specific. Therefore, the second column (Mid-Term 2021-2028) focuses more on technologies—less on defined systems.

The columns can be treated as capability tiers for specified time frame OPFOR. Please note: ***No force in the world has all systems at the most modern tier.*** The OPFOR, as with all military forces worldwide, is a mix of legacy and modern systems. Thus the emerging OPFOR force comprises a mix of COE time frame Tier 1-4 systems and newer systems. One would expect that some Near- or Mid-term adversaries with lower military technology capabilities could move up one or two capability tiers from (for instance) current COE capability Tier 4, to COE Tier 2. The most likely upgrade for emerging OPFOR used in most training simulations would be to move the OPFOR from COE Tier 2 to Tier 1, with added niche emerging systems.

We have previously stated that an OPFOR force can portray a diverse force mix by separating brigades and divisions into different tiers. The OPFOR also has the option of incrementally adding higher tier systems to lower tier units, as selective upgrades. Because most of the below systems in the 2015-2020

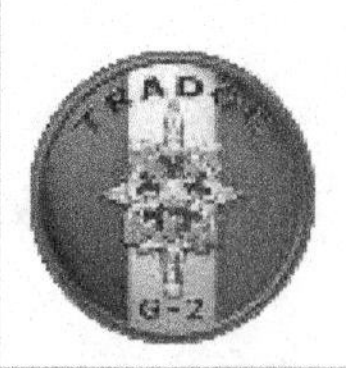

column are currently fielded, an adversary might also incrementally upgrade COE Tier 1 or 2 units by adding fielded assets from 2015-2020 as described in that column. However, until that time frame, we cannot assure beforehand when all of those technologies will appear. Again, the tables are not predictive. The OPFOR force designer may choose a middle road between current Tier 1-4 and future systems; in many countries they are upgrading legacy and even recent systems to keep pace with state-of-the-art systems. Thus they may look to subsystem upgrades such as noted in Chapter 15.

If a specialized system for specific role is missing from the table below, continue to use the OPFOR system noted in Tiers 1-4. Please remember that these projections reflect "possible" technology applications for future systems. They incorporate current marketed systems and emerging technologies and subsystems, may be combined in innovative ways. The table below is not a product of the US intelligence community, and is not an official US Army forecast of future "threats". It is approved only for use in Army training applications and simulations.

Future OPFOR (2028 and after) is described in various portrayals. But it is generally FOUO or classified and is not included in the WEG.

OPFOR Capabilities: Near-And Mid-Term

SYSTEM	NEAR-TERM OPFOR (FY 15-20)	MID-TERM OPFOR (FY 21-28)
	INFANTRY WEAPONS	
Infantry Assault Rifle	Rifle 6.8mm to 600 m day/night, w/EO LRF/pointer computer sight. Fire around corner sight EO link. Under-barrel grenades 600 m (CS gas, HEDP, EO recon, starburst, HE airburst, concussion). Rifle grenades 400m: HEAT, DP, smoke	On-bipod range 600 m. Sight on all weapons link to laptop/PDA/NVG/ helmet viewer w/real-time RF link. Multispectral smoke, TV/II recon/ atk rd, tandem HEAT grenades. Remote fire platform, 60m link.
Thermobaric grenades and Magazine grenade launcher	43-mm 4-round hand-held launcher for urban fight to 350m. Thermobaric grenades, also for hand throw, underbarrel.	Range 600 m for hand-held and under-barrel launchers, night sight. Add flechette, TV/II recon grenade.
AT/AP Hand Grenade	HEAT/Frag, 165-mm penetration, 20 m Frag radius, 20 m range, weighs 1.1 kg. Rifle grenades: HEAT 150mm to 300m	Hand grd to 40 m. Dual purpose bullet-thru rifle grd, no recoil, 150mm/Frag 20 m, 3 in belt pack.
Squad Machinegun	7.62x54 mm, frangible/sabot rds 1,300m. EO/3 gen II computer LRF sight 1,500m.	Add MMW radar, 5 km detection.
Combat Shotgun (replace one assault rifle)	12-gauge pump or semi-auto, 12 rds. Short and long change-out barrels, day/ night sights. Variable choke. Shells: HE, AP-sabot, door-buster, starburst, slug, concussion, frangible, flechette/anti-UAV	Time fuzed focused fragmentation airburst rd for use against dug in personnel, aircraft and UAVs. Multispectral smoke, CS grenades. TV/II recon rounds to 400 m.
Sniper Rifle Light	Bolt action, 7.62 mm rd, 15 lbs max weight with ammo. 10X optic w/2 gen II night channel. Range to 1000m.	Ballistic EO holographic LRF sight. Fused IR/FLIR channel 1,500m. Remote fire robot. Laser designator
Anti-Material Rifle (AMR) or Sniper Rifle (Heavy)	Semi-auto .50 cal. Weight 25 lbs. AMR/ anti-armor range 1,800 m. Armor pen 20 mm. As sniper rifle, range 1,000-1,500 m. Frangible multipurpose rd (AP 11 mm, incendiary 20 fragments). EO sight (20x) with 3 gen II night channel.	Ballistic EO holographic laser range-finder sight. Night sight fused IR/FLIR. Range 2,500+ m. Remote fire platform-60m link or weapon robot option. Laser designator.

SYSTEM	NEAR-TERM OPFOR (FY 15-20)	MID-TERM OPFOR (FY 21-28)
Automatic Grenade Launcher (AGL)-Light	35mm man portable launcher with 6/9/12-round drums. HEAT grenade range 600 m 80mm penetration. Frag-HE grenade range 1,500m. Buckshot grenade. EO day/ 3 gen II night sights. 1 per infantry squad	Air-burst munition (ABM), ballistic sights. EO and Fused IR/FLIR sight. Remote fire. Multispectral smoke grenades. Recon, HEAT/HE TV-guided atk grenades to 1,000 m
Automatic Grenade Launcher (AGL)-Heavy Weapons squads and vehicles	40 x 53 mm weight 17 kg. Range 2,200m. Ballistic fire control computer w/ EO sight. Dual-purpose grenade, HE with 60mm armor penetration. Buckshot round. Electronic fuzed HE air-burst munition (ABM). 32/48-round cans. Thermal night sight, range 2,200 m.	HEAT rd defeats 200+ mm armor. EO/ fused IR/FLIR sight. Multispectral smoke, unattended ground sensor (acoustic, seismic RF), and comms jam grenades. Robot option. Mount on all maneuver/recon vehicles. TV/IR attack grenades.
Multi-purpose Grenade Launcher (disposable)	76mm thermobaric HEAT, 250m range, 440 mm penetration. Reusable II sight.	Range to 400m. Fire from enclosed spaces. Nil smoke, little noise.
Antitank Grenade Launcher (disposable)	125mm tandem HEAT 300m range, 1000+ mm. Shoulder fired. Nil smoke.	Multipurpose DP effects, 500 m. Reduced recoil-enclosed spaces.
Antitank Grenade Launcher (ATGL - medium range) Mid-Term: Expand to AD/AT Missile Launcher	60mm launch tube, from enclosed spaces. Tandem warhead (1,150 mm to 600m), dual purpose 1700m. Ballistic LRF/3 gen II night sight to 1,500 m. Remote launch tripod. Nil smoke. High velocity 57-mm DP high vel rocket 1,000m, 300 mm pen	SAL-H, TV/IR-guided grenades to 1,000 m. Fused IR/FLIR night sight. ADAT KE dart fits converted launcher. Range 4 km. Laser designator 5 km, including artillery and mortar rounds.
Antitank Grenade Launcher (long range) Mid-Term: Expand to AD/AT Missile Launcher	125mm tandem HEAT 800+m range, 1100+ mm. HE-Thermobaric grenade to 1700 m. LRF computer sight. EO day/3 gen II Night sight. Nil smoke. Remote-fire platform option. Tripod and bipod.	SAL-H/TV/IR-guided: HEAT and HE grenades 1,200 mm. ADAT SAL/LBR KE dart to 4 km. EO and fused IR/FLIR sight, laser designator to 5 km for arty/mortar rds.
Remote-fire Platform and Weapon Robot or Laser Target Designator (LTD) Robot	Man-portable, <15 kg, 60m Laptop/PDA link. EO/3 gen II sight. MG/AGL/rifle. LTD robot TV/2nd gen FLIR, 10km range	Tracked, 24 kg, 2 hour charge, fused II/FLIR 10 km, 10 km rg RF link. LTD has 3rd gen FLIR, range 15km
Acoustic Targeting System (ATS)	Backpack/vehicle triangulates on aircraft, vehicle weapons to 6 km, MGs 2.5 km. Helmet mount to 800 m. Light display.	Increased range (10 veh weapon, 5 MG). Add auto-return fire for MG. Link to veh weapons/nets auto-slew
General Purpose and Air Defense Machinegun	12.7mm low recoil on ground tripod. Chain gun version on light vehicles, ATV, motorcycle, etc. TUV/LAV use RWS. Remote operated ground or robot version. Frangible rd 2 km, sabot 2.5 km. RAM/RAP/IR camouflage/ screens. TV/FLIR fire control. Lightweight MMW radar 5 km. Display link to AD azimuth warning net. Emplace in 10 sec. RF/radar DF set. ATS control option.	Stabilized gun and sights. Remote-operated computer FCS with PDA or laptop. Fused FLIR/ II to 5 km. Frangible, sabot rds to 3 km. Laser dazzler blinds enemy. Micro-recon/ heli atk UAVs. Robot version. Some light/AD vehicles replace w/ 30-mm recoilless gun on RWS. AHEAD round 4 km, FCS 10 km. Add-on ADAT missile launcher
Man-portable attack UAV (NLOS Backpack Munition)	2.5 kg tube launch with PDA, CCD/IR image, 10 to km and 155 m altitude, at 100-160 km/hr, with 10 min loiter, in-flight arm for HE charge, NLOS dive attack vs moving/static targets	Ranges to 20 km with 40 min loiter. Remote, air, ground, water craft, vehicle launch. 3rd gen thermal view. AT/AP remote sensor mines.
Infantry Flame Weapon	Reusable thermobaric 90-mm grenade (2/lchr) to 800 m. Effects = 152mm artillery rd. Targets personnel, bunkers, LAVs, etc. Nil smoke. EO/II night sight	Precursor (200 mm pen) DP grd. Computer LRF day/night sight. SAL-H guided. Remote fire and robot option. Use in enclosed space

SYSTEM	NEAR-TERM OPFOR (FY 15-20)	MID-TERM OPFOR (FY 21-28)
Vehicle/Man-portable Close Protection System (CPS)	Smoke grenade launchers can use multi-spectral smoke, CS smoke, Frag-HE grenades, range 3-40 meters, depending on angle. ATS control. Man-portable.	Man-portable remote control launcher. Quick load 3-6 grenades. Other Grds: CS gas, HE, AT/AP mines. 2-4 pods/vehicle.
Infantry Weapon Night Sight (Night Optical Device- NOD)	3d gen II night vision goggles/sights/ IR pointers for riflemen range 1,000. AGLs, MGs, sniper rifles/AMRs, ATGLs to 1,500m. FLIR recon sensors 3,000 m.	Uncooled 3rd gen FLIR (thermal and II combined) NVGs and weapon sights infantry 600m. Priority weapon sights 2,000+ m.
	INFANTRY WEAPONS	
Armored Personnel Carrier (APC)	8x8 wheeled chassis. Add ERA. 30-mm gun (and imp rd), coax MG. FOG NLOS ATGM lchr 4 km. Thermobaric ATGM. FLIR. 2 remote 7.62-mm MGs and 40 mm ABM AGL. CPS and ATS. Attack UAV launch	10x10 wheeled hybrid drive. Box ERA. CPS. Fused FLIR/II sight 13 km. 30-mm recoilless chain gun, RWS. Air-burst rds. ADAT KE missile and NLOS ATGM to 8 km. TV/IR attack grenades.
APC Fire Support Vehicle (Weapons Squad APC or Infantry Support Vehicle [1/ pltn or company], or Company Command Vehicle in Mech APC Bn)	Wheeled 8x8 chassis with ERA. 100mm & 30mm guns, 40 mm ABM AGL, auto-tracker, hunter-killer FCS. Gun-launch ATGM NLOS (SAL) 8 km fire on move. 30 and 100-mm HE elec fuzed rd 7 km. Imp 30-mm rd. 12.7 mm AD MG, 2 remote 7.62 MG. ADAT KE msl lchr 7+ km. Laser designator 10 km. CPS, ATS.	Above chassis & drive, ERA, fused FLIR/II sight. 100mm KE/600 CE protection. Cased telescoped gun 45-mm. ADAT KE dart rd 4 km, SAL/LBR ATGM 8-12 km. CPS. Micro-UAVs recon/attack. Tunable laser designator to 15 km. Radar/ MMW radar. SATCOM. Atk grds
APC Air defense/Antitank (ADAT) Vehicle	APC Bn and Bde MANPADS btry, selected other units	See AIR DEFENSE
Infantry Fighting Vehicle	2-man turret, amphib tracked. Add ERA. 30mm gun (sabot, 110+mm pen). Frag-HE Electronic-fuzed ammo 5 km. Buckshot rd for UAVs. 40-mm ABM AGL, 4 x fiber-optic guided ATGM 8 km launch on move, 2nd gen FLIR. Auto-track, hunter-killer FCS. Remote MGs 12.7mm, 2 x 7.62. Laser designator 15 km. CPS/ATS	Hybrid drive. Box ERA 100mm KE /600 CE. 45-mm CTG. Fused FLIR /II sight 13 km. ADAT dart rd 4 km. SAL/LBR ATGM 8-12 km. MMW radar. Micro-UAVs recon/atk. Radar warner, laser radar. Tunable LTD 15 km. CPS. 2 remote MGs, 1x 12.7. TV/IR attack grenades
IFV ADAT Vehicle IFV Bn/Bde MANPADS	IFV chassis and APC ADAT weapons and upgrades	See AIR DEFENSE, APC ADAT for weapons and upgrades
Heavy Infantry Fighting Vehicle (Heavy IFV in Heavy Bn, Infantry Fire Support Vehicle, or IFV Company Command Vehicle, as Required)	2-man turret, amphib tracked, Box ERA. Auto-track, hunter-killer FCS, ATGM lch on move. 100 and 30mm guns. 100 mm HEAT, DPICM rounds. 40mm ABM AGL, NLOS (LBR/SAL) ATGM 8+km lch-on-move. 30/100-mm HE electronic fuzed rd 7 km. 30-mm buckshot rd for UAVs. AD 12.7mm MG, 2 remote 7.62 MG. Laser designator 15 km. CPS/ATS	Hybrid drive. Armor and box ERA protects 300mm KE/800 CE. 45-mm CTG, KE, HE, ADAT rds. KE missile 8 km. Micro-UAVs recon/ atk. CPS. Fused FLIR/II sight 13 km. ATGM 8-12 km. Tunable laser designator to 15 km. Radar/MMW warners. AGL, 2 remote MGs, 1x 12.7. TV/IR atk grds
HIFV ADAT Vehicle HIFV and Amphib Bn/Bde	HIFV chassis with APC ADAT weapons and upgrades	See AIR DEFENSE, APC ADAT for weapons and upgrades

SYSTEM	NEAR-TERM OPFOR (FY 15-20)	MID-TERM OPFOR (FY 21-28)
Main battle tank	Welded turret, 3rd gen ERA, more armor. 125mm gun, bigger sabot (800+mm pen), LBR ATGM 6 km. SAL/IR-homing rd to 5 km in 1 sec, SAL-H ATGM 8 km.. Improved 2nd gen FLIR (7 km) and 50X Day/night sights. ATGM fire on move. Auto-tracker, laser radar, laser dazzler blind sights. Focused frag HE rd for heli, light AT targets. HEAT-MP, DPICM sub munitions rds. IR/MMW CM. Active suspension. CPS/ATS. Controls robot	Reduced remote turret, compartmented crew, electronic/ceramic armor, 500 mm top/mine armor. Laser/radar warners. Hybrid drive. CPS/ATS/APS. Sabot defeats 1000 mm KE. KE ATGM to 12 km. Tunable LTD to 15 km. ADAT msl 8 km. Fused 3rd gen FLIR/II sight 100 X to 13 km. MMW FC. Atk/recon micro-UAV, atk grds. Controls a robot tank.
Tank Robot (Near Term) Robotic Tank (Mid-Term)	Tracked LTD tank robot fits on platoon cmd tank when unused. Light armor, MMW/ IR screens - no signature. It designates SALH ATGMS and rds.	1/2 size MBT. Driver seat for pre-battle. Armor, weapons, mobility/ survivability (CPS/ATS/APS) same as tank. ATGM launch veh version
Tank ADAT Vehicle Tank Bn/Bde MANPADS	Tank chassis and APC ADAT weapons and upgrades	See AIR DEFENSE, APC ADAT for weapons and upgrades
Armored Tactical Utility Vehicle (TUV)	4x4 swims, 1/4 mt amphib trailer, Remote 12.7-mm MG and 40-mm AGL). Multirole (mech/recon/C4/AD/AT/security/ log). Run-flat, central tire inflation. CPS/ATS	6x6 hybrid drive, mine protection. 30-mm gun, RWS (see APC). Recon masted radar/fused FLIR/II sights. Smoke, recon/atk grenades. CPS.
Armored TUV ADAT Vehicle Infantry, SF, other units	12.7-mm MG, 2x lchr FOG/ IR-homing ATGM, EO/FLIR sight, manpack ADAT lchr. AD net azimuth warning. CPS/ATS	See above. Tunable laser designator, range 15 km. Radar warning receiver. MMW radar.
MANPADS Vehicle	Bn/Bde, insurgents. Truck, TUV, ATV. Remote launch, EO/thermal sight. Azimuth warner. Smoke/ATS	See Air Defense
Light Strike Vehicle	4x4 rear engine, 4-person, 2 m ford. 35-mm AGL, 12.7-mm MG, and 40-mm ATGL. ATS	Light armor/mine shields. Hybrid drive. Amphib (Bladders). 30-mm gun RWS (see APC). ATGM 8 km.
Light Strike Vehicle ADAT	4x4 rear engine, 4-person, 2 m ford. 35-mm AGL, 12.7-mm MG, KE LBR Msl. FOG/IR-hom ATGM 4 km. ATS	Light armor/mine shields. Hybrid drive. Amphib (Bladders). 30-mm gun RWS (see APC). ATGM 8 km.
Tactical Motorcycle Motorcycle ADAT version	Low noise diesel engine, 35-mm AGL Swim sacks. MMW/IR camouflage and screen. ATS	Continuous rubber track. FOG/ IR-homing ATGM, imp MANPADS. Track conversion in snow/swamp.
All-Terrain Vehicle (ATV) and ATV ADAT	6x6, 4-person capacity, 3.5 mt payload. Swim. Has 12.7-mm MG, 35-mm AGL. ADAT, AT, other roles. Amphib trailer. Track conversion in snow/swamp. ATS	8x8. Mine protection. Hybrid electric/diesel drive. Snap-on cab for cold weather etc. 23-mm light chain gun on pintle mount.
RECONNAISSANCE, INTELLIGENCE, SURVEILLANCE, TARGET ACQUISITION		
Binocular Laser rangefinder and Goniometer	Handheld 20km detection, 5-7km recognition, GPS. Thermal channel (below) goniometer, computer - digital transmit	See Thermal Binoculars (below). Heads-up display links to terminal. Transmit images to net.
Helmet Cam	Soldier camera link to laptop/PDA 2 km. NVG feed. Remote mast-mount.	Improved night viewer with 3 gen II or thermal. Nigh range 2 km.
Thermal Binoculars	Uncooled 2 gen FLIR. 2x electronic zoom (EZ), image stabilization. Detect 9 km (13 EZ), recognition 3.5 km (5.5 EZ)	Add LRF, laser pointer, internal GNSS. Fused FLIR/II camera. FOs call indirect fires 10-13 km, 6+ with precision, direct fire 5.5 km+. IDs heli at 7 km w EZ, detects at 13 km

SYSTEM	NEAR-TERM OPFOR (FY 15-20)	MID-TERM OPFOR (FY 21-28)
Laser Target Designator/ Rangefinder (Manportable)	Man-portable, encoded, designate SAL-H rounds, bombs, ATGMs to 10 km. 2 gen thermal sight. Mounts on sensor robot	Tunable laser designator with encoded pulse to 15 km. Mounts on sensor robot
Observer Sensor Suite For Recon, SPF, Security, Anti-tank, Air Defense, Artillery (Dismount, ATV, Motorcycle, Vehicle)	Goniometer/laser designator base. Laptop or radio link. GPS, thermal laser range-finder binoculars, manpack radar. Aircraft azimuth warner. Net with UGS, remote camera, micro-UAVs.	Mount on Sensor Robot. Increased range, encryption, SATCOM. Fused FLIR/II night sight. Tunable encoded LTD to 15 km designates for all SAL-H munitions.
Laptop Computer for Digital Sensor Network	System accesses sensor links: video cameras tactical units, UGS monitor, maps/unit status displays, azimuth and alert nets. Digital data links, microphones for discussion, ground station terminal. Access encrypted internet links, long-range cordless and SATCOM phones. Terminal to remote-detonate mines and control minefields.	Personal data assistant for dismount use or for mounting in or linking to weapon FCS. Solar rechargeable batteries, extended range on links with retransmission UAVs. Use for hand-off control of UAVs, in-flight munition retargeting. Fuse UAV, weapons, cameras, TV recon grde image, battle management data.
Surveillance radar	Man-portable low probability of intercept GS radar to detect/classify vehicles 30km, detect personnel 18km. Netted digital/graphic display.	Remotely operated, on a mast, with man-portable day/night EO sensor suite or from concealed base.
Mortar and Grenade Recon Rounds TV/IR attack grenades: Mid-Term	82 mm mortar round with a CCD TV camera to 5,700 m, aerial NLOS zoom view to laptop for 20 sec. Rifle/hand-held/AT grenades with TV cameras send video to PDA or laptop on descent.	Mortar rds (81/120), grenades with slewable fused FLIR/II and zoom. 40mm AGL grd 2,200 m. Shotgun grds. Recon, TV/IR attack grenades (HEAT/HE) from vehicle 82-mm smoke grenade launchers to 1,000m
Unattended Ground Sensor Set	Netted, acoustic, seismic, magnetic, IR. Acoustic sensor UGS array extends 12 km, for accuracy within 3m.	Robotic sensors with sleep mode, underground concealed hide position (self-relocate, dig in). Nil visual/IR/MMW signature.
Remote Cameras and Sensors	Motorized, masted, with constant-on, command-on or acoustic/seismic wakeup. 20-30km link range. CCD measures and in-ground mount. 2 gen FLIR day/night passive scan.	Robotic sensor entrenched and concealed. On wake-up, mast rises to RISTA mode. Integrated net digital display, link to sensor robots and robotic weapons.
Smart Dust	Rocket/UAV/aircraft scattered crush sensors emit for 1/2 hour.	Scatterable, attach to metal. Acoustic/crush/seismic. Emit 1 hour.
Sensor Robot	Man-portable tracked robot w/cameras in multi-sensor pods (acoustic/EO/ seismic) w/wake-up. Transmits image to monitor. Camera range 3 km. RAM. Laser designator direct munitions 10 km	Solar charge and vehicle quick charge, longer charge capability. Camera/link range 20-30 km. Self-entrench. Composition chassis and RAM is undetectable to sensors.
Acoustic sensor vehicle	Vehicle mounts microphones or dismount array, DFs/acquires aircraft, vehicles, or artillery. Rapid queuing and netted digital display. Range 10 km, accuracy 200m. Three vehicle set can locate artillery to 30 km with 1-2% accuracy in 2-45 sec. DF/ cueing rate 30 targets/min.	Range extends to 20-30 km with 10 m accuracy. Micro-UAVs with microphones to supplement the network in difficult terrain. Track and engage multiple targets. Range and accuracy SAB. Hybrid electric/diesel drive.

SYSTEM	NEAR-TERM OPFOR (FY 15-20)	MID-TERM OPFOR (FY 21-28)
Wheeled Armored Reconnaissance Vehicle (ARV)	4x4 and 4 aux wheels, low profile. 12.7-mm AD MG. NLOS FOG ATGM 8 km Multi-sensor mast, 2nd gen FLIR. GS radar classify vehicles 30 km, detect person 18km, laser designator 15 km, UGS, laser radar, MANPADS, ATGM. CPS/ ATS. Conformal MMW-IR net, MMW/IR grds. Canister UAV 10 km.	Hybrid drive. IFF. Fused FLIR/II to 24 km. Micro-UAV range 35 km. 30 mm recoilless chain gun on RWS (see APC). SAL/LBR ATGM 8-12 km. Tunable LTD 15 km. Multi-spectral smoke launcher and recon and TV/IR attack grenades to 1,000 m. Sensor robot. CPS.
Tracked Reconnaissance Vehicle	2-man turret, 30-mm gun, 12.7-mm AD MG, MANPADS, ATGM. Masted multi-sensor suite, 2 gen FLIR, laser radar, auto-tracker, laser target designator direct arty /mortar rds/bombs, ATGMs 15 km. GPS/ inertial nav, digital data. Radar detects vehicles 30 km, personnel 18. UGS net. Canister UAV. CPS/ATS	Hybrid drive. Fused FLIR/II to 24 km. IFF, Micro-UAVs to 35 km. 45 mm CT gun. ADAT KE round 4 km. SAL/LBR ATGM 8-12 km. Multi-spectral smoke launcher and recon and TV/IR atk grenades. Tunable laser designator 15 km. Sensor robot. CPS.
Long-range sensor vehicle	Tracked vehicle with elevated sensor suite on pod. Day/night TV, MMW radar detect to 45 km vehicle, 20 km personnel. 2 gen FLIR Net to UGS, UAVs, etc. Digital links to arty, AT, AD, recon, etc. 12.7-mm AD MG. Laser target designator to 15 km. CPS/ATS.	Longer range, increased target handling/transmission capacity. Manpack AD/AT LBR missile to 8 km. Fused FLIR/II to 24 km. Tunable laser designator to 15 km. Hybrid electric/diesel drive. CPS. Recon and TV/IR atk grenades.
Ground or Vehicle Launch Mini-UAV	2-backpack system. Man-portable ground launcher, and laptop terminal. Vehicle-launch from rail or canisters. TV/FLIR. Range 35 km, 3-hr endurance.	IR auto-tracker. Laser designator. Cassette launcher for vehicles. Signal retransmission terminal. Bus dispense micro-UAVs, UGS, mines
Micro-UAV	Hand-launch 4-rotor, 4 kg, 5 km/1 hr, GPS map/view on PDA/netbook. Atk grenade	< 1 kg for dismount sqd/tm, 2 km range. Add grenade for atk UAV
Heliborne MTI Radar	Range 200 km, endurance 4 hrs.	Range 400 km. Add SAR mode.
Commercial Satellite Imagery	Resolution 5 m for IR, SAR also available. <2 days for request. Terminal on tactical utility vehicle at division. Can be netted to other tactical units.	Response time reduction (to <6 hours). 1-m resolution.
ANTI-TANK		
Manpack Air Defense and Antitank (ADAT) Kinetic-Energy Missile Launcher (also listed in Air Defense)	Co/Bn substitute for ATGMs and AD. Targets helicopters and LAVs. Shoulder launch missile with 3 KE LBR submissiles 8 km, 0 m altitude. Submissiles have 25-mm sabot/HE warhead. Nil smoke. Mount on robotic launcher (below). FLIR night sight.	Fits in 45-100mm guns. Defeats all targets up to 135 mm KE. Range 8 km, time of flight 6 sec. Fused FLIR/II sight 10 km. Launch from enclosed spaces. Can mount on robotic ADAT launcher or ADAT Robot vehicle (below).
Man-portable ATGM Launcher (Also pintel/vehicle dismount)	SACLOS guided to 3 km. Tandem warhead defeats 1,200mm. Thermal sight. Jam-proof low noise/smoke. Fire from enclosed spaces. Can mount on robotic launcher (below)/vehicles.	Twin ATGM remote ground veh/ launch station with auto-tracker. Fused FLIR/II sight 5 km. NLOS /IIR homing missile to 4 km. Can use ADAT missile. Laser dazzler
Ground Turret	Ready-made hole mount turret for hoist installation, w/12.7 mm MG, 4 km ATGM launcher, thermal night sight, and radar absorbent/IR reflective paint on cover. Invisible until activated.	Add remote/unmanned pop-up turret. FOG-M top-attack or IIR-homing attack 8 km. Tandem warhead 1,300 mm. Fused FLIR/II sight to 10 km. CPS

SYSTEM	NEAR-TERM OPFOR (FY 15-20)	MID-TERM OPFOR (FY 21-28)
Ground/ Vehicle Pintle Mount ATGM Launcher	Combat support vehicle with portable robotic twin launcher (below). FOG-M top-attack or IIR-homing direct attack 4 km. Tandem warhead defeats 1,000+ mm. Thermal sight 5 km range. Low noise/smoke, countermeasure-resistant.	Range increase to 8 km and 1300 mm penetration. Thermobaric ATGM. Fused FLIR/II sight to 10 km. Launch from enclosed spaces. Laser dazzler. ADAT robot vehicle.
Robotic ADAT Launcher ADAT Robot Vehicle	Pintle mount shoulder/ground/ATV/ vehicle launch. Robotic launcher-60 m link. Twin auto-tracker. Operator in cover/spider hole. MMW/IR absorbent screen and net for operator, launcher and surrounding spall. CPS/ATS.	Masted 4-launcher, hybrid drive to self-entrench, then move to launch point. Fused FLIR/II sight to 10 km. Remote link 10 km. Most AD and AT host vehicles have 2 control stations and 2 robots. ATGM same as above. CPS.
Towed Antitank Gun	125mm gun, larger sabot (700+mm), LBR/SAL-homing ATGM 8 km. Stabilized FCS sights, auto-tracker. Auxiliary propulsion unit. TV day sight with (32x). Combined MMW radar and 2nd gen thermal night sight (5-7 km). Add SAL-H/IR HEAT rd 5 km in 1 sec, HEAT-MP, DPICM submunition round. ATS.	Remote unmanned gun with cassette, towed, dug into position, netted into AT net. Concealed position (retractable base and IR/MMW concealed). Fused FLIR/II sight to 10 km. KE ATGM (8 km), direct link to micro-UAVs and UGVs). Laser dazzler
Heavy Recoilless Gun, 106 mm and Recoilless Gun Vehicle (RGV)	TOW or RGV on TUV. Tandem HEAT round 700+mm 3 km. SAL-H, tandem ATGM (1,000+ mm), 8 km dive attack. .50-cal spotter rifle to 2,500 m. Laser designator. Computer sight, 2gen FLIR. HE, flechette rounds. RGV CPS/ATS.	HEAT rd 900+mm. Remote weapon system mount for APC, IFV, and TUV chassis. Fused FLIR/II sight to 10 km. Nil smoke/noise. Tunable laser designator for SAL munitions 15 km. Hybrid drive for RGV.
Self-Propelled Antitank Gun	Amphibious airborne tracked, 125 mm gun, larger sabot (700+ KE), SAL ATGM to 8 km. SAL-H/IR HEAT rd 5 km in 1 sec, DPICM submunition round, focus frag HE rd. Stabilized TV day sight (32x), 2 gen FLIR 5 km, auto-tracker. Laser designator 15 km. CPS/ATS.	Hybrid drive. MMW FC radar, NLOS ATGMs (8/12 km), direct link to micro-UAVs, UGVs). Fused FLIR/II sight 10 km. Micro-UAVs recon/atk. Laser dazzler. Tunable LTD 15 km. CPS, TV/IR attack grenades
Tracked ATGM Launcher Vehicle	Box ERA 300mm. NLOS/IIR ATGM launcher on IFV. 1,300 mm dive attack, 8 km. HE Thermobaric ATGM. Low noise/ smoke signature. 12.7-mm AD MG. Laser designator to 10 km. CPS/ATS. Manport ADAT KE missile launcher.	Imp ERA (300mm KE, 600 CE). Hybrid drive. NLOS/KE ATGMs LBR/SAL defeats 1,300mm at 8/12 km. EMP option. Fused FLIR/II 13 km. 2-target auto-track. Launch on move. Laser dazzler. Micro-UAV atk/ recon. 2 robots. Atk grds.
Wheeled ATGM Vehicle	4x4 Armored TUV with same launcher system as above. CPS/ATS.	Same launcher as above. Hybrid drive. Robot vehicle.
Airborne Infantry ATGM Launcher Vehicle	Airborne/amphib tracked light armored. Same launcher as above. CPS/ATS.	Same launcher as above. Hybrid drive, ERA, atk grds. Robot veh.
Heavy ATGM launcher Vehicle	Tracked, 6 lchrs, SAL-homing ATGM 1,400mm dive attack, 10 km. Warheads HEAT, Multi-purpose (HEAT/Frag-HE). 12.7-mm MG. Jam-proof auto-tracker, Laser designator 15 km. CPS/ATS.	Hybrid drive. Add IIR homing, 12 km range, EMP, and thermobaric. Warheads. Fused FLIR/II 13 km. Laser dazzler. Designator UAV 30 km range and 3-hr loiter. Atk grds

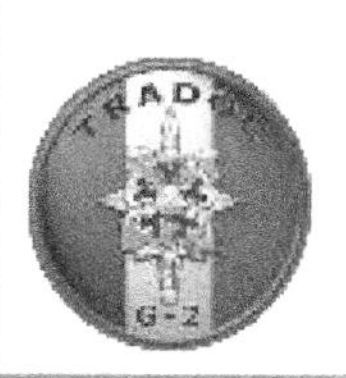

SYSTEM	NEAR-TERM OPFOR (FY 15-20)	MID-TERM OPFOR (FY 21-28)
Heavy ATGM Launcher Vehicle (and Land Attack Cruise Missile - LACM)	Tracked vehicle with 16 x SAL-homing ATGMs, Hybrid drive. RF-guided phase, 40 km. Fused FLIR/II acq to 10 km. MMW TA radar to 40 km. Warhead: 28-kg Frag-HE=1,300 mm penetration. UAV to 40 km with LTD (15 km range). FW and boat mounts. Anti-heli radar guided or SAL-homing. Atk grds.	Hybrid drive. Guidance adds radar or IIR homing. Warheads: Multi-purpose (HEAT/ Frag-HE) defeats 1,400mm), Bus for sensor-fuzed sub munitions, EMP warhead. Laser designator UAV range 100 km, 3-hr loiter time.
Attack UAV	Hit-to-kill system. Day/night 60+ km, up to 2 hours. GNSS/inertial navigation, TV/FLIR, Frag-HE warhead. They include an anti-radiation variant.	Cargo UAV 100 km dispenses IR/ MMW/SAL DP (600mm HEAT) sub munitions, EMP munitions, SAL ATGMs – UAV LTD 30 km.
Attack UAV Launcher Vehicle	Hit-to-kill UAV launch from modular launcher, 18 UAVs. GNSS/inertial nav, to 500 km. First version anti-radiation homing. Added TV guided and multi-seeker attack (hit-to-kill) UAV. Laser designator range 15 km. CPS/ATS.	Hybrid drive. Bus reusable UCAV with 4 ATGMs to 10 km, SAL-H bombs, or bus dispensing 16 terminally-homing sub munitions (with MMW/ IR seekers, or laser-homing DP sub munitions). CPS. LTD
Micro-Attack UAV	Hand or canister-launch UAV with TV and FLIR guidance to 10 km, 100-600 m altitude, with .25-.5 kg warhead.	Cassette/smoke grenade launcher launch for tactical vehicles. Recon and attack (top-attack) UAVs.
Mini-Attack UAV	Hand or vehicle canister-launch UAV with TV and FLIR guidance to 35 km, 100-600 m altitude, 1-4 kg warhead.	Cassette launcher launch for tactical vehicles. Recon and attack (DP with tandem 600 mm top-attack).
	FIRE SUPPORT	
Man-portable Mortar	Conventional munitions, 82mm FRAG-HE 6.7 km, RA 13.0 km. SAL-H 6.7 km. Day/night direct/indirect fire sight. GPS. Prox fuze. Tandem ATGM 7 km.	Increased range and accuracy. Ballistic computer sight. Fused FLIR/ II 10 km.Self-lay. Dual guided (diff GNSS course correct/SAL) 13 km
Towed Mortar Upgrade	120-mm FRAG-HE – 9 km. Prox fuze. ADHPM: SAL-H and IR-homing HEAT – 9 km, Sensor-fuzed – 7 km. Night capable direct/indirect fire sight, self-lay.	Improved range/precision. Ballistic computer sight. Fused FLIR/II 10 km. Dual guided round (differential GPS corrected, SAL) to 12 km.
Towed Combination Gun	GPS gun lay/nav system. Frag-HE range 8.1 km (and prox), RAP 12.8, HEAT 1 km, SAL-H 12.8. Mortar rds SAB.	Automated fire control, Fused II/ FLIR 13 km. Autonomous lay, diff GNSS. Auxiliary propulsion unit.
Self-Propelled Combination Gun	120-mm gun/mortar system. GPS gun lay. Cannon Frag-HE (prox fuze option) 13 km, -RAP 18, HEAT 1 km. All mortar rounds. ADHPM: Mortar SAL-H and IR-homing 9 km, Sensor-Fuzed 7 km. Cannon SAL-H rd 9 km. CPS/ATS.	IFV chassis. Hybrid drive. Laser designator 15 km, diff GPS, automated FCS, autonomous lay. Fused FLIR/II 13 km. SAL-H, GPS rounds 12 km. SAL tandem HEAT ATGM to 20 km. APS
Towed Medium Gun-Howitzer	FRAG-HE - 30 km, FRAG-HE BB - 39 km, Artillery delivered high precision munitions (ADHPM): SAL-H - 25 km, Sensor-Fuzed – 27 km. GPS 40 km	Autonomous lay/fire direction. Enhanced lethality, differential GPS corrected munitions (and sensor-fuzed) 60 km.
Self-Propelled Medium Gun-Howitzer	45-cal gun. GNSS/inertial land nav, self emplace, FC. Munitions: FRAG-HE – 30 km, FRAG-HE base bleed - 39 km. ADHPM: SAL-H - 25 km, Sensor-Fuzed – 27 km, GPS-corrected 40 km.	Automated fire control. Barrel cooling, thermal warning systems. Autonomous lay/fire direction. Differential GNSS corrected rds (and sensor-fuzed) 60 km.

SYSTEM	NEAR-TERM OPFOR (FY 15-20)	MID-TERM OPFOR (FY 21-28)
Self-Propelled Medium Gun-Howitzer Tracked	Ford depth 5.5 m. 40-cal gun. GPS/ inertial land nav, self emplace and FC. FRAG-HE 23 km, FRAG-HE rocket ast 31.5 km. SAL-H rd 25 km, Sensor-Fuzed rd 27 km, GPS corrected 40 km	Automated FC. Autonomous lay/FD. Barrel cooling, thermal warning systems. Differential GNSS course corrected rds (and sensor-fuzed) 60 km.
Self-Propelled Medium Gun	Conventional munitions, FRAG-HE-BB – 30.5 km, FRAG-HE-RA–40 km. ADHPM: SAL-H - 25 km, Sensor-Fuzed – 24 km. GPS corrected 40 km.	Automated FC, barrel cooling and thermal warning, autonomous fire direction. Diff GNSS corrected rounds (and sensor-fuzed) 60 km.
Manportable Single Round Rocket Launcher	122mm FRAG-HE – 10.8 km. SAL-H, Sensor fuzed 10.8 km. On tripod	Increased range and accuracy. Enhanced lethality.
Rocket Launcher Pod (107mm) For Use on Improvised/ Modified Launch Platforms	6-tube (2x3 rockets) pod mounts on cart, vehicle (e.g., amphibious/airborne APC), or ground stand. Remote launch fire control. Cart/vehicle 1-3 pods. GNSS. Range 8.5 km. Limited lateral launcher adjustment (move vehicle). Mines and DPICM warhead option.	Improved launcher mount with servo-motors and remote computer FCS and in-view GNSS data. Munitions include: EMP, smoke, UGS, SAL-homing HE, tandem HEAT, recon, chem. Use with laser designator. Range 10 km.
SP Medium Rocket Launcher (100mm to 220mm)	122mm 50-tubes. Self-emplace (GNSS/ inertial nav). Onboard FCS. Munitions: Frag-HE 90° precision fall 40 km, GPS course-corrected DPICM and Frag-HE 36 km, RF jammer rd 18.5, SAL-H rkt 32 km, Sensor fuzed 33 km.	Extended range. Increased accuracy and lethality. Course corrected diff GPS/ inertial) in DPICM, multi-role (HEAT, HE, incendiary). Motorized spades for quick displace.
SP Heavy Rocket Launcher (220-240 mm)	Self-locating launcher, 16 tubes. GNSS/ inertial nav. Onboard fire direction. Rockets: 220mm FRAG-HE –43 km, DPICM, Chemical. Thermobaric – 43 km ADHPM: Sensor-Fuzed – 43 km	Increased accuracy. Enhanced lethality. MRL can launch cruise missiles, UAVs. Diff GNSS Course corrected munitions (DPICM, sensor fuzed, mines) to 70 km.
SP Heavy Rocket Launcher (240-300 mm and larger)	Self-emplace 300-mm 12-tube launcher. GNSS/inertial nav, onboard FCS. Inertial course-corrected rockets 100 km: mines, DPICM, Chemical, and Thermobaric. Sensor-Fuzed 90 km, UAV rocket 90 km	Range (100+ km). Differential GPS lnchr, GNSS course-corrected rkts. Enhanced lethality. Launch cruise missiles (attack UAVs) and recon UAVs to 470 km.
Weapon Locating Radar Vehicle (Counter Mortar/ Counter-Battery Radar)	Detection range with low error rate Mortar: 30 km, Cannon artillery: 20-25km, Rocket: 40km, Tactical Missile: 55km.	Faster computer processors with digital links, differential GNSS, and decreased radial error
ENGINEER		
Improvised Explosive Device (IED)	Command (RF, wire) arm/detonate. Also sensor armed/fuzed. Large shaped charge, EFP, daisy chain arty rds, large IED, mine converted to cmd/SF. Defeat RF jammers, magnetic detectors.	Fuzes and radio links which can convert explosive devices and materials into intelligent IED fields (see intelligent minefield)
Minelayer, Towed	Lays 10 to 12 mines per min. Lines 20/40 m apart. Can also lay controllable minefields.	Advanced sensors. Target discrimination. Can lay intelligent mines.
Minelayer Vehicle	Armored chassis w/7.62mm MG, lays 1,000 m AT field with 5m between mines. Lay controllable mines. CPS/ATS	Add vehicle mount mine launchers. Also lays intelligent minefields. Hybrid electric/diesel drive.
Infantry Portable Scatterable Minelaying System	Remotely lays AT/AP mixed minefield 200-400m square from a distance up to 1090m. At platoon. 6 lb, 5 min set-up. Controllable mines.	Add intelligent mines. ATGL and AGL-delivered mines.

SYSTEM	NEAR-TERM OPFOR (FY 15-20)	MID-TERM OPFOR (FY 21-28)
Scatterable Mines	Deliver by artillery, cruise missile, UAV, rotary or fixed-wing aircraft. Non-metallic case, undetectable fill, resistant to EMP and jammers, w/self-destruct.	Advanced multi-sensor mines with wake-up and target discrimination. Prox fuze mines. Controlled minefields and intelligent mines.
Artillery Scatterable Mine Rounds and Rockets	Cannon, MRL, mortar, gun/mortar. 122-mm MRLs can fire AT and/or AP mines and covers 24-81 hectares.	Extended range. Controlled minefields (RF link)
Remote Mine Launcher Pod System (Vehicle, trailer, ground)	APC w/180 x 140-mm pods, scatters mines, UGS, jammers, CS gas, and smoke grenades, 30-60m from pod. Can lay field AT/AP 1-1.2km x 30-120m. CPS/APS	Multi-sensor mines with wake-up, target discrimination. Controlled minefields, intelligent mines. Prox fuze mines (up to 540) 2 km 10 sec
Off-Route Mines (Side-Attack and Top-Attack)	Autonomous weapons that attack vehicles from the side as the vehicles pass. 125-mm Tandem HEAT (900+ mm). Target speed 30-60 km/h, range 150m acoustic and infrared sensors.	Sensor-fuzed EFP 600mm KE top attack. Remote or sensor actuated (controller turn-on/off), 360-degree multi-sensor array. Hand/ heli/ UAV/arty/ATGL mortar emplace.
Controlled Mines and Minefield	AT/AP, machine emplaceable. Armed, disarmed, detonated by RF command. Chemical fills and non-metallic cases are undetectable. With CM and shielding, negate jammers/pre-detonating systems.	Control may be autonomous, based on sensor data and programmed in decision logic, or by operators monitoring with remote nets.
Smart Mines	Wide-area munitions (WAM) smart autonomous, GPS, seismic/acoustic sensors. AT/AV top-attack, stand-off mine. Lethal radius of 100 m, 360°. Hand-emplace	Discriminate targets. Reports data to a monitor, evaluate target paths, built-in logic. Use GPS to artillery/ heli-emplace. Non-nuclear EMP or HPW options
Unexplosive Ordnance (UXO)	Artillery cannon or rocket DPICM sub munitions in impact pattern.	Unused blue remote-launch precision munition pods may be seized and used against them.
Intelligent Minefields (including Non-nuc EMP, Jam, and HP Microwave)	Developmental programs and not proliferated	Self-healing, autonomous monitoring of obstacle integrity. Advanced sensors, target discrimination, built-in logic. Non-nuclear EMP or HPW.
Engineer Reconnaissance Vehicle	Tracked IFV chassis. Amphibious- recon equip: sonar, NODs, rangefinder, soil analyzer, gyrocompass, underwater mine detection. CPS/ATS	Hand-held and vehicle-mounted ground-penetrating radars for mine detection. Hybrid electric/diesel drive. CPS
Obstacle Clearing vehicle	Tank chassis, NBC-protected, dozer (3.8m), crane (2mt), scoop/ripper, and mine detonator. CPS/ATS	Hybrid electric/diesel drive.
Vehicle or Towed Line Charge Mineclearing System	Mounted on truck, IFV, APC, TUV or tank. Rocket launch 10 tubes HE or FAE, to 3km. Breach lanes 10x60m.	
Line-Charge Mineclearing Vehicle	Clears lane 6x9 m. 2 line charges. CPS/ATS.	Hybrid electric/diesel drive.
INFORMATION WARFARE		
Lightweight Mobile ESM/DF	0.7-40 GHz, ESM/DF	SATCOM intercept capabilities
Electronic Warfare Radio Intercept/DF /Jammer System, VHF	Intercept, DF, track & jam FH; identify 3 nets in non-orthogonal FH, simultaneous jam 3 fixed freq stations (Rotary/fixed wing/UAV capable)	Integrated intercept/DF/jam for HF/VHF/UHF

SYSTEM	NEAR-TERM OPFOR (FY 15-20)	MID-TERM OPFOR (FY 21-28)
Radio Intercept/DF HF/VHF/UHF	Intercept freq range 0.1-1000 MHz. (Rotary/fixed wing/UAV capable)	Wider Freq coverage. SATCOM intercept. Fusion/cue w/other RISTA for target location/ID
Radio HF/VHF/UHF Jammer	One of three bandwidths; 1.5-30/20-90/100-400 MHz, intercept and jam. Power is 1000W. (Rotary/fixed wing/UAV capable)	Increased capability against advanced signal modulations. UAV and mini-UAV Jammers.
Portable Radar Jammer	Power 1100-2500W. Jam airborne SLAR 40-60km, nav and terrain radars 30-50km. Helicopter, manpack.	UAV and long range fixed wing jammers.
High-Power Radar Jammer	Set of four trucks with 1250-2500 watt jammers at 8,000-10,000 MHz. Jams fire control radars at 30-150 km, and detects to 150 km.	UAV jammer and airship jammer. Hybrid electric/diesel drive.
Portable GNSS jammer	4 -25 W power, 200-km radius. Man-portable, vehicle & airborne GNSS jammers, airship-mounted jammers	Man portable, vehicle & airborne (UAV) GNSS jammers-increased range and power, and improvements in antenna design
Arty-delivered and ATGL-launch Jammer	HF/VHF (1.5-120 MHz), 700m Jamming radius, est. (1-hr duration). 300 m for ATGL-launched version	Increased capability against advanced signal modulations
Missile and UAV-delivered EMP Munition	Cruise missiles and ballistic missile unitary warhead and submunition.	Increased capability against advanced signal modulations
Artillery-delivered and Manpack EMP Munition	Cannon (152/155-mm), rocket (122/220 /300-mm), and mortar (82/120-mm).	Increased power, capability, and range.
Cruise Missile Graphite Munitions and Aircraft "Blackout Bombs"	400-500 kg cluster bombs/warheads with graphite strands to short out transmission stations and power grids.	Rocket precision and UAV-delivered munitions.
EMP Mine	Larger EMP mine. Effective radius 350 m, irregular/disruptive 500 m.	See intelligent minefields and smart mines
COMMAND AND CONTROL		
Radio, VHF/FM, Frequency-hopping	30-88 MHz, 100 hps, channels: 2,300, Mix of analog and digital radios, tactical cellular/digital phone, all nets digitally encrypted. Burst trans. UAV Retrans	Digital radios, tactical cellular/digital phone, and satellite phones, all nets encrypted
Radio Relay Station, VHF/UHF,	60-120/390-420 MHz, range 30-40km per hop LOS	Digital communications networks. Network management station, automated battlefield management system
Command Post Vehicle, Division (wheeled and tracked versions)	4xHF/VHF high power, 1x VHF, 75-2000km. Digital comms, graphics, voice back-up. SATCOM digitally encrypted.	Completely digital comms net thru all levels, fiber-optic cables. Networked automated, secure, and integrated battle management system
DECEPTION & COUNTERMEASURE SYSTEMS		
Armored Vehicle Decoy, Mobile	Towed trailers & decoy heater units, and flares. Used in concert with obscured target vehicle for positioning near target to divert homing munitions. Radar (and motorized) corner reflectors. Inflatables, tethered, move w/air currents.	Acoustic decoys w/seismic effects. Multi-spectral (high-fidelity) decoys powered for acoustic and IR signatures. Linked to vehicle warning systems
Armored Vehicle Decoy, Stationary	Multi-spectral (high-fidelity) erectable/ inflatable vehicle mock-ups, w/heaters & motorized radar corner reflectors	Acoustic decoys w/seismic effects. Multi-spectral decoys powered acoustic/IR signatures

SYSTEM	NEAR-TERM OPFOR (FY 15-20)	MID-TERM OPFOR (FY 21-28)
Vehicle and Weapon System Camouflage and Concealment	Tactical vehicles have MMW/IR paint and conformal nets, multi-spectral grenades, side skirts, thermal blankets, Thermal screens, laser/radar warners, acoustic engine & track noise modifiers.	Add mist thermal image concealment systems.
Camouflage and Concealment for Dismounts	Thermal screens and pop-up stands conceal from overhead, front, side visual/ thermal day/night vs MMW & IR. Face masks/gloves. Foxhole blast devices.	Ready-made spider hole covers, invisible to visual/ MMW/ IR sensors. Remote control option
Air Defense System Decoy	Manufactured and improvised decoys used with decoy emitter. Covered by AD systems in air defense ambushes.	Multispectral simulators of varied gun and missile systems mounted on robotic chassis.
Air Defense System Decoy RF Emitter	Expendable RF remote emitters with signal to match specific nearby radars, to trigger aircraft self-protection jammers.	Mounted on robotic chassis.
Non-Lethal (or Less Lethal) Weapons	Acoustic directed energy system, sticky foam, rubber bullets, acoustic disrupters	RF crowd disruption emitter. Water cannons. Laser dazzlers
ROTARY WING AIRCRAFT		
Attack Helicopter	30-mm auto-cannon, 8 NLOS FOG/IIR-homing ATGMs, range 8 km. Two pods semi-active laser homing (SAL-H) rockets 80mm (20x8 km) or 122mm (5x9 km). 2x LBR KE ADAT msl (warhead w/3 KE submissiles, 8 km range). Laser designator 15 km. UAVs to 30 km. 2nd gen FLIR auto-tracker. Radar and IR warners and jammers, chaff, flares	Tandem cockpit, coax rotor, 30-mm auto-cannon. 8 x RF/SAL-H ASMs to 18 km (28+kg HE=1300+mm), 2x SAL-H rocket pods (80mm or 122mm), 2 ADAT KE msl 8 km, and 2x MANPADs. 1/3 have ASM to 100 km. Fire control with fused II/ FLIR to 30 km, and MMW radar, link to ground LTD. Radar jammer. Atk and LTD UAVs to 30 km.
Multi-role Medium Helicopter and Gunship	24 troops or 5000kg internal. Medium transport helicopter. Range 460km. 30-mm auto cannon, 8 FOG-M/IIR ATGMs to 8 km, 40 x 80 mm laser-homing rockets, 4 AAMs. ATGM launchers can launch mini-UAVs and more AAMs. Mine pod option. Day/night FLIR FCS.	Fused FLIR/II to 15 km. 6x SAL-H ATGMs 18 km, 2 AAMs, 2 x 80/ 122-mm SAL-H rocket pods (20 or 5 ea). Laser designator to 15 km, and links to ground LTD. Aircraft survivability equipment (radar jammers and IR countermeasures).
Multi-role Helicopter and Gunship	12 troops (Load 400 kg internal, 1,600 external. Range 860 km. 23 mm cannon, 2 AAM, 4 SACLOS ATGMs to 13 km, TV/FLIR, day/night. Mine delivery pods	Launch 6x SAL-H ATGM to 18 km, 28+kg HE warhead. 2 x AAM Air-to-surface missile to 100 km. Pod w/7x SAL-H 90-mm rockets. Fused FLIR/II to 15 km. ASE
Light Helicopter and Gunship	3 troops (Load 750 kg internal, 700 external). Range 735 km. 20 mm cannon, 1 x 7.62mm MG, 6 SAL-H ATGMs to 13 km, 2 AAMs. FLIR night sight. Laser target designator. Mine pods	Launches 4x SAL-H ATGMs, to 18 km range. Fused FLIR/II to 15 km.
Helicopter and Fixed-Wing Aircraft Mine Delivery System	Light helicopter pod scatters 60-80 AT mines or 100-120 AP mines per sortie. Medium helicopter or FW aircraft scatters 100-140 AT mines or 200-220 AP mines per sortie.	Controllable and intelligent mines for aircraft delivery. Larger aircraft can hold multiple pods.
FIXED WING AIRCRAFT		
Intercept FW Aircraft	30-mm auto-gun, AAM, ASM, ARMs TV/laser guided bomb. 8 pylons Range 3,300 km. Max attack speed: Mach 4.	Stealth composite. ASE. Max G12+ All weather day/night. Unmanned option

SYSTEM	NEAR-TERM OPFOR (FY 15-20)	MID-TERM OPFOR (FY 21-28)
Multi-Role Aircraft	30-mm gun, AAM, ASM, ARM pods, guided, GPS, sensor fuzed bombs, 14 hard points. Thrust vectoring. FLIR	Improved weapons, munitions. Unmanned option. ASE all radars. Max G12+ All weather day/night
Ground-Attack Aircraft	Twin 30-mm gun, 8 x laser ATGMs 16 km 32 kg HE, 40 SAL-H 80mm rockets, ASMs, SAL-H and GPS sensor fuzed bombs, AA-10 and KE HVM AAM. 10 hard points. Range 500+km. FLIR	Stealth composite design. ASE. Unmanned option. Max G12+ 80-mm/122-mm rockets SAL-H, SAL-H ASM (28+kg HE=1300+ mm), to 40 km, 2 gen FLIR, radar jammer, day/night
	OTHER MANNED AERIAL SYSTEMS	
High-altitude Precision Para-chute and Ram-air Parachutes	High-altitude used with oxygen tanks. Ram-air parachute includes powered parachute with prop engine.	Increased range and portability. Reduced signature. Increased payload.
Ultra-light Aircraft.	Two-seat craft with 7.62-mm MG, and radio. Folds for carry, 2 per trailer.	Rotary-winged, two-seat, MG, 1/ trailer. Auto-gyro, more payload.
	UNMANNED AERIAL VEHICLES	
UAV (Brigade) It may also be employed in other units (e.g., artillery, AT missile, and naval)	Rotary wing, TV/FLIR/auto-tracker, with LRF and LTD designates targets to 15 km. Flies 180 km/6 hours, 220 km/hr, 2- 5,500 m alt, 100 kg payload. Can carry 2 AD/anti-armor missiles+MG for atk	Range extends to 250 km. Increased payload. Attack version can carry 2 SAL-H ATGMs (12 km range) or 1+ 4 70-mm SAL-H rockets (7 km, defeats 200 mm).
UAV (Divisional)	Day/night recon to 250 km. GNSS/inertial nav, digital links, retrans. SLAR, SAR, IR scanner, TV, ELINT, ECM suite, jammer/ mine dispensers. Laser designator 15 km.	Increased range, endurance. Diff GNSS. Composite materials, low signature engine. SATCOM Retrans relay links. Attack sub munitions.
UAV (Operational)	Day/night recon to 400+km. GNSS/ inertial nav with digital links. SLAR, SAR, TV, IR scanner, ELINT, ECM suite. Jammer option. Mine dispense. Laser target designator 15 km. Retrans/relay	Increased ranges, endurance. Diff GNSS. High altitude ceiling (35 km) option. Retrans/ relay/SATCOM links. UAV attack sub munitions. Laser target designators.
Unmanned Combat Aerial Vehicle (on Operational UAV platform)	Medium UAV with 4 ATGMs (fly out 10 km), laser guided bombs. Laser designator 15 km. Mine dispensers. GPS jammer, EW jammers. Range 400+ km.	Stealth composite design. ASE. Twin dispensers (pylons) with 16 terminally-homing sub munitions, MMW/IR seekers. Range 500+ km
	THEATER MISSILES	
Short-Range Ballistic Missile Transporter-Erector Launcher (TEL) and Cruise Missile (CM) Launcher	Twin launch autonomous vehicle (GPS/ inertial nav, self-emplace and launch). Range 450 km. Non-ballistic launch, separating GPS corrected reentry vehicle (RV) with decoys, CCD, 10-m accuracy. ICM, cluster, nucs. EMP warhead. CM option. TEL may convert to 6 x CM TEL (500 km, 3-m accuracy, below radar). Vehicle decoys. Vehicle has visual/MMW/ IR signature of a truck.	Missile improve range (TBM 800 km, cruise 1,000), with 1-m accuracy. TBM has GNSS-corrected maneuvering RV. Warheads for both include terminal-homing sub munitions, precision cluster munitions, EMP. Cruise missiles pre-program or enroute waypoint changes. Countermeasures include penetration aid jammers.
Medium-Range Ballistic Missile	Autonomous vehicle. Separating maneuvering warhead to 1300 km. GNSS,10-m CEP. Warheads include ICM, cluster, EMP, nucs. Penaids include decoys, jammers. Visual/MMW/IR signature of a truck.	Range 2,300 km, 1-m CEP. Differential GNSS, terminal homing, separating warhead. Warheads include EMP, terminal-homing cluster munitions. Non-ballistic launch and trajectory
Land-attack SAM system (secondary role for system)	The SAM system uses its EO sight and LRF (short/med range, strat "hittiles")	Range extends with SAM ranges. Passive operation with TV/FLIR.

SYSTEM	NEAR-TERM OPFOR (FY 15-20)	MID-TERM OPFOR (FY 21-28)
Cruise Missile Launcher Vehicle (Multi-role) Category includes specialized cruise missiles, long-range ATGMs, and SAM systems to engage targets at 12+ km.	Includes truck with 24 missile launchers. Range is 40 km. 28-kg Frag-HE warhead =1,300 mm penetration. Pre-program phase GNSS/inertial nav is used. LTD to 25 km range. Thermal night camera to 10 km. Support UAV with LTD is used. FW, RW, and sea-launch options.	Range 100 km. Penetration aids (countermeasures). IR Terminal-homing warhead or IR-homing submunitions can be used. Armored/ tracked launcher will mount 16 x 40 km missile launchers.
Cruise Missile Cassette launcher Vehicle	Off-road truck, GPS nav for autonomous ops. 16/lchr. Range 470 km; preprogram GNSS inertial guidance, with in-course correction, 10 CEP. Munitions include cluster munitions, thermobaric, chemical, DPICM/mine submunition scatter.	Launcher fire direction. Supersonic missile Diff GNSS/ inertial nav, 1-m CEP. Range 900 km. EMP warhead option. Warheads include homing cluster munitions. Penetration aids-countermeasures.
AIR DEFENSE		
General Purpose and Air Defense Machinegun	12.7mm low recoil for ground tripod. Chain gun light strike vehicle, ATV, motorcycle, etc, on pintle. TUV/LAV use RWS. Remote operated ground or robot option. Frangible rd 2 km, sabot 2.5 km. RAM/RAP/IR camouflage/ screens. TV/FLIR fire control. Lightweight MMW radar 5 km. Display link to AD azimuth warning net. Emplace 10 sec. RF/radar DF set. ATS control option.	Stabilized gun and sights. Remote-operated computer FCS with PDA/ laptop. Fused II/ FLIR 5 km. Frangible, sabot rds to 3 km. Laser dazzler blinds sights. Micro-recon/ heli atk UAVs. Robot mounts MG. Some light and AD vehicles replace gun with 30-mm recoilless chain gun on RWS, fires AHEAD round 4 km, plus Add-on ADAT missiles.
Improvised Multi-role Man-portable Rocket Launcher (AD/Anti-armor)	4-tube 57-mm launcher with high-velocity dual-purpose rockets. EO day/ night sight. Blast shield. Range 1,000 m. Penetration 300 mm, 10 m radius.	Prox fuze, 1,500 m range. Penetration 400 mm, 20 m radius.
Man-portable SAM launcher	6 km day/night range/ 0-3.5 km altitude all aircraft, velocity mach 2.6. Thermal night sight. Proximity fuze, frangible rod warhead (for 90% prob hit and kill). Approach/ azimuth link to AD warning net. Twin launcher vehicle quick mount. Nil smoke. Mount on robotic AD/AT launcher. RF/radar DF set on helmet.	Warhead/lethal radius increased air/ground targets. Improved seekers - not be decoyed by IR decoys/jammers. Fused II/ FLIR 10 km. Launch from enclosed spaces. Laser dazzler. Optional AD/AT LBR KE warhead missile – 8 km. Mount on AD/AT robot vehicle
MANPADS Vehicle Conversion Kit (Lt Stk Veh, Van, recon TUV, truck, etc)	Twin launcher and ADMG on improvised IR SAM vehicle. Day/night IR auto track FCS, MMW radar. Display link AD net. RF/radar DF set to 25 km. Camouflage	Replace launcher with 3-missile launcher: 2x ADAT KE SAMs, 1x IR SAMs. Total 6 missiles, (3+3)
Manpack Air Defense and Antitank (ADAT) Kinetic-Energy Missile Launcher (also listed in Anti-tank)	At company/Bn, can replace ATGMs and SAMs. Targets heli and LAVs. Missile has 3 KE LBR darts (submissiles) 8 km, 0 m altitude. Camo screen. Dart is 25-mm sabot with HE sleeve. Nil smoke. Fits on robotic ADAT launcher. Helmet RF/radar DF.	Larger sabot kills all targets up to 200 mm (KE) armor. Range 8 km, time of flight 5 sec. Fused II/ FLIR 10 km. Launch from enclosed spaces. Can mount on 3x remote launcher w/ IR auto-tracker, which. fits on AD/AT robot vehicle

SYSTEM	NEAR-TERM OPFOR (FY 15-20)	MID-TERM OPFOR (FY 21-28)
Towed/Portee/Vehicle Mount AA Short Range gun/missile system	2x23mm gun. MMW/IR Camou/screen. Frangible rd to 3,000 m (17mm pen). Onboard radar/TV fire control, ballistic computer, 5 km MMW radar, thermal night sight, auto-tracker, net azimuth warner. Add twin MANPADS. RF/radar DF set, 25 km. RWS on veh hull/turret. CPS/ATS.	Replace with twin 30-mm recoilless chain gun, range. Frangible, sabot, AHEAD rds to 4 km. TV/fused II/FLIR auto-tracker 10 km. MMW radar, Twin MANPADS/ADAT KE missile 8 km) lchr. APU to 15-kph self relocation. Robotic option. Laser dazzler.
Air Defense System Decoys (visual decoy, decoy emitter)	See DECEPTION & COUNTERMEASURE SYSTEMS	
Brigade gun/missile turret for mount on tracked mech IFV, wheeled mech APC, truck (motorized) chassis	Twin 30-mm gun, APFSDS/frangible rds 4 km. 30-mm buckshot rd for UAVs. Mounts 4x hyper-velocity LBR-guided SAMs to 8 km, 0 m min altitude. Passive IR auto-tracker, FLIR, MMW RADAR. 2/battalion. Track/launch on move. Targets: air, LAVs, other ground targets. RF/radar DF set with 25 km range. CPS/ATS.	Dual mode (LBR/radar guided) high velocity missile, 12 km, 0 m min altitude. Auto-tracker, to launch and fire on move. Phased array radars. Fused II/FLIR to 19 km. Twin 30-mm recoilless chain gun with AHEAD-type rds to 4 km. Micro recon/heli atk UAVs. TV/IR attack grenades.
Divisional gun/missile system on tracked mech IFV, wheeled mech APC, truck (motorized) chassis	Target tracking radar 24km. TV/FLIR. 8 x radar/EO FCS high velocity missiles to 18 km/12 at 0 m min altitude. Auto-track and IR or RF guided. 2 twin 30mm guns to 4 km. 30-mm buckshot rd for UAVs. RF/radar DF. CPS/ATS	Hybrid drive. Missile 18 km at 0 m, can kill LAVs. Fused II/FLIR auto-tracker 19 km, launch on move. Radar 80 km. Home on jam. Twin 30-mm recoilless chain gun with electronically fuzed air-burst rds to 4 km. Micro-recon/heli-atk UAVs. TV/IR attack grenades.
APC Air defense/AT Vehicle in APC Bn (Company Command Vehicle, MANPADS Vehicle in Bn/Bde)	1-man turret on 8x8 chassis. 30mm gun, 30-mm buckshot rd for UAVs. 100-X TV, 2 gen FLIR. 2x LBR ATGM lchrs 6 km, 2x veh MANPADS lchrs. Two dismount teams. 1x MANPADS lchr, 1x ADAT KE lchr. Total 18 msls. 12.7-mm MG. RF/radar DF to 25 km. CPS/ATS.	10x10 chassis, hybrid drive, box armor. Gun 30-mm recoilless gun on RWS. Ammo includes AHEAD-type to 4 km. Add 2 veh launchers for 5 HVM AD/AT (KE LBR) missiles, 8 km. Anti-helicopter surveillance/attack micro-UAVs. Fused II/FLIR 10 km. MMW radar. TV/IR attack grenades.
IFV ADAT Vehicle IFV Bn/Bde MANPADS	IFV chassis with features noted above. APC ADAT weapons and upgrades	See AIR DEFENSE, APC ADAT for weapons and upgrades
HIFV ADAT Vehicle HIFV Bn/Bde MANPADS	HIFV chassis with features noted above. APC ADAT weapons and upgrades	See AIR DEFENSE, APC ADAT for weapons and upgrades
Tank ADAT Vehicle Tank Bn/Bde MANPADS	Tank chassis with features noted above. APC ADAT weapons and upgrades	See AIR DEFENSE, APC ADAT for weapons and upgrades
Towed Medium Range AA gun/missile system	35mm revolver gun 1,000 rd/min. Gun rds: frangible, HE prox, electronic-fuzed. 4 SAMs/lchr, 45 km, 0 min altitude. Radar 45 km for 4 tgts. Resists all ECM. 2 gen FLIR auto-tracker to 20 km. RF/radar DF 25 km. SAM modes include active homing, home-on-jam. RAP/RAM/IR camo. CPS/ATS	Hybrid-drive auxiliary power units for local moves. Improved FCS, phased array radar, low probability of intercept, and acq to 80 km. Fused II/3rd gen FLIR auto-tracker to 35 km in day/night all-weather system. Ability to track and engage 8 targets per radar.
Medium-range ground SAM system	Tracked lchr. Radar to 150 km. 4 x radar-homing SAMs to 45km, 0 m min altitude (4 targets at a time). Home on jam. Use as cruise missile - priority ground tgts to 15 km, water 25 km. Fused 3rd gen FLIR auto-tracker. RF/radar DF. CPS/ATS	Hybrid drive. Improved FCS with radars and EO fused II/3rd gen FLIR day/night all-weather system to range 50 km. Radar range 200 km.

SYSTEM	NEAR-TERM OPFOR (FY 15-20)	MID-TERM OPFOR (FY 21-28)
Strategic SAM System	Cross-country truck launchers, 1 x track-via- missile SAMs 400 km, at Mach 7. 1x ATBM/high maneuver missile to 200 km. Also 8 x "hittile" SAMs to 120 km. Modes are track-via-missile and ARM (home-on-jam). All missiles 0 m to 50 km altitude vs stealth aircraft/UAVs/ ASMs. All strat/op missiles in IADS. Local IADS all AD. Battery autonomous option. Over-the-horizon TA radar veh to 400 km. Mobile radar to 350 km. Site CM, decoys.	Off-road trucks and tracked with hybrid drive. All missiles Mach 7. 1 x "big missile to 500 km. OTH radar to 600 km range with 5-min emplace-displace. Targets include all IRBMs. Increased target handling capacity (100/ battery in autonomous operations).
Operational-Strategic SAM System	Same as above on tracked chassis. Mobile FOs all batteries. AD radars on airships.	Same as above on tracked chassis.
Anti-helicopter Mines (Remote and Precision Launch)	In blind zones force helos upward or deny helo hides and landing zones. Range 150m. Acoustic and IR fuse, acoustic wake-up, or cmd detonation. Directed fragmentation. Precision-launch mines use operator remote launch, proximity fuze for detonation. RF/radar DF.	Stand-alone multi-fuse systems. Remote actuated hand-emplaced mines with 360-degree multi-sensor array, pivoting/orienting launcher, 4-km IR-homing missile. Operator monitors targets and controls (turns on or off) sections, mines or net.
Helicopter Acoustic Detection System	Early warning of helicopters. Acoustic sensors to 10km, 200m CEP. IR sensors can also be linked to air defense net.	Range 20 km, 50 m CEP. Track and engage multiple targets. Digital link to AD net, AD unit, IADS.

Military Technology Trends 2028

Year 2028 is a demarcation line for focusing on future military technologies. Even with the "Revolution in Military Affairs", most major technology developments are evolutionary, requiring one or more decades for full development. Subsystem upgrades can be added in less time. Most of the technologies noted below are in conceptual or early developmental stage or fielded at this time. Many exist in limited military or commercial applications, and can be easily extrapolated to 2028 and the near future time frame. Over this period and beyond, military forces will see some legacy systems fade to obsolescence and be replaced, or be relegated to lesser roles or lower priority units. Most will be retained and updated several times. New systems and technologies will emerge, be developed, become widely implemented, mature, and reach evanescence, requiring updates.

Infantry
• Infantry with improved weapons/sensors as primary lethal agent for combined arms • Weapon-delivered remotely-guided sub munitions and sensors for infantry weapons • Day/night sensors integrated, netted, with UAVs, robotics, and direct links to fire support • Visual/IR/MMW materials with signature management to avoid detection • Increased lethality weapons and precision for man-portable and vehicle weapons, robotic weapons • Increased range and effectiveness for use in Beyond Line-Of-Sight (BLOS) and MOUT operations • Tube launch UAVs, UCAVs, and remote overhead camera munitions for vehicles and dismounts limit collateral damage
Armor

<table>
<tr><td>

- Tank crew in hull, with insensitive ammunition, electromagnetic armor, and active protection systems
- Hybrid (diesel/electric) drive, and MMW/IR signature management
- Overhead guns and missiles, electro-thermal chemical gun, and cased telescoped rounds
- BLOS precision 12+km, 1,500+ mm lethality, KE missiles, and sensor/attack UAVs and robots
- Infantry carrier remote weapons manned by passengers for 360° all-aspect protection
- Heavy combat support vehicles/Heavy IFVs option to accompany tanks/IFVs
- Micro-UAVs, attack UAVs, and UCAVs for vehicle launch

</td></tr>
<tr><th>Anti-Armor</th></tr>
<tr><td>

- Increased penetration (1,500+ HEAT/1,000+ KE), including lightweight capability for infantry
- Infantry homing grenades with top-attack EFP or tandem
- KE hypervelocity missiles/missile rounds 10+ km vehicles. KE ATGMs for infantry
- Laser designators on AT grenade launchers, also used for precision artillery/air/naval rounds/ATGMs
- Attack UAVs and laser target designator UAVs for precision strikes throughout the battle zone

</td></tr>
<tr><th>Artillery</th></tr>
<tr><td>

- Autonomous operation/rapid self-emplace/displace with integrated netted FCS
- Precision munitions: laser/IR/MMW homing, EFP multi-sensor fuzed
- Inertial/GPS/muzzle-velocity radar course-correction on conventional rounds/rockets
- Combination guns integrate tactical unit BLOS fires and strikes

</td></tr>
<tr><th>Sensors</th></tr>
<tr><td>

- Multi-spectral immediate all-weather sensor transmission with real-time display
- Remote unmanned sensors, weapon-launch and robotic sensors and manned sensors
- Sensor nets integrated and netted from team to strategic and across functional areas
- Micro-UAVs and remote overhead camera munitions for vehicles and dismount teams

</td></tr>
<tr><th>Aircraft</th></tr>
<tr><td>

- Continued but selective use of FW and rotary wing for stand-off weapons, sensors
- Aircraft critical for transport, minelaying, jamming, other support missions
- Laser designators on AT grenade launchers, also used for precision artillery/air/naval rounds/ATGMs

</td></tr>
<tr><th>Other Aerial Systems</th></tr>
<tr><td>

- High-altitude UAVs, long-endurance UAVs, and UCAVs seamlessly integrated with other intelligence and support systems
- Recon/attack low-signature UAVs and UCAVs and stand-off munitions at all levels down to squads
- Ballistic missiles with non-ballistic trajectories, improved GNSS/homing re-entry vehicles, precision sub munitions, EMP
- Shift to canister launchers of tactical cruise missiles with precision homing and piloted option, cluster warheads, EMP
- Laser designators on AT grenade launchers, also used for precision artillery/air/naval rounds/ATGMs
- Airships and powered airships for long-duration and long-range reconnaissance, and variety of other roles
- Increased use of ultra-lights and powered parachutes

</td></tr>
<tr><th>Air Defense</th></tr>
<tr><td>

- Integrated Air Defense System with day/night all-weather RISTA access for all AD units
- Improved gun rounds (AHEAD/guided sabot) and missiles (anti-radiation homing, jam-resistant)
- Autonomous operation with signature suppression, counter-SEAD radars and comms
- Shoulder-launch multi-role (ADAT) hypervelocity missiles/weapons immune to helicopter decoys and jammers,
- Micro-UAVs and airships for multi-role use includes air defense recon and helicopter attack
- Acquisition/destruction of stealth systems and aerial munitions and ground rockets to 500+ km

</td></tr>
<tr><th>Information Warfare</th></tr>
<tr><td>

- Jammer rounds most weapons, electro-magnetic pulse rounds, weapons of mass effects
- UAVs, missiles and robots carry or deliver jammers/EMP/against point targets and for mass effects
- Multi-spectral decoys for most warfighting functions
- Computer network attack and data manipulation

</td></tr>
</table>

Access Denial
Non-lethal Weapons
• EMP/graphite/directed energy weapons to degrade power grid, information networks, and military systems • Space-based data manipulation to deny adversary use of satellite systems • Population control effects (acoustic devices, bio-chemical and genetic weapons, resources attack, dirty bomb) • Anti-materiel agents and organisms (microbes, chemicals, dust, and nanotech) • Countermeasures, tactical and technical, in all units to degrade enemy sensor and weapon effectiveness.

Worldwide Equipment Guide
Chapter 11: Irregular Forces

Distribution Statement: Approved for public release; distribution is unlimited.

Chapter 11: Irregular Forces

The conflict spectrum in the Complex Operational Environment includes not only modernized systems in upper tier forces but also older systems in less developed forces. Forces all across the capability spectrum use older, cheaper or improvised weapons because that is what they are able to procure, afford or require to blend in with their environment. Generally, Irregular forces are lightly armed attack troops and their equipment is not based on what they want or in some cases need but what is available. The equipment of an Irregular threat actor tends to improve over time as they increase their lines of supply, which complements their usual strategy of lassitude.

Irregular forces are armed individuals or groups who are not members of the regular armed forces, police, or other internal security forces (JP 3-24). Irregular forces can be insurgent, guerrilla, or criminal organizations or any combination thereof. Any of those forces can be affiliated with mercenaries, corrupt governing authority officials, compromised commercial and public entities, active or covert supporters, and willing or coerced members of a populace.

Arms Procurement. To maintain a force that is prepared to deal with its security challenges requires an equipment program of acquisition and procurement. Irregular forces do not usually have the luxury of the acquisition methods available to states such as internal manufacture or purchase through international defense agreements. There are laws and regulations that govern the control of military equipment such as the Arms Export Control Act and International Traffic in Arms Regulations (ITAR) specifically to stop Irregular forces and large criminal organizations from obtaining military grade equipment. Despite all the obstacles Irregular forces can acquire weapons from regional military sources, through purchase on the black market or fabricate them internally.

Regional Procurement. An Irregular force has a number of ways in which it can procure weapons and explosives regionally. The most dangerous Irregular forces are those that are supplied directly from an external country. Hezbollah are a recipient of support from Iran and Syria, and what began with caution and relatively small amounts of weapons in the 1980's has become a strategic alliance that supplies short-range precision guided munitions such as the Fateh-100 or the M-600.

Areas of instability, corruption, state weakness and long running conflict can be awash with uncontrolled weapons that are easy to procure. In 2003, Iraqi forces abandoned their positions as the coalition forces advanced on Baghdad, they left military bases and storage depots unguarded. Millions of tons of weapons and explosives were left for the taking by anybody. These weapons fueled the insurgency that followed.

Weapons and explosives can be taken using force or guile from legitimate government sources in the Irregular actors AO. Much of Boko Haram's (BH) military hardware is stolen from the Nigerian Army. BH fighters have conducted raids against remote military outposts and looted ammunition bunkers. In addition BH sympathizers in the Army have been accused of leaving armory doors unlocked which has left the militants well-armed.

Illicit Arms Trade. Insurgents, Guerrillas, armed gangs, and terrorists can all multiply their force through the use of illegally acquired firepower but an alternative to seizing weapons by force is to purchase them unlawfully through the Black Market. It is important not to underestimate the magnitude of the Black Market as

it is possible to purchase top of the line tanks and radar systems with the latest technology. The illicit circulation of small arms, light weapons and their associated ammunition alone values around $1bn a year.

The arms market is a good example, for all the wrong reasons of increasing globalization. The illicit arms trade takes maximum advantage of all the open trade developments including minimized custom regulations and relaxed border controls which leads to easier movement. The faking of shipping manifests or end-use certificates, bribing officials and concealing arms as humanitarian aids are common practice. The structure of the small arms black market is complex and stretches across the globe and the activities of the black market arms dealer's stretch to other trans-national criminal organizations, like drug and human traffickers.

Sources of small arms supplies to the black market are varied but most weapons start off the legal side and then get diverted to the illegal sphere. Small arms can enter illegal circulation through theft, leakage and divergence. The management of government's stockpiles is an acute problem and a prominent source of the illegal weapons in circulation. Stockpiles consist of obsolete and surplus weapons that are often collected as part of a disarmament program and not destroyed. Much of the international focus and funding is on the destruction or containment of chemical weapons stockpiles and the destruction of post conflict and legacy mines and thus small arms stockpiles remain comparatively under the radar and uncontrolled.

Improvised Weapons. Improvised weapons can be everyday objects made from non-military materials utilized without alteration, such as machetes, pocket knives or baseball bats. These weapons are characterized as primitive but continue to be effective. Systems encountered in Vietnam such as punji stakes, Malayan Gates and blowguns are an example of using natural materiel in an innovative way for the accomplishment of a military task such as an obstacle to movement.

Some seemingly ubiquitous military systems can be missing from an Irregular threat actor's arsenal because they are not suited to the fight. In Iraq heavy mortars were absent because coalition counter battery fire could quickly and accurately destroy the large caliber hard-to move equipment. Instead the Insurgents chose to repeatedly fire Type 63 107mm rockets at coalition bases. The Type 63 was self-stabilizing and could be fired from a simple ramp, it was easy to conceal and could be set on a timer thus reducing the danger to the shooter.

Improvised weapons include also include flame and incendiary devices to ignite fuels and ammunition supplies. Fuel-air-explosive IEDs or "Molotov cocktails" have been employed in almost all conflicts, and an air droppable version of incendiary devices known as "barrel bombs" have been seen in a recent conflict. These improvised bombs include large containers filled with flammable and shrapnel producing material and are pushed from a helicopter over a variety of military and civilian targets.

Improvised weapons are most associated with less robust forces, but they can also be the most effective method of accomplishing the mission. Military or non-military materials could be used to trigger major disasters such as forest or urban fires, breached dams or levees to initiate floods. The list of improvised weapons available and there methods of employment are limited only by human imagination.

The most populous class of improvised weapons is the improvised explosive device (IED). Any explosive devices can be used to make a type of IED, they can be of various design with differing amounts of explosive fill and different detonation mechanisms. Historically, the most numerous IEDs encountered on the battlefield use hand grenades, these can be rigged by wedging them into objects to act as camouflage or by tying them to trip wires. Often grenades are hidden on bodies, weapons, or objects to be picked up by soldiers. Artillery rounds are also favored for IED construction for their larger size, they can be placed in vehicles and delivered to target with

devastating accuracy, examples of Vehicle Bourne IED's range in size from a bicycle to a dump truck. Mines have long had capability for sophisticated fuzing and remote control units seen in some IEDs. Both mines and IEDs can be converted for command-arming and detonation, and for precision sensor fuzing.

Many IEDs are not made from military munitions. Bulk explosives (such as Dynamite, TNT, C-4, etc.) are used in IEDs. Terrorists such as the shoe bomber and anti-Israeli groups used Triacetone Triperoxide (TATP), precisely because it is highly sensitive. The most common explosive in the US is ammonium-nitrate fuel oil (ANFO, an insensitive slurry mixed onsite with the bulk of the mix as common fuel oil) for mining and road construction. The slurry can then be poured or pumped deep into spaces where other explosives cannot fit. Explosives can be improvised from common materials. In 1995 domestic terrorist Timothy McVeigh created a home-made variant of ANFO in a VBIED (vehicle-borne IED, with "volumetric explosive" effects) to blow up the Murrah Building in Oklahoma City. Questions and comments on data listed in this chapter should be addressed to:

WO2 Matthew Tucker (U.K Army)

DSN: 552-7994 Commercial (913) 684-7994

E-mail address: matthew.j.tucker28.fm@mail.mil

Belgian Fabrique Nationale 7.62x51mm NATO Main Battle Rifle, FAL

SYSTEM	SPECIFICATIONS	AMMUNITION	VARIANTS
Alternative Designation: None. Date of Introduction: 1954 Proliferation: Widespread; like the Hk G3, it has been used by 70+ nations, worldwide, at some time or another, and is still found in many Third World militaries/security services.	Weight: FAL 50.00: 4.3 kg (9.48 lb) FAL 50.61: 3.90 kg (8.6 lb) FAL 50.63: 3.79 kg (8.4 lb) FAL 50.41: 5.95 kg (13.1 lb) Length: FAL 50.00 (fixed stock): 1,090 mm (43 in) FAL 50.61 (stock extended): 1,095 mm (43.1 in) FAL 50.63 (stock extended): 998 mm (39.3 in) FAL 50.41 (fixed stock): 1,125 mm (44.3 in) Barrel length: FAL 50.00: 533 mm (21.0 in) FAL 50.61: 533 mm (21.0 in) FAL 50.63: 436 mm (17.2 in) FAL 50.41: 533 mm (21.0 in) Action: Gas-operated, tilting breechblock Rate of fire: 650–700 rds/min Muzzle velocity: FAL 50.00: 840 m/s (2,756 ft/s) FAL 50.61: 840 m/s (2,755.9 ft/s) FAL 50.63: 810 m/s (2,657.5 ft/s) FAL 50.41: 840 m/s (2,755.9 ft/s) Effective firing range: 400–600 m sight adjustments Feed system: 20 or 30 round detachable & 50-round drum magazines. Sights : Aperture rear sight, post front sight; Sight radius: FAL 50.00, FAL 50.41: 553 mm (21.8 in) FAL 50.61, FAL 50.63: 549 mm (21.6 in) Various telescopic and night visions sights are available	7.62x51mm NATO; ball, tracer, armor piercing. All known variants.	Numerous, particularly among those manufactured under license from FN in other nations. Most common variants are the standard FN infantry weapon with a fixed stock. Also the para models with a 17.2 inch barrel and folding stocks.

NOTES
LIKE THE HK G3, ONE OF THE MOST PROLIFIC MAIN BATTLE RIFLES EVER PRODUCED; CAN BE FOUND VIRTUALLY AROUND THE WORLD, STILL IN SERVICE WITH SOME THIRD WORLD MILITARIES AND SECURITY ORGANIZATIONS.

Russian 7.62-mm Assault Rifle AK-47/AKM

SYSTEM	SPECIFICATIONS	AMMUNITION	VARIANTS
Alternative Designations: AK, Kalashnikov Date of Introduction: 1949 (AK) 1961(AKM) Proliferation: Widespread (over 50 million) Feed: 30-round curved box magazine Fire Mode: Selective, automatic or semi-automatic Operation: Gas SIGHTS: Type: Fore, pillar; Rear, U-notch Magnification: None Night Sights Available: Yes	Description: Weight (kg): Loaded (with magazine): 3.8 Empty (w/o magazine): 4.3/3.14 Length (mm): 870/880 Rate of Fire (rd/min): Cyclic: 600 Practical: Automatic: 100 Semiautomatic: 40	M1943 (57N231S) Caliber/length: 7.62x39-mm Type: Ball, steel core Range (m): Effective: 300 Maximum: 800 Armor Penetration: 6 mm mild steel plate at 300 Steel helmet at 1,000 m Flak vest at 60 m Muzzle Velocity (m/s): 718 M1943 (T-45 or 57N231P) Caliber/length: 7.62x39-mm Type: Ball-Tracer Range Effective: 300 Maximum: 800 Trace (m): 800 Muzzle Velocity (m/s): 718	Numerous. Many countries manufacture clones of the AK-47 or weapons using the basic AK action. Some of these are made in different calibers. AKS: Folding stock AK-47. AKM: Improved AK-47, sights, magazine, and stock. AKMS: Folding stock variant of AKM.

NOTES

PHOTO IS OF AN AKM. ALL 7.62X39 MM KALASHNIKOV ASSAULT RIFLES ARE VERY DEPENDABLE WEAPONS. THEY PRODUCE A HIGH VOLUME OF FIRE AND ARE SIMPLE TO MAINTAIN AND PRODUCE. THE PRIMARY DIFFERENCE BETWEEN THE AK-47 AND THE IMPROVED AKM IS THE RECEIVER. THE RECEIVER OF THE AK-47 IS FORGED AND MACHINED WHILE THE RECEIVER OF THE AKM IS STAMPED METAL FACILITATING EASIER AND LESS COSTLY MANUFACTURING. BOTH THE AK-47 AND THE AKM CAN MOUNT A 40-MM UNDER-BARREL GRENADE LAUNCHER. THE AK-47 AND AKM HAVE BEEN REPLACED IN MANY ARMIES BY THE NEWER AK-74. THE AK-74 IS BASICALLY AN AKM RE-CHAMBERED TO FIRE A 5.45X39 MM CARTRIDGE. THE 7.62X39 MM RPK LIGHT MACHINEGUN IS BASED ON THE AK/AKM DESIGN WHILE THE RPK-74 IS A LIGHT MACHINEGUN VERSION OF THE AK-74. BOTH ARE INFANTRY SQUAD LEVEL SUPPORT WEAPONS.

Handheld Shaped Charge Grenade RKG-3 / RGK-3M

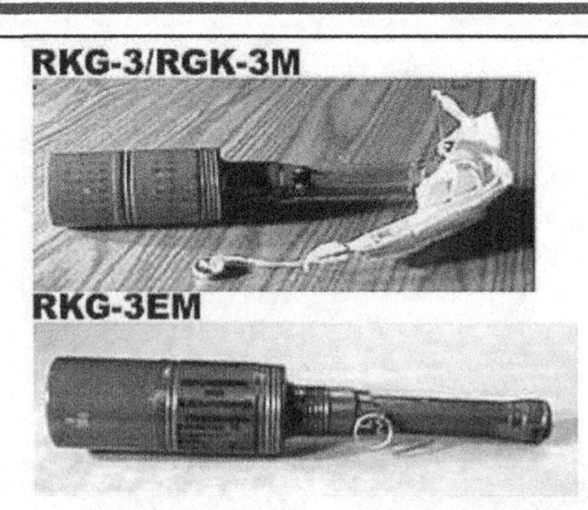

SYSTEM	SPECIFICATIONS
Alternative Designations	M79, PR Type 3, HEAT
Date, Country of Origin	1950, RUS
Proliferation	Soviet Bloc, Vietnam, Syria, Iraq, Insurgent groups
Crew	1
Weight Firing (kg)	1.1
Length Firing (m)	0.38
Diameter (m)	0.065
Rifling	No
Emplacement Time (min)	1
Fire from Inside Building	Yes

SIGHTS	SPECIFICATIONS
Name	N/A
Type	N/A
Sight Range Direct (m)	Line Of Sight

AMMUNITION	SPECIFICATIONS
Missile Type, Name	Shaped Charge Grenade, RKG-3
Max Aimed Range (m)	20
Penetration (mm)	125
Warhead Type	Hollow Charge with Drogue Parachute
Fuze Type (mm)	Point Detonating
Explosive Quantity (g)	384
Missile Type, Name	Shaped Charge Grenade, RKG-3M
Max Aimed Range (m)	20
Penetration (mm)	165
Warhead Type	Hollow Charge with Drogue Parachute
Fuze Type (mm)	Point Detonating
Explosive Quantity (g)	560

VARIANTS

RKG-3E – Steel Liner for hollow charge and an increased 170mm armor penetration.

RKG-3EM – Copper liner for hollow charge and an increased 220mm armor penetration.

UPG-8 – Training grenade

Notes

DUE TO IMPROVEMENT IN MODERN TANK ARMOR, ANTITANK GRENADES ARE GENERALLY CONSIDERED OBSOLETE IN CONVENTIONAL WARFARE. HOWEVER, THEY CAN BE AND HAVE BEEN USED EFFECTIVELY BY GUERILLAS AND INSURGENTS AGAINST ARMORED PERSONNEL CARRIERS AND RECONNAISSANCE VEHICLES WHICH LACK THE HEAVIER MODERN ARMOR.

Russian 7.62-mm Light Machinegun RPK

SYSTEM	SPECIFICATIONS	AMMUNITION	VARIANTS
Alternative Designation: None Date of Introduction: 1964 Proliferation: Widespread	Weight (kg): Empty (w/o magazine) (kg): 4.9 Loaded (with magazine): 5.67 w/40-rd mag Mount: Bipod Length (mm): Overall: 1,035 Barrel: 591 Quick Change Barrel: No Rate of Fire (rd/min): Cyclic: 650 Practical (auto): 150 (80 sustained, see note) Practical (semi): 50 Fire Mode: Selective Operation: Gas Feed: 40 round-curved box or 75-rd drum magazine. Can also use the 30-round curved box magazine used by the AKM. Sight Type: Leaf sights Magnification: None Night Sights Available: yes, (luminous front/rear)	M1943 (57N231S) Caliber/length: 7.62x39-mm Type: Ball, steel core Range (m): Effective: 800 Maximum: 800 Armor Penetration: 6 mm mild steel plate at 300 m Steel helmet at 1,000 m Flak vest at 60 m Muzzle Velocity (m/s): 718 M1943 (T-45 or 57N231P) Caliber/length: 7.62x39-mm Type: Ball-Tracer Range Effective: 800 Maximum: 800 Trace (m): 800 Muzzle Velocity (m/s): 718 M1943 Caliber/length: 7.62x39-mm Type: API Armor Penetration (mm @ 0° obliquity @ 500m): 8 Muzzle Velocity (m/s): N/A	RPKS: Folded stock version (820 mm in length)

NOTES

THE RPK IS THE LIGHT MACHINEGUN VARIANT OF THE AKM AND AS SUCH IS AN EXTENDED VERSION OF THE AKM. IT HAS A LONGER, HEAVER BARREL THAN THE AKM (591 MM VS 414 MM). MOST MOVING PARTS ARE INTERCHANGEABLE WITH THE AK-47 OR AKM ASSAULT RIFLES. THE SUSTAINED RATE OF FIRE CANNOT EXCEED 80 RDS PER MINUTE DUE TO "COOK OFF". IT HAS BEEN REPLACED BY THE 5.45-MM RPK-74 IN MANY ARMIES. THE RPK FILLS THE ROLE OF A SQUAD LEVEL SUPPORT WEAPON.

Russian 7.62-mm General Purpose Machinegun PKM and Pecheng (PKP)

SYSTEM	SPECIFICATIONS	AMMUNITION	VARIANTS
Alternative Designation: See Variants. Date of Introduction: (PKM/PKT/PKP): 1971/1968/2001 Proliferation: Widespread (PKM/PKT) NOTE: Picture, above, is of a PKP (Pecheneg).	Crew: 2 Weight (kg): Empty (w/o magazine) (PKM/PKT) (kg): 8.4/10.66 Ammo box (only) with 100/200-rd belt (kg): 3.9/8.0 Tripod (lightweight) (kg): 4.75 Length (mm): Overall (PKM/PKT): 1,160/1,080 On tripod (PKS): 1,267 Barrel: 658. Barrel Change: Yes Mount Type: Pintle, coaxial, bipod or tripod (Stepanov) Mounted On: (see VARIANTS) Rate of Fire (rd/min): Cyclic: 650 Practical: 250 (PKM) Up to 600 for Pecheneg/PKP Fire Mode: Automatic Operation: Gas Feed: Belt, 100-rd belt carried in a box fastened to the right side of the receiver. 25-rd belts can be joined in several combination lengths (100/200/250) Type: Open iron sights Sighting range (PKM/PKT) (m): 1,500/2,000 Night Sights Available: Yes	57-N-323S Caliber: 7.62x54-mm rimmed Type: Ball Max Range (PKM/PKT) (m): 3,800/4,000 Practical Range (PKM/PKT) (m): Day: 1,000/2,000 Night: 300/INA Armor Penetration @ 0° obliquity, and 500 range (mm): 8 steel plate @ 520 m (mm): 6 Flak vest: 110 m Muzzle Velocity (PKM/PKT) (m/s): 825/855 7BZ-3 Caliber and Length: 7.62x54-mm rimmed Type: Armor piercing incendiary Max Range (PKM/PKT) (m): 3,800/4,000 Practical Range (PKM/PKT) (m): Day: 1,000/2,000 Night: 300/INA Armor Penetration @ 200 range (mm): 10 Muzzle Velocity (PKM/PKT) (m/s): 808	PKM: Squad machinegun PKT: Vehicle mounted MG with solenoid electric trigger, remote sight, and a longer heavier barrel. It lacks a stock and, bipod. Some are coaxial to a main gun and use its sights. Others operate separately. They generally do not dismount for ground use. PKS: Lightweight tripod-mounted infantry weapon PKMS: Lightweight tripod-mounted variant of the PKS PKB (PKBM): Pintle-mounted on APCs, SP guns, BRDM, BTRs, has butterfly trigger rather than solenoid, double spade grips, and front and rear sights

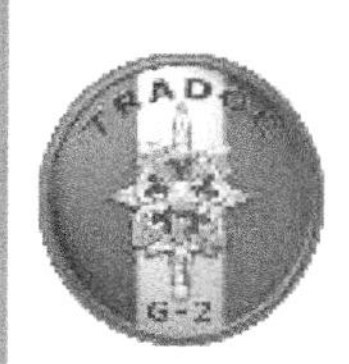

NOTES
THE 7.62-MM GENERAL-PURPOSE MACHINEGUN (PKM) IS A GAS-OPERATED, BELT-FED, SUSTAINED-FIRE WEAPON. THE BASIC PKM IS BIPOD-MOUNTED BUT CAN ALSO FIT IN VEHICLE FIRING PORTS. IT IS CONSTRUCTED PARTLY OF STAMPED METAL AND PARTLY OF FORGED STEEL. THE NEWER VARIANT PKP (PECHENEG) FEATURES IMPROVED COOLING SYSTEM, AND A HEAVY, FIXED BARREL THAT DOES NOT REQUIRE CHANGING NOR CAN IT BE CHANGED BY THE CREW. IT IS DESIGNED TO FILL THE ROLE OF A TRUE, SQUAD LEVEL GPMG FOR SUPPORT IN RUSSIAN INFANTRY AND SPETSNAZ UNITS.

Russian 12.7mm Heavy Machinegun NSV/NSVT/KORD

SYSTEM	SPECIFICATIONS	AMMUNITION	VARIANTS
Alternative Designation: NSVS (when mounted on tripod) Date of Introduction: 1974 Proliferation: Widespread. The original plant is in Kazakhstan (NSV/NSVT). After dissolution of the USSR, a Russian plant now produces the KORD HMG. NOTE: The HMG in the above picture is of a Finnish NSV.	Weight (kg): Total System (w/6T7): 43 Empty: 25 Loaded: INA Tripod (6T7 tripod): 16 Length (mm): Overall: 1,560 On 6T7 Tripod: 1,900 Width (on 6T7 tripod) (mm): 860 Height (on 6T7 tripod) (mm): 380 Barrel Life (rds): 5,000 Barrel Change Time (sec): 5 Barrel Weight (kg): 9.2 Mount Type: 6T7 (infantry) tripod or 6U6 (w/seat) universal tripod Mounted On: (see VARIANTS) Traverse (°): 360 Elevation (°): -5 to +75 Rate of Fire (rd/min): Cyclic: 680-800 Practical: 100 Fire Mode: Automatic; short bursts long bursts (10 to 15), or continuously Operation: Gas Feed: Left or right from metal link belt from 50-rd boxes	12.7-mm cartridge API (B-32) API-T (BZT-44) HEI Typical Combat Load: 300 rds Name: B-32 Caliber and Length: 12.7x108-mm Type: Armor Piercing Incendiary Max Range (grd) (m): 7,850 Effective Range (m): AA: 1,000 Ground: 2,000 Armor: 800 Night (w/1PN52-1): 1,000 Armor Penetration @ 0° obliquity @ 500/1,000m range (mm): 20/13.2 Muzzle Velocity (m/s): 860	NSVT: Tank-mounted version (See NOTES) A tripod-mount (6T7) version is available for infantry use in a ground role. However, the NSVT appears more commonly mounted on the turrets of tanks as an antiaircraft machinegun. Russian NSV/NSVT: The Russian version can produce the guns for either Russian 12.7x108 or NATO 12.7 x 99 (.50-cal) ammunition. Kord: A Russian modernized version of the NSV/NSVT. Improvements include reduced weight (50% for hand-carry 6P57), reduced recoil, increased barrel life, improved reliability, improved accuracy, increased burst rate capacity, and improved reliability and maintenance. Reduced weight and recoil permits use with the 6T19 light machinegun bipod. Like the above Russian MGs, the Kord can be produced in either ammunition version. Vehicle version is 6P49. Swivel mount hand-operated versions are 6P58 and 6P59.

NOTES

ON THE T-72 AND THE T-80, IT HAS A ROTATING MOUNT AND CAN BE FIRED FROM WITHIN THE TANK. THE TANK COMMANDER EMPLOYS THE K10-T REFLEX SIGHT TO ENGAGE AIRCRAFT. ON THE T-72/T-80 MOUNT HE ENGAGES GROUND TARGETS WITH METALLIC SIGHTS ON THE GUN ITSELF. THE T-64 TANK MOUNTS A MODIFIED VERSION WITH A FIXED MOUNT ON THE COMMANDER'S CUPOLA. IT FIRES BY MEANS OF AN ELECTRICAL SOLENOID WHEN THE TANK IS BUTTONED UP AND AN OPTICAL SIGHT INSIDE THE CUPOLA IS USED. INSTEAD OF THE NORMAL 50-ROUND AMMUNITION BELT CONTAINER, THE NSVT ON THE T-64 MAY USE A LARGER BELT CONTAINER HOLDING 200 ROUNDS.

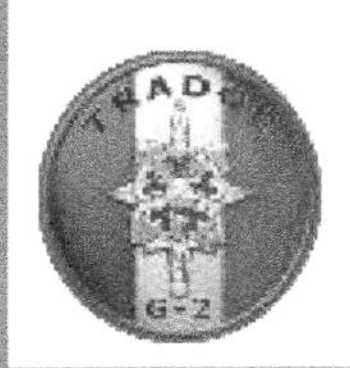

Russian 7.62-mm Sniper/Marksmen Rifle SVD

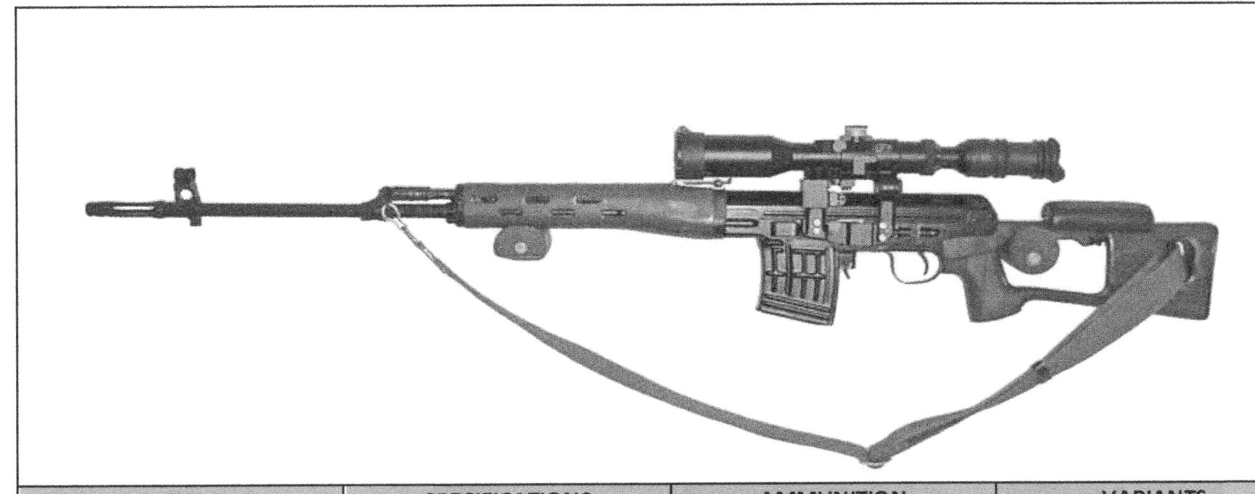

SYSTEM	SPECIFICATIONS	AMMUNITION	VARIANTS
Alternative Designation: SVD, Dragunov Date of Introduction: 1963 Proliferation: Widespread Fire Mode: Semi-automatic only SIGHTS: Name: PSO-1 Type: Infrared detection capability for night firing Magnification: 4x Field of View (°): 6 Sighting Range (m): 1,300 Night Sights Available: Yes. NSPU-3. The NSPU-3 increases accuracy to 1,000 m at night or during poor visibility.	Description: Weight (kg): Loaded (with magazine): 4.5 Empty (w/o magazine): 4.3 Length (mm): Overall: 1,230 With Bayonet: 1,370 Barrel: 620 Rate of Fire (rd/min): 30 Operation: Gas Feed: 10-rd detachable box magazine (15-rd available for the SVD-S)	Name: 57-N-323S Caliber and Length: 7.62x54-mm rimmed Type: Ball, standard steel-core Range (m): Effective: 600 Effective Night: 300 sight INA Armor Penetration (mm): Steel plate: 6 @ 520 m Flak vest: Yes @ 110 m Muzzle Velocity (m/s): 828 Name: Sniper (7N1) Caliber/length: 7.62x54R-mm rimmed Type: Steel core Range (m): Effective With Scope: 1,000 Effective W/O Scope: 800 Armor Penetration: INA Muzzle Velocity (m/s): 823 Name: 7N13 Caliber/length: 7.62x54R-mm rimmed Type: Enhanced penetration (steel core) Range (m): Effective With Scope: 1,000 Effective W/O Scope: 800 Armor Penetration (mm): Steel Plate: 6 @ 660 m Flak Vest: 800 m	SVD-S: Folding stock, 15-rd magazine. SVU: Bullpup (trigger forward of magazine). OTs-03AS: SVU w/PSO-1 sight. 6V1: SVD with PSO-1 sight. 6V1-N3: SVD with NSPU-3 night sight.

		Muzzle Velocity (m/s): 828 Name: 7B2-3 Bullet: B-32 Caliber/length: 7.62x54R-mm rimmed Type: AP-I Range (m): Effective With Scope: 1,000 Effective W/O Scope: 800 Armor Penetration: 10-mm armor plate @ 200 m Muzzle Velocity (m/s): 808 Name: 7T2m Bullet: T-46 Caliber/length: 7.62x54R-mm rimmed Type: Tracer Range Range (m): Effective With Scope: 1,000 Effective W/O Scope: 800 Trace (m): 1,200 Time of Trace (sec): 3 Muzzle Velocity (m/s): 798	

NOTES

THE BOLT MECHANISM AND GAS RECOVERY SYSTEM OF THE SVD ARE SIMILAR TO THOSE OF THE AK AND AKM. THE 7.62X54-MM RIMMED CARTRIDGE OF THE SVD IS NOT INTERCHANGEABLE WITH THE 7.62X39-MM RIMLESS ROUND OF THE AK-47/AKM. THE SVD PERFORMS BEST WHEN USING TARGET GRADE AMMUNITION, HOWEVER STANDARD (PKM/PKT) 7.62X54-MM RIMMED ROUNDS MAY ALSO BE FIRED. EVERY OPFOR INFANTRY SQUAD HAS AN SVD EQUIPPED DESIGNATED MARKSMAN (DM).

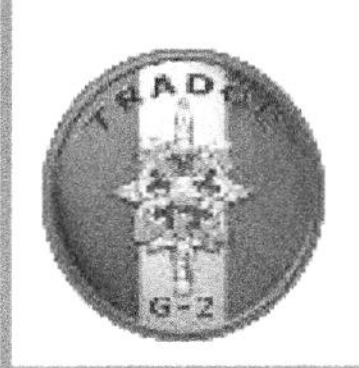

United States .50-cal Anti-Materiel Rifle M82A1A

SYSTEM	SPECIFICATIONS	AMMUNITION	VARIANTS
Alternative Designation: None Date of Introduction: 1984 Proliferation: Widespread (45+ nations) SIGHTS Name: Unertl Type: Optical (matches trajectory of .50-cal Raufoss Grade A) Magnification: 10x Name: Swarovski Type: Optical (with ranging reticle) Magnification: 10x42 Night Sights Available: yes Magnification: 10x42 Name: Barrett Optical Ranging System (BORS) Type: Ballistic Computer. This add-on device couples to the telescope, in place of the rear scope ring. It can then adjust for range, air temperature, round used, and other factors.	Weight (kg): Empty (w/o magazine): 14.75 Length (mm): Overall: 1,448 Barrel: 736 Rate of Fire (rd/min): 20 Operation: Recoil Feed: 10-rd detachable box magazine Fire Mode: Semi-automatic only Typical Combat Load: 30 rounds Ammunition Types: (.50-cal cartridge) Raufoss Grade A Ball (M2/M33) AP (M2) AP-I (M8) API-T (M20) Tracer (M10/21) SLAP (M903) MP (MK211 Mod 0)	Name: Raufoss Grade A (match)(DODIC A606) (USMC) Caliber/length: .50-cal BMG/12.7-mm x 99-m (NATO) Type: Standard operating round Range (m) (equipment-size targets): Maximum (w/scope): 1,800 Muzzle Velocity (m/s): 854 Name: MP NM140 (Nammo) MK211 Mod 0 Caliber/length: .50-cal BMG/12.7-mm x 99-m (NATO) Range (m) (equipment-size targets): Maximum (w/scope): 1,800 Armor Penetration: 11 mm @45° @1,000 m Fragmentation: 20 fragments after hitting 2 mm steel Incendiary Effect: Ignition of JP4 and JP8 Accuracy: <15 cm @ 550 m Muzzle Velocity (m/s): 915 Name: AP-S NM173 (Nammo) Caliber/length: .50-cal BMG/12.7-mm x 99-m (NATO) Type: Armor piercing Range (m) (equipment-size targets): Maximum (w/scope): 1,800 Armor Penetration: 11 mm @30° @1,500 m Accuracy: <15 cm @ 550 m Muzzle Velocity (m/s): 915 Name: M903 (Olin)	M107-A1 is a product improvement of the M82A1A. Improvements include: reduction in weight by 5 pounds; cylindrical titanium muzzle brake; titanium barrel key/recoil buffer system in order to operate with a Barrett suppressor; functional modifications to increase durability and ease of operation. Barrett introduced this variant in 2013.

A commonly associated scope is the Leupold Mark 4; but it can also work with other scopes. Night Sights Available: yes		Caliber/length: .50 cal BMG/12.7-mm x 99-m (NATO) Type: Saboted Light Armor Penetrator (SLAP) (actual bullet is tungsten .30 inch penetrator wrapped in a .50-cal plastic sabot) Range (m) (equipment-size targets): Maximum (w/scope): 1,500 Armor Penetration: 19 mm (.75 in) @1,500 m Accuracy: INA Muzzle Velocity (m/s): 1,014 Name: M8 Caliber/length: .50-cal BMG/12.7-mm x 99-m (NATO) Type: Armor piercing incendiary Range (m) (equipment-size targets): Maximum (w/scope): 1,800 Armor Penetration: 20 mm @ 100 m Accuracy: <25 cm @ 550 m Muzzle Velocity (m/s): 881 Name: M20 Caliber/length: .50-cal BMG/12.7-mm x 99-m (NATO) Type: Armor piercing incendiary-Tracer Trace (m): 91 to 1,463 Armor Penetration: 20 mm @ 100 m Accuracy: <25 cm @ 550 m Muzzle Velocity (m/s): 887	

NOTES

THE M82A1A PROVIDES MANEUVER COMMANDERS WITH THE TACTICAL OPTION OF EMPLOYING SNIPERS WITH AN ANTI-MATERIEL WEAPON TO AUGMENT PRESENT 7.62-MM ANTI-PERSONNEL SNIPER RIFLES. RECOIL EQUALS 7.62X51-MM LEVELS. THE USMC USES RAUFOSS GRADE A AMMUNITION, BUT THE RIFLE IS CAPABLE OF FIRING ANY STANDARD 12.7X99-MM BROWNING MACHINEGUN AMMUNITION.

Chinese 60-mm Lightweight Long Range Mortar Type WX-90

SYSTEM	SPECIFICATIONS
Alternative Designations	W90
Date, Country of Origin	1990, CHI
Proliferation	INA
Crew	3
Weight Firing (kg)	23.6
Weight Travel (kg)	14
Weight Bipod (kg)	9
Length Barrel (cm)	1200
Height Firing (m)	1200
Rifling	No
Feed	Manual
Breech Mechanism Type	NA
Emplacement Time (min)	INA
Fire from Inside Building	No

SIGHTS	SPECIFICATIONS
Name	
Type	
Sight Range Direct (m)	
Sight Range Indirect (m)	

ARMAMENT	SPECIFICATIONS
Caliber(mm)	60.75
Rate of Fire (prac, cyclic)	30
Fire on Move	No
Elevation (deg min, max)	-0, +80
Traverse (deg)	7

AMMUNITION	SPECIFICATIONS
Caliber(mm), Type, Name	60mm, Mortar, Western
Max Eff Range (m)	4700
Max Range, Extended (m)	5,500Number
Min Range (m)	72
Muzzle velocity (m/s)	314
Combat Load (ready, stow)	NA

VARIANTS	
Name	W89t
Barrel Length (cm)	1200
Barrel weight (kg)	9
Range (m)	Number
Name	WW90-60Lt
Barrel Length (cm)	1300
Barrel weight (kg)	11.7
Range (m)	5775
Name	WW90-60M
Barrel Length (cm)	1080
Barrel weight (kg)	9.7
Range (m)	4400

Notes

THIS MORTAR WAS PRODUCED FOR THE PEOPLES LIBERATION ARMY BUT IS ALSO AVAILABLE ON THE EXPORT MARKET. IT IS RELATIVELY CHEAP AND FIRES WESTERN 60MM AMMUNITION.

Russian ATGM Launcher for AT-4 and AT-5 9P135

SYSTEM	SPECIFICATIONS	AMMUNITION	VARIANTS
Ground mounted portable launcher for AT-4 and AT-5 family of missiles. Alt designations: AT-4B/ AT-5B, 9P135M Firing Post, Fagot/Fagot M Date of introduction: 1973 Proliferation: At least 25 countries	Crew: 3 Weight, excluding missile (kg): 22.5 Length, in firing position (m): 1.1/1.3 Launcher: 9P135 (AT-4 only), 9P135M (AT-4/AT-5) Launch method: Disposable launch canister Rate of launch (min): 2-3, range dependent Ready/Stowed – 4/0 dismounted, 4/4 veh Fire Control System: Name: 9S451M1 Guidance control box Guidance: SACLOS Command Link: Wire Beacon Type: Incandescent Infrared bulb Tracker Type: IR, 9S451M1 Susceptible to: EO Jammers, smoke, counter-fire Counter-countermeasures: Encoded pulse beacon, EO jammers Counter-dazzler adjustments to 9S451M1 Filter can be mounted on reticles Sights w/magnification: Day: 9Sh119M1, 4x Field of view (deg): 4.5 Acquisition Range (m): 4,000+ Night: 1PN86/Multi thermal sight Acquisition Range (m) 3,600+	Name: Konkurs Alt: AT-5/Spandrel-A Weight (kg) 25.2 (in tube) Type: Shaped Charge (HEAT) Penetration (mm): 650 Min/Max Range (m): 75/4000 Probability of hit(%): 90 Velocity (m/s): 200 Name: Konkurs-M Alt: AT-5b/Spandrel-B Weight (kg) 26.5 (in tube) Type: Tandem Shaped Charge (HEAT) Penetration (mm): 925 Min/Max Range (m): 75/4000 Probability of hit(%): 90 Velocity (m/s): 208 Name: Fagot Alt: AT-4/Spigot-A Weight (kg) 13 (in tube) Type: Tandem Shaped Charge (HEAT) Penetration (mm): 480 Min/Max Range (m): 70/2000 Probability of hit(%): 90 Velocity (m/s): 186 Name: Factoria (Fagot-M) Alt: AT-4b/Spigot-B Weight (kg) 12.9 (in tube) Type: Shaped Charge (HEAT) Penetration (mm): 550 Min/Max Range (m): 75/2500 Probability of hit(%): 90 Velocity (m/s): 180	P135M3: Konkurs-M Complex. Launcher with 1PN65 thermal sight and AT-5B missiles. Night range is 2,500m Tosan-1: Iranian version of AT-5 Launcher can be modified to launch other missiles such as the Indian Nag (TV/IR/SAL-homing) and the AT-3E/Malyutka-2 TPVP/1PN65 thermal sight is available with a range of 2,500. It weighs 13kg Slovienian TS-F sight has a 3,600-meter detection range

NOTES
THE RUSSIANS CATEGORIZE THE AT-4/4B SYSTEM AS PORTABLE RATHER THAN MAN PORTABLE. FOR DISMOUNT CARRY LOAD IT IS DIVIDED AMONG THREE PACKS. DUE TO ITS GREATER WEIGHT AT5/5B FITS INTO THE HEAVY CLASS AND SHOULD ONLY BE CARRIED SHORT DISTANCES FROM VEHICLE <500M.

Russian 40-mm Antitank Grenade Launcher RPG-7V

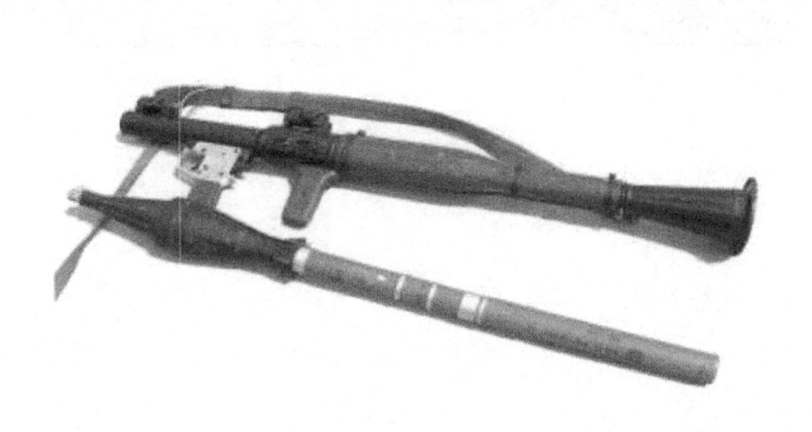

SYSTEM	SPECIFICATIONS	AMMUNITION	VARIANTS
Alternative Designation: N/A Date of Introduction: 1962 Proliferation: 70+ countries	Crew: 2 RPG-7V is light enough to be carried and fired by one person if needed. With a crew of 2, an assistant grenadier normally deploys left of the gunner to protect him from small arms fire. The full set has two bags: one has two grenades, spare parts, tools and accessories. The other has three more grenades. Caliber Launcher (mm): 40 The grenade warhead is forward of tube. Thus grenade diameter can be 105 mm or more. Weight (kg): 7.9 empty, loaded varies with grenade Length (mm): 950 Rate of Fire (rd/min): 4-6 Fire From Inside Building: No	40-mm grenade PG-7V PG-7VM PG-7VS PG-7VL PG-7VR TBG-7V OG-7V OG-7VM Combat load: 5 rockets Grenade Components: Warhead, rocket motor, tail assembly See Infantry Weapons for further details.	This is the most widely proliferated infantry AT system in the world. There are dozens of copies and variants of this launcher. RPG-250: Prototype and test base for the RPG-7V. RPG-7B1N3, -7N, and -7N1: Night site variant RPG-7V1: Upgrade w/bipod and improved PGO-7V3 sight. This is the standard production ATGL version since the late 1990s. RPG-7D, RPG-7DV1, and RPG-7D2N3 (night): Folding variants used by airborne troops. RPG-7D3 is the airborne counterpart to RPG-7V1. Type 69-1, II, III: Chinese upgrades variants with lighter weight, a wide range of munitions, and 3.0 x longer range sights.

NOTES

RPG-7V IS THE STANDARD (TIER 4) SQUAD ANTITANK WEAPON IN USE BY THE OPFOR. IT REQUIRES A WELL-TRAINED GUNNER TO ESTIMATE RANGES AND LEAD DISTANCES FOR MOVING TARGETS. THE RPG-7V HAS BEEN USED TO SHOOT DOWN HELICOPTERS IN SEVERAL CONFLICTS.

Russian 72.5-mm Antitank Disposable Launcher RPG-22

SYSTEM	SPECIFICATIONS
Alternative Designations	
Date, Country of Origin	1985, RUS
Proliferation	At least 9 countries
Crew	1
Weight Firing (kg)	2.8
Weight Travel (kg)	
Weight Tripod (kg)	
Length Firing (m)	0.85
Length Travel (m)	0.75
Rifling	Yes or No
Feed	Disposable
Breech Mechanism Type	NA
Emplacement Time (min)	1
Fire from Inside Building	No
SIGHTS	**SPECIFICATIONS**
Name	
Type	Iron
Sight Range Direct (m)	50, 150, 200, 250
Sight Range Indirect (m)	
Night Sights	No

ARMAMENT	SPECIFICATIONS
Caliber(mm)	**72.5**
Rate of Fire (prac, cyclic)	1, 1
Fire on Move	No
Elevation (deg min, max)	Shoulder held
Traverse (deg)	Shoulder held
AMMUNITION	**SPECIFICATIONS**
Missile Type, Name	**HEAT,**
Max Aimed Range (m)	250
Max Effective Range(m)	250
Penetration (mm)	390, C
Missile Diameter (mm)	72.5mm
Missile Weight (kg)	1.48
Initial Muzzle Velocity (m/s)	133
Maximum Velocity (m/s)	300

NOTES

THE RPG-22 IS A LIGHTWEIGHT, SHOULDER-FIRED, PRELOADED, DISPOSABLE ANTI-ARMOR WEAPON INTENDED FOR FIRING ONE ROUND, AFTER WHICH THE TUBE IS DISCARDED. IT IS BASICALLY A SCALED-UP VERSION OF THE RPG-18 (SIMILAR TO THE US LAW) AND HAS NO DEDICATED GRENADIER; HOWEVER, ALL SOLDIERS TRAIN TO USE THE SQUAD-LEVEL DISPOSABLE WEAPON.

Chinese 107mm Improvised Rocket Launcher Type 63

SYSTEM	SPECIFICATIONS
Alternative Designations	None
Date, Country of Origin	1960's, CHI
Proliferation	At least 20 countries
Crew	2
Weight Firing (kg)	3
Weight Travel (kg)	
Weight Tripod (kg)	10
Length Firing (m)	
Length Travel (m)	
Height Firing (m)	1.5
Width Firing (m)	1
Rifling	No
Feed	Manual
Breech Mechanism Type	Electric Battery
Emplacement Time (min)	3
Fire from Inside Building	No

SIGHTS	SPECIFICATIONS
Name	N/A
Type	Estimation
Sight Range Direct (m)	Line of Sight
Sight Range Indirect (m)	
Night Sights	

ARMAMENT	SPECIFICATIONS
Caliber(mm)	**107**
Rate of Fire (prac, cyclic)	1, 2
Fire on Move	No
Elevation (deg min, max)	-0, + est 60
Traverse (deg)	0

AMMUNITION	SPECIFICATIONS
Caliber(mm), Type, Name	**107, Rocket, Type 63**
Max Aimed Range (m)	8000
Warhead Type	Frag HE Spin stabilized
Weight (Kg)	18.8
Effect	1600 Steel Balls
Caliber(mm), Type, Name	**107, Rocket, Type 63 II**
Max Aimed Range (m)	8500
Warhead Type	Controlled Frag HE
Weight (Kg)	18.84
Effect	1214 Frag
Caliber(mm), Type, Name	**107, Rocket, Type 63 SS**
Max Aimed Range (m)	8500
Warhead Type	HE Incendiary White Phosphorous
Weight (Kg)	18.74
Effect	1600 Frag
Caliber(mm), Type, Name	**107, Rocket, Type 81 DP**
Max Aimed Range (m)	8000
Warhead Type	DPICM
Weight (Kg)	8.4
Effect	Sub munitions with HE effect 80mm penetration

NOTES

THE TYPE 63 CAN ALSO REFER TO A TOWED 12-TUBE MBRL PRODUCED IN CHINE IN THE EARLY 1960'S. IT WAS WIDELY USED BY THE PLA UNTIL THE 1980 AND IS VERY SIMILAR TO THE SOVIET BM-14. ALTHOUGH THE MBRL'S HAVE BEEN USED BY IRREGULAR ACTORS THEY ARE MORE LIKELY TO HAVE ACCESS TO THE MUNITIONS AND BE FORCED TO IMPROVISE THE LAUNCHER.

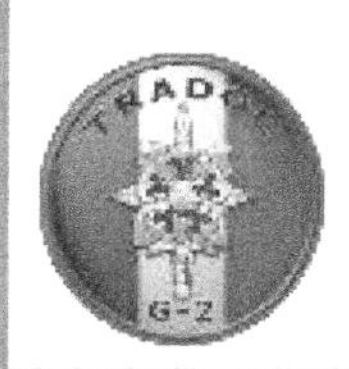

Insurgent 57-mm Improvised Rocket Launchers C-5K

SYSTEM	SPECIFICATIONS
Alternative Designations	S-5K
Date, Country of Origin	2000, INS
Proliferation	At least 2 Insurgent groups
Crew	1
Weight Firing (kg)	8
Length Firing (m)	1.42
Rifling	No
Feed	Manual
Breech Mechanism Type	Open
Emplacement Time (min)	1
Fire from Inside Building	INA
Launch control	Trigger, wire connecting to a battery in the stock

SIGHTS	SPECIFICATIONS
Name	Post, PGO-7
Type	Iron, Optical
Sight Range Direct (m)	Line of Sight
Aiming Limitation	Heat and Ash distract user
Night Sights	No

VARIANTS

Launcher tubes are extracted from UB32 Helicopter launch air to surface rocket pods. Tubes can be fabricated from pipe and some launchers have been seen with a blast shield to protect the user.

Early Chechen launchers were all welded, including grips and appeared to be sturdy.

In Tikrit a pedestal-mounted launcher with 4 unused rockets was found in the street.

ARMAMENT	SPECIFICATIONS
Caliber(mm)	57
Rate of Fire (prac, cyclic)	1, 2
Fire on Move	No
Elevation (deg min, max)	Est -10, est +65
Traverse (deg)	360

AMMUNITION	SPECIFICATIONS
Caliber(mm), Type, Name	57, Rocket, S5K
Free Flight Range (m)	2000
Max Eff Range, Day (m)	400
Rocket Weight (Kg)	3.65
Warhead Weight (Kg)	1.13
Rocket Length (m)	0.83
Fuze Type	Point Detonating
Penetration (mm)	150, C
Caliber(mm), Type, Name	57, Rocket, S5
Free Flight Range (m)	4000
Max Eff Range, Day (m)	400
Rocket Weight (Kg)	5.1
Warhead Weight (Kg)	1.1
Rocket Length (m)	1.42
Fuze Type	Point Detonating
Penetration (mm)	Damage light armored vehicles

NOTES

MOST LAUNCHERS USE A SINGLE TUBE BUT VARIANTS HAVE BEEN MADE WITH TWO THREE OR FOUR TUBES. VERSIONS HAVE BEEN SEEN WITH WOOD, METAL, AND PLASTIC HELD WITH A METAL STRAP, TAPE, AND WELDS THE S-5K ROCKETS FLAT TRAJECTORY OFFERS A MORE ACCURATE MUNITION FOR USE OVER DISTANCE BEYOND 200M THAN THAT OF THE HIGH BALLISTIC ARC OF AN ATGL, SUCH AS RPG-7V

Russian Surface to Air Missile Launcher SA-7 (Grail) / 9P54M

SYSTEM	SPECIFICATIONS	AMMUNITION	VARIANTS
Alt designations: 9K32M, Strela-2M Date of introduction: 1972 Proliferation: Worldwide Target: Low flying FW or Heli	Crew: 1, 2 with loader Launcher: 9P54M Length (m): 1.47 Diameter (mm): 70 Weight (kg): 4.71 Reaction time (sec): 5-10 Reload time (sec): 6-10 Fire on the move: Yes, short halt Fire Control: Sights w/magnification Target acquisition indicator Gunner visually identifies and acquires the target. IFF: Yes, can be fitted to the operator's helmet. A supplementary early warning system, passive RF antenna and headphones can be used to cue approach and direction.	Missile Name: 9M32M Range(m): 500-5,000 Max Altitude (m): 4,500 Min Altitude (m): 18 Length (m): 1.4 Diameter (mm): 70 Weight (kg): 9.97 Speed (m/s): 580 Propulsion: Solid fuel booster, and solid fuel sustainer motor Guidance: Passive 1-color IR homing (med IR range) Seeker field of view (deg): 1.9 Tracking rate (deg/sec): 6 Warhead type: HE Warhead weight (kg): 1.15 Fuze type: Contact Prob of Hit (%): 30 FW/ 40 Heli Self-destruct (Sec): 15 Countermeasures resistance: The seeker is fitted to reduce effectiveness of decoy flares and to block IR emissions.	SA-7b has improved propulsion over the older SA-7. SA-7b has better speed and range over the SA-7. SA-N-5: Naval version HN-5A: Chinese version Strela 2M/A: Yugoslavian upgrade Sakr Eye: Egyptian upgrade Strela-2M2: SA-7/7b and Strela 3/SA14 missiles converted with a Lomo upgrade 2-color IR seeker for detection/IRCM resistance similar to SA-18 SA-7b can be mounted in various vehicles, boats and vessels in four, six and eight tube launchers. It can also mount on helicopters including the Mi-8/17, Mi-24/35 and s-342 Gazelle

NOTES

THE MISSILE IS A TAIL CHASING HEAT (IR) SEEKER THAT DEPENDS ON ITS ABILITY TO LOCK ON TO HEAT SOURCES OF USUALLY LOW FLYING FIXED AND ROTARY WING AIRCRAFT. WHEN LAUNCHED TOWARDS A RECEDING AIRCRAFT, THE MANPADS CAN BE USED TO SCAN THE DIRECTION AND LOCK ON WITHOUT THE TARGET BEING VISUALLY ACQUIRED IN THE SIGHTS. A GUNNER MAY HAVE AN OPTIONAL 1L15-1 PORTABLE ELECTRONIC PLOTTING BOARD WHICH WARNS OF LOCATION AND DIRECTION OF APPROACHING TARGETS WITH A DISPLAY RANGE OF 12.5 KM. A VARIETY OF NIGHT SIGHTS ARE AVAILABLE INCLUDING 1 GEN (2-3,000M), 2 GEN (4,500M) AND THERMAL (5-6,000M).

Improvised Explosive Device Anti-Personnel Fragmentation Pipe Bomb

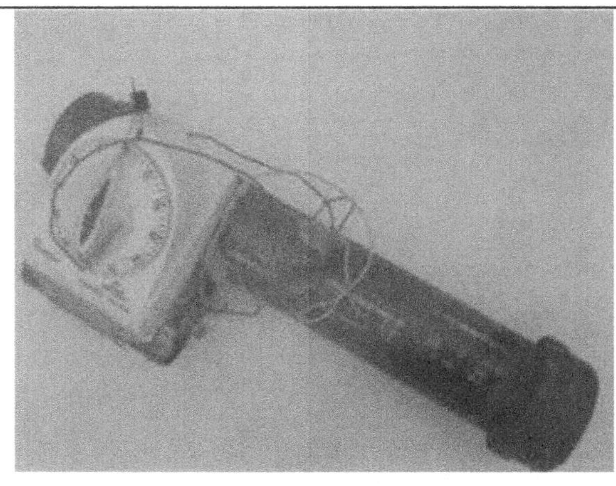

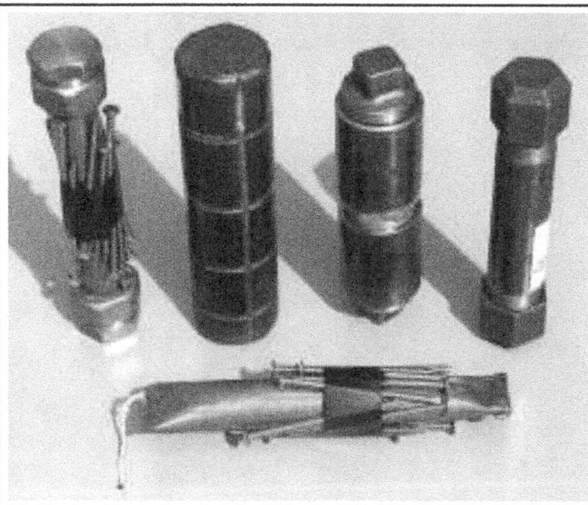

SYSTEM	SPECIFICATIONS
Alternative Designations	
Date, Country of Origin	NA
Proliferation	Worldwide
Crew	1
Weight Firing (kg)	Usually <2
Weight Travel (kg)	
Weight Tripod (kg)	
Length Firing (m)	Usually < 0.30
Feed	Single Use
Breech Mechanism Type	NA
Emplacement Time (min)	1
Fire from Inside Building	Yes

SIGHTS	SPECIFICATIONS
Name	
Type	Line Of Sight
Sight Range Direct (m)	50
Sight Range Indirect (m)	
Night Sights	No

ARMAMENT	SPECIFICATIONS
Caliber(mm)	NA
Rate of Fire (prac, cyclic)	1,1
Fire on Move	Yes
Elevation (deg min, max)	Hand Thrown
Traverse (deg)	360

EXPLOSIVE

Usually low grade without the need for a detonator. Black powder, fireworks or chlorate mixture are popular.

VARIANTS

Virtually any type of container that would provide confinement for an explosive material, glass jars, plastic pipe and appliances.

Some common variants are the Pressure Cooker bomb or the Letter Bomb

NOTES

A CRUDE DEVICE, OFTEN A PIPE CAPPED AT BOTH ENDS AND FILLED WITH EXPLOSIVE. CAN BE PACKED WITH NAILS OR SCREWS TO INCREASE DAMAGE. PIPE BOMBS CONCENTRATE PRESSURE AND RELEASE IT THROUGH THE FAILURE OF THE OUTER CASING. THEY ARE EASY TO MAKE IN ANY SHAPE OR SIZE BUT THE SIMPLE UNSOPHISTICATED NATURE OF THE DEVICE MAKES THEM DANGEROUS IF INCORRECTLY HANDLED.

Improvised Explosive Anti-Armor Device Explosive Formed Projectile

SYSTEM	SPECIFICATIONS
Alternative Designations	Dragon (Taliban)
Date, Country of Origin	2006, Insurgents
Proliferation	Iran, Lebanon, Hezbollah, Iraq, and Afghanistan
Crew	1
Weight Firing (kg)	Various
Length Firing (m)	>0.25
Rifling	Yes or No
Feed	Text
Breech Mechanism Type	Text
Emplacement Time (min)	Various
Fire from Inside Building	Yes
SIGHTS	**SPECIFICATIONS**
Name	
Type	NA
Sight Range Direct (m)	Line of Sight
Sight Range Indirect (m)	Remoted and unmanned
Night Sights	

AMMUNITION

Usually High Explosive

EFP components have been found with:

C4 Explosive / Gunpowder / TNT / HME / RDX

VARIANTS

EFPS were reportedly provided to Hezbollah proxy groups fighting in Iraq in kit form.

Machinery for forming the copper cones was discovered in Sadr City, Iraq indicating another variant.

Improvised EFP devices have been found cased in foam and painted to look like rocks

NOTES

EFP TECHNOLOGY WAS INVENTED IN THE 1930'S BY THE OIL INDUSTRY. IT HAS BEEN USED IN ANTI-ARMOR WEAPONS SINCE WW2.

EFPS WERE USED WITH DEVASTATING EFFECT IN IRAQ BY IRANIAN BACKED GROUPS AND IN AFGHANISTAN BY THE TALIBAN. A NORMAL IED WOULD CAUSE DAMAGE TO A HUMVEE AND EFP WOULD COMPLETELY DESTROY IT.

TOYOTA (DOUBLE CAB) HILUX "Technical" Multi-Role Vehicle

SYSTEM	SPECIFICATIONS
Alternative Designations	Pick-up Truck
Date, Country of Origin	1968, JPN
Proliferation	Worldwide
Crew	1
Troop Capacity	10
Combat Weight (kg)	2810
Length (m)	5.26
Height (m)	1.86
Width (m)	1.83
Ground Pressure (kg/cm^2)	INA
Drive Formula	4x4
AUTOMOTIVE	**SPECIFICATIONS**
Engine Type	Water-cooled, in-line, 6-cylinder gasoline Text
Cruising Range (km)	
Max On-Road (km/h)	
Max Off-Road (km/h)	100
Max Cross-Country (km/h)	INA
Max Swim (km/h)	NA
Fording Depth (m), note	0.7t

PROTECTION	SPECIFICATIONS
Radio	
Armor, Turret Front (mm):	No
Applique Armor (mm)	No
Reactive Armor (mm):	No
Mobility (mine clearing, self-entrenching)	No
NBC Protection System	No
Smoke Equipment	No

VARIANTS

The Toyota Hiluz can be up armored however most technical will be regular civilian vehicles used by an irregular force. There are reports that drug cartels have used armored pick-ups in the past.

A Hilux can be made to accommodate rockets, air defense guns or heavy machine guns. The photos above are from Libya and Afghanistan and illustrate examples fitted with DShK heavy machineguns.

A technical can be almost any civilian truck or vehicle that can be armed with different weapon systems (rockets, air defense guns, or heavy machineguns).

NOTES

THE TOYOTA HILUX IS A GOOD EXAMPLE OF A TECHNICAL BASED ON REPORTING FROM LIBYA, SYRIA, AFGHANISTAN, AND MALI. IRREGULAR FORCES USE THESE PARTICULAR TYPE OF VEHICLE BECAUSE OF THE VEHICLE'S RELIABILITY IN HOSTILE CONDITIONS. MOST ARE 4X4 VEHICLES TO ALLOW MOBILITY IN RESTRICTIVE TERRAIN. THEY ALLOW THE THREAT ACTOR TO BE AGILE, FAST AND BLEND INTO THE CIVILIAN POPULATION. THE DOWNSIDE OF HAVING LIMITED ARMOR IS OFFSET BY HIDING IN PLAIN SIGHT.

Sudanese 4x4 Light Tactical Vehicle Karaba VTG01

SYSTEM	SPECIFICATIONS
Alternative Designations	Safir
Date, Country of Origin	2008, SUD
Proliferation	Iran, Libya, CAR
Crew	1
Troop Capacity	5
Combat Weight (mt)	1500
Length (m)	3.51
Height (m)	1.88
Width (m)	1.9
Ground Pressure (kg/cm²)	INA
Drive Formula	4 x 4
AUTOMOTIVE	**SPECIFICATIONS**
Engine Type	Diesel
Cruising Range (km)	500
Max On-Road (km/h)	130
Max Off-Road (km/h)	Number
PROTECTION	**SPECIFICATIONS**
Radio	Yes
NBC Protection System	No
Smoke Equipment	No

ARMAMENT	SPECIFICATIONS
Caliber(mm), Type, Name	73, Recoilless Gun, SPG-9 RR
Rate of Fire (prac, cyclic)	6
Fire on Move	No
Elevation (deg min, max)	-3, +7
Caliber(mm), Type, Name	0, ATGM, AT-3 or AT-5
Rate of Fire (prac, cyclic)	2,3
Fire on Move	INA
Elevation (deg min, max)	INA

VARIANTS
Radio Station
Command Post
Ambulance

NOTES
THE KARABA VTG01 IS BASED ON THE IRANIAN-MADE SAFIR, ALTHOUGH SHOWN OPEN TOPPED IT CAN HAVE AN OPTIONAL SOFT COVER. IT IS AN EXAMPLE OF A MODERN TACTICAL UTILITY VEHICLE THAT COULD BE PROCURED BY IRREGULAR ACTORS.

Worldwide Equipment Guide

Chapter 12: Chemical, Biological, Radiological, and Nuclear (CBRN) Weapons

Chemical, Biological, Radiological, and Nuclear (CBRN) Weapons

This section provides a basic primer for threat characteristics for selected CBRN Weapons/agents/platforms. This portion also discusses the following topics: overview of OPFOR's rational on CBRN weapons, CBRN Threats, and WEG sheets representative of blister agents, nerve agents, choking agents, biotoxins, and decontamination platforms. These types of threats discussed in this segment are either in the real world and or readily available and therefore likely to be encountered by US forces in varying levels of conflict in the future. CBRN weapons can be used by a hybrid threat and is not limited to regular actors, but also irregular and criminal elements.

The list of CBRN systems/agents within this chapter is not meant to be encyclopedic. This chapter will be further developed with additional agents in upcoming editions. This edition of the CBRN chapter provides the US training community with a list of representative capabilities that allow scenario developers and the rest of the training community to create a dynamic threat to prepare today's warfighter for tomorrow's battlefield.

The section is divided into two major categories—***The CBRN Primer and WEG Sheets on CBRN assets/systems***. The CBRN primer provides insight into how the OPFOR composite views CBRN weapons. The second section of the primer address current CBRN threats. The WEG sheets (section) examine types of agents, and decontamination systems.

Questions and comments on data listed in this chapter should be addressed to:

Mr. Kristin Lechowicz

DSN: 552-7922 Commercial (913) 684-7922

E-mail address: Kristin.d.lechowicz.civ@mail.mil

This portion of the WEG is broken into two distinct but connected narratives. The first section discusses the OPFOR's rationale with regards to usage of CBRN systems. This OPFOR segment will not go into considerable depth on OPFOR tactics due to the fact that this subject is addressed in detail in Training Circular (TC) 7-100.2 chapter 13 (CBRN and Smoke). The second piece of this narrative discusses CBRN threats and has supporting CBRN related WEG sheets.

OPFOR and CBRN Issues

This section consists of a number of significant excerpts from TC 7-100.2 in order to provide a basic context for OPFOR CBRN related topics (for additional information on the subject of OPFOR CBRN and smoke tactics see the link in the above paragraph).

Key points on the OPFOR's CBRN issues:

- The OPFOR maintains a capability to conduct chemical, nuclear, and possibly biological or radiological warfare.
- The OPFOR is most likely to use chemical weapons against even a more powerful enemy.
- The OPFOR is equipped, structured, and trained to conduct both offensive and defensive chemical warfare.
- The OPFOR is continually striving to improve its chemical warfare capabilities.
- The OPFOR views chemical defense as part of a viable offensive chemical warfare capability.
- The OPFOR use the threat of numerous methods of CBRN delivery systems as an intimidating factor.
- The OPFOR could use CBRN against a neighbor as a warning to any potential enemy.
- The OPFOR uses the fact that CBRN weapons place noncombatants at risk as a positive factor.
- The OPFOR may threaten to use CBRN weapons as a way of applying political, economic, or psychological pressure by allowing the enemy no sanctuary.
- The OPFOR might use CBRN weapons either to deter aggression or as a response to an enemy attack.
- The OPFOR will use CBRN weapons on own troops in order to reach overarching objectives.
- The use of INFOWAR at every echelon is a key component in the OPFOR's CBRN program.
- The OPFOR may develop and employ radiological weapons.
- The OPFOR has nuclear capabilities.

CBRN Delivery Systems

The OPFOR has surface-to-surface missiles (SSMs) capable of carrying nuclear, chemical, or biological warheads. Most OPFOR artillery is capable of delivering chemical munitions, and most systems 152-mm and larger are capable of firing nuclear rounds. Additionally, the OPFOR could use aircraft systems and cruise missiles to deliver a CBRN attack. The OPFOR has also trained special-purpose forces (SPF) as alternate means of delivering CBRN munitions packages.

The Effects of CBRN on the Battlefield

The use of CBRN weapons can have an enormous impact on the battlefield and in peacetime and wartime operational environments. These types of weapons are a subset of weapons of mass destruction (WMD). WMD are weapons or devices intended for or capable of causing a high order of physical destruction or mass casualties (death or serious bodily injury to a significant number of people). The casualty-producing elements of WMD can continue inflicting casualties on the enemy and exert powerful psychological effects on the enemy's morale for

some time after delivery. Conventional weapons e.g., precision weapons or volumetric explosives) can also take on the properties of WMD.

Real World Threats CBRN Primer

Classifying of chemical agents

- Lethal agents
- Nonlethal agents
- Persistent: Agent that remains able to cause casualties for more than 24 hours to several days or weeks.
- Non-Persistent: dissipates and/or loses its ability to cause casualties after 10 to 15 minutes.

Subcategories of agents as the following:

- **Nerve:** Occupational Safety & Health Administration (OSHA) defines as highly toxic chemicals called "organophosphates" that poison the nervous system and disrupt bodily functions which are vital to an individual's survival.

Types and Characteristics Chemical Agents

		PERSISTENCE	PERSISTENCE		ENTRANCE	
TYPE OF AGENT	SYMBOL	SUMMER	WINTER	RATE OF ACTION	VAPOR/AEROSOL	LIQUID
NERVE	GA, GB, GD	10 min-24 hr	2 hr-3 days	Very Quick	Eyes, Lungs	Eyes, Skin, Mouth

*ARMY FIELD MANUAL NO. 8-10-7. Health Service Support in a Nuclear, Biological, and Chemical Environment.

- **Blister agents:** OSHA defines blister agents or "vesicants" are chemicals which have severely irritating properties that produce fluid filled pockets on the skin and damage to the eyes, lungs and other mucous membranes. Symptoms of exposure may be immediate or delayed until several hours after exposure.

Types and Characteristics Chemical Agents

		PERSISTENCE	PERSISTENCE		ENTRANCE	
TYPE OF AGENT	SYMBOL	SUMMER	WINTER	RATE OF ACTION	VAPOR/AEROSOL	LIQUID
BLISTER	HD, HN	3 days-1 wk	Weeks	Slow	Eyes, Skin, Lungs	Eyes, Skin
	L, HL	1-3 days	Weeks	Quick	Eyes, Skin, Lungs	Eyes, Skin, Mouth
	CX	Days	Days	Very Quick	Eyes, Lungs, Skin	Eyes, Skin, Mouth

*ARMY FIELD MANUAL NO. 8-10-7. Health Service Support in a Nuclear, Biological, and Chemical Environment.

- **Biotoxins:** OSHA defines as biological agents include bacteria, viruses, fungi, other microorganisms and their associated toxins. They have the ability to adversely affect human health in a variety of ways, ranging from relatively mild, allergic reactions to serious medical conditions, even death.
  Properties of Selected Biological Agents

BW Agents[1]	Likely Methods of Dissemination	Transmissibility Person-to-Person	Infectivity	Lethality[2]	Stability[2]
Anthrax (Inhalation)	Spores in aerosols	None	Moderate	High	Spores are highly stable
Brucellosis	1. Aerosol 2. Sabotage (food supply)	None	High	Low	Long persistence in wet soil and food
Cholera	1. Sabotage (food/water supply) 2. Aerosol	Negligible	Low	Moderate to high	Unstable in aerosol and pure water; more so in polluted water
Glanders	Aerosol	DNA	DNA	DNA	DNA
Melioidosis	Aerosol	Negligible	High	Variable	Stable
Plague (Pneumonic)	1. Aerosol 2. Infected Vectors	High	High	Very high	Less important because of high transmissibility
Psittacosis	Aerosol	Negligible	Moderate	Very low	Stable
Shigellosis	Sabotage (Food/Water Supply)	DNA	DNA	DNA	DNA
Tularemia	Aerosol	Negligible	High	Moderate if untreated	Not very stable
Typhoid Fever	1. Sabotage (food/water supply) 2. Aerosol	Negligible	Moderate	Moderate if untreated	Unknown
Q Fever	1. Aerosol 2. Sabotage (food supply)	None	High	Very low	Stable
Rocky Mountain Spotted Fever	1. Aerosol 2. Infected Vectors	None	High	High	Not very stable
Trench Fever	1. Aerosol 2. Vector	None	DNA	Low	DNA
Typhus Fever	1. Aerosol 2. Infected vectors	None	High	High	Not very stable
Chikungunya	Aerosol	None	High	Very low	Relatively stable
Crimean-Congo Hemorrhagic Fever	Aerosol	Moderate	High	High	Relatively stable
Dengue Fever	Aerosol	None	High	Low	Relatively unstable
Eastern Equine Encephalitis	Aerosol	None	High	High	Relatively unstable
Western Equine Encephalitis	Aerosol	None	High	Low	Relatively unstable
Ebola Fever	Aerosol	Moderate	High	High	Relatively unstable
Far Eastern Tick-borne Encephalitis	1. Aerosol 2. Milk	None	High	Moderate	Relatively unstable
Hantaan Virus (Korean HFV)	Aerosol	None	High	Moderate	Relatively stable
Juinn Hemorrhagic Fever	Aerosol	DNA	DNA	DNA	DNA

- **Choking agents:** The Center for Disease Control and Prevention (CDC) defines choking agents or pulmonary agents as chemicals that cause severe irritation or swelling of the respiratory tract (lining of the nose, throat, and lungs). Chart not available.
- **Systemic/Blood Agents:** The Center for Disease Control and Prevention (CDC) defines blood agents as poisons that affect the body by being absorbed into the blood. Chart not available.

OTHER TOXIC CHEMICALS

In addition to traditional chemical warfare agents, the OPFOR may find creative and adaptive ways to cause chemical hazards using chemicals commonly present in industry or in everyday households. In the right combination, or in and of themselves, the large-scale release of such chemicals can present a health risk, whether caused by military operations, intentional use, or accidental release.

Toxic Industrial Chemicals

Toxic industrial chemicals (TICs) are chemical substances with acute toxicity that are produced in large quantities for industrial purposes. Exposure to some industrial chemicals can have a lethal or debilitating effect on humans. They are a potentially attractive option for use as weapons of opportunity or WMD because of—

- The near-universal availability of large quantities of highly toxic stored materials.
- Their proximity to urban areas.
- Their low cost.

- The low security associated with storage facilities.

Employing a TIC against an opponent by means of a weapon delivery system, whether conventional or unconventional, is considered a chemical warfare attack, with the TIC used as a chemical agent. The target may be the enemy's military forces or a civilian population.

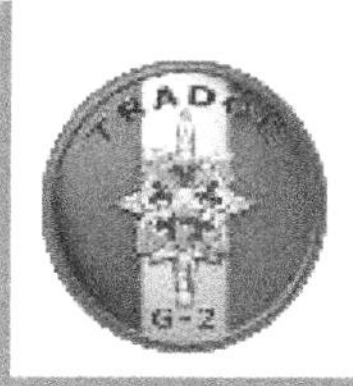

SULFUR MUSTARD: BLISTER AGENT

Sulfur Mustard

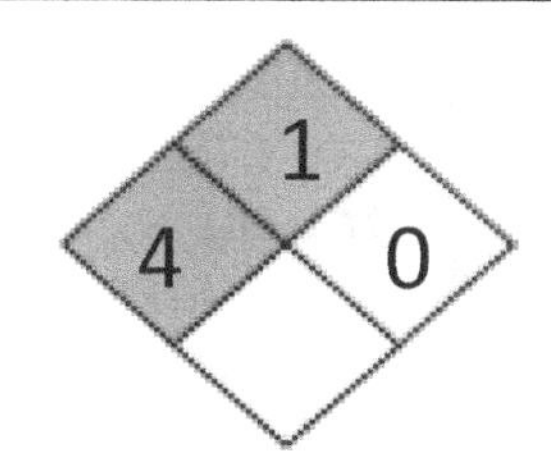

NFPA 704 Signal

AGENT CHARACTERISTICS	SPECIFICATIONS		ROUTES OF EXPOSURE	SPECIFICATIONS
Alternative designations:	HD (distilled sulfur mustard), Mustard, Mustard gas		Inhalation:	Yes
Military designation:	HD		Ingestion:	Yes
Chemical abstracts service number:	505-60-2		Skin/eye absorption:	Yes
UN Number	2810		METHODS OF DISSEMINATION	
Date of introduction:	1822		Indoor air:	Yes
Chemical Formula:	$C_4H_8Cl_2S$		Water:	Yes
Appearance:	Liquid	Food:	Yes	
Color:	Yellow or brown	Outdoor air:	Yes	
Description type agent:	Blister		Agricultural:	Yes
Description:	Blister agent (vesicant) that causes severe, delayed burns to the eyes, skin, and respiratory tract.		Indoor air:	Yes
Lethality:	Yes		Agricultural:	Yes
Nonpersistent:	No		Historic usage*:	Yes
Flammability*:	1			
Health*:	4			
Instability/reactivity*:	0			
Flashpoint° (F):	219.2			
Boiling Point ° (F):	422.6			

NOTES:

DESCRIPTION: MUSTARD GAS IS A VESICANT THAT WAS FIRST USED IN CHEMICAL WARFARE IN WORLD WAR I (EVEN THOUGH DISCOVERED PREVIOUSLY). IT CONSISTS OF AN OILY LIQUID OR CLEAR VAPOR. THE

UN REPORTS 'AFTER WWI' THE FOLLOWING USAGES OF HD : RUSSIA (1919), MOROCCO (1923-1926), ABYSSINIA (1935-1940), MANCHURIA (1937-1945), YEMEN (1963-1967), IRAQ/IRAN (1982-1988), POSSIBLE SYRIA (2014). UN #: 2810 (GUIDE 153). CDC REPORTS HD DAMAGES CELLS WITHIN MINUTES OF CONTACT; HOWEVER, PAIN AND OTHER HEALTH EFFECTS ARE DELAYED UNTIL HOURS AFTER EXPOSURE. HD IN LARGE DOSES CAN BE FATAL. HD IS 2 TO 5 TIMES MORE PERSISTENT IN WINTER THAN IN SUMMER. UN #: 2810 (GUIDE 153). THE CDC STATES THAT HD AFFECTS THE NERVOUS SYSTEM CAUSING "CHOLINERGIC TOXICITY" WITH THE FOLLOWING SYMPTOMS: EXCESSIVE SALIVA, TEARS AND URINE; GASTROINTESTINAL (GI) CRAMPING AND DIARRHEA; VOMITING (EMESIS); AND CONSTRICTED OR PINPOINT PUPILS (MIOSIS). HD HAS AN ODOR OF GARLIC, ONION, HORSERADISH, OR MUSTARD.

SARIN: CHEMICAL NERVE AGENT

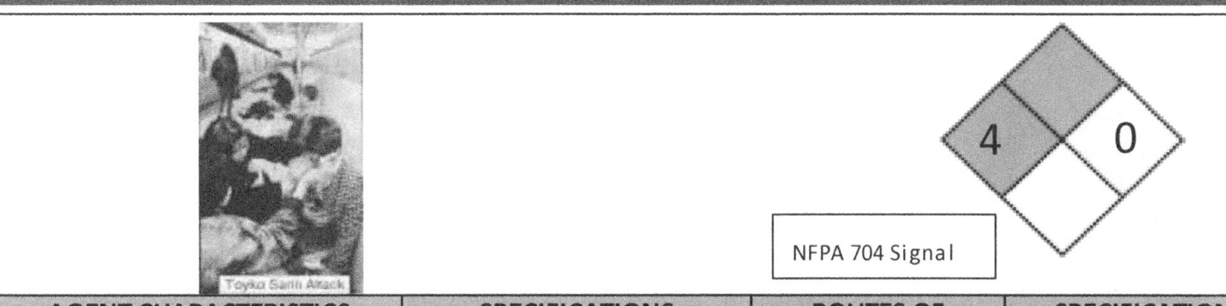

AGENT CHARACTERISTICS	SPECIFICATIONS	ROUTES OF EXPOSURE	SPECIFICATIONS
Alternative designations:	Trilone, Zarin	Inhalation:	Yes
Military designation:	GB	Ingestion:	Yes
Chemical abstracts service number:	107-44-8	Skin/eye absorption:	Yes
UN Number	2810	Ingestion:	Possible
Date of introduction:	1938	METHODS OF DISSEMINATION	
Chemical Formula:	$C_4H_{10}FO_2P$	Indoor air:	Yes
Appearance:	Liquid	Water:	Yes
Color:	Clear	Food:	Yes
Description type agent:	Nerve	Outdoor air:	Yes
Description:	Chemically similar to Organophospate	Agricultural:	Yes
Lethality*:	Yes	Indoor air:	Yes
Nonpersistent:	Yes	Agricultural:	Yes
Flammability*:	1	Historic usage*:	Yes
Health*:	4		
Instability/reactivity*:	0		
Flashpoint° (F):	>536		
Boiling Point ° (F):	297		

NOTES:

SARIN HAS BEEN USED BY REGULAR AND CRIMINAL ELEMENTS. MARCH 1995, AUM SHINRIKYO RELEASED SARIN GAS IN A TOKYO SUBWAY, 12 DEAD. IRAQI MILITARY USED SARIN DURING IRAN-IRAQ WAR (1980-1988) KILLING 26,000. LETHALITY: DEATH WITHIN MINUTES = 1 TO 10 MILLILITER (ML) ON SKIN ABSORPTION. INDICTORS OF ATTACK: ENEMY MOVING DECONTAMINATION UNITS FORWARD. ENEMY PUTTING ON CHEMICAL PROTECTIVE EQUIPMENT. SEVERE SIGNS OF EXPOSURE: LOSS OF CONSCIOUSNESS; SEIZURES, PARALYSIS; DEATH. SYMPTOMS: SKIN EXPOSURE: PROFUSE SWEATING, MUSCULAR TWITCHING, NAUSEA, VOMITING, DIARRHEA, AND WEAKNESS. INHALATION EXPOSURE: PINPOINT PUPILS, RUNNY NOSE, SHORTNESS OF BREATH.

VX: Nerve Agent

VX nerve agent

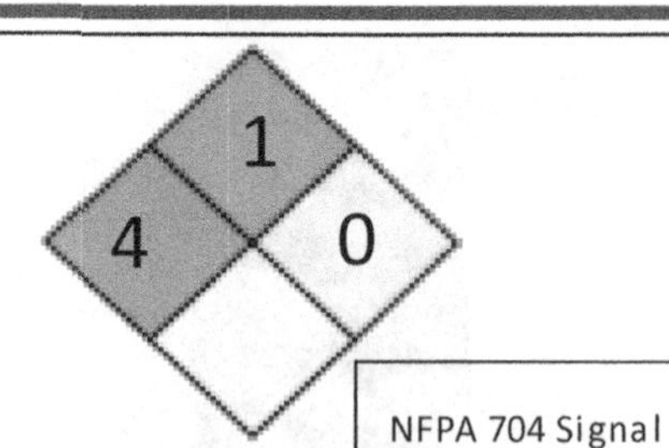

NFPA 704 Signal

AGENT CHARACTERISTICS	SPECIFICATIONS	ROUTES OF EXPOSURE	SPECIFICATIONS
Alternative designations:	Methylphosphonothioic acid, O-ethyl S-(2-diisopropylaminoethyl) methylphosphonothiolate	Inhalation:	Yes
Military designation:		Ingestion:	Yes
Chemical abstracts service number:	50782-69-9	Skin/eye absorption:	Yes
UN Number	2810	METHODS OF DISSEMINATION	
Date of introduction:	Early 1950s	Indoor air:	Yes
Chemical Formula:	$C_{11}H_{26}NO_2PS$	Water:	Yes
Appearance:		Outdoor air:	Yes
Color:	Clear, Amber	Agricultural:	Possible
Description type agent:	Nerve	Historic usage:	
Description:	One of the most toxic tasteless and odorless human-made chemical warfare agents		
Lethality:	Yes		
Nonpersistent:	No		
Flammability:	1		
Health:	4		
Instability/reactivity:	0		
Flashpoint° (F):	318.2		
Boiling Point ° (F):	568.4		

NOTES:

SYMPTOMS AND EFFECTS: BLURRED VISION, CONFUSION, DROOLING, EXCESSIVE SWEATING, COUGH, NAUSEA, DIARRHEA, SMALL PINPOINT PUPILS, CONVULSIONS, LOSS OF CONSCIOUSNESS, PARALYSIS, RESPIRATORY FAILURE POSSIBLY LEADING TO DEATH.

CHLORINE: CHOKING (LUNG DAMAGING) AGENT

WWI Chlorine Attack

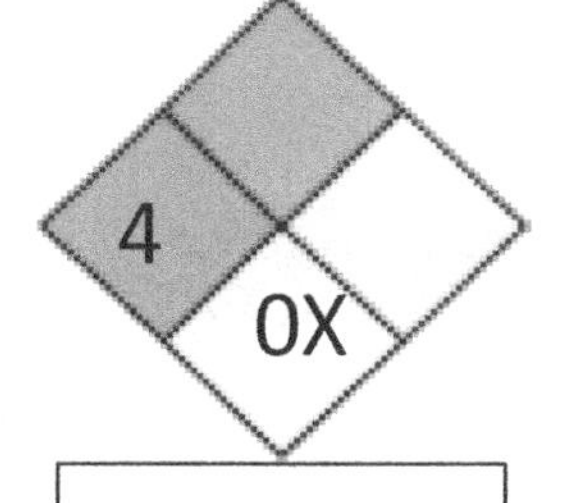

NFPA 704 Signal

AGENT CHARACTERISTICS	SPECIFICATIONS	ROUTES OF EXPOSURE	SPECIFICATIONS
Alternative designations:	Molecular chlorine	Inhalation:	Yes
Military designation:		Ingestion:	Yes
Chemical abstracts service number:	7782-50-5	Skin/eye absorption:	Yes
UN Number	1017	METHODS OF DISSEMINATION	
Date of introduction:		Indoor air:	Yes
Chemical Formula:	Cl_2	Water:	Yes
Appearance:	Gas	Food:	Unlikely
Color:	Greenish Yellow	Outdoor air:	Yes
Description type agent:	Choking	Agricultural:	Unlikely
Description:	Toxic gas with corrosive properties.	Indoor air:	Yes
Lethality:	Yes	Historic usage*:	Yes
Nonpersistent:	Yes		
Flammability:	0		
Health*:	4		
Instability/reactivity*:	0		
Flashpoint° (F):	INA		
Boiling Point ° (F):	-30.3		

NOTES:

CHLORINE IS USED AS AN INDUSTRIAL/HOUSEHOLD CLEANER/DISINFECTANT. IT IS USES RANGE FROM WATER TREATMENT TO CHEMICAL WARFARE (DATING BACK TO WORLD WAR I). PERSISTENCE: MINUTES TO HOURS. UN NUMBER: 1017. SYMPTOMS: BLURRED VISION. BURNING SENSATION IN THE NOSE, THROAT, AND EYES, COUGHING CHEST TIGHTNESS, DIFFICULTY BREATHING OR SHORTNESS OF BREATH, NAUSEA AND VOMITING, WATERY EYES, WHEEZING, LUNG PROBLEMS AND DEATH. BBC REPORTS POSSIBLE CHLORINE ATTACKS IN SYRIA IN APRIL/MAY 2014 AND MARCH 2015. UN SUPPORTED THE ABOVE STATEMENT WITH THE FOLLOWING: "HIGH DEGREE OF CONFIDENCE, THAT CHLORINE HAD BEEN USED AS A WEAPON IN THREE VILLAGES IN NORTHERN SYRIA FROM APRIL TO AUGUST 2014."

CHLOROACETOPHENONE: Riot Control/Tear Agent

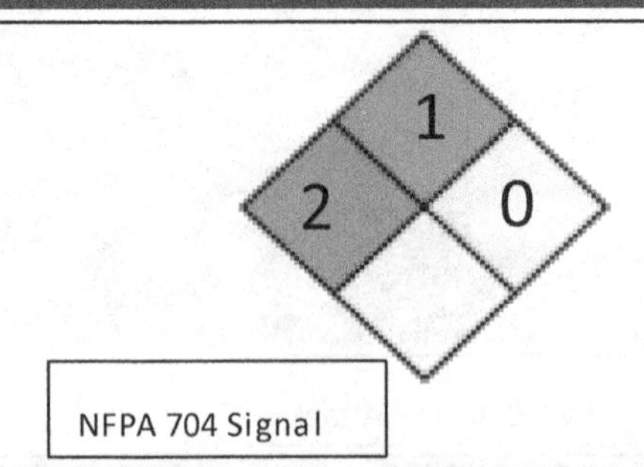

NFPA 704 Signal

AGENT CHARACTERISTICS	SPECIFICATIONS	ROUTES OF EXPOSURE	SPECIFICATIONS
Alternative designations:	2-Chloro-1-phenylethanone, 2-Chloroacetophenone, alpha-Chloroacetophenone, Chemical mace, Chloromethyl phenyl ketone, Phenyl chloromethylketone, Tear gas	Inhalation:	Yes
Military designation:	CN	Ingestion:	Yes
Chemical abstracts service number:	532-27-4	Skin/eye absorption:	Yes
UN Number	1697	METHODS OF DISSEMINATION	
Date of introduction:		Indoor air:	Yes
Chemical Formula:	C_8H_7ClO	Water:	Yes
Appearance:	Gas	Outdoor air:	Yes
Color:	Gray, white, or colorless	Agricultural:	Possible
Description type agent:	Riot Control/Tear Agent	Indoor air:	Yes
Description:	Military and law enforcement use CN for riot control.	Historic usage:	Yes
Lethality:	Yes*		
Nonpersistent:	Yes		
Flammability:	1		
Health*:	2		
Instability/reactivity*:	0		
Flashpoint° (F):	244		
Boiling Point ° (F):	472-473		

NOTES:
EXPOSURE (MILD TO MODERATE): RUNNY NOSE, EYE IRRITATION, COUGHING, SNEEZING, DIFFICULTY BREATHING, SHORTNESS OF BREATH, CHOKING, BURNING SENSATION, NAUSEA, VOMITING. (SEVERE): FLUID IN LUNGS, NARROWING OF AIRWAYS, FAINTING, INCREASED HEART RATE, LONG TERM EXPOSURE IN CONFINED SPACE CAN IN RARE INSTANCES CAUSES DEATH.

RICIN: Biotoxin

Caster Beans

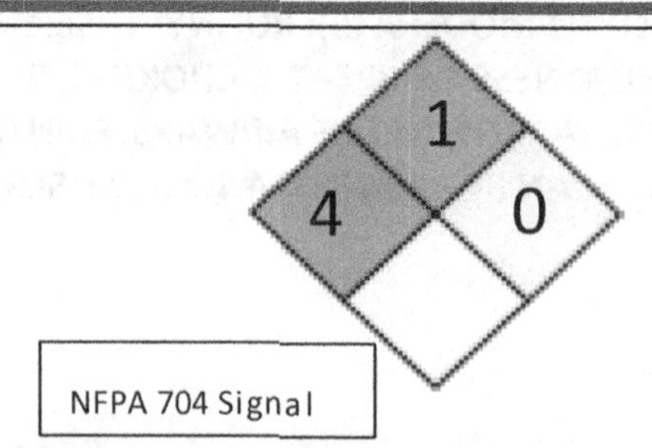

NFPA 704 Signal

AGENT CHARACTERISTICS	SPECIFICATIONS	ROUTES OF EXPOSURE	SPECIFICATIONS
Alternative designations:	Ricine, Ricins	Inhalation:	Yes
Military designation:		Ingestion:	Yes
Chemical abstracts service number:	9009-86-3	Skin/eye absorption:	Possible
UN Number	3462	METHODS OF DISSEMINATION	
Date of introduction:	INA	Indoor air:	Yes
Chemical Formula:	INA	Water:	Yes
Appearance:	Powder, Liquid, Crystalline	Outdoor air:	Yes
Color:	White	Agricultural:	Possible
Description type agent:	Biotoxin	Historic usage:	Yes
Description:	Toxic protein from Castor Bean plant		
Lethality:	Yes		
Nonpersistent:			
Flammability:	1		
Health*:	4		
Instability/reactivity*:	0		
Flashpoint° (F):	INA		
Boiling Point ° (F):	INA		

NOTES:

SYMPTOMS AND EFFECTS: RICIN INGESTION GENERALLY APPEARS WITHIN A FEW HOURS AFTER CONSUMPTION; HOWEVER, SOME PATIENTS CAN BE ASYMPTOMATIC FOR A PERIOD OF TIME. FEVER, COUGH, RESPIRATORY DISTRESS, BURING OF THE THROAT, ACCUMULATION OF FLUID IN THE LUNGS. RICIN ATTACKS THE LIVER, CENTRAL NERVOUS SYSTEM, KIDNEYS, AND ADRENAL GLANDS (2 TO 5 DAYS AFTER EXPOSURE). 3-5 DAYS AFTER CONTACT MAY LEAD TO DEATH.

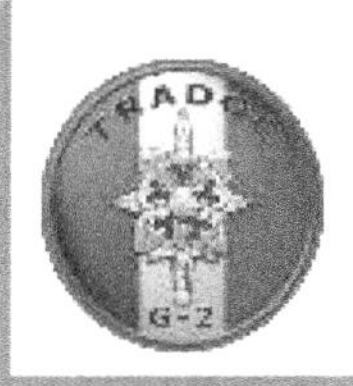

HYDROGEN CYANIDE: SYSTEMIC AGENT

NFPA 704 Signal

AGENT CHARACTERISTICS	SPECIFICATIONS	ROUTES OF EXPOSURE	SPECIFICATIONS
Alternative designations:	Formonitrile, Hydrocyanic acid, Prussic acid	Inhalation:	Yes
Military designation:	AC	Ingestion:	Yes
Chemical abstracts service number:	74-90-8	Skin/eye absorption:	Yes
UN Number	1051	Historic usage*:	Yes
Date of introduction:		METHODS OF DISSEMINATION	
Chemical Formula:	HCN	Indoor air:	Yes
Appearance:		Water:	Yes
Color:	Colorless or pale blue	Food:	Yes
Description type agent:	systemic chemical asphyxiant	Outdoor air:	Yes
Description:	interferes with the normal use of oxygen by nearly every organ of the body	Agricultural:	Possible
Lethality*:	Yes	Indoor air:	Yes
Nonpersistent:	No	Historic usage:	Yes
Flammability*:	4		
Health*:	4		
Instability/reactivity*:	2		
Flashpoint° (F):	0		
Boiling Point ° (F):	78		

NOTES:

THE CDC REPORTS THAT AFTER INHALATION EXPOSURE, SYMPTOMS BEGIN WITHIN SECONDS TO MINUTES; DEATH MAY OCCUR WITHIN MINUTES. AFTER SKIN EXPOSURE, ONSET OF SYMPTOMS MAY BE IMMEDIATE OR DELAYED FOR 30 TO 60 MINUTES. INGESTION OF HYDROGEN CYANIDE (AC) SOLUTIONS OR CYANIDE SALTS CAN BE RAPIDLY FATAL.

RUSSIAN DECONTAMINATION/SMOKE GENERATOR VEHICLE TMS-65M

SYSTEM	SPECIFICATIONS	FEATURES (CONT.)	SPECIFICATIONS
System		Max Swim:	INA
Alternative Designations:	None	Fording Depths (M):	1.5
Date Of Introduction:	INA	Radios, Frequency, And Range:	INA
Proliferation:	At Least 1 Country	Nozzle Movement:	Yes
Description:		Traverse:	90
Crew:	2	Elevation	12
Troop Capacity:	INA	Depression	23
Chassis:	Ural 4320, 6x6	Operating RPM's of Turbine Engine:	INA
Combat Weight (Mt):	INA	Idle Speed (rpm):	INA
Length Overall (M):	7.3	Max Effective Speed (rpm):	INA
Height Overall (M):	2.5	Ability to Generate on the Move:	Yes
Width Overall (M)	2.8	Performance Characteristics:	
Ground Pressure (Kg/Cm2):	INA	Engine Name:	VK-1A modified gas turbine
Automotive Performance:		Fuel Type:	INA

Engine Type:	YaMZ-238M2; V-8 liquid cooled diesel	Fuel Capacity (l):	INA
HP:	240	Operating Time (min):	60
Cruising Range (Km):	1040	Time to Initiate Smokescreen (min):	10
Speed (Km/H):		Generator Efficiency (%):	96
Max Road:	82	Smoke Screening System:	Yes
Max Off-Road:	INA	Emplacement Time:	10-12 minutes
		Displacement Time:	10-12 minutes

NOTES

SMOKE GENERATOR: PHYSICAL CHARACTERISTICS: NAME: INA. LENGTH (M): INA. HEIGHT (M): INA. WIDTH (M): INA. NUMBER OF NOZZLES: 6 LOCATION OF NOZZLE: INA. NOZZLE ORIFICE SIZE (M): INASMOKE AGENT: GOST 305-73. NOMENCLATURE: GOST-305-82 PETROLEUM OBSCURANTS: MANMADE AGENTS. CLOUD COLOR: WHITE. CAPACITY OF SMOKE AGENT (L): 1500. SMOKE AGENT CONSUMPTION RATE (L/HR): 1500. ODOR: FAINT SULFUR. PROTECTION REQUIRED: NONE. THE TMS-65M IS PRIMARILY USED FOR THE DECONTAMINATION OF VEHICLES AND EQUIPMENT WHETHER STATIONARY OR ON THE MOVE. ADDITIONALLY, THE SYSTEM CAN BE USED TO DECONTAMINATE GROUND AND HARD SURFACE ROADS AS WELL AS PRODUCE PROTECTIVE SMOKESCREENS. THE VK-1 TURBINE ENGINE (A MODIFIED MIG-17 ENGINE) ALONG WITH A SEALED CONTROL CAB IS MOUNTED ON AN URAL 4320, 6X6 TRUCK CHASSIS. THE TURBINE ENGINE CAN BE MOVED VERTICALLY OR HORIZONTALLY BY CONTROLS LOCATED IN THE OPERATOR'S CONTROL CAB. THE OPERATOR'S CONTROL CAB IS FIXED TO THE LEFT SIDE OF THE TURBINE ENGINE AND IS EQUIPPED WITH FLOODLIGHTS FOR LIMITED VISIBILITY OPERATIONS. TWO 1500-LITER TANKS ARE MOUNTED ACROSS THE FRONT OF THE TRUCK BED BETWEEN THE DRIVER'S CAB AND THE TURBINE ENGINE. ONE TANK IS USED FOR OBSCURANT, DECONTAMINANTS, OR WATER AND THE OTHER TANK IS USED FOR THE FUEL TO POWER THE TURBINE ENGINE. THE OBSCURANT, DECONTAMINANT, OR WATER IS FED THROUGH PIPES INTO THE TURBINE ENGINE EXHAUST STREAM. DEPENDING UPON THE SIZE AND LEVEL OF CONTAMINATION THE TMS-65M IS CAPABLE DECONTAMINATING EQUIPMENT AT A RATE OF 30 SECONDS TO 3 MINUTES PER ITEM.

Slovakian Decontamination/Smoke Generator Vehicle TZ-74

SYSTEM	SPECIFICATIONS	FEATURES (CONT.)	SPECIFICATIONS
System		Max Swim:	
Alternative Designations:	None	Fording Depths (M):	
Date Of Introduction:	INA	Radios, Frequency, And Range:	
Proliferation:	3	Nozzle Movement:	
Description:		Traverse:	120
Crew:	2	Elevation	30
Troop Capacity:	INA	Depression	20
Chassis:	Tatra 148 PPR 15, 6x6	Operating RPM's of Turbine Engine:	13,000 rpm (equals 84%)
Combat Weight (Mt):	21.9	Idle Speed (rpm):	5400
Length Overall (M):	8.49	Max Effective Speed (rpm):	1400
Height Overall (M):	2.5	Ability to Generate on the Move:	No
Width Overall (M)	3.2	Performance Characteristics:	
Ground Pressure (Kg/Cm2):	INA	Engine Name:	Type M701 C-500
Automotive Performance:	Tatra 2-298-1	Fuel Type:	Diesel
Engine Type:	V-8 air cooled diesel	Fuel Capacity (l):	2,000
HP:	1200	Operating Time (min):	22

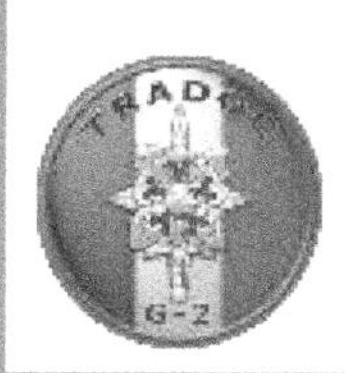

Cruising Range (Km):	400	Time to Initiate Smokescreen (min):	1
Speed (Km/H):	INA	Generator Efficiency (%):	98
Max Road:	71	Smoke Screening System:	INA
Max Off-Road:	INA	Emplacement Time:	approximately 10-12 minutes
		Displacement Time:	approximately 10-12 minutes

NOTES

THE TZ-74 IS PRIMARILY USED FOR THE DECONTAMINATION OF VEHICLES AND EQUIPMENT WHETHER STATIONARY OR ON THE MOVE. ADDITIONALLY, THE SYSTEM CAN BE USED TO DECONTAMINATE GROUND AND HARD SURFACE ROADS AS WELL AS PRODUCE PROTECTIVE SMOKESCREENS. THE TZ-74 IS BASED ON THE CHASSIS OF THE TATRA 148 PPR 15, 6X6 TRUCKS. THE SYSTEM USES A GAS TURBINE ENGINE TO GENERATE A SMOKE SCREEN OVER A LARGE AREA. THE GAS TURBINE IS MOUNTED ON THE REAR OF THE VEHICLE WITH THE OPERATOR'S CABIN ON THE LEFT-HAND SIDE, AND AN OBSCURANT TANK ON THE RIGHT-HAND SIDE. THE REMAINDER OF THE REAR OF THE VEHICLE CONSISTS OF STORAGE TANKS FOR LIQUIDS AND FUEL FOR THE TURBINE ENGINE. A SMOKESCREEN IS CREATED AS A RESULT OF THE INTRODUCTION OF OIL, DIESEL FUEL, OR A COMBINATION OF BOTH INTO THE EXHAUST NOZZLE OF THE TURBINE ENGINE. THE INTRODUCTION OF THE MIXTURE INTO THE TURBINE ENGINES EXHAUST CAUSES AN IMMEDIATE DISPERSION AND EVAPORATION OF THE MIXTURE THAT CONDENSES IN THE COOLER PART OF THE STREAM AFTER EXPULSION FROM THE MOTOR.

THE TZ-74 CAN PRODUCE SMOKE IN ABOUT A MINUTE AFTER THE TURBINE ENGINE IS TURNED ON. ALTHOUGH THE JET ENGINE HAS ENOUGH FUEL TO OPERATE FOR 3 HOURS, ITS MISSION TIME IS SEVERELY LIMITED BY THE CAPACITY OF THE INTERNAL LIQUID SOURCES AVAILABLE. HOWEVER, IF EXTERNAL SOURCES ARE USED, THE MISSION TIMES ARE DRAMATICALLY INCREASED. THE MAIN DECONTAMINANT TANK HAS A 5,000-LITER CAPACITY. THE FOG OIL TANK HAS A 200-LITER CAPACITY. THE FUEL TANK FOR THE GAS TURBINE HAS A 2,000-LITER CAPACITY.

Worldwide Equipment Guide

Volume 2: Air and Air Defense Systems

DEPARTMENT OF THE ARMY
UNITED STATES ARMY TRAINING AND DOCTRINE COMMAND
TRADOC G-2 LEAVENWORTH
801 HARRISON DRIVE
FORT LEAVENWORTH, KS 66027-1323

REPLY TO
ATTENTION OF:

ATIN-ZAL

16 March 2016

MEMORANDUM FOR: Distribution unlimited

SUBJECT: Worldwide Equipment Guide (WEG) Update 2015

1. In today's complicated and uncertain world, it is impossible to predict the exact nature of the next conflict that may involve U.S. joint forces. We must be ready to meet the challenges of any type of conflict, in all kinds of places, and against all types of threats in all Complex Operational Environments. As a training tool, the opposing force (OPFOR) must be a challenging, uncooperative sparring partner capable of stressing any or all warfighting functions and mission-essential tasks of the U.S. force.

2. The Army Training Circular 7-100 series describes the doctrine, organizations, TTP, and equipment of such an OPFOR and how to combine it with other operational variables to portray the qualities of a full range of conditions appropriate to Army training environments.

3. The WEG was developed to support the TC 7- 100 series and all OPFOR portrayal in training simulations (live, virtual, constructive, and gaming). The equipment portrayed in the WEG represents military systems, variants, and upgrades that US forces may encounter now and in the foreseeable future. The authors continually analyze real- world developments, capabilities, and trends to guarantee the OPFOR remains relevant.

4. Published in three volumes, (Ground; Airspace & Air Defense Systems; and Naval & Littoral Systems) the WEG is the approved document for OPFOR equipment data used in U.S. Army training. Annual updates are posted on the ATN website. Therefore it is available for downloading and local distribution. Distribution restriction is unlimited. This issue replaces all previous issues.

5. For comments or questions regarding this document, contact Mr. Cantin DSN 552- 7952, (913) 684-7952, john.m.cantin.ctr@mail.mil. If he is not available (or for specific issues), contact the POCs noted in the chapter introductions.

Gary E. Phillips
Assistant TRADOC G-2

Preface

The WEG was developed to support the TC 7-100 series and all OPFOR portrayal in training simulations (constructive, virtual, live, and gaming). The equipment portrayed in this WEG represents military systems, variants, and upgrades that U.S. forces may encounter now and in the foreseeable future. The authors continually analyze real-world developments, capabilities, and trends to guarantee that the OPFOR remains relevant.

This series of TC/FM and guides outline an OPFOR capable of portraying the entire spectrum of military and paramilitary capabilities against which the U.S. Military must train to ensure success in any future conflict.

Applications for this series of handbooks include field training, training simulations, and classroom instruction throughout the U.S. Military. All U.S. Military training venues should use an OPFOR based on these handbooks, except when mission rehearsal or contingency training requires maximum fidelity to a specific country-based threat. Even in the latter case, trainers should use appropriate parts of the OPFOR handbooks to fill information gaps in a manner consistent with what they do know about a specific threat.

Unless this publication states otherwise, masculine nouns or pronouns do not refer exclusively to men.

Introduction

This Worldwide Equipment Guide (WEG) describes the spectrum of worldwide equipment and system trends in the Complex Operational Environment (COE). Tier Tables provide baseline examples of systems with counterparts in other capability tiers. Other systems are added to offer flexibility for tailoring the force systems mix. Substitution Tables offer other system choices versus baseline examples.

The OPFOR in the COE should also include options for portraying "hybrid threat". Hybrid threat is defined as:

...the diverse and dynamic combination of regular forces, irregular forces, terrorist forces, and/or criminal elements, all unified to achieve mutually benefitting effects.

The OPFOR may use conventional weapons; however regular and irregular forces may also employ improvised systems, as described throughout this guide. Upgrade tables are included to capture WEG systems changes reflecting contemporary upgrade trends. Systems and technologies in Chapter 10, Countermeasures, Upgrades, and Emerging Technology, can be used in simulations for Near-Term and Mid-Term scenarios.

The pages in this WEG are designed for use in electronic form or for insertion into loose-leaf notebooks. This guide will be updated as often as necessary, in order to include additional systems, variants, and upgrades that are appropriate for OPFOR use.

The 2015 version of the WEG has changes in the format and presentation of information. We have attempted to make the information available more user friendly, easily accessible, and concise. Therefore, much of the narrative on some systems has been updated and edited. If you have questions on the presentation of information or anything else included in this guide, contact Mr. John Cantin at DSN 552-7952, commercial (913) 684-7952, email: john.m.cantin.ctr@mail.mil.

HOW TO USE THIS GUIDE

The WEG is organized by categories of equipment, in chapters. The format of the equipment pages is basically a listing of parametric data. This permits updating on a standardized basis as data becomes available. For meanings of acronyms and terms, see the Glossary. Please note that, although most terms are the same as in U.S. terminology, some reflect non-U.S. concepts and are not comparable or measurable against U.S. standards. For example, if an OPFOR armor penetration figure does not say RHA (rolled homogeneous armor) do not assume that is the standard for the figure. If there are questions, consult the Glossary, or contact us.

System names reflect intelligence community changes in naming methods. Alternative designations include the manufacturer's name, as well as U.S./NATO designators. Note also that the WEG focuses on the complete weapon system (e.g., AT-4/5/5B antitank guided missile launcher complex or 9P148 ATGM launcher vehicle), versus a component or munition (9P135 launcher or AT-4/5 ATGM).

Many common technical notes and parameters are used in chapters 3 through 6, since the systems contained in those chapters have similar weapon and automotive technologies. Chapters 2 (Infantry Weapons), 7 (Engineer and CBRN) and 8 (Logistics), offer systems with many unique parameters and therefore may not be consistent with those in other chapters.

The authors solicit the assistance of WEG users in finding unclassified information that is not copyright-restricted, and that can be certified for use. Questions and comments should be addressed to the POC below. If he is not available, contact the designated chapter POC.

Mr. John Cantin
DSN: 552-7952 Commercial (913) 684-7952
e-mail address: john.m.cantin.ctr@mail.mil

AKO PATH TO OPFOR COE PRODUCTS

To access WEG and other COE training products at our site, use either of these two methods. The AKO direct link is https://www.us.army.mil/suite/files/21872221.

Direct link to each volume is as follows:

Volume 1 Ground Forces	https://www.us.army.mil/suite/doc/25963538
Volume 2 Air and AD	https://www.us.army.mil/suite/doc/25963539
Volume 3 Naval Littoral	https://www.us.army.mil/suite/doc/25963540

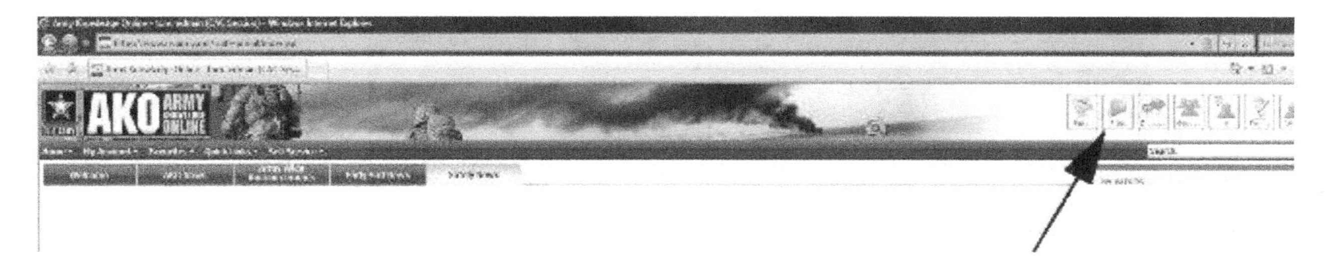

Or, navigate to the site as follows:

(1) Go to the AKO home page and click on Files (upper right).
(2) Then go to DOD Organizations (left)
(3) Then click on prompts per the sequence in the box.

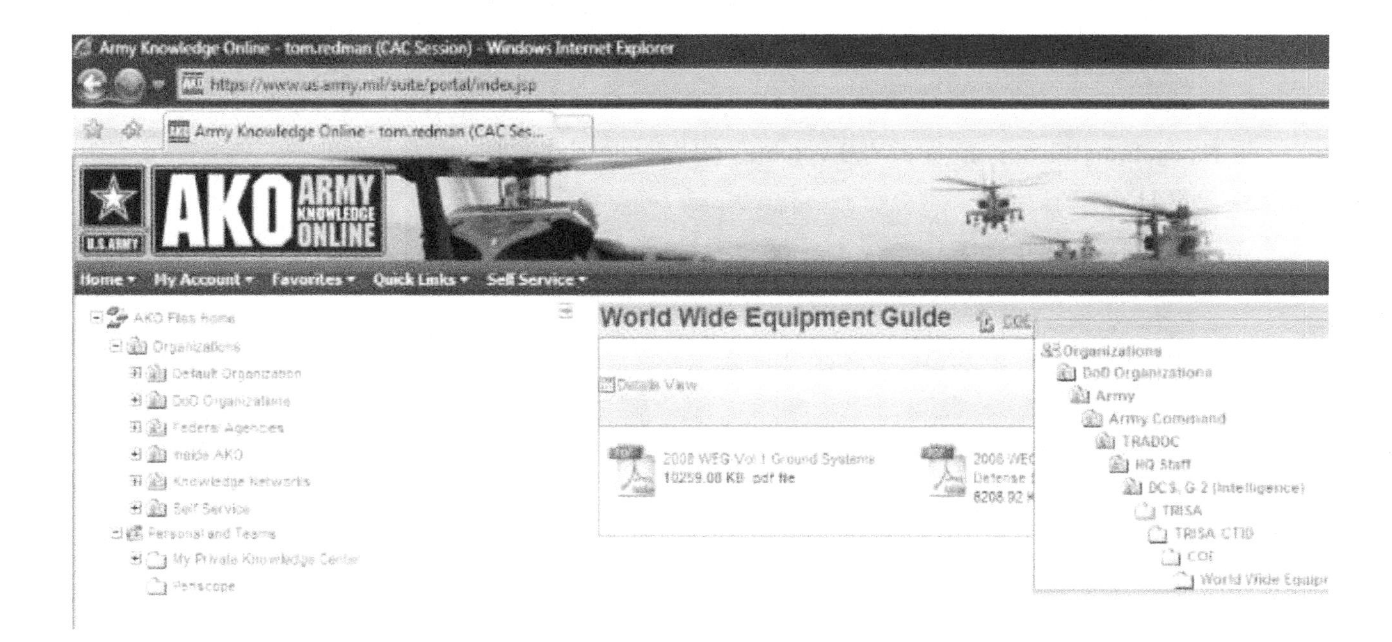

ATN PATH TO OPFOR COE PRODUCTS

To access WEG and other COE training products at our ATN go to the TRISA CTID page at the following link https://atn.army.mil/dsp_template.aspx?dpID=311.

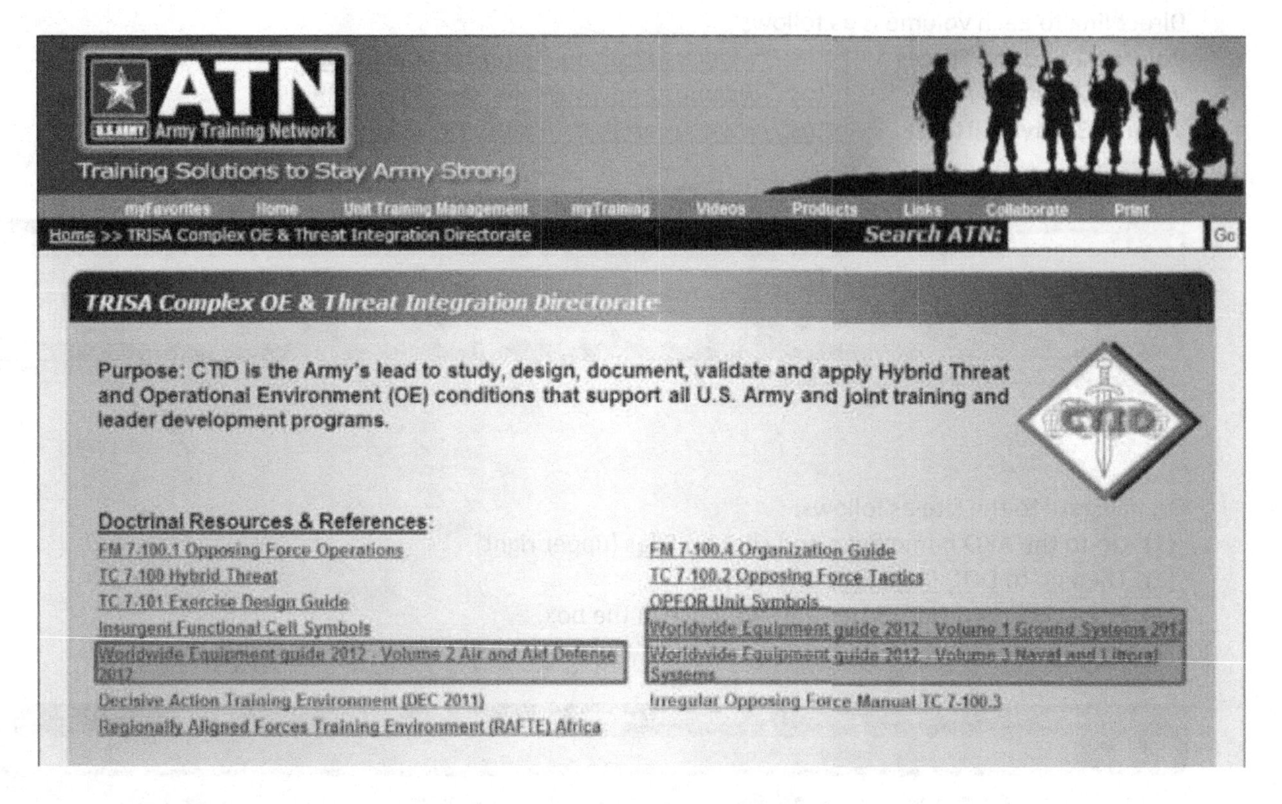

Navigate to the appropriate WEG Volume and download the PDF.

Opposing Force: Worldwide Equipment Guide
Chapters Volume 2

Changes to the 2015 Worldwide Equipment Guide

Many chapters have significant changes. Changes include specific changes in text and data, photos, equipment name changes, as well as added or deleted pages. For clarity, functional classifications of aircraft and some designators and names for specific models have been adjusted.

In these times of reduced economic resources for military force improvements, most forces are focusing more on upgrading existing systems, with reduced numbers of new fielded systems. Thus, many older systems are being upgraded to be more effective against even the most modern forces. Therefore, the number of variants for systems described in the WEG continues to expand. Some system names have been changed to add key upgrade variants which are featured on the data sheets. A red ink edition is available for users who want to know detailed changes to text and data. Major changes can be found on the following pages:

Units of Measure

The following example symbols and abbreviations are used in this guide.

Unit of Measure	Parameter
(°)	degrees (of slope/gradient, elevation, traverse, etc.)
GHz	gigahertz—frequency (GHz = 1 billion hertz)
hp	horsepower (kWx1.341 = hp)
Hz	hertz—unit of frequency
kg	kilogram(s) (2.2 lb.)
kg/cm^2	kg per square centimeter—pressure
km	kilometer(s)
km/h	km per hour
kt	knot—speed. 1 kt = 1 nautical mile (nm) per hr.
kW	kilowatt(s) (1 kW = 1,000 watts)
liters	liters—liquid measurement (1 gal. = 3.785 liters)
m	meter(s)—if over 1 meter use meters; if under use mm
m^3	cubic meter(s)
m^3/hr	cubic meters per hour—earth moving capacity
m/hr	meters per hour—operating speed (earth moving)
MHz	megahertz—frequency (MHz = 1 million hertz)
mach	mach + *(factor)* —aircraft velocity (average 1062 km/h)
mil	milliradian, radial measure (360° = 6400 mils, 6000 Russian)
min	minute(s)
mm	millimeter(s)
m/s	meters per second—velocity
mt	metric ton(s) (mt = 1,000 kg)
nm	nautical mile = 6076 ft (1.152 miles or 1.86 km)
rd/min	rounds per minute—rate of fire
RHAe	rolled homogeneous armor (equivalent)
shp	shaft horsepower—helicopter engines (kWx1.341 = shp)
μm	micron/micrometer—wavelength for lasers, etc.

OPFOR Air and Air Defense Systems – Tier Tables

The OPFOR organization and equipment must support the entire spectrum of Contemporary Operational Environment in U.S. forces training. The COE OPFOR includes "hybrid threats", and represents rational and adaptive adversaries for use in training applications and scenarios. The COE time period reflects current training as well as training extending through the Near Term. This chapter deals with current time frame systems. Lists of equipment on these tables offer convenient baseline examples arranged in capability tiers for use in composing OPFOR equipment arrays for training scenarios. For guidance on systems technology capabilities and trends after 2014, the user might look to Chapter 10, Countermeasures, Upgrades, and Emerging Technology. Those tables offer capabilities tiers for Near and Mid-Term.

OPFOR equipment is broken into four "tiers" in order to portray systems for adversaries with differing levels of force capabilities for use as representative examples of a rational force developer's systems mix. Equipment is listed in convenient tier tables for use as a tool for trainers to reflect different levels of modernity. Each tier provides an equivalent level of capability for systems across different functional areas. The tier tables are also another tool to identify systems in simulations to reflect different levels of modernity. The key to using the tables is to know the tier capability of the initial organizations to be provided. Tier 2 (default OPFOR level) reflects modern competitive systems fielded in significant numbers for the last 10 to 20 years.

Systems reflect specific capability mixes, which require specific systems data for portrayal in U.S. training simulations (live, virtual, and constructive). The OPFOR force contains a mix of systems in each tier and functional area which realistically vary in fielded age and generation. The tiers are less about age of the system than realistically reflecting capabilities to be mirrored in training. Systems and functional areas are not modernized equally and simultaneously. Forces have systems and material varying 10 to 30 years in age in a functional area. Often military forces emphasize upgrades in one functional area while neglecting upgrades in other functional areas. Force designers may also draw systems from higher or lower echelons with different tiers to supplement organizational assets. Our functional area analysts have tempered depiction of new and expensive systems to a fraction of the OPFOR force. The more common modernization approach for higher tier systems is to upgrade existing systems.

Some systems are used in both lower and higher tiers. Older 4x4 tactical utility vehicles which are 30 to 40 years old still offer effective support capability, and may extend across three tiers. Common use of some OPFOR systems also reduces database maintenance requirements.

Tier 1 systems are new or upgraded robust state-of-the-art systems marketed for sale, with at least limited fielding, and with capabilities and vulnerabilities representative of trends to be addressed in training. But a major military force with state-of-the-art technology may still have a mix of systems across different functional areas at Tier 1 and lower tiers in 2016.

Tier 2 reflects modern competitive systems fielded in significant numbers for the last 10 to 20 years, with limitations or vulnerabilities being diminished by available upgrades. Although forces are equipped for operations in all terrains and can fight day and night, their capability in range and speed for several key systems may be somewhat inferior to U.S. capability.

Tier 3 systems date back generally 30 to 40 years. They have limitations in all three subsystems categories: mobility, survivability and lethality. Systems and force integration are inferior. However, guns, missiles, and munitions can still challenge vulnerabilities of U.S. forces. Niche upgrades can provide synergistic and adaptive increases in force effectiveness.

Tier 4 systems reflect 40 to 50 year-old systems, some of which have been upgraded numerous times. These represent Third World or smaller developed countries' forces and irregular forces. Use of effective strategy, adaptive tactics, niche technologies, and terrain limitations can enable a Tier 4 OPFOR to challenge U.S. force effectiveness in achieving its goals. The tier includes militia, guerrillas, special police, and other forces.

Please note: ***No force in the world has all systems at the most modern tier.*** Even the best force in the world has a mix of state-of-the-art (Tier 1) systems, as well as mature (Tier 2), and somewhat dated (Tier 3) legacy systems. Many of the latter systems have been upgraded to some degree, but may exhibit limitations from their original state of technology. Even modern systems recently purchased may be considerably less than state-of-the-art, due to budget constraints and limited user training and maintenance capabilities. Thus, even new systems may not exhibit Tier 1 or Tier 2 capabilities. As later forces field systems with emerging technologies, legacy systems may be employed to be more suitable, may be upgraded, and continue to be competitive. ***Adversaries with lower tier systems can use adaptive technologies and tactics, or obtain niche technology systems to challenge advantages of a modern force.***

A major emphasis in an OPFOR is flexibility in use of forces and in doctrine. This also means OPFOR having flexibility, given rational and justifiable force development methodology, to adapt the systems mix to support doctrine and plans. The tiers provide the baseline list for determining the force mix, based on scenario criteria. The OPFOR compensates for capability limitations by using innovative and adaptive tactics, techniques, and procedures (TTP). Some of these limitations may be caused by the lack of sophisticated equipment or integration capability, or by insufficient numbers. Forces can be tailored in accordance with OPFOR guidance to form tactical groups.

An OPFOR force developer has the option to make selective adjustments such as use of niche technology upgrades such as in tanks, cruise missiles, or rotary-wing aircraft, to offset U.S. advantages (see WEG Chapter 15, Equipment Upgrades). Forces may include systems from outside of the overall force capability level. A Tier 3 force might have a few systems from Tier 1 or 2. The authors will always be ready to assist a developer in selecting niche systems and upgrades for use in OPFOR portrayal. Scenario developers should be able to justify changes and systems selected. With savvy use of TTP and systems, all tiers may offer challenging OPFOR capabilities for training. The Equipment Substitution Matrices can help force designers find weapons to substitute, to reflect those best suited for specific training scenarios.

Mr. John Cantin
DSN: 552-7952 Commercial (913) 684-7952
E-mail address: john.m.cantin.ctr@mail.mil

OPFOR Tier Tables, Airspace and Air Defense Systems

The OPFOR forces and equipment must support the entire spectrum of the Complex Operational Environment in U.S. forces training. The COE includes "hybrid threats", and represents rational and adaptive adversaries for use in training applications and scenarios. The COE time period reflects current training (2013), as well as training extending through the Near Term. This chapter deals with current time frame systems. Lists of equipment on these tables offer convenient baseline examples arranged in capability tiers for use in composing OPFOR equipment arrays for training scenarios.

OPFOR equipment is broken into four "tiers" in order to portray systems for adversaries with differing levels of force capabilities for use as representative examples of a rational force developer's systems mix. Equipment is listed in convenient tier tables for use as a tool for trainers to reflect different levels of modernity. Each tier provides an equivalent level of capability for systems across different functional areas. The tier tables are also another tool to identify systems in simulations to reflect different levels of modernity. The key to using the tables is to know the tier capability of the initial organizations to be provided. Tier 2 (default OPFOR level) reflects modern competitive systems fielded in significant numbers for the last 10 to 20 years.

Systems reflect specific capability mixes, which require specific systems data for portrayal in U.S. training simulations (live, virtual, and constructive). The OPFOR force contains a mix of systems in each tier and functional area which realistically vary in fielded age and generation. The tiers are less about age of the system than realistically reflecting capabilities to be mirrored in training. Systems and functional areas are not modernized equally and simultaneously. Forces have systems and material varying 10 to 30 years in age in a functional area. Often military forces emphasize upgrades in one functional area while neglecting upgrades in other functional areas. Force designers may also draw systems from higher or lower echelons with different tiers to supplement organizational assets. Our functional area analysts have tempered depiction of new and expensive systems to a fraction of the OPFOR force. The more common modernization approach for higher tier systems is to upgrade existing systems.

Some systems are used in both lower and higher tiers. Older 4x4 tactical utility vehicles which are 30 to 40 years old still offer effective support capability, and may extend across three tiers. Common use of some OPFOR systems also reduces database maintenance requirements.

Tier 1 systems are new or upgraded robust state-of-the-art systems marketed for sale, with at least limited fielding, and with capabilities and vulnerabilities representative of trends to be addressed in training. But a major military force with state-of-the-art technology may still have a mix of systems across different functional areas at Tier 1 and lower tiers in 2016.

Tier 2 reflects modern competitive systems fielded in significant numbers for the last 10 to 20 years, with limitations or vulnerabilities being diminished by available upgrades. Although forces are equipped for operations in all terrains and can fight day and night, their capability in range and speed for several key systems may be somewhat inferior to U.S. capability.

Tier 3 systems date back generally 30 to 40 years. They have limitations in all three subsystems categories: mobility, survivability and lethality. Systems and force integration are inferior. However, guns, missiles, and munitions can still challenge vulnerabilities of U.S. forces. Niche upgrades can provide synergistic and adaptive increases in force effectiveness.

Tier 4 systems reflect 40 to 50 year-old systems, some of which have been upgraded numerous times. These represent Third World or smaller developed countries' forces and irregular forces. Use of effective strategy, adaptive tactics, niche technologies, and terrain limitations can enable a Tier 4 OPFOR to challenge U.S. force effectiveness in achieving its goals. The tier includes militia, guerrillas, special police, and other forces.

Please note: ***No force in the world has all systems at the most modern tier.*** Even the best force in the world has a mix of state-of-the-art (Tier 1) systems, as well as mature (Tier 2), and somewhat dated (Tier 3) legacy systems. Many of the latter systems have been upgraded to some degree, but may exhibit limitations from their original state of technology. Even modern systems recently purchased may be considerably less than state-of-the-art, due to budget constraints and limited user training and maintenance capabilities. Thus, even new systems may not exhibit Tier 1 or Tier 2 capabilities. As later forces field systems with emerging technologies, legacy systems may be employed to be more suitable, may be upgraded, and continue to be competitive. ***Adversaries with lower tier systems can use adaptive technologies and tactics, or obtain niche technology systems to challenge advantages of a modern force.***

A major emphasis in COE is flexibility in use of forces and in doctrine. This also means OPFOR having flexibility, given rational and justifiable force development methodology, to adapt the systems mix to support doctrine and plans. The tiers provide the baseline list for determining the force mix, based on scenario criteria. The OPFOR compensates for capability limitations by using innovative and adaptive tactics, techniques, and procedures (TTP). Some of these limitations may be caused by the lack of sophisticated equipment or integration capability, or by insufficient numbers. Forces can be tailored in accordance with OPFOR guidance to form tactical groups.

An OPFOR force developer has the option to make selective adjustments such as use of niche technology upgrades such as in tanks, cruise missiles, or rotary-wing aircraft, to offset U.S. advantages (see WEG Chapter 9, Equipment Upgrades). Forces may include systems from outside of the overall force capability level. A Tier 3 force might have a few systems from Tier 1 or 2. The authors will always be ready to assist a developer in selecting niche systems and upgrades for use in OPFOR portrayal. Scenario developers should be able to justify changes and systems selected. With savvy use of TTP and systems, all tiers may offer challenging OPFOR capabilities for training. The Equipment Substitution Matrices (starting at pg 1-6) can help force designers find weapons to substitute, to reflect those best suited for specific training scenarios.

Mr. John Cantin
DSN: 552-7952 Commercial (913) 684-7952
e-mail address: john.m.cantin.ctr@mail.mil

	Tier 1	Tier 2	Tier 3	Tier 4
Fixed Wing Aircraft				
Fighter/Interceptor	Su-35	Su-27SM	Mirage III, MiG-23M	J-7/FISHBED
High Altitude Interceptor	MiG-31BS	MiG-25PD	MiG-25	--
Ground Attack	Su-39	Su-25TM	Su-25	Su-17
Multi-Role Aircraft	Su-30MKK	Su-30, Mirage 2000, Tornado IDS	Mirage F1, SU-24	MiG-21M
Bomber Aircraft	Tu-22M3/BACKFIRE-C	Tu-22M3/BACKFIRE-C	Tu-95MS6/BEAR-H	Tu-95S/BEAR-A
Command & Control	IL-76/MAINSTAY	IL-76/MAINSTAY	IL-22/COOT-B	IL-22/COOT-B
Heavy Transport	IL-76	IL-76	IL-18	IL-18
Medium Transport	AN-12	AN-12	AN-12	AN-12
Short Haul Transport	AN-26	AN-26	AN-26	AN-26
RW Aircraft				
Attack Helicopter	AH-1W/Supercobra	Mi-35M2	HIND-F	HIND-D
Multi-role Helicopter	Z-9/WZ-9	Battlefield Lynx	Lynx AH.Mk 1	Mi-2/HOPLITE
Light Helicopter	GAZELLE/SA 342M	GAZELLE/SA 342M	BO-105	MD-500M
Medium Helicopter	Mi-17-V7	Mi-171V/Mi-171Sh	Mi-8(Trans/HIP-E Aslt)	Mi-8T/HIP-C
Transport Helicopter	Mi-26	Mi-26	Mi-6	Mi-6
Other Aircraft				
Wide Area Recon Helicopter	Horizon (Cougar heli)	Horizon (Cougar heli)		
NBC Recon Heli	HIND-G1	HIND-G1	HIND-G1	--
Jamming Helicopter	HIP-J/K	HIP-J/K	HIP-J/K	HIP-J/K
Naval Helicopter	Z-9C	Ka-27/HELIX	Ka-27/HELIX	--
Op-Tactical Recon FW	Su-24MR/FENCER-E	Su-24MR/FENCER-E	IL-20M/COOT	--
EW Intel/Jam FM	Su-24MP/FENCER-E	Su-24MP/FENCER-E	IL-20RT and M/COOT	--
Long Range Recon	Tu-22MR/BACKFIRE	Tu-95MR/BEAR-E	Tu-95MR/BEAR-E	IL-20M/COOT
Long Range EW	Tu-22MP/BACKFIRE	Tu-95KM/BEAR-C	Tu-95KM/BEAR-C	--

	Tier 1	Tier 2	Tier 3	Tier 4
Air Defense				
Operational-Strategic Systems				
Long-Range SAM/ABM	Triumf/SA-21, SA-24	SA-20a w/SA-18	SA-5b w/SA-16	SA-5a w/S-60
LR Tracked SAM/ABM	Antey-2500, SA-24	SA-12a/SA-12b	SA-12a/SA-12b	SA-4b w/S-60
LR Wheeled SAM/ABM	Favorit/SA-20b, SA-24	SA-20a w/SA-18	SA-10c w/SA-16	SA-5a w/S-60
Mobile Tracked SAM	Buk-M1-2 (SA-11 FO)	Buk-M1-2(SA-11 FO)	SA-6b w/ZSU-23-4	SA-6a w/ZSU-23-4
Towed Gun/Missile System	Skyguard III/Aspide2000	Skyguard II/Aspide2000	SA-3, S-60 w/radar	SA-3, S-60 w/radar
Tactical Short-Range Systems				
SR Tracked System (Div)	Pantsir S-1-0	SA-15b w/SA-18	SA-6b w/Gepard B2L	SA-6a w/ZSU-23-4
SR Wheeled System (Div)	Crotale-NG w/SA-24	FM-90 w/SA-18	SA-8b w/ZSU-23-4	SA-8a w/ZSU-23-4
SR Gun/Missile System (Bde)	2S6M1	2S6M1	SA-13b w/ZSU-23-4	SA-9 w/ZSU-23-4
Man-portable SAM Launcher	SA-24 (Igla-S)	SA-24 (Igla-S)	SA-16	SA-14, SA-7b
Airborne/Amphibious AA Gun	BTR-ZD Imp (w/-23M1)	BTR-ZD with ZU-23M	BTR-ZD/SA-16	BTR-D/SA-16, ZPU-4
Air Defense/Antitank				
Inf ADAT Vehicle-IFV	BMP-2M Berezhok/SA-24	BMP-2M w/SA-24	AMX-10 w/SA-16	VTT-323 w/SA-14
Inf ADAT Vehicle-APC	BTR-3E1/AT-5B/SA-24	BTR-80A w/SA-24	WZ-551 w/SA-16	BTR-60PB w/SA-14
ADAT Missile/Rocket Lchr	Starstreak II	Starstreak	C-5K	RPG-7V
Air Defense ATGM	9P157-2/AT-15 and AD missile	9P149/Ataka and AD missile	9P149/AT-6	9P148/AT
Anti-Aircraft Guns				
Medium-Heavy Towed Gun	Skyguard III	S-60 with radar/1L15-1	S-60 with radar/1L15-1	KS-19
Medium Towed Gun	Skyguard III	GDF-005 in Skyguard II	GDF-003/Skyguard	Type 65
Light Towed Gun	ZU-23-2M1/SA-24	ZU-23-2M	ZU-23	ZPU-4
Anti-Helicopter Mine	Temp-20	Helkir	MON-200	MON-100

	Tier 1	Tier 2	Tier 3	Tier 4
AD Spt (C2/Recon/EW)				
EW/TA Radar Strategic	Protivnik-GE and 96L6E	64N6E and 96L6E	TALL KING-C	SPOON REST
EW/TA Rdr Anti-stealth	Nebo-SVU	Nebo-SVU	Nebo-SV	BOX SPRING
EW/TA Radar Op/Tac	Kasta-2E2/Giraffe-AMB	Kasta-2E2/Giraffe AMB	Giraffe 50	LONG TRACK
Radar/C2 for SHORAD	Sborka PPRU-M1	Sborka-M1/ PPRU-M1	PPRU-1 (DOG EAR)	PU-12
ELINT System	Orion/85V6E	Orion/85V6E	Tamara	Romona
Unmanned Aerial Vehicles				
High Altitude Long Range	Hermes 900	Hermes 900	Tu-143	Tu-141
Med Altitude Long Range	ASN-207	ASN-207 / Hermes 450	--	--
Tactical	Skylark III/Mohadjer 4B	Skylark II/Mohadjer 4	Shmel-I	FOX AT2
Vertical Take Off/ Landing	Camcopter S-100	Camcopter S-100	--	--

Vehicle/Man-Portable	Spylite	Spylite/Skylite-B	Skylite-A	--
Man-Portable	Skylark-IV	Skylark	--	--
Hand-Launch	Zala 421-12	Zala 421-08/421-21/Hexarotor VTOL		--
Artillery Launch	R-90 rocket	R-90 rocket		
Attack UAVs/UCAVs	Hermes 450	Hermes 450	Mirach-150	--
Theater Missiles				
Medium Range (MRBM)	Shahab-3B	Shahab-3A	Nodong-1	SS-1C/SCUD-B
Short-Range (SRBM)	SS-26 Iskander-M	SS-26 Iskander-E	M-9	SS-1C/SCUD-B
SRBM/Hvy Rkt < 300 km	Lynx w/EXTRA missile	Tochka-U/SS-21 Mod 3	M-7/CSS-8	FROG-7
Cruise Missile	Delilah ground, air, sea	Harpy programmed/piloted	Mirach-150 programmed	--
Anti-ship CM	BrahMos ground, air, sea	Harpy programmed/radar	Exocet	Styx
Anti-radiation	Harpy programmed/ARM	Harpy programmed/ARM	--	--

SYSTEMS SUBSTITUTION MATRIX VOLUME 1

This table provides a list of Vol 1 systems for users to substitute other systems versus OPFOR systems listed in guidance documents. Systems in italics are Tier 2 baseline systems used in the OPFOR Organization Guide. Systems are listed by type in tier order, and can substitute to fit a scenario. Some systems span between the tiers (e.g., 3-4). Also, systems can be used at more than one tier (e.g., 3-4).

	Tier
1. ROTARY-WING AIRCRAFT	
Light Helicopters	
SA-342M Gazelle	1-2
BO-105	3
MD-500MD/Defender	4
Attack Helicopters	
Ka-50/HOKUM and Ka-52	1
Mi-28/HAVOC	1
AH-1W/Supercobra	1
Mi-35M2	2
AH-1F/Cobra	2
Mi-24/HIND D/F	3-4
Medium Multi-role Helicopters	
Z-9/Haitun and WZ-9 Gunship	1
Battlefield Lynx	2-3
Mi-2/HOPLIGHT	4
Utility Helicopters	
Mi-17/Mi-171V	1-2
Mi-8/HIP-C	3-4
AS-532/Cougar	2-4
Ka-27/HELIX	3
Transport Helicopters	
Mi-26/HALO	1-2
Mi-6/HOOK	3-4
Reconnaissance Helicopters	
Horizon	1-2
2. FIXED-WING AIRCRAFT	
Fighter/Interceptor Aircraft	
MiG-31/FOXHOUND	1
Su-27/FLANKER-B and FLANKER-C	2
MiG-25/FOXBAT-B	2

	Tier
F-5/Freedom Fighter (Tiger)	3
Mirage III/5/50	3
J-7/FISHBED	3-4
J-8/FINBACK	3-4
J-6/F-6	4
Jaguar	4
J-6 (Jian-6)/F-6	4
Ground Attack Aircraft	
Su-39/FROGFOOT	1
Su-25TM/FROGFOOT	2
L-39/Albatros	4
Su-17/FITTER	4
Multi-role Aircraft	
EF-2000/Eurofighter	1
JAS39/Gripen	1
Rafale	1
Su-30M and Su-30MKK	1
Su-35/Su-27BM	1
MiG-29/FULCRUM	1-2
Mirage 2000	2
Tornado IDS	2
AJ37/Viggen	2-3
KFIR (Lion Cub)	2-3
F-4/Phantom	3
MiG-23/MiG-27 FLOGGER	3
Mirage F1	3
Su-24/FENCER	3
Q-5/FANTAN	3-4
MiG-21/FISHBED	4
Transport Aircraft	
An-12/CUB	1-2
An-2/COLT	3-4
An-26/CURL	1-4
IL-76/CANDID	1-2
IL-18/COOT	3-4
Bomber Aircraft	
H-5/Hongzhaji-5	4
H-6/Hongzhaji-6	4
Tu-22M3/BACKFIRE-C	1-2
Tu-95MS-6 and Tu-95S/BEAR	3-4

	Tier
Command and Control Aircraft	
A-50E/MAINSTAY	1-2

3. UNMANNED AERIAL VEHICLES

Skylark IV	1
Skylite-B	1
ASN-207	1
Vulture	1
Hermes 450S	1
Hermes 900	1
Skylark II	1-2
Zala 421-08	1-2
Zala 421-12	1-2
Camcopter S-100	1-2
Skylark	2
Skylite-A	2
Hermes 450	2
Pustelga	3
AT1	3
AT2 (200)	3
ASN-105 (D-4)	3
Shmel-1 and Pchela-1K	3

4. AVIATION COUNTERMEASURES, UPGRADES, EMERGING TECHNOLOGY

No Substitution Platforms

5. UNCONVENTIONAL AND SPF ARIAL SYSTEMS

No Substitution Platforms

6. THEATER MISSILES

Ballistic Missiles

Iskander-E, -M/SS-26	1-2
Shahab-3B	1
SS-21 Mod-3/Tochka-U	1-2
Shahab-3A	2
SCUD-B Mod 2/SS-1c Mod	2
M-11/DF-11/CSS-7	3
M-9/DF-15/CSS-6	3
M-7/CSS-8/B610	3
DF-3/CSS-2	3
SCUD-C/SS-1d	3-4
Nodong-1	3-4
SCUD-B/SS-1c	4

Cruise Missiles

BrahMos Supersonic Cruise Missile	1
Lynx Rocket/Missile System and Delilah	1
Iskander-E/-M/-K Cruise Missile Systems	1

	Tier
Nimrod 3 Long-Range ATGM/Atk UAV	1
Harpy/CUTLASS ARM/Attack UAV	1-2
Nimrod Long-Range ATGM/Atk UAV	2
Mirach-150 Attack UAV/Cruise Missile	3

7. AIR DEFENSE

Air Defense Command Vehicles and Radars

Giraffe AMB Radar/Cmd Veh	1
Sborka AD ACV (w/DOG EAR Radar)	1-2
Giraffe 50AT Radar/Cmd Veh	2
PPRU-1/PU-12M ACV (DOG EAR Radar)	3-4
Long Track Mobile AD Radar Vehicle	3-4

Electronic Warfare Systems for Air Defense

Orion ELINT System	1-2
Avtobaza Ground ELINT System	1-2

Anti-helicopter Mines

Helkir	2

Towed Antiaircraft (AA) Guns

GDF-005 Retrofit (Skyguard) III 35-mm	1
ZU-23M1 23-mm	1
ZU-23M 23-mm	2
GDF-005 35-mm (Skyguard Mk2)	2
GDF-003 35-mm (Skyguard)	3
KS-19M2 100-mm	3-4
S-60 57-mm	3-4
Type 65 37-mm	3-4
ZU-23 23-mm	3-4
M1939 37-mm	4
ZPU-4 14.5-mm Heavy Machinegun	4

Self-Propelled AA Gun System

BTR-ZD Imp 23-mm SP AA Gun	3
Gepard 35-mm SP AA Gun	3
BTR-ZD 23-mm SP AA Gun	3-4
ZSU-23-4 23-mm SP AA Gun	4
ZSU-57-2 57-mm SP AA Gun	4

Manportable Surface-to-Air Missiles (SAMs)

SA-18S/Igla-Super	1-2
Starstreak II High Velocity Missile	1
Starstreak High Velocity Missile	1-2
Stinger	1-2
Albi MANPADS Launcher Vehicle/Mistral	1-2
SA-18/GROUSE and SA-24/Igla-S	2

	Tier
SA-16/GIMLET	3
SA-7b/GRAIL	4
SA-14/GREMLIN	4
SHORAD SAM Systems	
Pantsir-S1-0/SA-22E Gun/Missile System	1
Crotale-New Generation	1
Tor-M2E (SA-15b Imp)	1
2S6M1 30-mm Gun/Missile System	1-2
FM-90 (Crotale Imp)	1-2
SA-15b/GAUNTLET	2
Crotale 5000	2
SA-8P/Osa-AKM-P1	2-3
SA-8b/GECKO Mod 1	3
SA-13b/GOPHER	3
SA-9/GASKIN	4
Medium Range Air Defense (MRAD)	
Aspide 2000 with Skyguard III	1
SA-11 FO/Buk-M1-2	1 -2
Aspide 2000 (Skyguard Mk 2)	1 -2
SA-11/Buk-M1/GADFLY	2-3
Pechora-M (SA-3 Imp)	2-3
Pechora-2M (SA-3 Imp Mobile)	2-3
SA-2/GUIDELINE	3-4
SA-3/GOA	3-4
SA-4b/GANEF Mod 1	3-4
SA-6b/GAINFUL Mod 1	3-4
SA-6a/GAINFUL 4	
Long Range Air Defense (LRAD)	
Antey 2500/SA-23, S-300V4	1
SA-21b/S-400/Triumf	1
SA-20b/Favorit/S-300PMU2	1
SA-20a/GARGOYLE/S-300PMU1	2
SA-12a/GLADIATOR, SA-12b/GIANT	2-3
SA-10C/GRUMBLE/S-300PMU	3
SA-10b/GRUMBLE/S-300PM	3-4
SA-5b/GAMMON	3
SA-5a/GAMMON	4

Worldwide Equipment Guide

Chapter 1: Rotary Wing Aircraft

Distribution Statement: Approved for public release; distribution is unlimited.

Chapter 2: Rotary Wing Aircraft

This chapter provides the basic characteristics of selected rotary-wing aircraft readily available to the OPFOR. The sampling of systems was selected because of wide proliferation across numerous countries or because of extensive use in training scenarios. ***Rotary-wing aircraft*** covers systems classified as light, attack, multirole, transport, and reconnaissance aircraft. Rotary wing aircraft can be used for a variety of roles, including attack, transport, direct air support, escort, target designation, security, reconnaissance, ambulance, anti-submarine warfare (ASW), IW, airborne C2, search and rescue (SAR), and anti-ship.

Because of the increasingly large numbers of variants of each aircraft, only the most common variants produced in significant numbers were addressed. If older versions of helicopters have been upgraded in significant quantities to the standards of newer variants, older versions may not be addressed. Helicopters can be categorized into capability tiers. Upgrades may designate different configurations of the same aircraft in different tiers. Technology priorities include multirole capability, more lethal weapons with longer range, ability to operate in all terrains, survivability/countermeasures, and sensors for day/night all-weather capability.

Helicopters can be configured for various combat missions (attack, direct air support, escort, target designation, etc.). The best armed combat helicopters are ***attack helicopters***, which may be used for all combat missions (including attack, direct air support, escort, anti-ship, etc), and some non-combat missions (transport, reconnaissance, SAR, etc). ***Helicopter gunships*** (combat configurations of multirole helicopters) can be used for all combat and non-combat missions, but are less suitable for attack missions against well-defended targets. Some of these missions can be executed by armed multirole helicopters.

The weapon systems inherent to the airframe are listed under Armament. They use various weapon mounts, including fuselage or turret nose gun, external mounted pylons (or hardpoints), and cabin weapons, including door guns. Pylons can mount single munitions, launchers or pods, sensor pods, or fuel tanks.

Munitions available to each aircraft are noted, but not all may be employed at the same time. Munition selection is based on mission and flight capability priorities. Munitions include bombs, missiles (ATGMs, air-to-surface missiles/ASMs, air-to-air missiles/AAMs), or rockets (single or in pods), mine pods, and automatic grenade launchers. For helicopter missions, other weapons and more ammunition can be carried in the passenger compartment. The most probable weapon loading options are also given, but assigned mission dictates actual weapon configuration.

Tables on aircraft weapons and aircraft-delivered munitions (ADMs) are at pages 2-22 to 2-26.

Questions and comments on data listed in this chapter should be addressed to:

Mr. John Cantin
DSN: 552-7952, Commercial (913) 684-7952
E-mail address: john.m.cantin.ctr@mail.mil

EUROPEAN LIGHT HELICOPTER BO-105

BO-105AT1 National War College Photo

SYSTEM	SPECIFICATIONS	SYSTEM	SPECIFICATIONS
Alternative Designations:	INA	Speed (km/h) Max (level):	242
Date of Introduction:	1972	Cruise:	205
Proliferation:	At least 40 countries	Ceiling: Service:	3050
Description:	Variants in "( )"	Hover (out of ground effect):	457
Crew:	1 or 2 (pilots)	Hover (in ground effect):	1,525
Transports	3 troops or 2 litters, or cargo.	Vertical Climb Rate (m/s):	7.5
Main rotor:	4	Speed (km/h) Maxi (level):	242
Tail rotor:	2	Cruise:	205
Engines	2x 420-shp Allison 250-C20B turboshaft	Ceiling: Service:	3050
Weight (kg) Maximum Gross:	2,500	Hover (out of ground effect):	457
Normal Takeoff:	2,000	Hover (in ground effect):	1,525
Empty:	1,301, 1,913 (PAH1)	Vertical Climb Rate (m/s):	7.5
Speed (km/h) Max (level):	242	Fuel (Liters) Internal:	570
Internal Aux Tank:	200 ea. (max 2x)	Cargo Compartment Dimensions (m): Floor Length:	1.9
Range(km): Normal Load:	555	Width: 1.4	1.4
With Aux Fuel:	961	Height: 1.3	1.3
Dimensions (m): Length (rotors turning):	11.9	Standard Payload (kg): Internal load: 690	690
Length (fuselage):	8.8	External on sling only: 1,200	1,200
Width: (m)			
Height:	3.0		
AMMUNITION	**SPECIFICATIONS**	**AMMUNITION**	**SPECIFICATIONS**
7.62-mm or 12.7-mm MG pods	2	HOT ATGM	6
2.75-in rocket pods (7 or 12)	2	AS-12 ASM pods (2 ea pod)	4
8-mm SNEB rocket pods (12)	2	Stinger AAM pod (4 ea pod)	4
50-mm SNIA rocket pods (28)	2	BO-105P/PAH1	6x HOT AT missiles, or rocket pods

TOW ATGM (4 ea pod)	8	AT Guided Missiles: HOT 3 Missile Weight (kg):	32 (in tube)
Rate of fire (missiles/min):	3-4, depending	Rate of fire (missiles/min):	3-4, depending
VARTIANTS	SPECIFICATIONS	VARIANTS	SPECIFICATIONS
BO-105CB:	The standard civilian production variant.	BO-105LS:	Upgraded to 2x 550-shp Allison 250-C28 turboshaft engines for extended capabilities in high altitudes and temperatures. Produced only in Canada.
BO-105CBS:	VIP version with a slightly longer fuselage to accommodate 6 passengers, some used in a SAR role.	BO-105M VBH:	Standard reconnaissance (light observation) version. Others are built in Chile, the Philippines, Indonesia (NBO-105), and Spain.
BO-105P:	German military variant	BO-105/ATH:	Spanish CASA assembled variant rigidly mounts 1x Rh 202 20-mm cannon under the fuselage.
BO-105 PAH-1:	Standard antitank version	BO-105M VBH:	Standard reconnaissance (light observation) version. Others are built in Chile, the Philippines, Indonesia (NBO-105), and Spain.
BO-105AT1:	Variant with 6 x HOT ATGMs		

NOTES

EXTERNAL STORES ARE MOUNTED ON WEAPONS "OUTRIGGERS" OR RACKS ON EACH SIDE OF THE FUSELAGE. EACH RACK HAS ONE HARDPOINT. THIS HELICOPTER IS PRODUCED BY THE EUROCOPTER COMPANY. IT WAS FORMED AS A JOINT VENTURE BETWEEN AEROSPATIALE OF FRANCE, AND DAIMLER-BENZ AEROSPACE OF GERMANY. OTHER MISSIONS INCLUDE: DIRECT AIR SUPPORT, ANTITANK, RECONNAISSANCE, SEARCH AND RESCUE, AND TRANSPORT. CLAMSHELL DOORS AT REAR OF CABIN AREA OPEN TO ACCESS CARGO AREA. CARGO FLOOR HAS TIE-DOWN RINGS THROUGHOUT.

UNITED STATES LIGHT HELICOPTER MD-500MD/DEFENDER

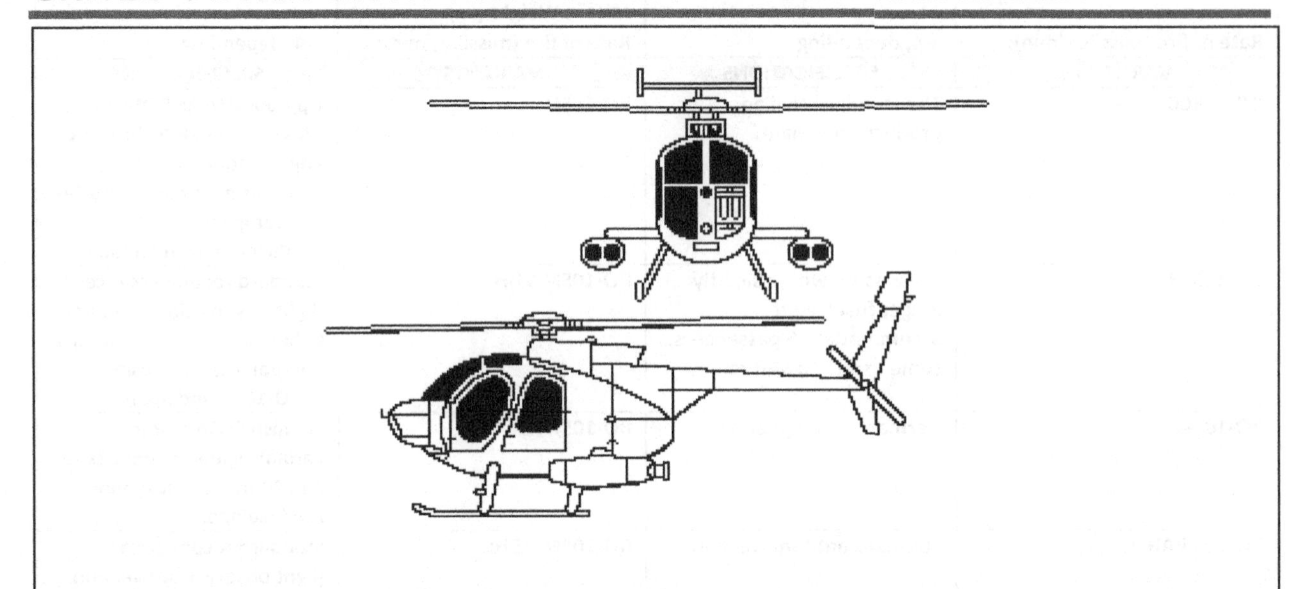

SYSTEM	SPECIFICATIONS	SYSTEM	SPECIFICATIONS
Alternative Designations:	Hughes model 369, Cayuse, Loach	Ceiling: Service:	4,635 (500), 4,875 (530)
Date of Introduction:	1977 (MD-500 MD)	Hover (out of ground effect):	1,830 (500), 3,660 (530)
Proliferation:	At least 22 countries	Hover (in ground effect):	2,590 (500), 4,360 (530)
Crew:	1 or 2 (pilots)	Vertical Climb Rate (m/s):	8.4 (500), 10.5 (530)
Transports	2 or 3 troops/cargo, or 6 on external platforms in lieu of weapons.	Fuel (Liters) Internal:	Internal: 240
Main rotor:	4 or 5 (see VARIANTS)	Internal Aux Tank:	80
Tail rotor:	2 or 4 (see VARIANTS)	Range(km): Normal Load:	485 (500), 430 (530)
Engines	2x 420-shp Allison 250-C20B turboshaft	With Aux Fuel:	961
Weight (kg) Maximum Gross:	1,361 (500), 1,610 (530)	Dimensions (m): Length (rotors turning):	9.4 (500), 9.8 (530)
Normal Takeoff:	1,090	Length (fuselage):	7.6 (500), 7.3 (530)
Empty:	896	Width: (m)	1.9
Speed (km/h) Maximum (level):	241 (500), 282 (530)	Height:	2.6 (500), 3.4 (530 over mast sight)
Cruise:	221 (500), 250 (530)		

AMMUNITION	SPECIFICATIONS	AMMUNITION	SPECIFICATIONS
BO-105P/PAH1	6x HOT AT missiles, or rocket pods	Rate of fire (missiles/min):	3-4, depending
AT Guided Missiles: HOT 3 Missile Weight (kg):	32 (in tube)	Other Missile Types:	HOT 2 multi-purpose (HEAT and Frag warheads)
Armor Penetration (mm):	1250	AVIONICS/SENSOR/OPTICS	The BO-105P has a roof-mounted direct-view, daylight-only sight to allow

			firing of HOT ATGMs. Options exist to fit a thermal imaging system for night operations and a laser designator. Night/Weather Capabilities: Available avionics include weather radar, Doppler and GPS navigation, and an auto-pilot. It is capable of operation in day, night, and with instruments under adverse meteorological conditions.
VARTIANTS	**SPECIFICATIONS**	**VARIANTS**	**SPECIFICATIONS**
BO-105CB:	The standard civilian production variant.	BO-105 PAH-1:	Standard antitank version
BO-105CBS:	VIP version with a slightly longer fuselage to accommodate 6 passengers, some used in a SAR role.	BO-105AT1:	Variant with 6 x HOT ATGMs
BO-105P:	German military variant	BO-105LS:	Upgraded to 2x 550-shp Allison 250-C28 turboshaft engines for extended capabilities in high altitudes and temperatures. Produced only in Canada.

NOTES

SURVIVABILITY/COUNTERMEASURES:
SOME MODELS HAVE RADAR WARNING RECEIVERS. CHAFF AND FLARE SYSTEMS AVAILABLE. INFRARED SIGNATURE SUPPRESSORS CAN BE MOUNTED ON ENGINE EXHAUSTS.

ARMAMENT
MD-500MD/SCOUT DEFENDER: VERSION WITH MOST PROBABLE ARMAMENT: FITTED WITH GUNS, ROCKETS, GRENADE LAUNCHERS, OR COMBINATION ON 2 X FUSELAGE HARDPOINTS. FOR GENERAL USE RECOMMEND 12.7-MM MG AND A TWIN TOW ATGM POD.

MD-500MD/TOW DEFENDER: TWIN TOW MISSILE PODS ON 2X HARDPOINTS; MOUNTS MISSILE SIGHT IN LOWER-LEFT FRONT WINDSHIELD.

ANTITANK GUIDED MISSILES
NAME: TOW 2
ALTERNATIVE DESIGNATIONS: BGM-71D
MISSILE WEIGHT (KG): 28.1 (IN TUBE)
WARHEAD TYPE: TANDEM SHAPED CHARGE
ARMOR PENETRATION (MM): 900 EST
MAXIMUM RANGE (M): 3,750
RATE OF FIRE (/MIN): 3-4, BASED ON RANGE
PROBABILITY OF HIT (%): 90

OTHER MISSILE TYPES: TOW, ITOW, TOW 2A

AVIONICS/SENSOR/OPTICS

TIME OF FLIGHT TO MAX RANGE (SEC): 21AVIONICS/SENSOR/OPTICS
THE MD-500MD ALLOWS FOR MOUNTING A STABILIZED DIRECT-VIEW OPTICAL SIGHT IN THE WINDSHIELD. OPTIONS EXIST TO FIT A MAST-MOUNTED, MULTIPLE FIELD OF VIEW OPTICAL SIGHT, TARGET TRACKER, LASER RANGE FINDER, THERMAL IMAGER, 16X FLIR FOR NIGHT NAVIGATION AND TARGETING, AND AUTOPILOT.
NIGHT/WEATHER CAPABILITIES: OPTIONAL AVIONICS INCLUDE GPS, ILS AND FULL INSTRUMENT WEATHER CONDITIONS PACKAGES.
THE MORE ADVANCED VARIANTS ARE FULLY CAPABLE OF PERFORMING MISSIONS UNDER ANY CONDITIONS.

FRENCH LIGHT HELICOPTER SA-341/342 GAZELLE

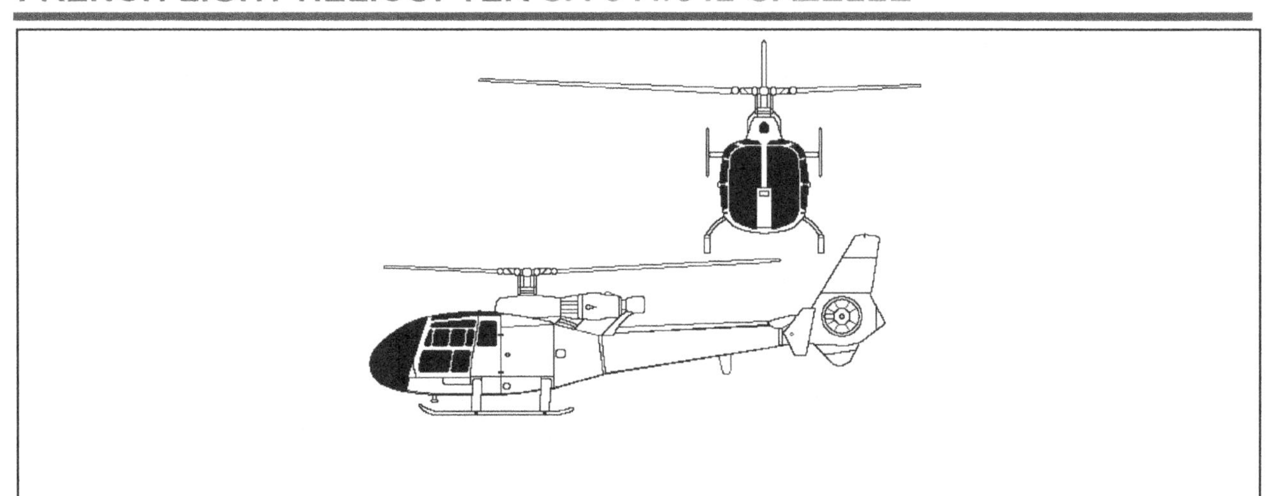

SYSTEM	SPECIFICATIONS	SYSTEM	SPECIFICATIONS
Alternative Designations:	Variants	Ceiling (m):	
Date of Introduction:	1961 SA-341, 1973 SA-342	Service:	Service: 4,100 (SA 341), 5,000 (SA 342)
Proliferation:	At least 23 countries	Hover (out of ground effect):	2,000 (SA 341)
Crew:	1 or 2 (pilots)		2,370 (SA 342)
Transports	3 troops or 1 litter, or cargo.	Hover (in ground effect):	2,850 (SA 341)
Blades - Main rotor:	3		3,040 (SA 342)
Tail rotor:	13 (fenestron in tail)	Vertical Climb Rate (m/s):	12.2
Engines:	1x 590-shp Turbomeca		
	Astazou IIIB turboshaft	Internal:	445
Weight (kg):		Internal Aux Tank:	90
Maximum Gross:	1,800 (SA 341), 1,900 (SA 342K), 2,000 (SA 342L/M)	Additional Internal Aux Tank:	200
Normal Takeoff:	1,800	Length (fuselage):	9.5
Empty:	998	Width:	2.0
Speed (km/h): Max (level):	310	Height:	3.1
Cruise:	270	Main Rotor Diameter:	10.5
Tail Rotor Diameter:	0.7	Height:	1.2
Cargo Compartment Dimensions (m):		Standard Payload (kg):	
Floor Length:	2.2	Internal load:	750
Width:	1.3	External on sling only:	700
AMMUNITION	**SPECIFICATIONS**	**AMMUNITION**	**SPECIFICATIONS**
SA 341H:	Can carry 4x AT-3 ATGMs, and 2x SA-7, or 128-mm or 57-mm rockets, and 7.62-mm machinegun in cabin.	SA 342M:	Armed version with 4 x HOT
SA 342L:	Export light attack variant with either rocket pods or Machine guns.	ATGMs	2x Mistral AAM, 7.62-mm MG.
SA 342K:	Armed antitank version with 4-6x HOT ATGMs and 7.62-mm MG.	HOT 3	Missile Weight (kg): 32 (in tube)

Missile Weight (kg):	32 (in tube)		Warhead: Tandem shaped Charge
7.62-mm Mini-TAT MG or 20-mm GIAT M.621 cannon or 2x 7.62-mm AA-52 FN MG pods	100 1,000		Armor Penetration (mm CE): 1250 Maximum Range (m): 75/4,000
2.75-in rocket pods (7 ea.)	2		Rate of fire (missiles/min): 3-4, depending on range
68-mm SNEB rocket pods (12 ea.)	2		
57-mm rocket pods (18 ea.)	2		
AT-3 SAGGER ATGM	4		
AS-12 ASM	4 or 2		
SA-7 GRAIL AAM	2		
MISTRAL AAM	2		
VARIANTS	**SPECIFICATIONS**	**VARIANTS**	**SPECIFICATIONS**
SA 341 GAZELLE:	SA 341 Gazelle: Developed by Aerospatiale in France. Others were built in the UK by Westland, and in Yugoslavia.	SA 341B/C/D/E:	Production versions for British military. Used in communications and training and roles.
SA 341F: SA 341F (CONT'D)	Production version for French Army. A GIAT M.621 20-mm cannon is installed on right side of some aircraft. Rate of fire is either 300 or 740 rpm. Upgraded engine to Astazou IIIC.	SA 341H:	Export variant.
SA 342K:	Armed SA 341F with Upgraded 870-shp Astazou XIVH engine, mostly exported to the Middle East	SA 342L:	Export light attack variant with Astazou XIVM engine.
SA 342M:	Improved ground attack variant for French Army, with 4-6 HOT ATGMs, possibly fitted with Mistral air- to-air missiles. Similar to SA 342L, but with improved instrument panel, engine exhaust baffles to reduce IR signature, navigational systems, Doppler radar, and other night flying equipment. Fitted with Viviane FCS with thermal sight for night attack. *This the OPFOR Tier 1 baseline light helicopter.*		

NOTES

MISSIONS INCLUDE: DIRECT AIR SUPPORT, ANTI-HELICOPTER, RECONNAISSANCE, ESCORT, SECURITY, TRANSPORT, AND TRAINING. EXTERNAL STORES ARE MOUNTED ON WEAPONS "OUTRIGGERS" OR RACKS ON EACH SIDE OF THE FUSELAGE. EACH RACK HAS ONE HARDPOINT. THE BENCH SEAT IN THE CABIN AREA CAN BE FOLDED DOWN TO LEAVE A COMPLETELY OPEN CARGO AREA. CARGO FLOOR HAS TIE DOWN RINGS THROUGHOUT.

UNITED STATES ATTACK HELICOPTER AH-1F/COBRA

SYSTEM	SPECIFICATIONS	SYSTEM	SPECIFICATIONS
Alternative Designations:	Bell 209	Ceiling (m):	
Date of Introduction:	By 1986	Service:	5,703
Proliferation:	At least 3 countries	Hover (out of ground effect):	915
Crew:	2 (pilots in tandem seats)		Hover (in ground effect): 4,270
Transports	N/A	Hover (in ground effect):	4,270
Blades – Main rotor:	2	Vertical Climb Rate (m/s):	4.0
Tail rotor:	2	Fuel (liters):	
Engines:	2 x 1,775-shp GE	Internal:	1,1,50
	T-700-GE-401 turboshaft	Range:	590 Normal LoadAux Fuel
Weight (kg):		Length (rotors turning):	17.7
Maximum Gross:	Maximum Gross: 6,700	Length (fuselage):	14.7
Normal Takeoff:	Normal Takeoff: 6,700	Width (including wing):	3.3
Empty:	Empty: 4,670	Height:	4.2
Speed (km/h): Max (level):	350	Main Rotor Diameter:	14.7
Cruise:	270	Tail Rotor Diameter:	3.0
Max “G” Force:	+2.5 to -0.5 g	Standard Payload (kg):	1,740
AMMUNITION	SPECIFICATIONS	AMMUNITION	SPECIFICATIONS
ARMAMENT	M197, 3x barrel 20-mm Gatling gun in chin turret. On 4 under wing hard points, it can mount 8 x TOW or Hellfire ATGMs (or four each), and 2 x 2.75-in FFAR rocket pods. AIM-9L/ Side winder provides air-to-air capability. Not all may be used at one time. Mission dictates weapon configuration.	MOST PROBABLE ARMAMENT: AH-1W:	A representative mix when targeting armor formations is eight Hellfire missiles, two2.75-in rocket pods and 750x 20-mm rounds. Gun is centered before firing under wing stores.
20-mm 3x barrel Gatling gun, M197:		Antitank Guided Missiles:	TOW 2
Range:	(practical) 1,500 m	Warhead Type:	Tandem Shaped Charge

Elevation:	21° up to 50° down	Armor Penetration (mm CE):	900+ estimated
Traverse:	220°	Maximum Range (m):	3,750
Ammo Type:	AP, HE	Rate of fire (missiles/min):	3-4 based on range
Rate of Fire:	Burst 16±4, continuous 730±50		HELLFIRE II
		Warhead Type:	Tandem Shaped Charge
		Armor Penetration (mm CE):	1,000+
		Maximum Range (m):	8,000+
		Rate of fire (missiles/min):	2-3
VARIANTS	**SPECIFICATIONS**	**VARIANTS**	**SPECIFICATIONS**
AH-1J:	Initial USMC twin engine AH-1 variant fielded in the early 1970s.	AH-1RO (Romania):	Construction of a variant, possibly called "Dracula", may occur in the near future.
AH-1T:	AH-1 variant with upgraded engines and powertrain for improved performance. This minimally expanded rotor system and overall dimensions of the AH-1J. Most older AH-1J Seacobra and AH-1Ts are still in operation, having been upgraded to the AH-1W standard.	AH-1Z/AH-1(4B)W:	Four-bladed variant called the "King Cobra" or "Viper", with better flight performance. It contains an integrated digital tandem cockpit and digital map display. Improved FCS includes helmet-mount sight system.

NOTES

AVIONICS/SENSOR/OPTICS: THE MISSILE TARGETING SYSTEM USES A TELESCOPIC SIGHT UNIT (TRAVERSE 110º, ELEVATION –60º/+30º) WITH TWO MAGNIFICATIONS/FIELDS OF VIEW, A LASER AUGMENTED TRACKING CAPABILITY, TV, AND VIDEO. ADDITIONAL MISSIONS INCLUDE: DIRECT AIR SUPPORT, ESCORT, TARGET DESIGNATION, SECURITY, RECONNAISSANCE, AIR TO AIR COMBAT, AND ANTI-SHIP. THIS AIRCRAFT COSTS APPROXIMATELY $10.7 MILLION, INEXPENSIVE COMPARED TO OTHER MODERN ATTACK HELICOPTERS; BUT ITS PERFORMANCE IS SIMILAR. THUS MANY NATIONS CONSIDER THIS AIRCRAFT AS A GOOD CANDIDATE FOR FIELDING IN ATTACK HELICOPTER SQUADRONS. <u>THIS IS THE OPFOR TIER 1 REPRESENTATIVE HELICOPTER SYSTEM.</u>

UNITED STATES ATTACK HELICOPTER AH-1W/SUPERCOBRA

SYSTEM	SPECIFICATIONS	SYSTEM	SPECIFICATIONS
Alternative Designations:	Bell 209	Ceiling (m):	
Date of Introduction:	By 1986	Service:	5,703
Proliferation:	At least 3 countries	Hover (out of ground effect):	915
Crew:	2 (pilots in tandem seats)		Hover (in ground effect): 4,270
Transports	N/A	Hover (in ground effect):	4,270
Blades – Main rotor:	2	Vertical Climb Rate (m/s):	4.0
Tail rotor:	2	Fuel (liters):	
Engines:	2 x 1,775-shp GE	Internal:	1,1,50
	T-700-GE-401 turboshaft	Range:	590 Normal LoadAux Fuel
Weight (kg):		Length (rotors turning):	17.7
Maximum Gross:	Maximum Gross: 6,700	Length (fuselage):	14.7
Normal Takeoff:	Normal Takeoff: 6,700	Width (including wing):	3.3
Empty:	Empty: 4,670	Height:	4.2
Speed (km/h): Max (level):	350	Main Rotor Diameter:	14.7
Cruise:	270	Tail Rotor Diameter:	3.0
Max "G" Force:	+2.5 to -0.5 g	Standard Payload (kg):	1,740
AMMUNITION	**SPECIFICATIONS**	**AMMUNITION**	**SPECIFICATIONS**
ARMAMENT	M197, 3x barrel 20-mm Gatling gun in chin turret. On 4 under wing hard points, it can mount 8 x TOW or Hellfire ATGMs (or four each), and 2 x 2.75-in FFAR rocket pods. AIM-9L/Sidewinder provides air-to-air capability. Not all may be used at one time. Mission dictates weapon configuration.	MOST PROBABLE ARMAMENT: AH-1W:	A representative mix when targeting armor formations is eight Hellfire missiles, two2.75-in rocket pods and 750x 20-mm rounds. Gun is centered before firing under wing stores.
20-mm 3x barrel Gatling gun, M197:		Antitank Guided Missiles:	TOW 2
Range:	(practical) 1,500 m	Warhead Type:	Tandem Shaped Charge
Elevation:	21° up to 50° down	Armor Penetration (mm CE):	900+ estimated
Traverse:	220°	Maximum Range (m):	3,750
Ammo Type:	AP, HE	Rate of fire (missiles/min):	3-4 based on range

Rate of Fire:	Burst 16±4, continuous 730±50		HELLFIRE II
		Warhead Type:	Tandem Shaped Charge
		Armor Penetration (mm CE):	1,000+
		Maximum Range (m):	8,000+
		Rate of fire (missiles/min):	2-3
VARIANTS	**SPECIFICATIONS**	**VARIANTS**	**SPECIFICATIONS**
AH-1J:	Initial USMC twin engine AH-1 variant fielded in the early 1970s.	AH-1RO (Romania):	Construction of a variant, possibly called "Dracula", may occur in the near future.
AH-1T:	AH-1 variant with upgraded engines and powertrain for improved performance. This minimally expanded rotor system and overall dimensions of the AH-1J. Most older AH-1J Seacobra and AH-1Ts are still in operation, having been upgraded to the AH-1W standard.	AH-1Z/AH-1(4B)W:	Four-bladed variant called the "King Cobra" or "Viper", with better flight performance. It contains an integrated digital tandem cockpit and digital map display. Improved FCS includes helmet-mount sight system.

NOTES

AVIONICS/SENSOR/OPTICS: THE MISSILE TARGETING SYSTEM USES A TELESCOPIC SIGHT UNIT (TRAVERSE 110º, ELEVATION –60º/+30º) WITH TWO MAGNIFICATIONS/FIELDS OF VIEW, A LASER AUGMENTED TRACKING CAPABILITY, TV, AND VIDEO. ADDITIONAL MISSIONS INCLUDE: DIRECT AIR SUPPORT, ESCORT, TARGET DESIGNATION, SECURITY, RECONNAISSANCE, AIR TO AIR COMBAT, AND ANTI-SHIP. THIS AIRCRAFT COSTS APPROXIMATELY $10.7 MILLION, INEXPENSIVE COMPARED TO OTHER MODERN ATTACK HELICOPTERS; BUT ITS PERFORMANCE IS SIMILAR. THUS MANY NATIONS CONSIDER THIS AIRCRAFT AS A GOOD CANDIDATE FOR FIELDING IN ATTACK HELICOPTER SQUADRONS. THIS IS THE OPFOR TIER 1 REPRESENTATIVE HELICOPTER SYSTEM.

RUSSIAN ATTACK HELICOPTER KA-50/HOKUM AND KA-52/HOKUM-B

SYSTEM	SPECIFICATIONS	SYSTEM	SPECIFICATIONS
Alternative Designations:	Black Shark, Werewolf , HOKUM-A	Ceiling (m):	
Date of Introduction:	Limited fielding by 1995. Ka-52 fielding starts in 2011.	Service:	5,500
Proliferation:	2 countries	Hover (out of ground effect):	4,000
Blades – Main rotor:	6 (2 heads, 3 blades each)	Hover (in ground effect):	Hover (in ground effect): 5,500
Tail rotor:	None	Vertical Climb Rate (m/s):	10
Engines:	2x 2,200-shp Klimov	Fuel (liters):	
Weight (kg):		Internal:	INA
Maximum Gross:	10,800	Range:	500 ea. (max 4 x)
Normal Takeoff:	9,800	Length (rotors turning):	16
Empty:	7,692	Length (fuselage):	15.0
Speed (km/h): Max (level):	310, 390 diving	Width (including wing):	7.34
Cruise:	270	Height:	(gear extended): 4.93 (gear retracted): 4
Sideward:	100+, Rearward: 100+	Main Rotor Diameter:	14.5 Cargo
Turn Rate:	Unlimited	Tail Rotor Diameter:	None
Max "G" Force: Survivability Countermeasures:	+3 to +3.5 g Main rotors and engines electrically deiced. Infrared signature suppressors can mount on engine exhausts. Pastel/L-150 radar warning receiver, laser warning receiver, IFF, chaff and flares. Armored cockpit. Self-sealing fuel tanks. Pilot ejection system.	Compartment Dimensions: Negligible Standard Payload:	External weapons load: 2,500 kg on 4 under-wing hard points.
AMMUNITION		**AMMUNITION**	**SPECIFICATIONS**
Most probable armament: HOKUM A/B/N:	Fuselage-mounted 30-mm cannon on right side, 40 x	Armor Penetration (mm):	1,200

	80-mm rockets,12 x Vikhr-M ATGMs, 2 x SA-24AAMs(ATGM pod can launch SA-24AAMs).		
Guided Missiles:	AT-16/Vikhr-M antitank missile	Rate of fire (missiles/min):	2-3 per range
Guidance:	Laser-beam rider, prox on/off	Range (m):	1,000- 10,000
Warhead:	Tandem shaped Chge (HEAT)	Other Missile Types:	AT-16 HE, Ataka 9M120-1 HEAT, HE

VARIANTS	SPECIFICATIONS
Ka-50A/HOKUM A:	Original Hokum. Due to poor performance, it will not be fielded.
Ka-50N/HOKUM N:	Night attack variant fitted with a nose-mounted FLIR from Thomson-CSF. The cockpit is fitted with an additional TV display, and is NVG compatible. These replace the Saturn pod on HOKUM-A. ATGM pods hold 6 AT-16/Vikhr missiles. Later, dual-seat versions were developed. Dual-seat arrangement can significantly improve effectiveness of a combat aircraft, because it frees up the pilot for precision flying, and provides a weapons officer who can give full attention to the combat mission.
Ka-52/Alligator/HOKUM-B: Ka-52/Alligator/HOKUM-B (cont'd):	Tandem, dual-seat cockpit variant of Ka-50, with 85% of its parts in commonality. Although performance is slightly inferior to Ka-50 in some areas (Max g 3.0, 3,600 m hover ceiling), it out-performs its predecessor in other areas (such as 310 km/h max speed), and has an equal service ceiling and range. An upgrade to the more powerful VK-2500 engine has begun. Ka-52 can be used as an air and ground attack. The fire control system employs a mast-mounted FH-01/Arbalet millimeter wave radar covering the front quadrant. The fire control system has a chin-mounted TV, FLIR, and laser in the UOMZ DOES stabilized ball mounted behind the cockpit. Also included is a Prichal laser range-finder/laser target designator (LTD), with a range of 18+ km. It can acquire, auto-track, and engage moving targets at a range of 15 km. Stationary targets can be engaged to 18+ km. The Ka-52 can launch AT-16/Vikhr ATGMs, with LBR guidance. However, there have been issues with that missile. A version of AT-9/Ataka, 9M120-1 now has added LBR guidance to its RF; so it could be used on the Ka-52, and supplement or replace Vikhr missile loads. Another option to replace or supplement Vikhr is Hermes-A. The aircraft has been displayed with 2 pods (12 multi-role missiles), and has been successfully tested. It is a 2-stage supersonic missile with a 170-mm booster stage and 130-mm sustainer. The aircraft can use its own LTD for guidance, or launch but defer to a remote LTD (man-portable, vehicle mounted, or UAV-mounted) for terminal phase, and shift to its next target. These multi-mode guided ASMs have a range of 18 (15-20) km, and a 28-kg HE warhead large enough to kill any Armored vehicle, and a wide variety of other air or ground targets.

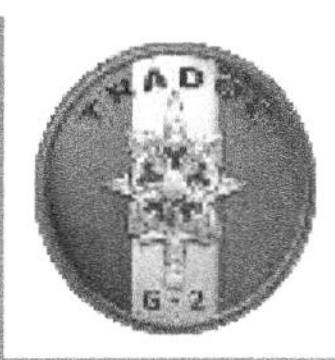

	This helicopter is also equipped with a Hermes-A multi-role missile with a Weight (kg): 32 (in tube) Guidance: Inertial/ MMW radar ACLOS or SAL-H with auto-tracker lock-on Warhead: HE, 28 kg Armor Penetration (mm): 1,30 0+ Rate of fire (missiles/min): 2 Range (m): 18,000 maximum A 40-km version of Hermes was tested and is due in the Near Term. A 100-km version (with a 210 mm booster, for 4 missiles per pylon) is featured at the KBP Tula site, and will be an option. Future versions will have an IR or radar-homing option. The Ka-52 adds workstation equipment for air battle management. It has 2 workstations with aircraft controls for mission hand-off. Russian forces have demonstrated operations with Ka-52s controlling flights of Ka-50N helicopters. It can also be used as a trainer for the Ka-50N.
Ka-50-2/Erdogan:	Russian/Israeli cooperative effort competing for the Turkey helicopter contract. The variant has Israeli avionics and a tandem dual seat cockpit similar to the Apache.

NOTES

AVIONICS/SENSOR/OPTICS: THE MISSILE TARGETING SYSTEM USES A TELESCOPIC SIGHT UNIT (TRAVERSE 110º, ELEVATION –60º/+30º) WITH TWO MAGNIFICATIONS/FIELDS OF VIEW, A LASER AUGMENTED TRACKING CAPABILITY, TV, AND VIDEO.
ADDITIONAL MISSIONS INCLUDE: DIRECT AIR SUPPORT, ESCORT, TARGET DESIGNATION, SECURITY, RECONNAISSANCE, AIR TO AIR COMBAT, AND ANTI-SHIP.
THIS AIRCRAFT COSTS APPROXIMATELY $10.7 MILLION, INEXPENSIVE COMPARED TO OTHER MODERN ATTACK HELICOPTERS; BUT ITS PERFORMANCE IS SIMILAR. THUS MANY NATIONS CONSIDER THIS AIRCRAFT AS A GOOD CANDIDATE FOR FIELDING IN ATTACK HELICOPTER SQUADRONS. <u>THIS IS THE OPFOR TIER 1 REPRESENTATIVE HELICOPTER SYSTEM.</u>

RUSSIAN ATTACK HELICOPTER MI-24/35 HIND

SYSTEM	SPECIFICATIONS	SYSTEM	SPECIFICATIONS
Alternative Designations:	Mi-25 or Mi-35 for exports	Hover (out of ground effect):	1,500
Date of Introduction:	1976 (HIND D)	Hover (in ground effect):	2,200
Proliferation:	At least 34 countries	Vertical Climb Rate (m/s):	15
Crew:	2 pilots in tandem cockpits		
	8 troops/4 litters	Internal:	1,840
Blades – Main rotor:	5	Internal Aux Tank (in cabin):	1,227
Tail rotor:	3	External Fuel Tank:	500 ea. x 2
Engines:	2x 2,200-shp Klimov TV3-117VMA turboshaft	Range (km):	
Weight (kg):		Normal Load:	450
Maximum Gross:	11,500	With Aux Fuel:	950
Normal Takeoff:	11,100	Dimensions (m):	
Empty:	8,500	Length (rotors turning):	21.6
Speed (km/h): Max (level):	335	Cargo Compartment Dimensions (m):	
Cruise:	295	Floor Length:	2.5
Max "G" Force:	1.75 g	Width:	1.5
Ceiling (m):		Height:	1.2
Service:	4,500	External weapons load:	1,500 kg (no weapons): 2,500 kg
AMMUNITION	**SPECIFICATIONS**	**AMMUNITION**	**SPECIFICATIONS**
Fuselage/nose mount gun/MG	1	57-mm S-5 rocket pods (32 ea.)	2-4
7.62/12.7-mm door MG	1	122-mm S-13 rocket pods (5 ea.)	2-4
AT-2/-6/-9 ATGMs	2	240-mm S-24 rocket pods (1 ea)	2-4
80-mm S-8 rocket pods (20 ea.)	2-4	250-kg bombs, including FAE	4
500-kg bombs, including FAE	2	Protection/Survivability/Countermeasures:	Armored cockpit and titanium rotor head defeat 20-mm rds. Overpressure system is used for NBC

			environment. Infrared signature suppressors on engine exhausts. Radar warning receivers, IFF. Infrared jammer, rotor brake. Armored cockpit. ASO-4 Chaff/flare dispenser. Auxiliary power unit for autonomous operation. Main and tail rotors are electrically deiced.
KMGU or K-29 Mine pods	2-4		
Gun/MG/AGL pods (See below)	2-4		
AA-8/R-90 or SA-24AAM	2-4		
ARMAMENT	Mi-24 has a fuselage or turret nose gun, and at least one door machinegun. It also has 6 pylons (hardpoints), on which it can mount bombs, missiles (ATGMs, ASMs, AAMs), rockets, and gun or grenade or mine pods. Mission dictates weapon configuration. Available munitions are shown above; not all may be employed at one time. As ammunition/payload weight is expended, more passengers can fit aboard the aircraft.	Most Probable Armament:	HIND D: Nose turret-mounted 4-barrel 12.7-mm Gatling type minigun, 1,470 rds, 4 pods of 57-mm rockets, and 4 x AT-2C/ SWATTER ATGMs. HIND E: Nose turret-mounted 4-barrel 12.7-mm Gatling type minigun, 40 x 80-mm rockets and 8 x AT-6C/SPIRAL ATGMs. HIND F: GSh-30K gun on fuselage, 40 x 80-mm rockets, 8 x AT-6C ATGMs, and 2x SA-24AAMs. Mi-35M2: Nose turret 23-mm twin gun 470 rds, 40 x 80-mm (or 10 x 122-mm) rockets, 8 AT-6c (or 8 AT-9), and 2 x SA-24 AAMs. For tank destroyer role, exchange rocket pods for 8 more ATGMs.
Fuselage-Mounted Guns/Machineguns:	Guns vary widely with different variants (see below). Some are fixed, providing accurate fires along the flight path. Nose turret guns offer more responsive fires against targets to sides, but may lack accuracy, range and ammo capacity of fixed guns. The gun is assisted by rear and	AVIONICS/SENSOR/OPTICS	The ATGM targeting system uses a low-level light TV, a laser target designator, PKV gunsight for pilot, air data sensor, and a missile guidance transmitter. Some versions and specific forces have upgraded FCS.

	side mount guns and arms operated by passengers. Onboard combat troops can fire personal weapons through cabin windows. For gunship missions, usually the only troop is a door gunner, thus permitting more ammo in the cabin. Also, to complement main gun fires, crews can add gun pods		
Guided Missiles:	AT-6b or AT-9/Ataka-M Guidance: Radio-guided Warhead: Tandem shaped Chge (HEAT) Armor Penetration (mm): 1,100, 800+ERA Rate of fire (missiles/min): 3-4 Range (m): 400-7,000 (6,000 AT-9) Other Missile Types: AT-6/Ataka HE, 9A2200 anti-helicopter w/prox fuze		

VARIANTS	SPECIFICATIONS
Mi-24A/HIND A/B/C:	The original -A helicopter had side-by-side seats, single-barrel 12.7-mm MG, 57-mm rocket pods, and AT-2a/b/SWATTER-A/B ATGMs. The export HIND A launched AT-3/SAGGER ATGMs. All of these missiles were manually controlled (MCLOS). The HIND B never entered production. HIND C was a trainer, without a gun pod. Nearly all of the older HIND A, B and C variants have been upgraded or modified to the HIND D or E standard.
Mi-24D/HIND D:	This represents an OPFOR Tier 4 helicopter capability. This gunship has a more powerful engine and improved fire control system. Other upgrades include a 4-barrel 12.7-mm Gatling type gun. Rocket pods can be mounted on the inner 4 pylons, and AT-2c/ SWATTER-C ATGMs can be mounted on wing pylons. These SACLOS missiles offer superior range and operational precision over earlier versions. There are NVGs and II sights, which permit night flying but virtually no night engagement capability, except in illuminated areas. Mi-25 is the export version.
Mi-24V/HIND E:	The most proliferated version. This variant represents OPFOR Tier 3 helicopter capability. It has the 4-barrel mini-gun and up to 8 AT-6/ Shturm-V series ATGMs (most recent is AT-6C). It can also launch Ataka/AT-9 series ATGMs. With its heads-up-display (HUD) fire control system, the aircraft can also launch AA-8 AAMs. Mi-35 is an export version of HIND E. Mi-35O night attack upgrade with an Agema FLIR ball.
Mi-24P/HIND F:	This gunship variant has A 30-mm twin gun affixed to right side. ATGMs are the AT-6 and AT-9 series. Mi-35P is an export version of the HIND F.

Mi-24PS:	Ministry of Internal Affairs version, with wingtip ATGM launchers, sensor ball with FLIR night sights and loud speakers.
Mi-24R/HIND G-1:	Mi-24V variant for NBC sampling. It has mechanisms for soil and air samples, filter air, and place marker flares.
Mi-24K/HIND G-2:	Photo-reconnaissance and artillery fire direction variant. It has a camera in the cabin, gun, and rocket pods, but no targeting system. Upgrades to the Mi-35M standard are the Mi-24VK-1 and Mi-24PK-2.
Mi-24PN/Mi-35PN:	Russian upgrade of Mi-24P/35P with Zarevo FLIR FCS.
Mi-24VP:	Mi-24VP is a Russian response to lack of satisfaction with the 30-mm gun. This variant replaces the gun with a twin 23-mm nose turret gun and 470-mm rounds. It has been fielded in limited numbers.
Mi-24VM/Mi-35M:	The program integrates a suite of compatible upgrades. It has main and tail rotors from Mi-28, and a new engine and transmission, with improved capability for nap-of-the-earth (NOE) flight. It includes: hardpoints reduced to 4, hover rise to 3,000 m, fiberglass rotor blades, fixed landing gear, scissors tail rotor, new nav, and stabilized all-weather FLIR ball FCS. Export Mi-24VP with FLIR sights is Mi-35M1 (NFI). Mi-35-PM is a Mi-35P upgraded to -M standard. Indian Mi-35s are upgrading to -M standard.
Mi-24VK-1 and Mi-24PK-2:	Upgrades for earlier helicopters to the Mi-35M standard.
The Mi-35M2:	This is the latest export version, and the most robust version of the Mi-24/35 HIND helicopter. This variant represents OPFOR Tier 2 helicopter capability. It has new 2,400- shp VK-2500 engines. Ceiling is increased to 5,700 m (4,000 hover). The French based FCS pod has a Chlio FLIR night sight. Armament is: twin barrel 23-mm nose turret gun, 12.7-mm NSV MG (at the cargo door), 16 x AT-6c (or AT-9) ATGMs, and 2 rocket pods. Other options include AA-8, AA-11, or SA-24 AAMs. A 30-mm nose gun is available. For tank destroyer role, exchange rocket pods for pods with 8 more ATGMs.
Mi-35D:	Export private venture upgrade with weapons systems from the Ka-50/Hokum helicopter. Changes include the Shkval FCS, Saturn FLIR, and up to 16 AT-16/Vikhr ATGMs. For AAM, the AA-18 would be replaced with AA-18S (SA-18S/Igla-Super).
Tamam Mi-24 HMSOP/ Mission 24:	Israeli upgrade program. It includes a TV FCS with FLIR, autotracker, and GPS. Contrary to other HINDs, The pilot sits in front, with the gunner in the rear. ATGM is the NLOS Spike-ER. The launcher can also launch Skylite UAVs, then hand them off to ground controllers.
Mi-24 Mk III:	South African upgrade. It has a 20-mm Gatling-type gun, and ZT-35/ Ingwe ATGM. The Ukrainian Super HIND Mk II would be similar, with Mokopa.

NOTES

ADDITIONAL MISSIONS INCLUDE: DIRECT AIR SUPPORT, ESCORT, TARGET DESIGNATION, SECURITY, RECONNAISSANCE, AIR TO AIR COMBAT, AND ANTI-SHIP. OPTIONAL UPGRADES INCLUDE THE MI-28'S AT-9/ATAKA 8-MISSILE LAUNCHER (16 TOTAL), OR ISRAELI SPIKE-LR ATGM LAUNCHER. A NEW UPGRADE IS ADDITION OF A LASER TARGET DESIGNATOR IN THE FCS, WHICH CAN GUIDE SEMI-ACTIVE LASER-HOMING BOMBS, AND LASER-GUIDED 57/80/122-MM ROCKETS FROM PODS.

RUSSIAN ATTACK HELICOPTER MI-28N/HAVOC

SYSTEM	SPECIFICATIONS	SYSTEM	SPECIFICATIONS
Alternative Designations:	N/A	Hover (out of ground effect):	3.600
Date of Introduction:	N/A	Hover (in ground effect):	INA
Proliferation:	Algeria, Kenya, Iraq, Venezuela	Vertical Climb Rate (m/s):	INA
Crew:	2	Fuel (liters):	
Blades – Main rotor:	5	Internal:	1,900
Tail rotor:	4	Internal Aux Tank (in cabin):	INA
Engines:	2x 2,200-shp Klimov TV3-117VMA turboshaft	External Fuel Tank:	INA
Weight (kg):		Range (km):	
Maximum Gross:	11,500	Normal Load:	475
Normal Takeoff:	10,400	With Aux Fuel:	1,100
Empty:	7,000	Dimensions (m):	
Speed (km/h): Max (level):	300	Length (rotors turning):	21.2
Cruise:	260	Cargo Compartment Dimensions:	Negligible
Sideward\Rearward:	100/100	Standard Payload:	3,640 kg on 4 under wing stores points.
Max "G" Force:	-.5 to +3.7 g	Width (including wing):	4.9
Ceiling (m):		Height:	4.7
Service:	6,000	Tail Rotor Diameter:	3.8

Survivability/Countermeasures:

Armored cockpit frame is made of titanium, steel and ceramic. It can withstand hits of 20-mm shells at a minimum. The cockpit glass is bulletproof to 12.7-mm rounds, and resistant to fragmentation from 20-mm shells. The HAVOC has a high altitude ejection system that jettisons wings and cockpit doors when the crew jumps to safety with parachutes. It has a "technical compartment" accommodating two persons, to evacuate the crew from downed aircraft. Main rotors and engines are electrically deiced. Self-sealing fuel tanks. Infrared signature suppressors mounted on engine exhausts. Radar warning receivers, pressurized cockpit, IFF, chaff, decoys

AMMUNITION	SPECIFICATIONS	AMMUNITION	SPECIFICATIONS
1x 2A42 30-mm cannon	250 Rds.	250/500-kg bombs	2-4
AT-6c or AT-9/Ataka pods (4 ea pod)	2-4	SA-24 AAM pod (2-4 ea)	2
S-8 80-mm rocket pod (20 ea) or S-13 122-mm rocket pod (5	2-4	KMGU scatterable mine pod	2-4

AS-12/KEGLER ASM	2	Mission dictates weapons configuration. Not all will be employed at the same time.	
23-mm gun pods (250 rds) Most Probable Armament:	2		
Mi-28A/N:	Chin turret-mounted 2A42 30-mm auto-cannon, 40 x 80-mm (or 10 x 122-mm) unguided or semi-active laser-homing rockets, 14 x AT-6c/Kokon-M ATGMs, and 2 x SA-24 AAMs. Note. The ATGM pods can launch other ATGMs and selected AAMs.	SENSOR/OPTICS	The HAVOC has optical magnification, a HUD, 2 FLIR sights, targeting radar, and a laser designator for target engagement. A helmet sighting system turns the cannon in the direction the pilot is looking. Rotor blade-tip pitot tubes give speed/drift data for targeting at low airspeed.

VARIANTS	SPECIFICATIONS
Mi-28A:	The original version, and is primarily a daylight only aircraft.
Mi-28N:	The Mi-28N has avionics upgrades. Use of night-vision goggles gives day/night, all-weather mission capability. The "Night version."
Mi-28NE(for export):	This aircraft features an integrated rotor-hub radar for targeting and navigation, autopilot, an inertial nav system, thermal night sight, and low-light level TV helmet targeting system for target engagement. It is probable that changes for the Mi-28M (below) will be applied to Mi-28N, and in fact, to all Mi-28s.
Mi-28M:	Next upgrade version currently in development. It includes 2x 2,400-shp Klimov VK-2500 (TV3-117SB3) turboshaft engines, improved transmission, and more efficient rotor blades. These compensate for added avionics weight, and increases in armament basic load. The aircraft's upgraded avionics offer better coordination of group combat actions through datalinks. A likely ATGM change will be to the Krizantema/AT-15, with 6,000-m range and 1,500+ mm penetration. A version of AT-9/Ataka, 9M120-1 now has RF and laser beam rider guidance as on Krizantema. Thus Ataka can be used to supplement AT-15missile loads.

NOTES
ADDITIONAL MISSIONS INCLUDE: DIRECT AIR SUPPORT, ESCORT, TARGET DESIGNATION, SECURITY, RECONNAISSANCE, AIR TO AIR COMBAT, AND ANTI-SHIP. ALTHOUGH THIS AIRCRAFT IS ROUTINELY COMPARED TO THE U.S. AH-64 APACHE, IT IS MUCH LARGER AND LESS MANEUVERABLE THAN ITS U.S. COUNTERPART.

BRITISH MEDIUM MULTIROLE HELICOPTER LYNX

SYSTEM	SPECIFICATIONS	SYSTEM	SPECIFICATIONS
Alternative Designations:	AH. Mk-1, 7, 9	Hover (out of ground effect):	3,230, 5,126
Date of Introduction:	1977	Hover (in ground effect):	3,660
Proliferation:	At least 11 countries	Vertical Climb Rate (m/s):	7
Crew:	2 pilots. Transports 9 troops, 6 litters, or cargo.	Fuel (liters):	
Blades – Main rotor:	4	Internal:	985
Tail rotor:	4	Internal Aux Tank (in cabin):	696
Engines:	2x 900-shp Rolls Royce Gem 42-1 turboshaft, 2x 1,260 LHTEC CTS800-4N turboshaft (Mk 9)	Range (km):	
Weight (kg):		Normal Load:	630
Maximum Gross:	4,535, 5,126 (Mk 9)	With Aux Fuel:	1,342
Normal Takeoff:	2,658, 3,496 (Mk 9)	Main Rotor Diameter:	12.8
Empty:	2,578	Tail Rotor Diameter:	2.2, 2.4 (Mk 9) Cargo
Speed (km/h): Max (level):	289	Floor Length:	2.1
Cruise:	259, 285 (Mk 9)	Width:	1.8
Sideward/Rearward:	Sideward:130/ Rearward:INA	Height:	1.4
Max "G" Force:	+2.3 to -0.5	Standard Payload (kg):	
Ceiling (m):		Internal load:	907
Service:	INA	External on sling only:	1,360, 2,000 (Mk 9)

SURVIVABILITY/COUNTERMEASURES:

Engine exhaust suppressors, infrared jammer, and flare/chaff dispensers are available. Rotor brake and self-sealing fuel tanks are used.

ARMAMENT

The Lynx employed by ground forces can be equipped with two 20-mm cannons mounted externally to permit 7.62-mm machineguns to be fired from the cabin. Two fuselage pylons allow for external stores.

AVIONICS/SENSOR/OPTICS

Army variants equipped for TOW missiles have a roof-mounted sight (over the left-hand pilot's seat) with IR and thermal capabilities for firing. Optional equipment allows for target magnification, LLLTV, cameras, and IR searchlight. Safire or other FLIR for night capability.

VARIANTS	SPECIFICATIONS
Developed under a partnership between predominantly Westland of the United Kingdom and Aerospatiale of France. Listed below are primary and most proliferated variants used by ground forces. Many others exist in small numbers for ground and naval forces.	
Lynx AH. Mk 1:	The basic army multirole and gunship version. This aircraft has skid-type landing gear. Most have been converted to Mk 7 format.
Lynx AH. Mk 7:	Also known as AH 1. Upgraded British army version, some with improved main rotor blades. Reverse-direction tail rotor to reduce noise signatures and improve performance. Aircraft has skid-type landing gear.
Lynx AH. Mk 9:	Aka Super Lynx or Light Battlefield Helicopter. Implemented tricycle-type landing gear, improved rotor blades, and upgraded engines to increase performance. Mostly used in tactical transport role, with no ATGM launch capability.
Battlefield Lynx:	Export version of Lynx AH. Mk 9 that can be armed with ATGMs.

NOTES

THIS AIRCRAFT WAS DESIGNED TO BE BOTH A TRANSPORT AND AN ATTACK AIRCRAFT. MISSIONS INCLUDE: DIRECT AIR SUPPORT, ANTI- HELICOPTER, RECONNAISSANCE, ESCORT, SECURITY, TRANSPORT, AND TRAINING. EACH FUSELAGE SIDE HAS ONE PYLON ALLOWING FOR A SINGLE GUN POD OR MISSILE RACK. LYNX IS CAPABLE OF SINGLE-ENGINE FLIGHT IN THE EVENT OF LOSS OF POWER BY ONE ENGINE (DEPENDING ON AIRCRAFT MISSION WEIGHT) WITH ITS ENGINE LOAD SHARING SYSTEM. IF AN ENGINE FAILS, THE OTHER'S OUTPUT INCREASES.

RUSSIAN MEDIUM MULTIROLE HELICOPTER MI-2/HOPLITE

SYSTEM	SPECIFICATIONS	SYSTEM	SPECIFICATIONS
Alternative Designations:	INA	Hover (out of ground effect):	1,000
Date of Introduction:	1965	Hover (in ground effect):	2,000
Proliferation:	Widespread	Vertical Climb Rate (m/s):	4.5
Crew:	1 (pilot)	Fuel (liters):	
Blades – Main rotor:	3	Internal:	600
Tail rotor:	2	External Fuel Tank:	238 ea.
Engines:	2x 400-shp PZL GTD-350 (series III and IV) turboshaft	Range Max Load (km):	170
Weight (kg):		Internal Fuel Load:	440
Maximum Gross:	3,700	With Aux Fuel:	790
Normal Takeoff:	3,550	Main Rotor Diameter:	14.6
Empty:	2,372	Tail Rotor Diameter:	2.7
Speed (km/h): Max (level):	220	Dimensions (m):	
Cruise:	194	Length (rotors turning):	17.4
Sideward/Rearward:	INA	Length (fuselage):	11.9
Max "G" Force:	INA	Width:	3.2
Ceiling (m):		Height:	3.7
Service:			

SURVIVABILITY/COUNTERMEASURES:
Main and tail rotor blades electrically deiced.

ARMAMENT
The Lynx employed by ground forces can be equipped with two 20-mm cannons mounted externally to permit 7.62-mm machineguns to be fired from the cabin. Two fuselage pylons allow for external stores.

AVIONICS/SENSOR/OPTICS
The cannon is pilot sighted, and fire is adjusted by controlling attitude of the aircraft.

ARMAMENT

23-mm Automatic Cannon, NS-23KM: Range: (practical) 2,500 m Elevation/Traverse: None (rigidly-mounted) Ammo type: HEFI, HEI, APT, APE, CC Rate of Fire (rpm): (practical) 550 7.62-mm or Pintle-mounted Machinegun: (may be mounted in left-side cabin door) Range: (practical) 1,000 m Ammo type: HEFI, HEI, APT, APE, CC Rate of Fire (rpm): (practical) 250 OR 12.7-mm or Pintle-mounted Machinegun: (may be mounted in left-side cabin door) Range: (practical) 1,500 m Ammo type: API, API-T, IT, HEI Rate of Fire (rpm): (practical) 100	
VARIANTS	SPECIFICATIONS
Mi-2B:	Upgrade with improved navigation and electrical systems
Mi-2R:	Ambulance version that carries 4x litter patients.
Mi-2T:	Transport version that carries 8 personnel.
Mi-2URN:	Armed reconnaissance variant, employs 57-mm unguided rockets, and mounts a gun sight in the cockpit for aiming all weapons.
Mi-2URP:	The antitank variant. Carries 4x AT-3C Sagger ATGMs on external weapons racks, and 4x additional missiles in the cargo compartment.
Mi-2US:	The gunship variant, employs an airframe modification that mounts a 23-mm NS-23KM cannon to the portside fuselage. It also employs 2x 7.62-mm gun pods on external racks, and 2x 7.62-mm pintle-mounted machineguns in the cabin.
PZL Swidnik:	A Polish-produced variant under license from Russia. It features minor design changes, but same performance, characteristics, and missions. Polish MOD officials will upgrade the gunship version with a new ATGM. Likely choice is between the Israeli 6 km FOG-M Spike-ER missile, and the 4 km HOT-3. The 4-missile launcher will also have a thermal night sight.
NOTES EXTERNAL STORES ARE MOUNTED ON WEAPONS RACKS ON EACH SIDE OF THE FUSELAGE. EACH RACK HAS TWO HARDPOINTS FOR A TOTAL OF FOUR STATIONS. ADDITIONAL MISSIONS INCLUDE; DIRECT AIR SUPPORT, RECONNAISSANCE, TRANSPORT, MEDEVAC, AIRBORNE COMMAND POST, SMOKE GENERATING, MINELAYING, AND TRAINING. THE CABIN DOOR IS HINGED RATHER THAN SLIDING, WHICH MAY LIMIT OPERATIONS. THERE IS NO ARMOR PROTECTION FOR THE COCKPIT OR CABIN. AMMO STORAGE IS IN THE AIRCRAFT CABIN, SO COMBAT LOAD VARIES BY MISSION. SOME MI-2USS CURRENTLY EMPLOY FUSELAGE-MOUNTED WEAPON RACKS RATHER THAN THE 23-MM FUSELAGE-MOUNTED CANNON, WHICH IS REMOVED. SOME VARIANTS HOWEVER, STILL EMPLOY THE CANNON.	

CHINESE MEDIUM MULTI-ROLE Z-9/HAITUN AND WZ-9 GUN SHIP

SYSTEM	SPECIFICATIONS	SYSTEM	SPECIFICATIONS
Alternative Designations:	50	Hover (out of ground effect):	1,020 Z-9A 1,600 Z-9B
Date of Introduction:	1994	Hover (in ground effect):	1,950 Z-9A 2,600 Z-9B
Proliferation:	At least 3 countries.	Vertical Climb Rate (m/s):	246
Crew:	1 for Z-9, 2 for WZ-9. Transports 9-12 troops, 4-8 litters or cargo.	Fuel (liters):	
Blades – Main rotor:	4	Internal:	1,140
Tail rotor:	13 Z-9A, 11 Z-9B/WZ-9	External Fuel Tank:	180
Weight (kg):		Normal Fuel Load:	860
Maximum Gross:	4100	With Aux Fuel:	1,000
Empty:	2050	Dimensions (m):	
Speed (km/h): Max (level):	315	Length (rotors turning):	13.7
Cruise:	280	Length (fuselage):	12.1 without rotors
Max "G" Force:	INA	Compartment Dimensions: (m)	
Ceiling (m):		Floor Length:	2.2
Service:	4,500 Z-9A, 6,000 Z-9B/WZ-9	Width:	1.9
		Internal load:	INA
		External on sling only	1,600
		Max:	2,038

SURVIVABILITY/COUNTERMEASURES:
Light armor panels. All composite rotors and fenestron, and composite body structure reduce signature. Nomex honeycomb in structure. Limited countermeasure capability.

ARMAMENT
Two fixed 23-mm guns or 12.7-mm MGs.
Two pylons permit mounting up to 8 ATGMs, or 4 plus 2 rocket pods.

MOST PROBABLE ARMAMENT

Combat versions (WZ-9 and Z-9G) have Twin 23-mm gun, four Red Arrow-8F ATGMs, 2x 7-round 90-mm rocket pods, and 2 TY 90 IR-homing AAMs.	
ROCKETS AND MISSILES Name: Red Arrow-8F Type: ATGM Warhead: Tandem Shaped Charge Armor Penetration (mm CE): 1,100 Min/Max Range (m): 100/4,000 Rate of fire (missiles/min): 3-4, depending on range. Name: Type 90-1 Type: Air-to-surface rocket Warhead: Frag-HE Max Range (m): 7,000	
AVIONICS/SENSOR/OPTICS WZ-9 has a day/night all-weather capability with gyro-stabilized TV/IRST FLIR chin pod gunsight, and SFIM autopilot. Transponder and weather radar is optional. Datalink for naval observation supports over-the-horizon attack	
NIGHT/WEATHER CAPABILITIES: The aircraft is NVG compatible, and through instruments, avionics, autopilot, and Doppler navigation system, is capable of operations day and night, and is instrumented for adverse meteorological conditions.	
VARIANTS	SPECIFICATIONS
Z-9A:	Military production version with some upgrades, such as Arriel 1C2 engine, upgrade instrument panel, and 150-kg payload increase.
Z-9A 100:	Indigenously produced version.
Z-9B:	Current production version for multi-role use, based on Dauphin 2 designs. Changes include 11-blade tail rotor.
Z-9C:	Naval version for ASW and SSM, with Sinatra HS-12 dipping sonar and torpedo. It has a datalink to support targeting for YJ-82 SSM. An expected near-term upgrade is the C-701 TV guided air-to-surface missile.
WZ-9:	Light attack version of Z-9B (see ARMAMENT, left). Poss aka Z-9W. Export version is Z-9G.
Z-9Z:	Reconnaissance prototype.
NOTES DESPITE STATEMENTS FROM SOME SOURCES, WZ-9 IS TOO LIGHTLY PROTECTED TO BE AN "ATTACK HELICOPTER". THE Z-9 WAS DESIGNED TO BE ADAPTABLE FOR A VARIETY OF ROLES, INCLUDING TRANSPORT, DIRECT AIR SUPPORT, ESCORT, SECURITY, RECONNAISSANCE, AMBULANCE, ANTI-SUBMARINE WARFARE, IW, AIRBORNE C2, SEARCH AND RESCUE, ANTI-SHIP, AND ANTI-SUBMARINE WARFARE. EACH FUSELAGE SIDE HAS ONE PYLON ALLOWING FOR A SINGLE POD OR MISSILE RACK. AN EXPECTED UPGRADE FOR WZ-9/Z-9G IS THE RED ARROW 9 LASER-BEAM RIDER/MMW GUIDED ATGM, WITH 1,200 MM PENETRATION AND 5 KM RANGE.	

EUROPEAN MULTIROLE HELICOPTER AS-532/COUGAR

SYSTEM	SPECIFICATIONS	SYSTEM	SPECIFICATIONS
Alternative Designations:	AS 332 Super Puma, SA 330 Puma	Fuel (liters):	
Date of Introduction:	1981	Internal:	1,497 (UC/AC), 2,000 (UL/AL), 2,020 (U2/A2)
Proliferation:	At least 38 countries	External Fuel Tank:	
Crew:	2 (pilots) Transports: 20-29 troops or 6-12 litters (variant dependent), or cargo.	With Aux Fuel:	1,017 (UC/AC), 1, 245 (UL/AL), 1,176 (U2/A2)
Blades – Main rotor:	4	Dimensions (m):	
Tail rotor:	5, 4 (U2/A2)	Length (rotors turning):	18.7-19.5 (U2/A2)
Weight (kg): Normal Takeoff:	8,600 (Mk I), 9,300 (Mk II)	Length (fuselage):	15.5 (UC/AC),
Maximum Gross:	9,000 (Mk I), 9,750 (Mk II)	Floor Length:	16.3 (UL/AL), 16.8 (U2/A2)
Empty:	4,330 (UC/AC), 4,460 (UL/AL), 4,760 (U2/A2)	Width:	3.6-3.8 (U2/A2)
Speed (km/h): Max (level):	275 (Mk I), 325 (Mk II)	Main Rotor Diameter	15.6-16.2 (U2/A2)
		Tail Rotor Diameter:	3.1-3.2 (U2/A2)
Ceiling (m):	270	Cargo Compartment Dimensions (m):	
Service:	4,100	Floor Length:	6.5 (AC/UC), 6.8 (UL/AL), 7.9 (U2/A2)
Hover (out of ground effect):	1,650 (Mk I) 1,900 (Mk II)	Width/Height:	1.8/1.5
Hover (in ground effect):	2,800 (Mk I), 2,540 (Mk II)	Standard Payload (kg):	
Vertical Climb Rate (m/s):	7	Internal load:	3,000
		External on sling only:	4,500

WEAPONS

7.65-mm MG (2)

Other Loading Options

20-mm twin gun pods (2), 68-mm rocket pods (22 each), (2), 2.75-in rocket pods (19 each), (2), External fuel tanks (600 liters).

Mission dictates weapons configuration. Not all will be employed at the same time.	
ARMAMENT The Mk I variants may employ 2x 7.65-mm machine guns on pintle-mounts in the cabin doors when employed in a transport role.	
MOST PROBABLE ARMAMENT The armed versions have side-mounted 20-mm machineguns and/or axial pods fitted with 68-mm rocket launchers.	
AVIONICS/SENSOR/OPTICS Night/Weather Capabilities: The aircraft is NVG compatible, and through its instruments, avionics, full autopilot, and navcomputer, is capable of operation in day, night, and instrument meteorological conditions.	
VARIANTS	**SPECIFICATIONS**
SA 330 Puma:	Developed in the late 1960s by Aerospatiale in France. Others were built in the UK, Indonesia, and Romania.
AS 332 Super Puma:	Differs from the SA 330 Puma through an improved rotor system, upgraded engines, stretched fuselage, and a modified nose shape.
The Cougar name was adopted for all military variants. In 1990, all Super Puma designations were changed from AS 332 to AS 532 to distinguish between civil and military variants. The "5" denotes military, "A" is armed, "C" is armed-antitank, and "U" is utility. The second letter represents the level of "upgrading".	
AS-532 Cougar UC/AC Mk I:	The basic version with a short fuselage to carry 20 troops.
AS-532 Cougar UL/AL Mk I:	This version has an extended fuselage, which allows it to carry 25 troops and more fuel. It is also capable of carrying an external load of 4,500 kg.
AS-532 Cougar U2/A2 Mk II:	This 1992 version is the longest variant of the Cougar line. It has an improved Spheriflex rotor system with only 4x tail rotor blades, and 2x 2,100-shp Turbomeca Makila 1A2 turboshaft engines that allow an increased cargo carrying capability. It can transport 29 troops or 12 litters, or an external load of 5,000 kg. Primarily used for combat search and rescue, and as an armed version. It may be armed additionally with a 20-mm cannon or pintle-mounted .50 caliber machine guns.
NOTES THIS HELICOPTER IS PRODUCED BY THE EUROCOPTER COMPANY. IT WAS FORMED AS A JOINT VENTURE BETWEEN AEROSPATIALE OF FRANCE, AND DAIMLER-BENZ AEROSPACE OF GERMANY. ADDITIONAL MISSIONS INCLUDE: VIP TRANSPORT, ELECTRONIC WARFARE, AND ANTI-SUBMARINE WARFARE.	

RUSSIAN PATROL/ANTI-SUBMARINE HELICOPTER KA-27/HELIX

SYSTEM	SPECIFICATIONS	SYSTEM	SPECIFICATIONS
Alternative Designations:	N/A	Fuel (liters):	
Date of Introduction:	1980	Internal:	4,720
Proliferation:	At least 6 countries	Range (km):	800
Crew:	2 (pilot, navigator) , 1-3 sensor operators	Dimensions (m):	
Blades – Main rotor:	6 (2heads, 3 blades each)	Length (rotors turning):	31.8
Tail rotor:	None	Length (fuselage):	11.3
Normal Takeoff:	11,000	Width:	5.65
Maximum Gross:	10,700	Height:	5.4
Empty:	6,400	Main Rotor Diameter:	15.9
Speed (km/h): Max (level):	250	Cabin Dimensions (m):	
Cruise:	230	Length:	4.52
Ceiling (m):		Width:	1.3
Service:	6,000	Height:	1.32
Hover:	3,500	Main Rotor Diameter: 15.9	Main Rotor Diameter: 15.9
Vertical Climb Rate (m/s):	12.5	Cabin Dimensions (m):	Cabin Dimensions (m):
Standard Payload (kg):		Length:	4.52
Internal load:	4,000	Width:	1.3
External load:	5,000	Height:	1.32

WEAPONS
7.62 mm machine gun (1)
PLAB 250-120 bombs (2)(rarely used)
AT-1MV 400 mm Torpedoes (2)
Mission dictates weapons configuration. Not all will be employed at the same time.

MOST PROBABLE ARMAMENT
Torpedoes

AVIONICS/SENSOR/OPTICS

Auto-hovering, automatic flight control system, 360 degree search radar, directional ESM, Doppler, dipping sonar, magnetic anomaly detector (MAD), sonobuoys stored internally.	
Night/Weather Capabilities: Designed to operate day and night in adverse weather.	
VARIANTS	**SPECIFICATIONS**
Ka-27PL Helix-A:	ASW version.
Ka-27PS Helix-D:	Ka-27PS Helix-D: SAR version. Fitted with 300 kg rescue hoist. Hooks under fuselage for loads up to 5,000 kg
Ka-28:	Export version of Helix-A. Max takeoff weight increased to 12,000 kg. Max fuel and range also increased.
Ka-29TB Helix-B:	Armored assault troop version operated from amphibious landing ships or aircraft carriers. Armed with single four-barrel 7.62 mm machine gun, can also fit a 30 mm Type 2A42 cannon. Four stores pylons for 80 mm rocket pods, 57 mm rocket pods, 23 mm gun pods, incendiary tanks, or anti-tank missiles.
Ka-31 AEW:	Airborne early warning version of Ka-29 fitted with rotating radar antenna underneath the aircraft.
Ka-32A2:	Ka-32A2: Paramilitary transport version used by police. Pintle mounted guns in window, hydraulic hoist, loudspeakers, and searchlights. Can carry 11 passengers.
Ka-32A7: .	Armed version of Ka-27PS. 13-passenger capacity. Two GSh-3L 23mm cannons, B-8V-20 rocket pods, two AS-20 Kayak anti-ship missiles or AS-10 Karen air-to-air missiles.
NOTES	
THE HELIX IS PRIMARILY A NAVAL HELICOPTER, FOR MISSIONS SUCH AS SHIP-BASED ANTI-SUBMARINE WARFARE, DIRECT AIR SUPPORT, TRANSPORT, RESCUE, EW, ANTI-SHIP, AND AIR-TO-AIR. THE HELIX HAS THE DISTINCTIVE CONTRA-ROTATING MAIN ROTOR SYSTEM FAVORED BY THE KAMOV BUREAU. THE CONTRA-ROTATING DESIGN ELIMINATES THE NEED FOR A TAIL ROTOR.	

RUSSIAN MULTIROLE HELICOPTER MI-8/HIP-C AND VARIANTS

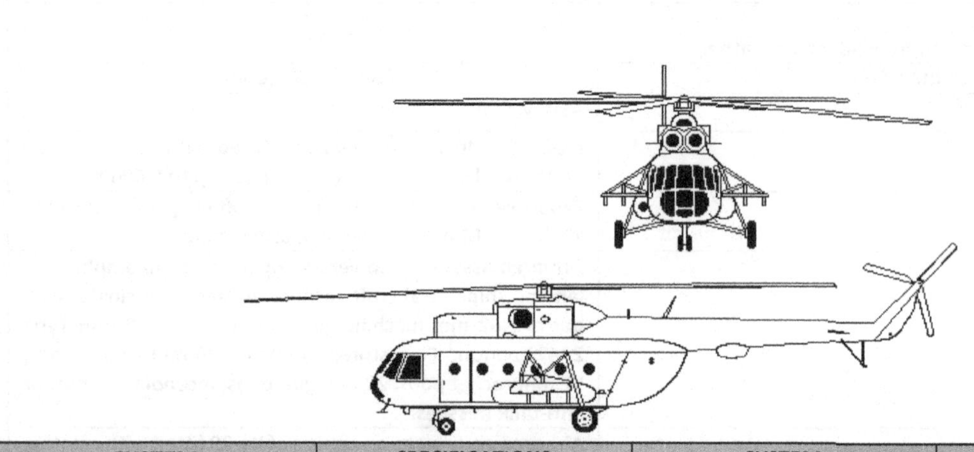

SYSTEM	SPECIFICATIONS	SYSTEM	SPECIFICATIONS
Alternative Designations:	Rana in India	Vertical Climb Rate (m/s):	9
Date of Introduction:	1967	Fuel (liters):	1,870 total, 3,700 max
Proliferation:	At least 54 countries	Internal:	445
Crew:	3 (2x pilots, 1x flight engineer) Transports: 24-26 troops (HIP-C, HIP-E)	Internal Aux Tank:	915 ea., up to 2
Blades – Main rotor:	5	Auxiliary Cabin Tank:	915 each, 1 or 2
Tail rotor:	3 right side, left on upgrades	Range (km):	
Engines:	2x 1,700-shp Isotov TV2-117A turboshaft. Upgrades use Mi-17 engines.	Maximum Load:	INA
Weight (kg):		Normal Load	690
Maximum Gross:	12,000	With Aux Fuel:	950
Normal Takeoff:	11,100	Dimensions (m):	Dimensions (m):
Empty:	6,990	Length (rotors turning):	25.4
Speed (km/h):		Length (fuselage):	18.2
Maximum (level):	250	Width:	2.5
Cruise:	240	Height:	5.6
Ceiling (m):		Main Rotor Diameter:	21.3
Service:	4,500	Tail Rotor Diameter:	3.9
Hover (out of ground effect):	850	Height	1.8
Hover (in ground effect):			

SURVIVABILITY/COUNTERMEASURES:
Can be fitted with armor. Main and tail rotor blades electrically deiced. Infrared jammer, chaff and flares. Armor on some variants.

ARMAMENT	
HIP C has four external hardpoints. HIP E -F have six; other variants have none. Weapons include fuselage/nose MGs, rockets, ATGMs, bombs, mines, and AAMs. Only a selected mix of munitions will fit. Mission dictates weapon configuration. Troops can fire their personal weapons from pintles and windows and doors. Assault versions may have fewer onboard troops to carry more ammunition. The K-29 dispenser can hold POM-2S or PTM-3 mines.	
AVIONICS/SENSOR/OPTICS	
Night/Weather Capabilities: The Mi-8 is equipped with instruments and avionics allowing operation in day, night, and is instrumented for bad weather conditions.	
VARIANTS	SPECIFICATIONS
The original civilian version produced at Kazan is called Mi-8. A civilian version produced at Ulan-Ude is called Mi-8T.	
Mi-8T/HIP C:	Initial fielded version for medium assault/transport, with 4 external hard points and noted engines and rotor. Probable assault armament mix is 7.62-mm MGs, 4x 57-mm or 2x 80-mm rocket pods.
Mi-8PS:	Military VIP transport variant of civilian HIP-C deluxe Mi-8 Salon.
Mi-8TVK/HIP E:	Assault or transport helicopter. Assault probable armament with 6x hard points: 12.7-mm nose turret MG, 4x AT-2 type ATGMs, and 2 x rocket pods or bombs.
Mi-8TV/HIP-F:	Export version uses AT-3 type ATGMs.
Mi-8SMV/HIP J:	Airborne electronic countermeasures (ECM) platform. R-949 jammer, and up to 32 dispensable jammers.
Mi-8PPA/HIP K:	Airborne IW comms intercept/jam platform characterized by 6x "X"-shaped antennas on the aft fuselage.
Mi-8VP/HIP D:	Comes in two variants. Mi-8VPK is an airborne communications platform with rectangular comms canisters mounted on weapons racks. Mi-8VzPU is an airborne reserve command post. Mi-9/HIP G: Airborne command relay post characterized by antennas, and Doppler radar on tailboom.
Mi-14/HAZE:	Naval HIP upgrade variant.
Mi-17/Mi-171/HIP H:	Mi-17/Mi-171/HIP H: Upgrade helicopters produced after 1977, with more powerful engines, left-side tail rotor, and a five blade rotor. Many Mi- 8 helicopters have been upgraded to the Mi-17/HIP-H standard.
NOTES	
MORE THAN 12,000 HIP HELICOPTERS HAVE BEEN PRODUCED. MISSIONS INCLUDE DIRECT AIR SUPPORT, TRANSPORT, RECONNAISSANCE, EW, MEDEVAC, SEARCH AND RESCUE, SMOKE GENERATING, AND MINELAYING. THERE ARE DOZENS OF VARIANTS AND A MORE THAN A DOZEN UPGRADES AND UPGRADE PACKAGES. INTERIOR SEATS ARE REMOVABLE FOR CARGO CARRYING. RESCUE HOIST CAN LIFT 150 KG. CARGO SLING SYSTEM CAPACITY IS 3,000 KG. THE MI-8 IS CAPABLE OF SINGLE-ENGINE FLIGHT IN THE EVENT OF LOSS OF POWER BY ONE ENGINE (DEPENDING ON AIRCRAFT MISSION WEIGHT) BECAUSE OF AN ENGINE LOAD SHARING SYSTEM.	

RUSSIAN MULTIROLE MI-17/HIP-H AND MI-171SH GUNSHIP

SYSTEM	SPECIFICATIONS	SYSTEM	SPECIFICATIONS
Alternative Designations:	Mi-8M for home use, Mi-17 for export. With Mil Plant design and Kazan, Ulan-Ude plant products, varied mission designs and upgrades, nomenclatures vary. Export nomenclatures vary from Russian military-use products.	Vertical Climb Rate (m/s):	9
Date of Introduction:	1977, 1981 as Mi-17	Fuel (liters):	1,870 total, 3,700 max
Proliferation:	At least 23 countries, with more than 5,000 in service worldwide.	Internal:	445
Crew:	3 (2x pilots, 1x flight engineer). Transports up to 26, 36 troops military seating, or 12 casualties.	Internal Aux Tank:	915 ea., up to 2
Blades – Main rotor:	5	Auxiliary Cabin Tank:	915 each, 1 or 2
Tail rotor:	3	Range (km):	
Engines:	2x 2,200-shp Isotov TV3-117VM	Maximum Load:	INA
Weight (kg):	13,000	Normal Load	Up to 580, 675 Mi-17-V5
Maximum Gross:	11,100	With Aux Fuel:	1,065
Normal Takeoff:	7100-7370 (variant dependent)	Dimensions (m):	See Mi-8/HIP-C
Empty:		CARGO COMPARTMENT DIMENSIONS (M):	
Maximum (level):	300	Width: 2.3, Height: 5.5 Others see Mi-8	
Cruise:	200	Standard Payload (kg):	
		Internal load: 4,000	
Service:	6,000	External sling: 4,000 (5,000 Mi-17-V5)	
Hover (out of ground effect):	1,670		
Hover (in ground effect):			

SURVIVABILITY/COUNTERMEASURES:
Armor plating (military versions), main and tail rotor blades electrically deiced. Infrared jammer, chaff and flares, exhaust diffusers. Missile warners include LIP. Shear-cutters. Like Mi-8 it has single-engine flight ability.

MOST PROBABLE ARMAMENT	
HIP H: Fitted with 1x 12.7mm MG or AG-17 30-mm AGL, aft 7.62-mm MG, 4x AT-2C/SWATTER and 40x 80-mm rockets.	
AVIONICS/SENSOR/OPTICS Night/Weather Capabilities: The Mi-17 is equipped with instruments, GPS nav, avionics, Doppler radar, autopilot for operation in day and night, map display screen, and instruments for meteorological conditions.	
VARIANTS	**SPECIFICATIONS**
Mi-17/HIP-H:	Original production HIP-H had 2x 1,950-shp Isotov TV3-117MT engines from Mi-14/HAZE, a new main rotor, and left-side tail rotor (distinguishing it from HIP-C). The reconfigured cab has rear clamshell doors. Many early HIP models are modified to the Mi-17 standard. Counterpart export and Russian-use variant weapons, sensors, and other features may differ to fit requirements
Mi-17T/Mi-8M:	Military variant added crew armor plating. The assault version has 1x 12.7mm MG or 30-mm AG-17 AGL, aft 7.62-mm MG, and 40x 80-mm rockets.
Mi-17P:	Descendent of the HIP K airborne jamming platform characterized by large rectangular antennas along aft fuselage.
Mi-17PG:	Variant with H/I-band pulse and continuous wave jamming system.
Mi-17PI:	Variant with D-band jammer, able to jam up to 8 sources simultaneously.
Mi-8MT:	Early "Hot and high" upgrade, with 2x 2,070-shp Klimov TV3-117VMA engines for greater rate of climb, higher hover ceiling Mi-19: Airborne CP on Mi-17 chassis. Mi-19R: Abn rocket artillery regiment CP. Many common versions now use 2,200-shp engines as noted at left. Kazan makes the Mi-17-1V export/Mi-8MTV multi-role, the Mi-17-V5/Mi-8MTV-5 multi-role (with APU and increased sling load), and Mi-172 passenger version. Ulan-Ude produces the Mi-171 export/Mi-8AMT multi-role, and the Mi-171Sh combat helicopter. Mi-171A is a civilian version.
Mi-17N/Mi-8MTO/Mi-8N:	Upgrade night assault variant tested in Chechnya, with FLIR sights. It led to the helicopter noted below.
MI-171-SH/MI-8AMTSH TERMINATOR (RUS): Better armored 2001 gunship, with upgrades, e.g., 2x 2,200-shp engines. The FCS includes Raduga-Sh ATGM day sight from	

Mi-35M, FLIR night sight.

MOST PROBABLE ARMAMENT:
2 x 7.62-mm MGs, 8x AT-6c/AT-9 ATGMs, and 40 x 80-mm rockets. Frangible rod AT-9 missiles can be used for air-to-air combat. Also, AA-18S/SA-18S AAMs (SAMs) can be used. The ATGM pod can also launch AAMs. IR warner and flares. For export, they can fit other sensors and/or munitions.

Newest variant is the Mi-17-V7 multi-role from Kazan, with VK-2500 engines
rated at 2,500 shp. It can operate at high altitude, and offers 14,000 max take-off weight, 5,000 kg internal payload, and 6,000 kg max external sling load. Gunship has a laser designator for semi-active laser-homing munitions (bombs, 80/ 122-mm rockets or ATGMs).

Israeli Peak-17 gunship upgrade for India has FLIR/CCD day/night FCS, either Spike-ER (8 km) or LAHAT ATGM (13 km, below), and can launch Skylite UAVs.

NOTES
MISSION DICTATES WEAPONS CONFIGURATION. NOT ALL WILL BE EMPLOYED AT THE SAME TIME.

RUSSIAN TRANSPORT HELICOPTER MI-6/HOOK

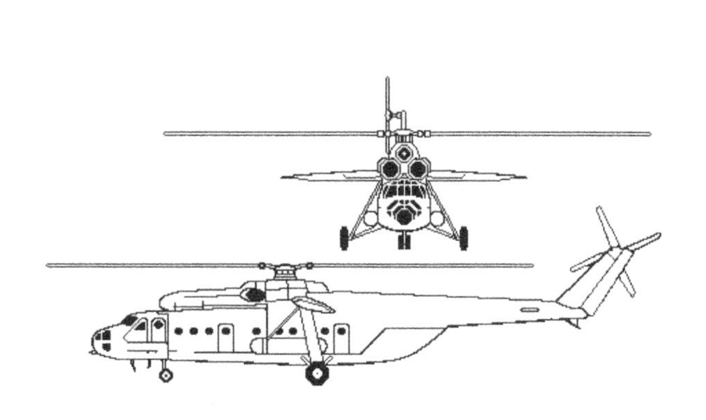

SYSTEM	SPECIFICATIONS	SYSTEM	SPECIFICATIONS
Alternative Designations:	INA	External Fuel Tank:	3,490
Date of Introduction:	1961	Range (km):	
Proliferation:	At least 15 countries	Max Load:	620
Crew:	5 (2 pilots, 1x navigator, 1x flt engineer, 1x radio operator)	With Aux Fuel:	1,000 km
Blades – Main rotor:	5	Dimensions (m):	
Tail rotor:	4	Length (rotors turning):	41.7
Engines:	2x 5,500-shp Soloviev D-25V (TV-2BM) turboshaft	Length (fuselage):	33.2
Weight (kg):		Width (including wing):	15.3
Maximum Gross:	42,500-46,800	Height:	9.9
Normal Takeoff:	40,500	Main Rotor Diameter:	35.0
Empty:	27,240	Cargo Compartment Dimensions (m): Floor Length: 12 Width: 2.65 Height: Variable from 2.0 to 2.5 Standard Payload: Internal: 12,000 kg with rolling takeoff External: 8,000 kg at hover Transports over 65 troops, or 41 litters, or 1x BRDM-2 scout car, or 1x BMD, or 1x GAZ truck, or 1x 7,500 liter POL truck or 12,000 liters in soft bladders.	
Maximum (level):	300		
Cruise:	250		
Ceiling (m):	4,500		
Internal:	6,315		
Internal Aux Tank:	INA		

AVIONICS/SENSOR/OPTICS

Night/Weather Capabilities:

The avionics and navigational package, and a fully functioning autopilot allow for day/night all-weather operation.

VARIANTS	SPECIFICATIONS
Mi-6A/-6T/HOOK A:	Basic civil and military transport version.
Mi-6VKP/HOOK B:	Airborne command post variant.
Mi-6VUS/HOOK C:	Developed airborne command post. Also known as Mi-22.
Mi-6AYaSh/HOOK D:	Airborne command post with possible side-looking airborne radar fairing.

Mi-6S:	MEDEVAC variant.
Mi-6TZ:	Tanker variant.
NOTES	
REMOVABLE STUB WINGS, WHEN INSTALLED, ARE FIXED AT A 15° INCIDENCE RELATIVE TO THE LONGITUDINAL AXIS. THEY PROVIDE 20% OF THE TOTAL LIFT IN FORWARD FLIGHT. AIRCRAFT PRODUCTION ENDED IN 1981. AIRCRAFT HAS HYDRAULICALLY ACTUATED REAR CLAMSHELL DOORS AND RAMP, PROVISIONS FOR INTERNAL CARGO TIE-DOWN RINGS, AN 800 KG CAPACITY INTERNAL WINCH SYSTEM IN CARGO COMPARTMENT, FLOOR CAPACITY IS 2,000 KG/M^2, AND A CENTRAL HATCH IN THE CABIN FLOOR FOR SLING LOADS	

RUSSIAN TRANSPORT HELICOPTER MI-26/HALO

SYSTEM	SPECIFICATIONS	SYSTEM	SPECIFICATIONS
Alternative Designations:	INA	Range (km):	
Date of Introduction:	1983	Max Load:	800
Proliferation:	At least 5 countries	With Aux Fuel:	1200
Crew:	5 (2x pilots, 1x navigator, 1x flt engineer, 1x loadmaster)	Dimensions (m):	
Blades – Main rotor:	8	Length (rotors turning):	40
Tail rotor:	5	Length (fuselage):	33.5
Engines:	2x 11,400-shp Lotarev D-136 turboshaft	Width:	8.2
Weight (kg):	56,000	Height	8.1
Maximum Gross:	49,500	Main Rotor Diameter:	32
Normal Takeoff:	28,240	Tail Rotor Diameter: 7.6	7.6
Empty:	28,240	Cargo Compartment Dimensions (m): Floor Length: 12 Width: 3.3 Height: variable from 2.9 to 3.2	
Maximum (level):	295		
Cruise:	255		
Ceiling (m):	4,500 Hover (out of ground effect): 1,800 Hover (in ground effect): 4,500	Standard Payload: Internal or external load: 20,000 kg Transports over 80 troops, 60 litters, or 2x BRDM-2 scout cars, or 2x BMDs, or 1x BMP or, 1x BTR-60/70/80 or, 1x MT-LB.	
Internal:	11,900		

AVIONICS/SENSOR/OPTICS

Night/Weather Capabilities:
The avionics and navigational package, Doppler weather radar, and a fully functioning autopilot allow for day/night all-weather operation.

VARIANTS	SPECIFICATIONS
Mi-26MS:	Medical evacuation version.
Mi-26T:	Freight transport.
Mi-26TZ:	Fuel tanker with an additional 14,040 liters of fuel in 4x internal tanks and 1,040 liters of lubricants, pumped through 4x 60-meter long refueling nozzles for refueling aircraft, and 10x 20-meter long hoses for refueling ground vehicles. Fuel transfer rate is 300 liters/minute for aviation fuel, and 75-150 liters/minute for diesel fuel. The refueling system can easily be removed to allow the aircraft to perform transport missions.

NOTES
THE HALO A HAS NO ARMAMENT. THE LOAD AND LIFT CAPABILITIES OF THE AIRCRAFT ARE COMPARABLE TO THE U.S. C-130 HERCULES TRANSPORT AIRCRAFT. THE LENGTH OF THE LANDING GEAR STRUTS CAN BE HYDRAULICALLY ADJUSTED TO FACILITATE LOADING THROUGH THE REAR DOORS. THE TAILSKID IS RETRACTABLE TO ALLOW UNRESTRICTED APPROACH TO THE REAR CLAMSHELL DOORS AND LOADING RAMP. THE CARGO COMPARTMENT HAS TWO ELECTRIC WINCHES (EACH WITH 2,500 KG CAPACITY) ON OVERHEAD RAILS CAN MOVE LOADS ALONG THE LENGTH OF THE CABIN. THE CABIN FLOOR HAS ROLLERS AND TIE-DOWN RINGS THROUGHOUT. THE HALO HAS A CLOSED-CIRCUIT TELEVISION SYSTEM TO OBSERVE POSITIONING OVER A SLING LOAD, AND LOAD OPERATIONS. THE MI-26 IS CAPABLE OF SINGLE-ENGINE FLIGHT IN THE EVENT OF LOSS OF POWER BY ONE ENGINE (DEPENDING ON AIRCRAFT MISSION WEIGHT) BECAUSE OF AN ENGINE LOAD SHARING SYSTEM. IF ONE ENGINE FAILS, THE OTHER ENGINE'S OUTPUT IS AUTOMATICALLY INCREASED TO ALLOW CONTINUED FLIGHT.

FRENCH HELIBORNE BATTLEFIELD SURVEILLANCE RADAR SYSTEM HORIZON

SYSTEM	SPECIFICATIONS
Alternative Designations:	Helicoptere d Observation Radar et d'Investigation sur zone
Date of Introduction:	1994
Proliferation:	At least one country
Crew:	4
Platform:	Mounted on AS-32UL/Cougar helicopter
Combat Weight (mt):	11.5
Antenna size (m):	3.5 x 5
Radio:	INA
RADAR	

Antenna:

Mount: Vertical post mount pointing downward from left rear. Radar stows under helicopter tail on take-off and landings, then lowers hydraulically during operation.
Antenna Type: Doppler, with MTI
Mode: Search
Scan Method: Antenna rotates horizontally for azimuth scan. Radar rotates 10°/sec, for a low pulse repetition frequency (PRF). Electronic for elevation.

SYSTEM

Transmitter:

Transmitter Type: Traveling Wave Tube fully coherent, agile frequency and adaptive burst mode.
Frequency band: I/J
RF maximum (GHz): 12.0
Power (kw): 50
Mode: Doppler MTI radar

Receiver and Processing Requirements:

Aircraft has onboard processing system. The processor is designed for a low false alarm rate. Ground station is mounted in a 7-mt truck. Each ground station holds 2 workstations. System receives 60° and 90° sector scans, independent of aircraft flight dynamics. Real-time digital data link can be integrated into French RITA communications net. Each moving target is automatically detected, located, analyzed, and classified. System can operate separately or as part of an intelligence network.

Protection and Electronic Counter-countermeasures:

Radar snapshot mode reduces vulnerability to anti-radiation missiles.

Very low side lobes reduce ECM effects.

The aircraft carries flares and decoys.

VARIANTS
System derived from the Orchidee system used in Desert Storm. Orchidee was compatible with the British Astor and US JSTARS systems.
PERFORMANCE
Surveillance range (km): 200 / 150 in rain clutter Surveillance rate: 20,000 km^2every 10 sec Target location accuracy (m): 40 Datalink range: 120 km, Agatha data link Surveillance targets: Wheeled or tracked vehicles, moving or hovering rotary wing aircraft, slow-flying FW aircraft, watercraft. Target speed (km/hr): 4-400, including nap-of-the-earth (NOE) Flight speed (km/hr): 130 Surveillance altitude (m): 2,000-4,000 Endurance (hrs): 4
NOTES
THE SYSTEM WAS DESIGNED TO OPERATE UNDER ARMY CONTROL AT DIVISION LEVEL. HORIZON SET CONSISTS OF 2 AIRCRAFT, ONE GROUND STATION, NAVIGATION EQUIPMENT, AND AGATHA DATA LINK.

ROTARY WING AIRCRAFT WEAPONS AND AIRCRAFT-DELIVERED MUNITIONS (ADM)

A wide variety of weapons and munitions can be employed on rotary-wing aircraft for use against aerial, ground, and waterborne targets. Weapons can be generally categorized as guns, launchers, and dispensers. Munitions are primarily rounds, rockets, missiles, bombs, grenades, mines, and torpedoes (see the tables below). However, new technologies continue to emerge, and are expanding the ability of aircraft to deliver lethality and execute other missions for and against military forces. Technology trends for more lethal air attack include abilities to: launch reconnaissance UAVs to support their missions in roles such as target selection and designation, launch attack UAVs, and add new weapons and munitions for long-range precision attack. The following weapons and munitions apply to RW systems in this chapter. Fixed-wing aircraft can use these munitions and a variety of heavier ones.

GUNS

Mount/Gun Name	Producing Country	Caliber or mm/Type	Barrels (if 2+)	Mount, Fixed or Turret/ Pod (Fixed)	# of Rounds/ Rds per Min	Munition Types (Other Than Ball-T, API-T, HEI-T)	Munition Range (m)/ Lethality (penetration-mm)
AA-52	France	7.62 MG *1	1	Pod	500+/900		1,200 heavy barrel
M134	U.S.	7.62 Mini-gun	6	M27or Mini-TAT turret, M18 pod	1500/2,000, 4,000		1,500 m
PKM	Russia	7.62 MG		Cabin, rear	Varies/250practical		1,000/ 8 at 500 m
PKT	Russia	7.62 MG		Nose fixed, rear, pod	3,800/250 practical		2,000/ 8 at 500 m
AN/M2	U.S./Others	.50-cal MG	1	Door pintle, or fixed, pod	/750-850	APFSDS-T, SLAP	1,800
NSV-T	Russia	12.7 MG	1	Door pintle or fixed, pod	/800	Incendiary, Duplex-T *2	2,000/20 at 500, 13.2 at 1,000
YakB-12.7	Russia	12.7 Gatling	4	USPU-24 chin turret GUV-8700 pod	1,470/4,500 750/4,500	Incendiary, Duplex-T *2	2,000
M197	U.S.	20 Gatling	3	Nose turret	/750		1,500
M 621	France	20 Cannon	1	THL-20 turret, pod, right side fixed *3	100+/650	APDS	1,500-2,000 m
9A669 GUV 9A624 9A622	Russia	23 Cannon 7.62 Mini-gun	2 4	Pod with 3 guns, the 23-mm, and 2 x 7.62 mini-guns	750/300 or 3,400 2200/	Frangible, APFSDS-T	2,500+/16 at 1,000 m for Frangible 2,000/ 8 at 500 m
GSh-23L Type 23-3	Russia China	23 Cannon	2	USPU-24 chin turret NPPU-24 right side *3 UPK-23-250 pod fixed *3	470/3,400 470/4,300 250/300 or 3,400	Frangible, APFSDS-T	
NS-23KM	Russia	23 Cannon	2	Right side fixed	550 practical	Frangible, Frag-HE, CC*4 APFSDS-T	2,500/19 @ 1000 m API-T
2A42	Russia	30 Cannon	1	NPPU-280 chin turret	460/250/200 or 600	Frangible, Frag-HE, CC*4 APFSDS-T	4,000/45 at 2,000 m for APFSDS-T
GSh-30K	Russia	30 Cannon	2	Right side fixed	250/varies to 2,600	Frangible, Frag-HE, CC*4 APFSDS-T	4,000/45 at 2,000 m for APFSDS-T

***1 Early versions of AA-52 were in 7.5 x 54 mm.**

***2 Duplex round has 2 cartridges, to double fire saturation in the beaten zone.**

***3 Gun (on fuselage or in a pod) has a fixed base mount, but can flex in elevation. An example is the UPK-23-250 flexible gun pod, which can depress guns to 30 degrees.**

***4 CC is a 30mm canister round with 28 sub-projectiles for use against soft targets and personnel with increased fire saturation in the beaten zone.**

AERIAL ROCKETS

Name	Producing Country	Caliber (mm)	Guidance No/Yes	Pod Name (# per pod)	Munition Nomenclature	Lethal Munition Type	Munition Range (m)/ Lethality (penetration-mm)	Comments
SNIA	France	50	No	/28				
S-5	Russia Others	57	No/SAL-H	UB-9 UB-16-57 UB-32	S-5K, KO, KP, KPB S-5, S-5M, S-5OM S-5Cor	HEAT-Frag, Frag-HE Frag-HE HEAT SAL-H	2,000/200 4,000 4,500 7,000/200	SAL-H: Semi-active Laser-Homing, on aircraft equipped with a laser target designator.
SNEB	France	68	No/SAL-H	Heli TDA 68-12C/12 Heli TDA 68-22C/22	Type 253 Type 26P Type 24, 26	HEAT-MP Frag-HE APERS	1,600/INA 1,600	There are reports of SAL-H capability - see above
S-8	Russia Others	80	No/SAL-H	B-8V7/7 B-8V20A/20 B-8M1/20	S-8KOM S-8T S-8DM S-8BM S-8ASM S-8Cor	HEAT-Frag Tandem HEAT Frag HE APHE Flechette HEAT SAL-H	4,000/400 antitank 4,000/600+ antitank 4,000/HE fuel-air 2,200/2 m concrete + HE INA 8,000/ 400	SAL-H see above. Other assets, such as aircraft or ground forces with LTD can laze rockets to target. S-8PM with jammer
Hydra-70/ 2.75 inch rkt	U.S. Others	70	No	M260/7, M261/19	M151 and M229 M261 M255A1	HE HE-MPSM Flechette	8.8/M151 10-lb Warhead, M229 17-lb 7,000/9 DP submunitions	MPSM is multipurpose, programmable time fuze. SAL-H in R&D.
S-13	Russia Others	122	No/SAL-H	B-13R/5 B-13L/5	S-13 S-13-OF S-13DF S-13T S-13Cor	HEAT Frag-HE HE thermobaric APHE HEAT SAL-H	4,000/3 m soil, 1 m concrete +HE 3,000/Frag-HE 6,000/equal to 40 kg of TNT 4,000/6 m soil, 1 m concrete + HE 9,000/700	SAL-H see above
S-24B	Russia	240	No/SAL-H Inertial	/1	V-24APD RV-24 S-24BMZ	Frag-HE PD fuze Frag-HE prox fuze Frag-HE	2,000/23.5 kg warhead	SAL-H option see above Fuze conversion kit with fins
S-25	Russia	340	No/SAL-H	O-25/1	S-25-OFME S-25L S-25LD	Frag-HE prox fuze HE SAL-H HE SAL-H	2-4,000/190 kg warhead 7,000/150 kg HE warhead 10,000/150 kg HE, 8 m CEP	SAL-H see above S-25LD can also use TV or IR-homing
Type 90-1	China	90	No	/7	Type 90-1	Frag-HE	7,000	Chinese

* Aerial rockets are also referred to as air-to-surface rockets (ASRs), or as fin-folding aerial rockets (FFARs).

ANTITANK GUIDED MISSILES (ATGMS)

Name	Producing Country	Rate of Fire (#/min, based on range)	Guidance	#/Pod	Munition Nomenclature (If different)	Munition Type	Munition Range (m)/ Penetration (mm)	Comments
AT-2c	Russia	3-4	RF SACLOS	2		HEAT, HE	4,000/650	
AT-3c and AT-3e	Russia Others	2-3	Wire SACLOS	1 or 3	AT-3c, AT-3E	HEAT (comments), HE	3,000/520, 800 AT-3e	AT-3e has Tandem HEAT. Other Countries make copies/variants.
AT-6/Shturm-V	Russia	3-4	RF SACLOS	4 *1,2		HEAT, HE	5,000/650	
AT-6b/Shturm-V1	Russia	3-4	RF SACLOS	4 *1,2		Tandem HEAT, HE	6,000/1,000	
AT-6c/Shturm-V2	Russia	3-4	RF SACLOS	4 *1,2		Tandem HEAT, HE	7,000/1,000	
AT-9/Ataka	Russia	3-4	RF SACLOS	4 *1,2		Tandem HEAT, HE, AA frangible rod	6,000/1,100	Expected upgrades include 8-km range, IR/radar homing. See *1.
Krizantema/ AT-15	Russia	4-6	RF ACLOS/LBR	4		Tandem HEAT	6,000/1,250+ERA (1,500+)	2 simultaneous, separate targets
AT-16/Vikhr-M	Russia	2-3	Laser-beam rider	8 *2, 3		Tandem HEAT/HE *2	10,000 /1,200 *3	Proximity fuze on/off per target.
Hellfire	U.S./UK	2-3	SAL-H *5	4 *3	Hellfire, Hellfire II	Tandem HEAT + HE *2	Hellfire II 8000/1300+ equiv	
Hermes-A	Russia	2-2	Inertial/RF/SAL-H *5	6		Tandem HEAT + HE *2	18,000/1300+ equiv	28 kg warhead, 40 km version due
HOT	Europe	3-4	Wire SACLOS	2, 3, 4	HOT-2, HOT-3	Tandem HEAT	HOT 3 4000/1250+	
LAHAT	Israel	2-4 *4	SAL-H *5	4		Tandem HEAT	13,000/1,000+ Dive attack	
Mokopa	South Africa	2-4 *4	SAL-H *5			Tandem HEAT	10,000/1,350+	Variant of Hellfire
Red Arrow-8F	China	3-4	Wire SACLOS	2 or 4		Tandem HEAT	4,000/1,100	

Spike-ER	Israel	2-3	Fiber-Optic *5 and IIR homing	2 or 4		Tandem HEAT	8,000/1,000+ Dive attack	AKA: NTD, Dandy. ER stands for Extended Range
TOW/BGM-71	U.S./Others	3-4	Wire SACLOS	2 or 4	TOW-2	Tandem HEAT	TOW 2 3750/900+	2-missile pod on MD-500. Other countries make copies/variants.

*1. AT-6 and variants, and AT-9 and variants, are interchangeable in launchers for each other.

*2. Launcher pods can also launch AA-16, AA-18, or AA-18S air-to-air missiles, decreasing the number of ATGMs in the pod for a given mission.

*3. AT-16 and Hellfire II have combined HEAT and HE warheads for multi-role use. The AT-16 also has proximity fuse that can be engaged in-flight for aircraft and materiel targets.

*4. With semi-active laser homing (SAL-H) guidance, launcher craft can hand off missile control to another designator, and launch other missiles without delays from missile flight time.

*5. Guidance modes such as SAL-H and fiber-optic can be categorized as non-line-of-sight, whereby the launcher craft can be outside of view of the target, and can avoid return fires.

6. For additional information on antitank and anti-armor missiles, see Vol 1 Chapter 6.

AIR-TO-AIR MISSILES (AAMS)

Name	Producing Country	Also SAM or ATGM *1	Guidance	Pod Name (# per pod)	Munition Type	Munition Range (km)/Warhead (kg)	Comments
AA-2C or D/ATOLL/R-13M	Russia		IR-homing	/1, 2	Frag-HE	8/7.4	AIM-9L upgrade phasing out
AA-8/APHID/R-60M	Russia		IR-homing	/1	HE Continuous rod prox	8 low altitude/3.5	Upgrade missile with DU rod
AA-11/ARCHER/R-73 RMD1	Russia		IR-homing	/1	HE Continuous rod prox	30/7.4	
AA-11/ARCHER/R-73 RMD2	Russia		IR-homing	/1	HE Continuous rod prox	40/7.4	
SA-7b/Strela-2M	Russia/Others	MANPADS SAM	IR-homing	/1	Frag-HE	5/1.15	
SA-14/Strela-3	Russia/Others	MANPADS SAM	IR-homing	/1, 2, 4	Frag-HE	6/1.0	
SA-16/Igla-1	Russia/Others	MANPADS SAM	IR-homing	/1, 2, 4	Frag-HE	5.2+/1.27	
SA-18/Igla	Russia/Others	MANPADS SAM	IR-homing	/1, 2, 4	Frag-HE	6/1.27	
SA-24 (SA-18S)	Russia	MANPADS SAM	IR-homing	/1, 2, 4	Continuous rod, prox fuze	6+/2.5	Aka: Igla-S/Igla-Super
AIM-9L/Sidewinder	U.S./Others	Veh/towed SAM	IR-homing		Frag-HE	17.7/9.5	
AT-6c and AT-9/Ataka	Russia	Veh ATGM	RFSACLOS	/4, 8 *1 *2	Tandem HEAT	7/7.4, 6/7.4 Ataka	Penetration 1,000-1,100 mm
Ataka 9A2200 Missile	Russia	Veh ATGM	RFSACLOS	4, 8 *1 *2	Continuous rod, prox fuze	6/	Also fit AT-6 launchers
AT-16/Vikhr-M	Russia	RW ATGM	Laser-beam rider	/8 *1 *2	HEAT/HE with prox on/off	10,000 /INA	Penetration 1,300+ mm
Mistral 2	France	Veh/pedestal SAM	IR-homing	ATAM/1, 2	Frag-HE, prox	6/3	On Gazelle
Spike-ER	Israel	Veh/man-port ATGM	FOG_M, IIR-homing	/4 *1 *2	Tandem HEAT	8.0/INA	Penetration 1,000+ mm
Starstreak	UK	AD/AT or multi-role	Laser-beam rider	ATAS/4 *1	3 x Sabots with Frag-HE	7/.9 kg per submissile	3 x high-velocity submissiles
Stinger	U.S./Others	Veh/MANPADS SAM	IR-homing	ATAS/4, 2	HE	4.5+/1.0	
TY-90/Yitian	China	Veh-launch SAM	IR-homing	/2, 1	HE, frangible rod	6/3	Too large for MANPADS use

*1. All ATGMs can be used to engage helicopters hovering or flying low and slow, esp. nap-of-the-earth mode (35 km/hr or less). These ATGMs can engage RW aircraft at all times.

2. ATGM launcher can substitute 1 or more SAMs.

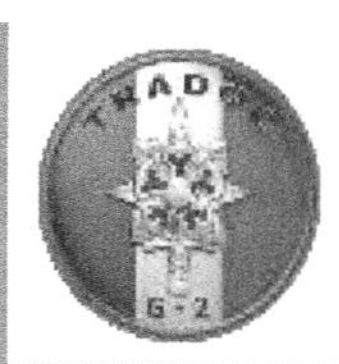

AIR-TO-SURFACE MISSILES (ASMS)

Name	Producing Country	Mission	Guidance	#/Pod	Warhead Type	Munition Range (km)/ Penetration (mm)	Comments
AS-10/KAREN/Kh-25ML **Kh-25-MR** **Kh-25-MT** **Kh-25MTP**	Russia	Tactical Tactical, AT Tactical, AT Tactical, AT	SAL-H RF-Guided TV-Guided Thermal-Guided	1	Frag-HE/90 kg Frag-HE/90 kg Frag-HE/90 kg Frag-HE/90 kg	20/ 10/ 20/ 20/	
AS-12/KEGLER/Kh-25MP	Russia	Anti-radar	Passive-homing	1	90 kg	40/	
AS-12/AS.12	France	Tactical, AT, Anti-ship	Wire SACLOS	2	SAPHE, 28 kg	7/	
AS-17/KRYPTON/Kh-31P	Russia	Anti-radar	Passive homing	1	90 kg	100/	
AS-17/KRYPTON/Kh-31A	Russia	Anti-ship	Active radar	1	90 kg	50/	
C-701	China	Anti-ship, land attack	TV, IR-homing	4	SAPHE, 29 kg	20/	MMW-homing tested
Hermes-A	Russia	Tactical, AT	Inertial/RF/SAL-H	6-8	Frag-HE, 28 kg	40/1300+	100 km version due
Sea Skua	UK	Anti-ship	Semi-active Radar	1	SAPHE, 28 kg	25/	
Guided Rockets see pg 2-23	Russia		SAL-H				

* Systems designed for use with laser guidance are generally called missiles. However, some rockets can be adapted with SAL-H modifications for near-ASM range and precision.

BOMBS[1]							
Name	**Weight (kg)**	**Guidance (if any)**	**Type**	**Nomenclature Specific Bomb**	**Warhead or Submunition/# if more than 1/Nomenclature/Type**	**Munition Range (m)/ Lethality (penetration-mm)**	**Comments**
GBU-100	120		ASW Depth Bomb		HE 100 kg		
SZV	94	Underwater Acoustic	ASW Depth Bomb		HEAT 19 kg	600 m in depth	Steers on glide fins
FAB-100	117		General Purpose	M80	HE 39 kg		
OFAB-100	100		Blast-Frag		Frag-HE 60 kg		
FAB-250	250		General Purpose	M79	HE 105 kg	30 radius	
OFAB-250	250		Blast-Frag		Frag-HE 210 kg		
RBK-250 Glide bomb (Dispenser)	273		Cluster Cluster	RBK-250-275AO- RBK-250AD-1	150 AO-1sch bomblets /60 AO-2.5 RT AP bomblets /30 PTAB-2.5KO HEAT bomblets Chemical bomblets	4,800 m^2 destructive area	Like MK-118
ZAB-250	250		Incendiary		200 kg Napalm		
KhB-250			Chemical		200 kg Sarin, VX, mustard, etc		
FAB-500/M62	500		General Purpose		HE 450 kg		
OFAB-500	515		General Purpose		Frag-HE 155 kg		
OFZAB-500	500		General Purpose		Frag-HE Incendiary 250 kg		
ODAB-500PM	520		Fuel-Air Explosive		193 kg		
KAB-500Kr	560	TV guided	Precision Attack		Concrete-piercing 380 kg, 200 kg chg	1500 m^2 destructive area	
KAB-500L	534	SAL-H	Precision Attack		HE 400 kg with 195 kg of charge	1500 m^2 destructive area	
RBK-500U Glide bomb (Dispenser)	504 500 520 427 334 525 525 525 500 467		Cluster	RBK-500AO OAB-2.5RT PTAB PTAB-1M ShOAB-0.5 BETAB-500ShP OFAB ZAB PPM SPBE-D	108/ AO-2.5 APAM ICM/bomblets 126/ 5RTM APAM 352/ PTAB HEAT bomblets 60/ PTAB-2.5KO HEAT bomblets 268/ PTAB-1M HEAT bomblets 565/ 0.5 ShOAB-0.5 AP bomblets 10/ BETAB-M concrete piercing bomblets 10/ OFAB APAM bomblets 168/ ZAB incendiary bomblets 48/ PPM mines 15 IR sensor-fuzed 14.5 kg bomblets Chemical bomblets	6,400 m^2 destructive area 210 m^2 destructive area 210 mm penetration top-atk 300 m x 400 m/210 mm top atk 300 m x 400 m Runway penetrators EFP top-attack[2]	Improvement over the RBK-500
ZAB-500	500		Incendiary		480 kg Napalm		

1. Only Russian RW aircraft in this chapter employ bombs. Thus, all bombs listed are Russian.
2. EFP - Explosively-formed penetrator

Other ordnance includes sub munition and mine dispensers, minelayer ramps, automatic grenade launchers, anti-ship torpedoes, anti-submarine mines, and torpedoes. Selected RW aircraft can launch UAVs; therefore a near-term capability will be ability to launch attack UAVs or UCAVs and guide them to engage targets.

Worldwide Equipment Guide

Chapter 2: Fixed Wing Aircraft

Distribution Statement: Approved for public release; distribution is unlimited.

Chapter 3: Fixed Wing Aircraft

This chapter provides the basic characteristics of selected fixed-wing aircraft readily available to COE OPFOR across the spectrum of joint operations. This sampling of systems was selected because of wide proliferation across numerous countries or because of already extensive use in training scenarios. Additional data sheets addressing other widely proliferated aircraft will be sent with further supplements to this guide. Many foreign militaries are leveraging advances in automated technologies in order to use increasing amounts of data across all warfighting capabilities. Increases in processing power and broadband technologies through commercial research and development make real time situational awareness and communications on the move a tangible objective for many foreign forces. The emergence of rudimentary Integrated Battlefield Management Systems (IBMS) in tier three forces represents this global trend. Net Centric operations are viewed worldwide as a key element of modern military operations, an IBMS is a system that integrates multiple command and control formats as well as sensor data into one display that improves situational awareness through multiple sources.

Because of the increasingly large numbers of variants of each aircraft, only the most common variants produced in significant numbers were addressed. If older versions of airplanes have been upgraded in significant quantities to the standards of newer variants, the older versions were not addressed.

Fixed-Wing Aircraft generally covers the systems that will affect the planning and actions of the ground maneuver force, aircraft commonly employed by the OPFOR when in close proximity to enemy ground forces, as well as strategic aircraft. This chapter classifies aircraft as fighter/interceptor, strike, ground-attack, multi-role, bombers, special-role, and transport aircraft. Multi-role aircraft are able to support missions across each of the categories. This chapter encompasses many aircraft which may have a dual civil/military application. It does not include, however, aircraft designed and used primarily for civil aviation.

The munitions available to each aircraft are mentioned, but not all may be employed at the same time. The weapon systems inherent to the airframe are listed under armament. The most probable weapon loading options are also given, but assigned mission dictates actual weapon configuration. Therefore, any combination of the available munitions may be encountered.

A wide variety of upgrade programs are underway. The FW aircraft variants noted are only a small representation of those available. For instance, application of GPS and commercial GPS map display units permits even the oldest aircraft to have precision location. Night vision systems coupled with the high level of night illumination existing in most areas of the world permit night use of older aircraft. Even though some weapons require linked effective night sights, many weapons, such as bombs (including sensor-fuzed), standoff GPS programmed cruise missiles, and munitions using remote guidance (such as semi-active laser-homing munitions guided by laser target designators) permit older aircraft to launch the munitions and rely on others to guide them to target. Other aerial systems can substitute for FW aircraft to execute what were FW missions. These include rotary-wing aircraft, unmanned aerial vehicles (including attack UAVs and UCAVs), improvised systems such as airships, and cruise missiles.

Many data sheets for joint systems were provided by Mr. Charlie Childress of JFCOM.

Questions and comments on data listed in this chapter should be addressed to:

Mr. John Cantin

DSN: 552-7952 Commercial (913) 684-7952

E-mail address: john.m.cantin.ctr@mail.mil

FIGHTER AIRCRAFT F-5 FREEDOM FIGHTER/TIGER

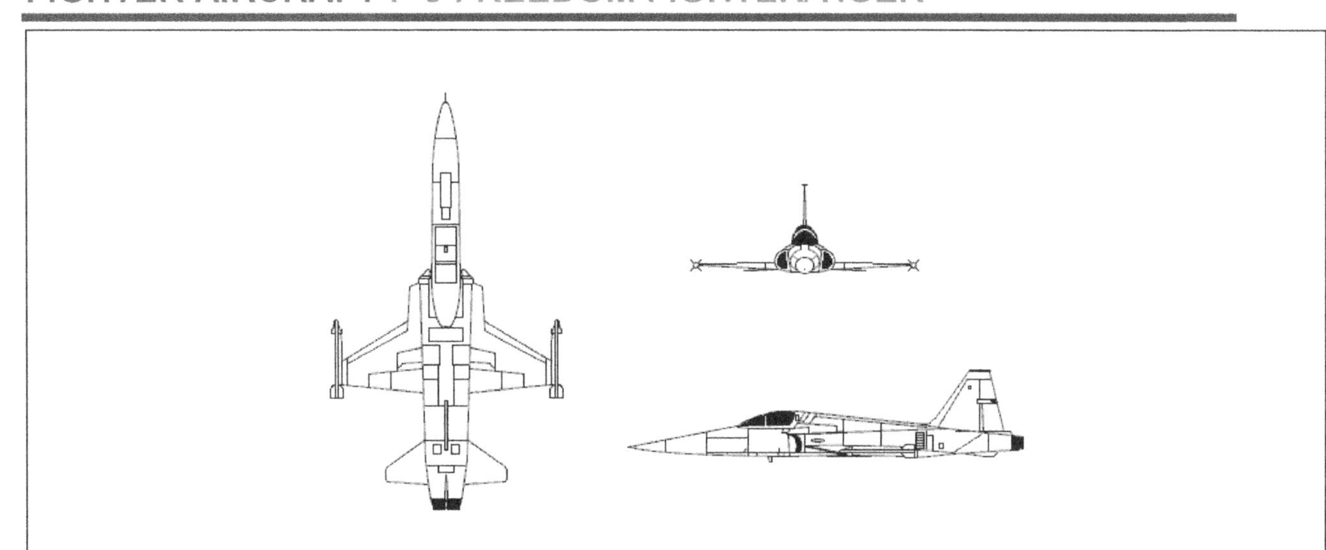

SYSTEM	SPECIFICATIONS	SYSTEM	SPECIFICATIONS
Weapon & Ammunition Types		Vertical Climb Rate (m/s):	146
2 x M239A2 20-mm Qty:	280	Vertical Climb Rate (m/s) (A/E):	175
Other Loading Options		Fuel (liters):	
AAMs:		Internal:	2,207
AIM-9 Sidewinders on wingtip launchers	2	Internal (A/E):	2,555
Pylons:		Range (km):	
Fuselage:	1	Ferry :	2,519
Underwing:	4	Ferry (A/E):	2,861
Max weapons:(kg):		Dimensions (m):	
F-5A	2,812	Length:	14.4
F-5E	3,175	Length (A/E):	14.6
900 kg	1	Wingspan:	7.7
227 kg	9	Wingspan (A/E):	8.1
AGM-65 submunitions dispensers rocket pods	1	Height:	4.1
GPU-5 30-mm gun pods	3	Standard Payload (kg):	
568-L or 1,041-L drop tanks	3	External (A):	2,812
Alternative Designations:	F-5A initial	External (E):	3,175

Date of Introduction:	1964	Hardpoints:	
Number of Countries Proliferated:	≥30	Centerline:	1
Description:		Wing Pylons:	4
Crew:		Survivability/Countermeasures:	
F-5A	1	Martin-Baker Mk10 F-5Es ejection seats:	Yes
F-5B/F	2	ECM systems:	Option
Engines:		RWR:	Option
5,000 lbs. thrust General Electric J85-21A turbojets w/afterburner (F-5E)	2	Chaff and Flare:	Option
Weight (kg):		ARMAMENT:	
Empty:	3,667	M239A2 20-mm cannon:	2
Empty (A/E):	4,410	AVIONICS/SENSOR/OPTICS	
Max Takeoff:	9,333	F-5A radar gun sight	Yes
Max Takeoff (A/E):	11,214	Pulse Doppler Radar (F-5E):	Yes
Speed (km/h):		Communications and Navigation (F-5E):	Yes
Maximum (at altitude):	1,489	Lead-Computing Optical Sight:	Yes
Maximum (at altitude) (A/E):	1,733	Central Air Data Compute:	Yes
Cruise:	904	Attitude and Heading reference system:	Yes
Max "G" Force (g):	INA	FLIR:	Yes
Ceiling (m):	15,789	Night/Weather Capabilities:	Yes

NOTES

THE F-5 IS A LIGHTWEIGHT, EASY-TO-FLY, SIMPLE-TO-MAINTAIN, AND RELATIVELY CHEAP SUPERSONIC FIGHTER. IT WAS ORIGINALLY OFFERED AS A CANDIDATE FOR THE U.S. LIGHTWEIGHT FIGHTER, BUT FOUND VIRTUALLY ALL ITS MARKET OVERSEAS.

APPEARANCE: WINGS: SMALL, THIN MOUNTED LOW ON THE FUSELAGE WELL AFT OF THE COCKPIT

ENGINES: TWO TURBOJETS ARE BURIED SIDE-BY-SIDE IN THE AFT FUSELAGE

FUSELAGE: LONG POINTED NOSE THAT SLOPES UP TO THE CANOPY, BEHIND THE CANOPY, A THICK DORSAL SPINE SLOPES DOWN TO THE TAIL

TAIL: DOUBLE-TAPER FIN HAS TO-SECTION INSET RUDDER. CROPPED DELTA TAIL PLANES ARE MOUNTED AT THE BOTTOM OF THE FUSELAGE IN LINE WITH THE FIN.

VARIANTS

F-5B FREEDOM FIGHTER: TWO-SEAT VERSION. FIRST EXPORT PRODUCTION VARIANT FLEW IN MAY 1964.

CF-5A/D: CANADIAN-BUILT VARIANT. POWERED BY 4,300 LBS. THRUST J85-CAN-15 TURBOJETS. CF-5AS ARE SINGLE SEAT FIGHTERS, AND CF-5DS ARE TWO-SEATERS.

NF-5A/D: CANADAIR BUILT AIRCRAFT FOR NETHERLANDS WITH MODIFIED WING INCLUDING LEADING-EDGE MANEUVERING SLATS AND LARGER DROP TANKS.

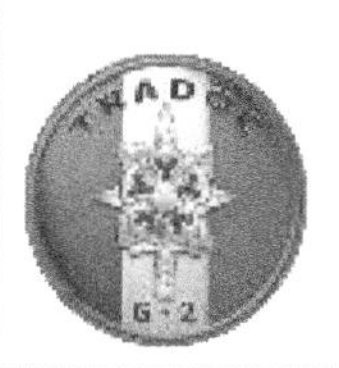

NORWEGIAN F-5A/B UPGRADE: PERFORMED ON 30 AIRCRAFT (17 A, 13 B). AS WERE FITTED WITH ALE-40 CHAFF/FLARE DISPENSERS. BS RECEIVED ALR-46 RWR, ALE-38 CHAFF/FLARE DISPENSERS, NEW RADIO, TACAN, IFF, AND LIS-600D ALTITUDE AND HEADING REFERENCE SYSTEM (AHRS).

F-5E TIGER II: SECOND GENERATION F-5 FIGHTER VERSION THAT REPLACED F-5A/B IN PRODUCTION.

F-5F: TWO-SEAT TRAINER RETAINS ONE CANNON WITH 140 ROUNDS, WEAPONS PYLONS, TIP RAILS; CAN BE FITTED WITH AVQ-27 LASER TARGET DESIGNATOR.

CHEGOONG-HO (AIR MASTER): SOUTH KOREAN NAME FOR F-5ES AND F-5FS ASSEMBLED BY KOREAN AIR.

CHUNG CHENG: TAIWANESE NAME GIVEN TO F-5ES AND F-5FS ASSEMBLED BY AIDC IN TAIWAN.

RF-5E TIGEREYE: PHOTO-RECONNAISSANCE VERSION WITH MODIFIED NOSE THAT ACCEPTS A VARIETY OF CAMERA-CARRYING PALLETS AND MOUNTING AN OBLIQUE FRAME CAMERA.

BRITISH/FRENCH LIGHT ATTACK AIRCRAFT JAGUAR

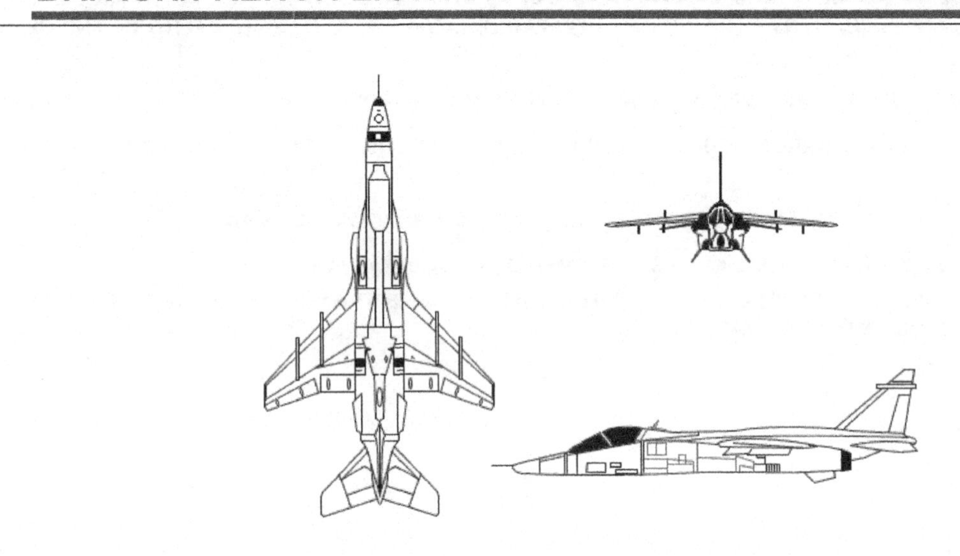

SYSTEM	SPECIFICATIONS	SYSTEM	SPECIFICATIONS
Weapon & Ammunition Types		Range (km):	
Two 30 mm Aden or DEFA 533 guns:	150	Combat Radius (km):	
Other Loading Options		Internal Fuel:	I537 - 852
400 kg or 445 kg:	8	External Fuel:	917 - 1,408
227 kg or 250 kg:	11	Dimensions (m):	
113 kg or 125 kg:	15	Length:	16.9
Rocket pods:	4-6	Wingspan:	8.7
Munitions dispensers:			
ECM pods:	4	Height:	4.9
Fuel drop tanks:	3	Standard Payload (kg):	
ATLIS laser designating pod (French):	1	Hardpoints:	5
Missiles:		Centerline:	1
AIM-9 Sidewinder/Matra/Magic R55:	2	Wing:	4
AS30L AGM:	2	Survivability/Countermeasures:	
Engines 8,040 lbs. thrust Rolls-Royce Turbomeca Adour Mk 104/804 turbofan with afterburner:	2	Martin-Baker zero/zero ejection seats:	Yes

Weight (kg):		ECM systems:	Yes
Maximum Gross:	15,700	Night Vision Goggles:	Yes
Normal Takeoff:	10,954	Bulletproof windscreen:	Yes
Empty:	7,000	ARMAMENT:	
Speed (km/h):		30 mm Aden or DEFA 533 guns:	2
Maximum (at altitude):	1,699, Mach 1.6	AVIONICS/SENSOR/OPTICS	
Maximum (sea level):	1,350, Mach 1.1	DARIN (display attack and ranging inertial navigation):	Yes
Landing Speed:	213	Nav/attack system:	Yes
Max "G" Force (g):	+8.6 g	ADF:	Yes
Ceiling (m):	14,000	Radar altimeter:	Yes
Vertical Climb Rate (m/s):	72	Central Air Data Compute:	Yes
Fuel (liters):	Fuel (liters):	Attitude and Heading reference system:	Yes
Internal:4,200	Internal:4,200	HUDWAC (head-up display and weapon aiming computer):	Yes
External:3,600	External:3,600	Night/Weather Capabilities:	Yes

NOTES

PRODUCED TO MEET A JOINT ANGLO-FRENCH REQUIREMENT IN 1965 FOR A DUAL-ROLE ADVANCED/OPERATIONAL TRAINER AND TACTICAL SUPPORT AIRCRAFT, THE JAGUAR HAS BEEN TRANSFORMED INTO A POTENT FIGHTER-BOMBER. THE RAF ORIGINALLY INTENDED TO USE THE AIRCRAFT PURELY AS AN ADVANCED TRAINER, BUT THIS WAS LATER CHANGED TO THE OFFENSIVE SUPPORT ROLE ON COST GROUNDS

APPEARANCE: WINGS: SHORT-SPAN, SWEPT SHOULDER-MOUNTED, ENGINES: TWO TURBOFANS IN REAR FUSELAGE, FUSELAGE: LONG AND SLEEK WITH LONG, POINTED, CHISELED NOSE, WIDENED AT AIR INTAKES.

NIGHT/WEATHER CAPABILITIES:

DAY/VFR MEDIUM AND LOW-LEVEL GROUND ATTACK/ RECONNAISSANCE AIRCRAFT. THE NIGHT VISION GOGGLES PROGRAM WILL ALLOW LIMITED NIGHT CAPABILITY.

VARIANTS

JAGUAR S/JAGUAR GR1: SINGLE-SEAT ATTACK VERSION DESIGNATED GR1 IN BRITISH SERVICE. FIRST EQUIPPED WITH ADOUR MK 102 ENGINES DEVELOPING 7,305 LBS. THRUST WITH AFTERBURNER.

JAGUAR A: FRENCH AIRCRAFT WITH TWIN-GYRO PLATFORM AND DOPPLER NAVIGATION, WEAPON-AIMING COMPUTER, MISSILE FIRE CONTROL FOR ANTI-RADAR MISSILE, FIRE CONTROL SIGHTING UNIT, AND LASER RANGER AND DESIGNATOR POD.

JAGUAR T2/JAGUAR E: TWIN-SEAT COMBAT-CAPABLE TRAINER VERSION: 35 AIRCRAFT DESIGNATED T2 IN BRITISH SERVICE AND E IN FRANCE AND OTHER NATIONS.JAGUAR INTERNATIONAL: EXPORT VARIANT, OFTEN WITH MORE EXTENSIVE AVIONICS FITS THAN BRITISH OR FRENCH AIRCRAFT.

SHAMSHER: JAGUAR INTERNATIONAL VARIANT SELECTED BY INDIA OVER THE MIRAGE F1 AND THE SAAB AJ37 VIGGEN AS THE DEEP PENETRATION STRIKE AIRCRAFT (DSPA).

CHINESE FIGHTER AIRCRAFT J-6 (JIAN-6)/F-6

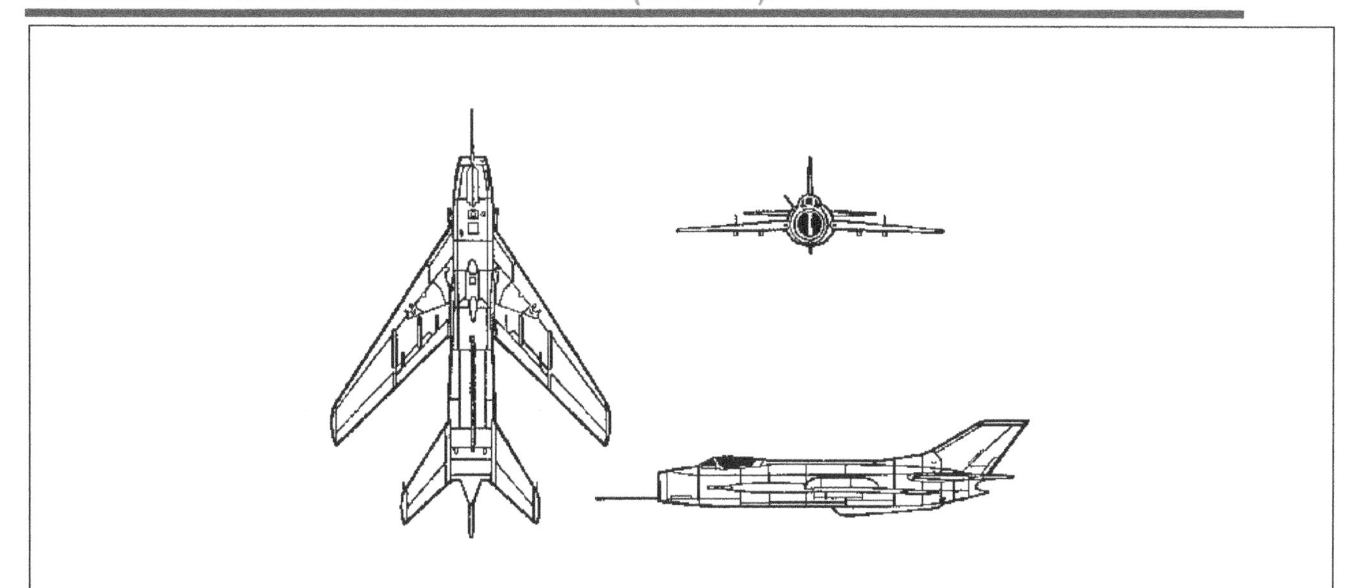

SYSTEM	SPECIFICATIONS	SYSTEM	SPECIFICATIONS
30 mm guns		Vertical Climb Rate (m/s):	152+
250 kg Bombs, or	2	Fuel (liters):	
400-L drop tanks, or	2	Internal:	2,170
760-L drop tanks, or	2	External (2 drop tanks):	800 or 1,520
CAA-1B AAM	2	Range (km):	
Inboard Stations:		Normal:	1,390
8 x 57-mm rockets, or	4	With 2 x 760 L drop tanks:	2,200
16 x 57-mm rockets, or	4	Dimensions (m):	
7 x 90-mm rockets, or	4	Length:	
Gun pods, or	4	Fuselage:	12.6
Practice bomb	4	With Nose Probe:	14.9
Alternative Designations:	see variants	Wingspan:	9.2
Date of Introduction:	1962	Height:	3.9
Proliferation:	10 countries	Hardpoints:	
Description:		Underwing:	6
Crew:	1 (pilot)	Survivability/Countermeasures:	
5,732 lbs. thrust Shenyang Wopen-6 turbojets (7,165 lbs. thrust with afterburner)	2	Martin-Baker zero/zero ejection seats:	Yes

Weight (kg):		Cockpit is pressurized:	Yes
Takeoff:		Fluid anti-icing system for windscreen:	Yes
Clean:	7,545	Tail warning system:	Yes
Typical:		ARMAMENT:	
with 2 AAMs and 760-L drop tanks:	8,965		
Max:	10,000	30-mm automatic cannons:	3
Empty:	5,760	AVIONICS/SENSOR/OPTICS	
Speed (km/h):		Airborne interception radar:	Yes
Maximum Clean:		VHF transceiver:	Yes
(at 11,000 m):	1,540, Mach 1.45	Blind-flying equipment,:	Yes
(at low level):	1,340, Mach 1.09	Radio compass:	Yes
Cruise:	950	Radio altimeter:	Yes
Max "G" Force (g):	+8	Night/Weather Capabilities:	No
Ceiling (m):	19,870		

NOTES

THE F-6 (JIAN-6 FIGHTER AIRCRAFT) IS THE CHINESE VERSION OF THE MIG-19, WHICH WAS STILL IN PRODUCTION IN CHINA IN THE MID-1990S.

APPEARANCE: WINGS: SHARPLY SWEPT, MOUNTED AT MID-FUSELAGE. ENGINES: TWO SMALL TURBOJETS ARE FITTED SIDE-BY-SIDE IN THE AFT FUSELAGE. FUSELAGE: RELATIVELY LONG AND SLENDER, SWELLING AFT FOR THE ENGINES WITH ENGINE NOSE INTAKE THAT HAS A CENTRAL SPLITTER PLATE. TAIL: THE SHARPLY SWEPT FIN HAS A SMALL DORSAL FILLET AND NEARLY FULL HEIGHT RUDDER.

VARIANTS

J-6: EQUIVALENT OF THE MIG-19S/SF DAYTIME FIGHTER WITH 3 X 30 MM GUNS, ONE AT EACH WING ROOT AND ONE ON THE FUSELAGE.

J-6A: EQUIVALENT OF THE MIG-19PF ALL-WEATHER FIGHTER. ARMED WITH STANDARD J-6 GUNS AND ROCKETS.

J-6B: EQUIVALENT OF THE MIG-19PM ALL-WEATHER FIGHTER. ARMED WITH THE AA-1 ALKALI RADAR HOMING MISSILES, NO GUNS.

J-6C: SIMILAR TO THE J-6, BUT WITH BRAKE CHUTE HOUSED IN BULLET-FARING AT THE BASE OF TAILFIN. SAME GUNS AS THE J-6A. EXPORT VARIANT WITH MARTIN-BAKER EJECTION SEATS AND AIM-9 SIDEWINDER MISSILES.

J-6XIN: SIMILAR TO J-6A, BUT WITH NOSE-MOUNTED INTERCEPTION RADAR. SAME GUNS AS THE J-6A.

JJ-6: TRAINER VERSION WITH TANDEM TWO-SEAT COCKPIT. EXPORT VERSIONS ARE FT-6. ARMED WITH ONLY THE FUSELAGE GUN.

JZ-6: A TACTICAL PHOTO-RECONNAISSANCE VERSION, ARMED WITH WING ROOT GUNS ONLY.

F-6: EXPORT VERSIONS.

CHINESE FIGHTER AIRCRAFT J-7 (JIAN-7)/FISHBED

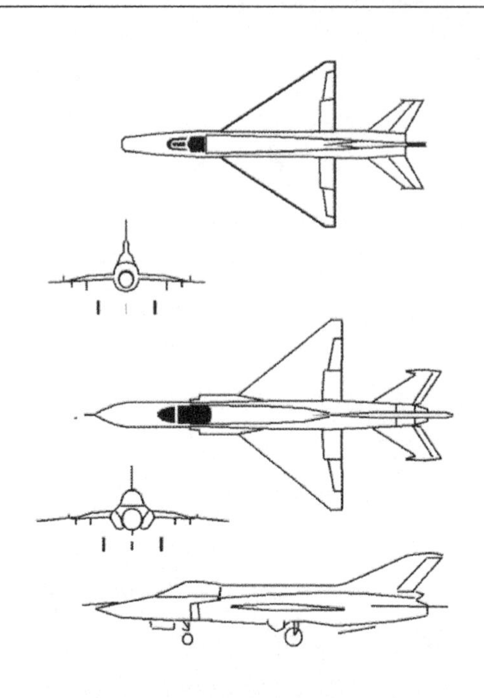

SYSTEM	SPECIFICATIONS	SYSTEM	SPECIFICATIONS
Weapon & Ammunition Types		Landing Speed:	310-330
23mm type 23- twin barrel (F-7M):	INA	Ceiling (m):	18,800
30mm Type 30-1 Cannons:	2	Vertical Climb Rate (m/s):	
AAMs:		(F-7B):	150
PL 5B (F-7M):	2 - 4	(F-7M):	180
PL-2/2A/5B/7 (J-711):	2	Fuel (liters):	
Rockets:		Internal:	2,385
12 round 57mm (F-7M):	4	Range (km):	
7 round 90mm (F-7M):	4	Low Alt:	370
18 round 57mm (J-7111)):	4	F-7B with 2 Pl-2 AAM:	
7 round 90mm (J-7111):	4	Internal fuel:	1,200
Bombs:		1 800-L drop tank:	1,490
100 kg (F-7M):	10	F-7M with 2 PL-7 AAM:	
250 kg (F-7M):	4	3 500-L drop tanks:	1,740
500 kg (F-7M)	2	Dimensions (m):	
50 / 150 kg (J-7111):	4	Length:	14.9

250 / 500 kg (J-7111):	2	Wingspan:	7.2
Fuel Tanks:		Height:	4.1
500 L (F-7M):	2	Standard Payload (kg):	1,800
800 L (F-7M):	1	Hardpoints:	
500 L (J-711):	3	Wing Pylons:	2
Alternative Designations:	F-7B, F-7M	Survivability/Countermeasures:	
Date of Introduction:	1965	Zero/130-850 km/h ejection seat:	Yes
Proliferation:	≥ 11	ECM systems:	
Crew:	1	Jammer:	Yes
Engines 9,700 lbs. thrust Wopen-7B turbofan, 13,500 lbs. thrust w afterburner:	1	ARMAMENT:	
Weight (kg):		M239A2 20-mm cannon:	2
Empty:		30-mm type 30-1 cannons with 60 rounds each in farings under front fuselage	2
(F-7B):	5,145	AVIONICS/SENSOR/OPTICS	
(F-7M):	5,275	Skyranger or Super Skyranger radar:	Yes
Max takeoff:		Heads-Up-Display and Weapons Aiming Computer	Yes
(F-7B):	7,372	ECM pod:	Yes
(F-7M):	7,531	Night/Weather Capabilities:	
Speed (km/h):		J-7111	Yes
Max:	2,175		

NOTES

THE SOVIETS LICENSED THE MANUFACTURE OF THE MIG-21F AND ITS ENGINE TO CHINA IN 1961, AND ASSEMBLY OF THE FIRST J-7 USING CHINESE-MADE COMPONENTS BEGAN EARLY1964. THE J-7 AIRCRAFT WAS THE MOST WIDELY PRODUCED CHINESE FIGHTER, REPLACING OLDER J-6 FIGHTERS, THE CHINESE VERSION OF THE MIG-19. IN 1995 IT WAS PROJECTED THAT J-7 PRODUCTION WOULD CONTINUE FOR AT LEAST ANOTHER DECADE, RESULTING IN A TOTAL INVENTORY OF NEARLY 1000 AIRCRAFT BY 2005, BUT THE PLAAF INVENTORY HAS REMAINED AT ABOUT 500 AIRCRAFT, SUGGESTING THAT PRODUCTION WAS EITHER SUSPENDED OR TERMINATED.
APPEARANCE: WINGS: MID-MOUNT, DELTA, CLIPPED TIP, ENGINES: ONE TURBOFAN IN FUSELAGE, FUSELAGE: CIRCULAR WITH DORSAL SPINE, TAIL: SWEPT-TAIL WITH LARGE VERTICAL SURFACES AND VENTRAL FIN
VARIANTS:
J-7 I/F-7: INITIAL PRODUCTION VERSION, SIMILAR TO MIG-21F FISHBED-C. THE 12,677-LBST WOPEN 7 ENGINE IS SAID TO BE MORE RELIABLE THAN THE TUMANSKY R-11 FROM WHICH IT WAS DERIVED. EXPORT MODELS ARE DESIGNATED F-7.

J-7 II/F-7B: UPRATED ENGINE, REDESIGNED INLET CENTER-BODY, INSTALLATION OF SECOND 30-MM CANNON, CENTERLINE DROP TANK HARDPOINT. ENTERED PRODUCTION IN EARLY 1980S.

JJ-7/FJ-7: TANDEM TWO-SEAT TRAINER VERSION DEVELOPED WELL AFTER THE SINGLE SEAT FIGHTERS. FIRST FLIGHT ON JULY 5, 1985.

F-7M AIRGUARD: CURRENT PRODUCTION VERSION AND EXPORT VERSION: RECOGNITION FEATURE IS RELOCATION OF THE PITOT TUBE FROM BELOW THE NOSE INTAKE TO ABOVE IT. FITTED WITH MARCONI SKYRANGER RADAR; GEC AVIONICS HEADS-UP-DISPLAY AND WEAPONS AIMING COMPUTER; INBOARD WING PYLONS FOR PL-2/2A/5B/7 OR MATRA MAGIC AAM, ROCKET PODS OR BOMBS UP TO 500 KG; ADDITIONAL OUTBOARD PYLONS WITH PLUMBING FOR 500-L DROP TANKS OR 50/150 KG BOMBS OR ROCKET PODS.

F-7P SKYBOLT: SIMILAR TO THE F-7M WITH SOME PAKISTANI EQUIPMENT: CANNON IS TWO NORINCO 30 MM CANNONS WITH 60 ROUNDS EACH. USUALLY CARRIES A 720-L CENTERLINE DROP TANK.

F-7M AIRGUARD: CURRENT PRODUCTION VERSION AND EXPORT VERSION: RECOGNITION FEATURE IS RELOCATION OF THE PITOT TUBE FROM BELOW THE NOSE INTAKE TO ABOVE IT. FITTED WITH MARCONI SKYRANGER RADAR; GEC AVIONICS HEADS-UP-DISPLAY AND WEAPONS AIMING COMPUTER; INBOARD WING PYLONS FOR PL-2/2A/5B/7 OR MATRA MAGIC AAM, ROCKET PODS OR BOMBS UP TO 500 KG; ADDITIONAL OUTBOARD PYLONS WITH PLUMBING FOR 500-L DROP TANKS OR 50/150 KG BOMBS OR ROCKET PODS.

F-7P SKYBOLT: SIMILAR TO THE F-7M WITH SOME PAKISTANI EQUIPMENT: CANNON IS TWO NORINCO 30 MM CANNONS WITH 60 ROUNDS EACH. USUALLY CARRIES A 720-L CENTERLINE DROP TANK.

CHINESE FIGHTER AIRCRAFT J-8/FINBACK

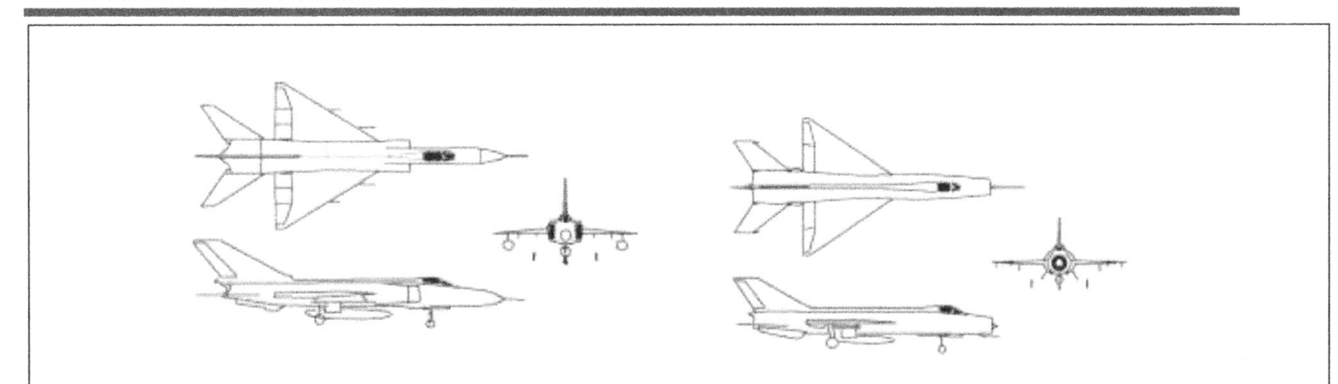

SYSTEM	SPECIFICATIONS	SYSTEM	SPECIFICATIONS
Weapon & Ammunition Types		Fuel (liters):	
23mm type 23- twin barrel (rnds):	200	Internal:	5,400
AAMs:		External:	1,760
PL-2B IR:	6	Range (km):	2,200
PL-7 medium range semi-active homing (optional):	6	Combat Radius:	800
Rockets:		Takeoff Run/Landing Roll (m):	
Quingan HF-16B 57 mm:	6	670/1,000 (w/afterburner and drag chute)	
90 mm AS rockets:	6	Dimensions (m):	
Bombs:	3	Length:	21.6
Fuel Tanks:	3	Wingspan:	9.4
Alternative Designations:	F-8	Height:	5.4
Date of Introduction:	1980	Standard Payload (kg):	
Proliferation:	1	External:	7
Crew:	1	Hardpoints:	6 under wing, 1 centerline
Engines14,815 lbs. thrust Wopen 13A-II turbojets with afterburner :	1	Survivability/Countermeasures:	
Weight (kg):		Pressurized cockpit with ejection seat:	Yes
Max Gross:	17,800	Radar warning receiver:	Yes
Normal Takeoff:	14,300	chaff and flares	Yes
Empty:	9,820	ARMAMENT:	
Speed (km/h):		23-mm Type 23-3 twin-barrel cannon:	1
Max (at altitude):	2,340	AVIONICS/SENSOR/OPTICS	

Max (sea level):	1,300	VHF/UHF and HF/SSB radios	Yes
Limit "G" Force (g):	+4.83	'Odd Rods' type IFF	Yes
Ceiling (m):	20,000	Monopulse nose-radar	Yes

NOTES

THE BEST THAT CAN BE SAID FOR THE J-8 IS THAT ONCE UPGRADED IT WILL BE NO MORE THAN AN ADVANCED OBSOLETE AIRCRAFT, COMPARABLE IN CONFIGURATION AND AERODYNAMIC PERFORMANCE TO THE SU-15/FLAGON. THE J-8 AND J-8-II AIRCRAFT ARE TROUBLE-PRONE AIRCRAFT WITH A POOR WEAPON SUITE AND AN INEFFICIENT ENGINE. AT BEST, THE J-8-II CAN BE COMPARED WITH AN EARLY MODEL (1960S) US F-4 PHANTOM. IN FACT, AFTER TWENTY-SIX YEARS THE J-8-II IS STILL IN THE DEVELOPMENT STAGE, HAS RESULTED IN ONLY ABOUT 100 FIGHTERS DEPLOYED, AND MEETS NONE OF THE REQUIREMENTS OF THE PLAN.

APPEARANCE: WINGS: SHARPLY SET DELTA WING, ENGINES: SIDE BY SIDE W OPEN TURBOJETS, FUSELAGE: SLENDER WITH NOSE ENGINE AIR INTAKE (J-8-I), SOLID CONICAL NOSE (J-8-II), TAIL: SWEPT WITH FULL-HEIGHT RUDDER.

VARIANTS

THIS AIRCRAFT IS AN ADAPTATION OF THE SOVIET MIG-21 FISHBED

J-8/F-8-I FINBACK-A: INITIAL PRODUCTION VERSION WITH WP-7P ENGINES AND NOSE AIR INTAKES. J-8 IS DESIGNATION FOR AIRCRAFT IN CHINESE SERVICE; F-8/F-8M DENOTES EXPORT VERSION. MORE THAN 100 J-8/F-8-IS WERE PRODUCED

J-8-II FINBACK-B: RADAR TYPE IS UNIDENTIFIED MONOPULSE RADAR, BUT MAY BE THE LEIHUA TYPE 317A IN SOLID NOSE HOUSING. SEVEN PYLONS FOR INCREASED WEAPONS INVENTORY AND NEW SIDE AIR INTAKES. OTHER CHARACTERISTICS SIMILAR TO F-8-II.

F-8-II FINBACK-B: IMPROVED VERSION WITH NEW 14,815 WOPEN-13A ENGINES, WING ROOT INTAKES, AND ALL-FLYING HORIZONTAL STABILIZERS, FOLDING VENTRAL FIN, 80%-COMPOSITE MATERIAL VERTICAL FIN AND IMPROVED AVIONICS.

F-8 ILM FINBACK-B: DESIGNATION FOR RUSSIAN MODIFIED F-8-IIS. INCLUDES: RUSSIAN AA-12 AND AA-10 MISSILES, A HEADS-UP-DISPLAY, GLOBAL POSITIONING SYSTEM RECEIVER, MULTI-FUNCTION COCKPIT DISPLAYS AND INTEGRATED ELECTRONIC COUNTERMEASURES. ALSO RUSSIAN PHAZOTRON ZHUK 8 II MULTI-FUNCTION PULSE DOPPLER FIRE-CONTROL RADAR.

RUSSIAN INTERCEPTOR AIRCRAFT MIG-25/FOXBAT-B

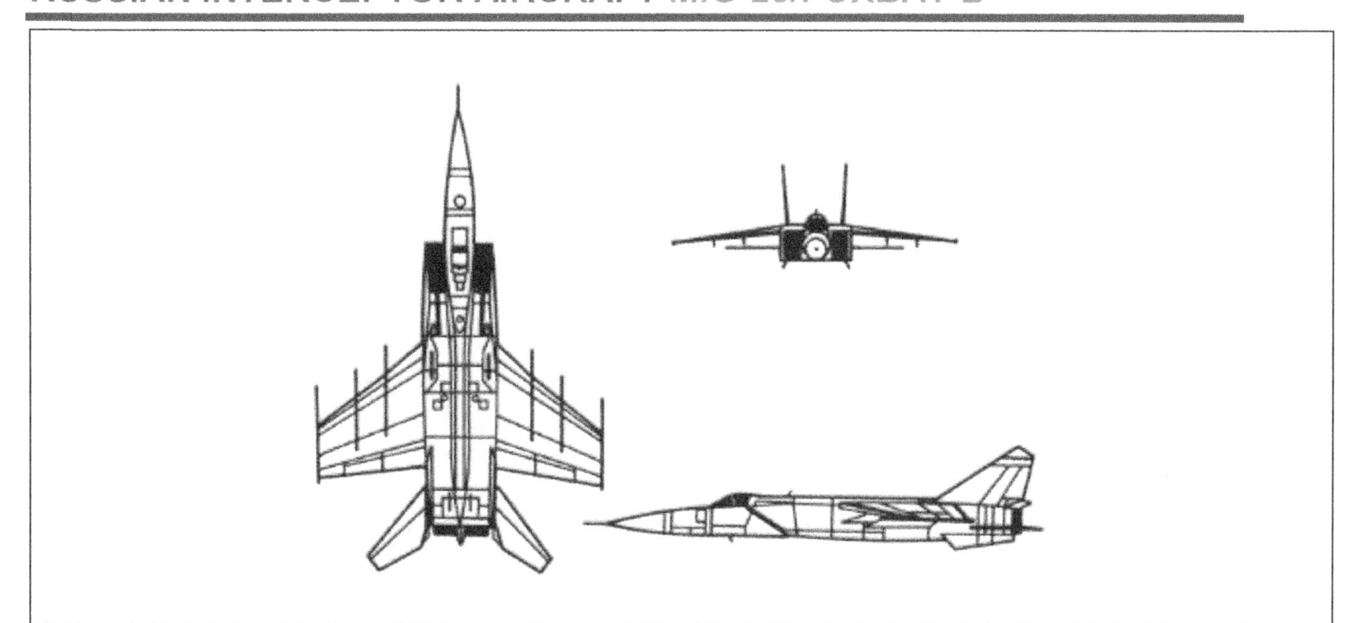

SYSTEM	SPECIFICATIONS	SYSTEM	SPECIFICATIONS
Air-to-Air Missiles:		Ceiling (m):	
AA-6 ACRID:	4	Service (clean):	
AA-7 APEX:	4	R Series:	23,000
AA-6 ACRID w/ AA-8 APHID/AA-11 ARCHER:	2/4	P Series:	20,700
AA-7 APEX w/ AA-8 APHID/AA-11 ARCHER:	2/4	With External Stores (R-Series):	20,700
Alternative Designations:	Mig-25 RB FOXBAT-B/ MiG-25PD FOXBAT-E	Vertical Climb Rate (m/s) (P Series):	208
Date of Introduction:	1967	Fuel (liters):	
Proliferation:	At least 10 countries	Internal:	17,470
Description:		External:	5,300
Crew:	1	Range (km):	
Engines 19,400 lbs. thrust Soyuz/ Tumansky R-15BD-300 turbojet (24,692 lbs. thrust with afterburner):	2	Dimensions (m):	
Weight (kg):		Length:	
Maximum Gross:		R Series:	21.6
R Series:	41,200	P Series:	23.8
P Series:	36,720	Wingspan:	

Clean Takeoff:	35,060 (R)	R Series:	13.4
Empty:	20,000 (P)	P Series:	14.0
Speed (km/h):		Height (gear extended):	
Maximum (at altitude):		R Series:	6.0
R Series:	3,000	P Series:	6.1
P Series:	3,390	Standard Payload (kg):	
Maximum (sea level):		External:	2,000 – 5,000
R Series:	1,200	Hardpoints (R Series):	
P Series:	1,050	Wing:	4
Cruise:		Fuselage:	6
R Series:	2,500	Hardpoints (P Series):	4
P Series:	3,000	Survivability/Countermeasures:	
Takeoff/Landing Speed:		pressurized cockpit with zero/130 – 1,250 km hour ejection seats	Yes
R Series:	360	Decoys:	Yes
P Series:	290	Radar jammer:	Yes
Max "G" Force (g):		radar and missile warning receivers:	Yes
P Series:	+4.5	ARMAMENT:	
Maximum with Max Internal fuel:	Supersonic:	Air-to-air missiles on four under-wing attachments:	Yes
With 5,300-litre Fuel Tank:	Supersonic:	AVIONICS/SENSOR/OPTICS:	
Subsonic (R Series):	2,400 (R)	Fire control radar in the nose Range (km):	
Takeoff Run/Landing Roll (m) (P Series):	1,250/800	Search:	100
		Tracking;	75

NOTES

THE FOXBAT IS A HIGH-PERFORMANCE, HIGH-ALTITUDE INTERCEPTOR. THIS FAST BUT MANEUVERABLE INTERCEPTOR HAS BEEN DEPLOYED AS A HIGH ALTITUDE RECONNAISSANCE PLATFORM. THOSE REMAINING IN RUSSIAN SERVICE ARE ALL RECONNAISSANCE VERSIONS. THE INTERCEPTORS PHASED OUT IN 1994.INTERCEPTOR VERSIONS REMAIN IN SERVICE WITH OTHER NATIONS.

APPEARANCE: WINGS: SHOULDER-MOUNTED, SWEPT-BACK, AND TAPERED WITH SQUARE TIPS, ENGINES: BURIED SIDE BY SIDE IN AFT FUSELAGE, FUSELAGE: LONG AND SLENDER WITH SOLID, POINTED NOSE. FLATS ARE MID- TO LOW MOUNTED ON FUSELAGE, SWEPT-BACK AND TAPERED WITH ANGULAR TIPS. TAIL: TWO SWEPTBACK, AND TAPERED VERTICAL FINS WITH ANGULAR TIPS

VARIANTS

MIG-25 FOXBAT-A: STANDARD INTERCEPTOR VERSION. WITHDRAWN FROM SERVICE IN RUSSIA IN THE 1990S.

MIG-25R FOXBAT-B: RECONNAISSANCE VERSION.

MIG-25RB FOXBAT-B: RECONNAISSANCE-BOMBER VERSION BUILT IN 1970.

MIG-25U FOXBAT-C: TWO-SEAT TRAINER VERSION.

MIG-25RBK FOXBAT-D: RECONNAISSANCE-BOMBER VERSION WITH SLAR.

MIG-25P/PD FOXBAT-E: INTERCEPTOR WITH IMPROVED RADAR THAT HAS LIMITED LOOK-DOWN/SHOOT-DOWN CAPABILITY, IR SENSOR UNDER THE NOSE, AND UPGRADED ENGINE.

MIG-25BM FOXBAT-F: FIGHTER/ATTACK VERSION WITH AS-11 KILTER ANT-RADAR MISSILES AND FREE-FALL BOMBS TO ATTACK GROUND BASED AIR DEFENSE FROM HIGH ALTITUDES.

RUSSIAN INTERCEPTOR AIRCRAFT MIG-31/FOXHOUND

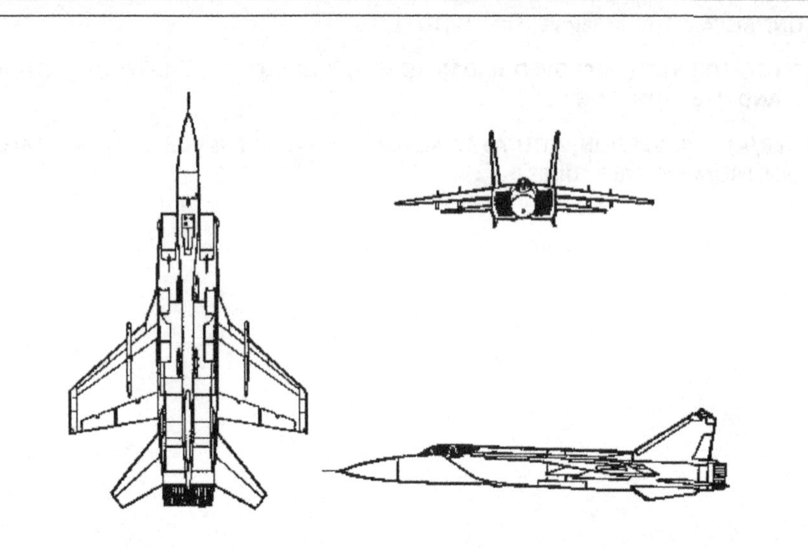

SYSTEM	SPECIFICATIONS	SYSTEM	SPECIFICATIONS
Weapon & Ammunition Types:		Range (km):	
GSh-23-6 23mm Gatling-type cannon (rnds):	260	Ferry without refueling:	3,300
Other Loading Options:		Combat Radius (km):	
Fuselage:		4 x AA-9 Amos, 2 x drop tanks, 1 in-flight refuel at Mach 0.8:	2,200
AA-9 AMOS AAMX:	4	4 x AA-9 Amos, 2 x drop tanks at Mach 0.85:	1,400
AA-6 ACRID:	2	4 x AA-9 Amos, no drop tanks at Mach 0.85:	1,200
AA-8 APHID:	4	4 x AA-9 Amos, no drop tanks at Mach 2.35:	720
AA-10 ALAMO (multirole):	8	Duration (hr.):	
AA-11 ARCHER (multirole):	8	internal and drop tanks only:	3.6
AA-12 ADDER (multirole):	8	drop tanks and in-flight refueling:	6 -7
Alternative Designations:		Takeoff Run/Landing Roll (m):	1,200/800
Date of Introduction:	1967	Dimensions (m):	
Proliferation:	$\leq$ 1	Length:	20.6
Description:		Wingspan:	13.5
Crew:	2	Height:	6.2

Engines 0,944 lbs. thrust Aviadvigatel D-30F-6 turbofan, 34,172 lbs. thrust with afterburner):	2	Standard Payload (pylons):	8
Weight (kg):		Survivability/Countermeasures:	
Maximum Gross:	46,200	Pilot and weapons system operator in tandem under individual rearward hinged canopies:	Yes
Normal Takeoff:	41,000	Active infrared and electronic countermeasures:	Yes
Empty:	21,820	Radar warning receiver:	Yes
Speed (km/h):		Wingtip ECM/ECCM pod:	Yes
Maximum (at altitude):	2,500	ARMAMENT:	
Maximum (sea level):	1,500	GSh-23-6 23mm Gatling-type cannon:	1
Maximum Attack Speed:	3,000	AVIONICS/SENSOR/OPTICS:	
Cruise:	1,010	N-007/S-800 Zaslon (Flash Dance) electronically scanned phased array look-down shoot-down fire control radar:	Yes
Max “G” Force (g):	+5	Long range nav system:	Yes
Ceiling (m):	24,400	Infrared search/track system:	Yes
Vertical Climb Rate (m/s):	42	Night/Weather Capabilities::	Yes
Internal:	20,250		
External:	5,000		

NOTES

THE MIG-31 IS AN ALL-WEATHER, TWO-SEAT INTERCEPTOR WITH ADVANCED DIGITAL AVIONICS. IT WAS THE FIRST SOVIET FIGHTER TO HAVE A TRUE LOOK-DOWN, SHOOT-DOWN CAPABILITY.

APPEARANCE: WINGS: SHOULDER-MOUNTED, MODERATE-SWEPT WITH SQUARED TIPS. ENGINES: TWO TURBOFANS. FUSELAGE: RECTANGULAR FROM INTAKES TO EXHAUSTS WITH A LONG, POINTED NOSE. TAIL: TAIL FINS ARE BACK-TAPERED WITH ANGULAR TIPS AND CANTED OUTWARD. LOW-MOUNTED FLATS ARE SWEPT-BACK AND TAPERED.

VARIANTS

THE MIG-31 FOXHOUND IS A SUBSTANTIALLY IMPROVED DERIVATIVE OF THE MIG-25 FOXBAT.

MIG-31 FOXHOUND-A INTERCEPTOR: ORIGINAL PRODUCTION VERSION.

MIG-31B/BS/E FOXHOUND-A INTERCEPTOR:

THE MIG-31B HAS FLASH DANCE A RADAR AND IMPROVED AA-9 AMOS AAMS, IN-FLIGHT REFUELING PROBE, AND NEW NAVIGATION SYSTEM.

MIG-31BS: SIMILAR TO THE MIG-31B, WITH RADAR ENHANCEMENT AND A-723 NAVIGATION.

MIG-31E: EXPORT VARIANT OF MIG-31B AIMED AT CHINA, INDIA, AND IRAN. NONE WERE SOLD.

MIG-31BM/FE FOXHOUND-A MULTIROLE FIGHTER: MID-LIFE UPGRADE FOR INTERCEPTORS. FITTED WITH ASMS, UPGRADED RADAR AND AA-11 AND AA-12 AAMS.MIG-31FE IS EXPORT VARIANT.

MIG-31M FOXHOUND-B MULTIROLE FIGHTER: UPGRADED LONG RANGE NAVIGATION SYSTEM AND IMPROVED PHASED ARRAY RADAR.

FRENCH FIGHTER AIRCRAFT MIRAGE III/5/50

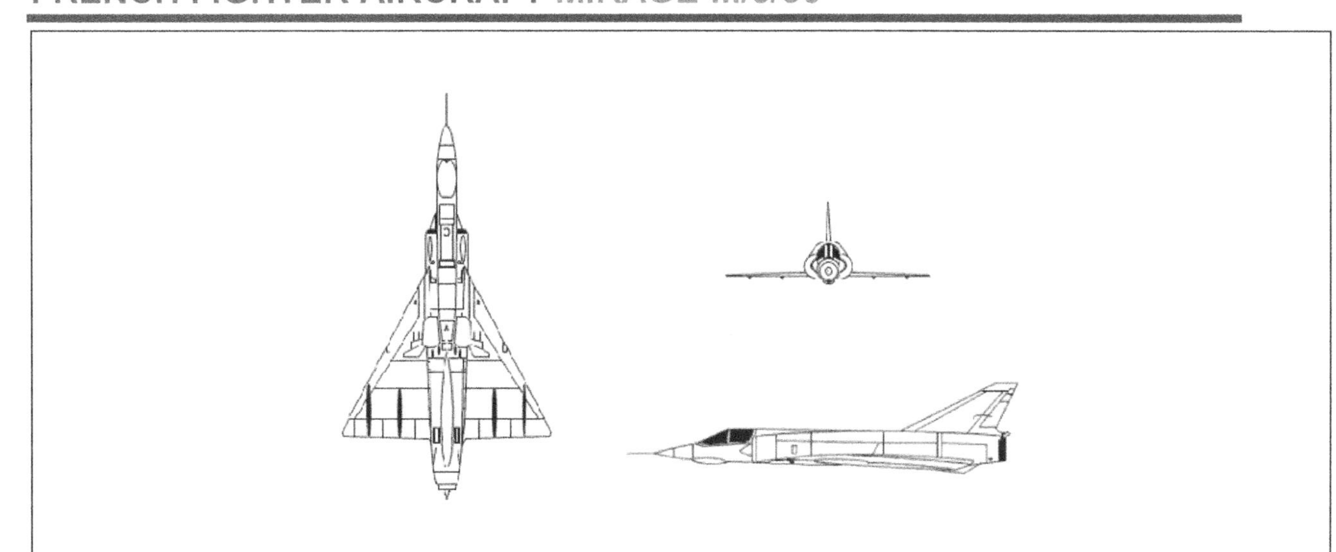

SYSTEM	SPECIFICATIONS	SYSTEM	SPECIFICATIONS
Weapon & Ammunition Types		Range (km):	
30-mm cannon:	125	Cruise:	
AAMs:		Mirage III:	1,670
Matra Magic 550:	2	Mirage 5:	1,930
AIM-9 Sidewinder:	2	Mirage 50:	2,133
Bombs:	12	Ferry:	4,000
125 kg /250 kg:	6	Takeoff Run/Landing Roll (m):	700-1,600/700
440 kg:	10	Dimensions (m):	
Rocket Pods:	1967	Length:	
68-mm or 100-mm:	2	Mirage III:	15.0
2 x 30-mm Cannon Pods (rnds ea.):	250	Mirage 5:	15.6
Alternative Designations:		Mirage 50:	15.6
Date of Introduction:	1959	Wingspan:	8.3
Proliferation:	>15	Height (gear extended):	4.3
Description:		Standard Payload (kg):	4,000
Crew:	1	Survivability/Countermeasures:	
Engines:		Martin-Baker zero/267 km/h ejection seat:	Yes

6,200 lbs. thrust SNECMA Atar 9C turbojet with afterburner (Mirage III/5):	1	Separate cockpit and avionics air conditioning systems:	Yes
7,200 lbs. thrust SNECMA Atar 9K50 turbojet, afterburner (Mirage 50):	1	Radar warning receiver:	Yes
Weight (kg):		ARMAMENT:	
Max Takeoff:	13,500	30-mm DEFA 552 (Mirage III):	2
Empty:	7,050	30-mm DEFA 553 (Mirage 5):	2
Speed (km/h):		30-mm DEFA 553 (Mirage 50):	2
Max (at altitude):	2,350; Mach 2.2	AVIONICS/SENSOR/OPTICS:	
Max (sea level):	1,390; Mach 1.1	Intercept or ground mapping radar:	Yes
Ceiling (m):		Fire-control radar in the nose:	Yes
Mirage III:	17,000	Navigation computer:	Yes
Mirage 5:	17,000	Automatic gun sight:	Yes
Mirage 50:	18,000	Night/Weather Capabilities:	
Vertical Climb Rate (m/s):	84	All-weather, day and night capable. (III/5)	
Fuel (liters):		Clear-weather day fighter. (50)	
Internal:	3,330		
External (III/5/50):	1,700/ 1,200/ 1,700		

NOTES

ONE OF THE MOST SUCCESSFUL AIRCRAFT PRODUCED FOR EXPORT TO BE PRODUCED OUTSIDE OF THE UNITED STATES AND THE FORMER SOVIET UNION. THE MIRAGE III/5/50 HAS PROVEN TO BE A COMPETENT GROUND ATTACK AIRCRAFT DESPITE ITS ORIGINAL DEVELOPMENT AS A HIGH ALTITUDE INTERCEPTOR. THE MIRAGE 5 AND 50 ARE SIMILAR TO THE III, BUT FITTED WITH SIMPLIFIED AVIONICS AND HAVE EXCLUSIVELY BEEN EXPORT VARIANTS.

APPEARANCE: WINGS: LOW-MOUNTED DELTA WINGS WITH POINTED TIPS. ENGINES: ONE TURBOJET INSIDE FUSELAGE. FUSELAGE: LONG, SLENDER, AND TUBULAR WITH A POINTED NOSE AND BUBBLE COCKPIT. TAIL: LARGE, SWEPT-BACK SQUARE TIP WITH A TAPERED FIN AND NO TAIL FLATS.

VARIANTS

MIRAGE IIIA: HIGH ALTITUDE INTERCEPTOR AND STRIKE AIRCRAFT FITTED WITH ROCKET MOTOR FOR TAKE-OFF.

MIRAGE IIIB: TWO-SEAT TRAINER VERSION OF IIIA WITH STRIKE CAPABILITY RETAINED NO INTERNAL CANNON.

MIRAGE IIIC: MAJOR PRODUCTION VARIANT OF IIIA. FITTED WITH ATAR 9B3 ENGINE, CYRANO II INTERCEPT AND GROUND-MAPPING RADAR.

MIRAGE IIID: TWO-SEAT TRAINER/STRIKE AIRCRAFT. NO INTERNAL CANNON.

MIRAGE IIIE: MAJOR PRODUCTION VARIANT. FITTED WITH ATAR 9C ENGINE, CYRANIO IIBIS RADAR, AND EXTRA AVIONICS BAY.

MIRAGE IIIEA/EO: AUSTRALIAN LICENSE-BUILT AIRCRAFT. (52 ATTACK AND 48 INTERCEPTOR AIRCRAFT)

MIRAGE 5: GROUND ATTACK VARIANT ORIGINALLY DEVELOPED FOR THE ISRAELI AIR FORCE. FITTED WITH TWO EXTRA FUSELAGE STORES STATIONS, FIRE CONTROL RADAR DELETED OR REPLACED BY RANGING RADAR. PERFORMANCE IDENTICAL TO MIRAGE III EXCEPT FOR LONGER RANGE ON INTERNAL FUEL; SOME FITTED WITH CYRANO OR AGAVE RADAR.

CZECH REPUBLIC TRAINER/LIGHT GROUND ATTACK AIRCRAFT L39 ALBATROSS

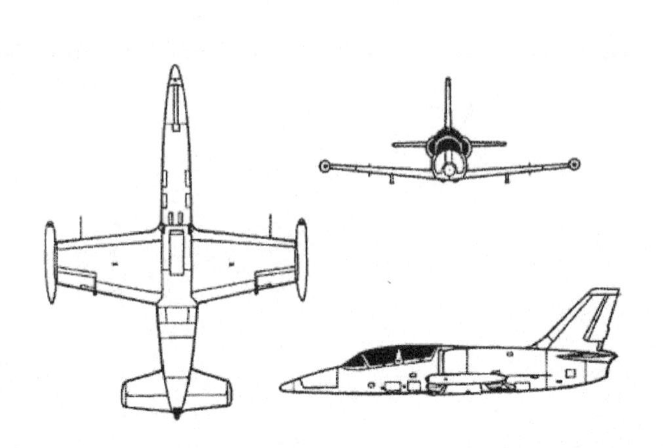

SYSTEM	SPECIFICATIONS	SYSTEM	SPECIFICATIONS
Weapon & Ammunition Types	Combat Load	Ceiling (m):	11,500
23-mm twin barrel GSh-23 cannon	150	Vertical Climb Rate (m/s):	22
Other Loading Options		Fuel (liters):	
Rocket Pods:	4	Internal:	1,255
Rocket Pods and 350 L drop tanks:	2	External:	8,40
IR Missiles and 350 L drop tanks:	2	Range (km):	
227 kg bombs:	4	With Max Fuel:	1,750
454 kg bombs:	2	Takeoff Run/Landing Roll (m):	530/650
113 kg bombs:	6	Dimensions (m):	
Dispensers and 350 L drop tanks:	2	Length:	12.2
350 L drop tank and Photo Recon Pod:	1	Wingspan:	9.5
SYSTEM		Height:	4.8
Alternative Designations:		Standard Payload (kg):	
Date of Introduction:	1974	External:	1,500
Proliferation:	22	Hardpoints:	

Description:		Fuselage:	1
Crew:	2	Wings:	4
Engines 3,792 lbs. thrust Ivanchenko AI-25TL turbofan:	1	Survivability/Countermeasures:	
Weight (kg):		Zero/150 km/hr. ejection seats:	Yes
Max Takeoff:	4,700	pressurized, heated, and air conditioned cockpit:	Yes
Clean Takeoff:	4,525	ARMAMENT:	
Empty:	3,455	23-mm GSh-23 twin barreled cannon:	Yes
Speed (km/h):		AVIONICS/SENSOR/OPTICS:	
Maximum (at altitude):	750	Weapon delivery and navigation system with HUD and video camera in front cockpit and monitor in rear cockpit	Yes
Maximum (sea level):	700	Gun/rocket/missile firing and weapon release controls in front cockpit only	Yes
Max "G" Force (g):	+8/-4	Night/Weather Capabilities:	Limited

NOTES

THE L39 ALBATROSS IS A VERY WIDELY FLOWN TRAINER/LIGHT ATTACK AIRCRAFT. THE DESIGN IS CZECHOSLOVAKIAN, THOUGH THERE ARE SIGNIFICANT SOVIET INPUTS AND THE AIRCRAFT IS IN SERVICE WITH VARIOUS SOVIET ALLIES.

APPEARANCE: WINGS: LOW, SLIGHTLY SWEPT. ENGINES: SINGLE TURBOFAN IN FUSELAGE. FUSELAGE: LONG, SLENDER, POINTED NOSE. TAIL: TALL, SWEPT VERTICAL WITH INSET RUDDER.

VARIANTS

L39C: BASIC FLIGHT TRAINER.

L39V: SIMILAR TO THE L39C, BUT WITH SINGLE SEAT COCKPIT AND MODIFIED TO ACT AS TARGET TOW AIRCRAFT.

L39ZO: ARMED VERSION OF L39C, ADDING FOUR UNDERWING HARDPOINTS FOR A VARIETY OF GROUND ATTACK STORES.

L39ZA: SIMILAR TO L39ZO, BUT WITH UNDER FUSELAGE GUN POD AND REINFORCED LANDING GEAR. USED FOR GROUND ATTACK AND RECONNAISSANCE MISSIONS.

L39MS: DEVELOPMENTAL VERSION INCORPORATING MORE ADVANCED AVIONICS AND NEW 4,852 LBS. THRUST ENGINE. ADDRESSED SOVIET AF REQUIREMENT TO TRAIN PILOTS FOR THE MIG-29 FULCRUM AND SU-27 FLANKER AIRCRAFT

L59:DEVELOPMENT OF THE L39MS WITH WESTERN ENGINE, AVIONICS, AND MARTIN-BAKER EJECTION SEATS

L-159: GROUND ATTACK VARIANT OF THE L59.

RUSSIAN GROUND-ATTACK AIRCRAFT SU-17/FITTER

SYSTEM	SPECIFICATIONS	SYSTEM	SPECIFICATIONS
Weapon & Ammunition Types		Ceiling (m):	
2x 30-mm NR-30 guns (rnds):	160	Service (clean):	
Other Loading Options		M2:	18,000
325-mm S-25 rockets:	1	M4:	15,200
80-mm S-8 rocket pods:	20	With External Stores:	INA
57-mm S-5 rocket pods:	32	Vertical Climb Rate (m/s):	230
AS-7/KERRY ASM:	1	Fuel (liters):	
AS-9/KYLE ASM:		Internal:	4,550
AS-10/KAREN ASM:		External 800 liter tanks:	≤ 4
AS-12/KEGLER ASM:		Range (km):	
AS-14/KEDGE ASM:		Max Load:	1,500
AA-2 ATOLL AAM:	2	With Aux Fuel:	INA
AA-8/APHID AAM:	2	Combat Radius:	330 to 685
AA-11/ARCHER AAM launchers:	2	Takeoff Run/Landing Roll (m):	
23-mm SPPU-22 Gun Pods		Prepared Surface:	900/950

External fuel tanks (liters)	800	Dimensions (m):	
100-kg, 250-kg, and 500-kg unguided and guided bomb		Length:	18.8
SYSTEM		Wingspan:	13.8 extended, 10.6 swept
Alternative Designations:	Su-20, Su-22, Strizh or Martlet	Height:	4.8
Date of Introduction:	1970	Standard Payload (kg):	
Proliferation:	19	External:	4,000
Description:		M4:	4,250
Crew:	1	Hardpoints:	8
Engines 28,660-hp Lyulka AL-21F-3(Su-17/20)/ or 1x 25,335-shp Tumansky:	1	Survivability/Countermeasures:	
R-29BS-300 (Su-22) turbojet with afterburner	1	Radar warning receiver:	Yes
Weight (kg):		chaff and flares:	Yes
Max Gross:		Armored cockpit:	
M2:	17,700	M3:	Yes
M4:	19,500	M4:	Yes
Normal Takeoff:		ARMAMENT:	
M2:	14,000	30-mm machinegun:	2
M4:	16,400	Range (practical) (m):	2,500
Empty:	10,000	Elevation/Traverse (rigidly mounted):	None
Speed (km/h):		Ammo Type:	
Max (at altitude):	Mach 2.1	HEFI:	Yes
Max (sea level):	Mach 1.1	APT:	Yes
Takeoff/Landing Speed:	265	CC:	Yes
Max "G" Force (g):	+7.0	Rate of Fire (rpm):	850
		AVIONICS/SENSOR/OPTICS:	Simple

NOTES

THE MID-WING PIVOT POINT OF THE SWEEP WINGS ALLOWS FOR POSITIONS OF 28, 45 OR 62 DEGREES. UP TO FOUR EXTERNAL FUEL TANKS CAN BE CARRIED ON WING PYLONS AND UNDER THE FUSELAGE. WHEN UNDER-FUSELAGE TANKS ARE CARRIED, ONLY THE TWO INBOARD WING PYLONS MAY BE USED FOR ORDNANCE. AVAILABLE MUNITIONS ARE SHOWN ABOVE; NOT ALL MAY BE EMPLOYED AT ONE TIME. MISSION DICTATES WEAPONS CONFIGURATION. EXTERNAL STORES ARE MOUNTED ON UNDERWING AND UNDERBODY HARDPOINTS. EACH WING HAS TWO POINTS, AND THE FUSELAGE HAS FOUR ATTACHMENT POINTS FOR A TOTAL OF EIGHT STATIONS. GUN PODS CAN BE MOUNTED TO FIRE REARWARD.

APPEARANCE: WINGS: LOW-MOUNT, VARIABLE, SWEPT AND TAPERED WITH BLUNT TIPS. ENGINES: ONE IN FUSELAGE, INTAKE IN NOSE. FUSELAGE: TUBULAR WITH BLUNT NOSE. TAIL: SWEPT-BACK AND TAPERED, FLATS MOUNTED ON FUSELAGE AND SWEPT-BACK.

VARIANTS

AIRCRAFT WAS DERIVED FROM SU-7 FITTER A BY INCORPORATING VARIABLE WINGS.

MANY VARIANTS ARE IN USE; HOWEVER, THE M3 AND M4 ARE THE MOST PROLIFERATED VERSIONS.

DOMESTIC AIRCRAFT USE NOMENCLATURE SU-17. EXPORT VERSIONS USE SU-20 AND SU-22.

SU-17/-17MK/-20/FITTER C: THE FIRST PRODUCTION VERSION. EXPORT IS CALLED SU-20.

SU-17M/ -17M2/ -17M2D FITTER D: EXTERNAL DOPPLER-NAV AND INTERNAL LASER RANGEFINDER. RECONNAISSANCE VERSION CALLED SU-17R.

SU-17UM/-22U/FITTER E: TWO-SEAT TRAINER WITH COMPONENTS OF SU-17M.

SU-17/FITTER G: COMBAT-READY TWO-SEAT TRAINER VARIANT OF FITTER H. EXPORT VERSION IS SU-22, WITH TUMANSKY ENGINE.

SU-17/-17M3/FITTER H: INCREASED PILOT VISIBILITY BY DROOPING THE AIRCRAFT NOSE, AND INCORPORATED AN INTERNAL DOPPLER-NAV AND LASER RANGEFINDER. RECONNAISSANCE VERSION CALLED SU-17M3R.

SU-17M4/-22M4/FITTER K: FIGHTER-BOMBER. ESSENTIALLY SAME AS ABOVE, BUT WITH AN ADDITIONAL AIR INTAKE. EMPLOYS DIGITAL NAVIGATION AND ATTACK AVIONICS.

SU-22/FITTER F: EXPORT VERSION OF FITTER D WITH TUMANSKY ENGINE.

SU-22/-22M3/FITTER J: SIMILAR TO FITTER H, BUT WITH INCREASED INTERNAL FUEL CAPACITY.

GEORGIAN/RUSSIAN MULTI-ROLE ATTACK AIRCRAFT SU-25TM AND SU-39

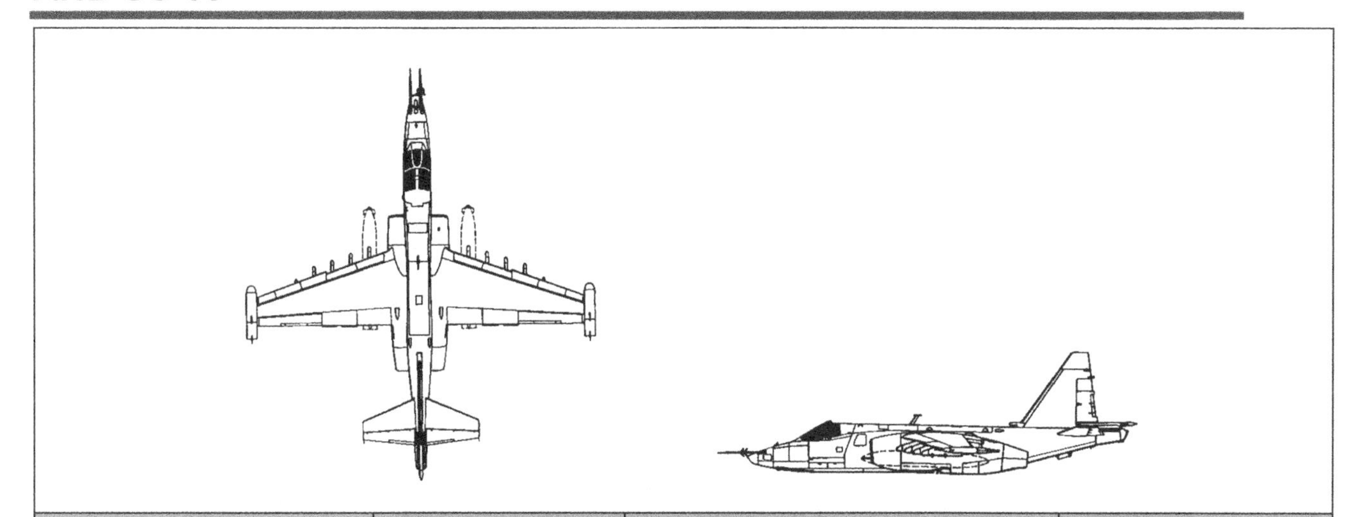

SYSTEM	SPECIFICATIONS	SYSTEM	SPECIFICATIONS
Weapon & Ammunition Types		Takeoff Run/Landing Roll (m):	
2x 30-mm Gsh-30 guns (rnds):	1200	Prepared Surface:	550/600
Other Loading Options		Unprepared Surface:	650/750
AT-16 Vikhr-M ATGM:	8	Max Load:	1,200
23- or 30-mm GSH gun pods (rnds):	260	Dimensions (m):	
UB-20 80/122/240/340-mm rockets w/ semi-active laser homing:	8	Length:	15.3
AS-10/KAREN ASM:	8	Wingspan:	14.5
AS-11/KILTER ASM:	8	Height (gear extended):	5.2
AS-14/KEDGE ASM:	8	Standard Payload (kg):	
AS-17/KRYPTON ASM:	8	External:	6,400
AA-8/APHID AAM:	2	Hardpoints under-wing, w/500 kg ea.:	8
AA-11/ARCHER AAM launchers:	2	+ 2 light outer (± 65 kg) for AAM	
AA-12 ADDER AAM:	2	Dimensions (m):	
SYSTEM		Length:	15.3
Alternative Designations:	Gratch, Rook,	Wingspan:	14.5
Date of Introduction:		Height (gear extended):	5.2
Su-25TM:	1995	Standard Payload (kg):	
Proliferation:	> 16	External:	6,400

Description:		Survivability/Countermeasures:	
Crew:		Armored Titanium cockpit and engines:	Yes
Su-25TM:	1	12-mm titanium plate added between engines:	Yes
Su-39:	2	Zero/100 km/ejection seat:	Yes
Engines:2 x 9,900 lbs. thrust R-195:	2	Self-sealing fuel tanks:	Yes
Weight (kg):		Strengthened flight control linkage:	Yes
Maximum Gross:	17,600	IFF:	Yes
Normal Takeoff:	14,500	Exhaust cooling:	Yes
Empty:	9,525	L166S1/ShokogruzEO infrared jammer:	Yes
Speed (km/h):		Sirena 3/Pastilradar warning receiver:	Yes
Maximum (at altitude):	880	Omul ECM pods with UV-26 flares:	Yes
Maximum (sea level):	950	ARMAMENT:	
Maximum Attack Speed:	690	Hardpoints;	10
Cruise:	700	AVIONICS/SENSORS/OPTICS:	
Takeoff/Landing Speed:	220	SUO-39 FCS pod with Shkval-M sight system and Mercury (LLLTV):	Yes
Max "G" Force (g):	+6.5 g	Laser radar:	Yes
Service Ceiling:	(m):	Khod thermal imager:	Yes
Vertical Climb Rate (m/s):	72	23X image magnification aiming system (to 25 km):	Yes
Fuel (liters):		Active bomb sight:	Yes
Internal:	3,840	Laser rangefinder/ designator 10-15 km:	Yes
External:		Kopyo-25 pulse Doppler multi-role radar:	Yes
800:	4	SAU-8 automated control system:	Yes
1150:	2	INS:	Yes
Range Max Load (km):	500	GPS:	Yes
Plus2 Aux Fuel tanks:	750 or 1250	Doppler Radar:	Yes
Ferry Range (Max Fuel):	2,500	Night/Weather Capabilities:	Yes
Combat Radius:	556		

NOTES

THE AIRCRAFT CAN CARRY A SELF-CONTAINED MAINTENANCE KIT IN 4 UNDER-WING PODS. THE LASER TARGET DESIGNATOR CAN GUIDE A VARIETY OF BOMBS, MISSILES, AND ROCKETS, INCLUDING S-24 SAL-H ROCKETS, S-25L ROCKETS TO 7 KM, AND S-25LD ROCKETS TO 10 KMMIG-25P/PD FOXBAT-E: INTERCEPTOR WITH IMPROVED RADAR THAT HAS LIMITED LOOK-DOWN/SHOOT-DOWN CAPABILITY, IR SENSOR UNDER THE NOSE, AND UPGRADED ENGINE.

THE ENGINES CAN OPERATE ON ANY TYPE OF FUEL TO BE FOUND IN THE FORWARD-OPERATING AREAS, INCLUDING DIESEL AND GASOLINE. THUS IT CAN OPERATE FROM UNPREPARED AIRFIELDS.

APPEARANCE: WINGS: HIGH-MOUNT, TAPERED BACK, ENGINES: BOTH ALONG BODY, UNDER WINGS.

REPRESENTATIVE MIX FOR TARGETING ARMOR IS: 30-MM GUN, 4 PODS (16) AT-16 ATGMS, AND 2 PODS OF SAL-H GUIDED ROCKETS. TWO OTHER PODS HOLD FUEL OR AS-10/12 MISSILES. MISSILES MAY REQUIRE A TV, RADAR OR IR POD FOR GUIDANCE. TWO OUTER MOUNTS HOLD SINGLE AA-8 MISSILE.

VARIANTS

THE SU-25 (FROGFOOT A) WAS THE ORIGINAL 1-SEAT AIRCRAFT FIELDED IN 1980, WITH SU-25K FOR EXPORT. EARLY SU-25S HAD 2X SOYUZ/ GAVRILOV

R95SH ENGINES. MOST ARE NOW UPGRADED.

SU-25B/-25UB/-25UBK/-UBP/: A TWO-SEAT COMBAT AIRCRAFT, NAVAL VERSION, AND TRAINER. THE SU-25UT/UTG TRAINERS ARE AKA FROGFOOT-B.

SU-39/SU-25TM (DOMESTIC): DEVELOPED FROM THE SU-25UB 2-SEAT TRAINER. FOR FCS SEE ABOVE. HEIGHT IS 5.2 M FOR AVIONICS AND EXTRA FUEL. NEW R-195 ENGINES OFFER MORE THRUST, RANGE, CEILING, AND LOAD. NEW COUNTERMEASURE SUITES ARE USED.

SU-25UBM: THE LATEST UPGRADE HAS THE SH013 NAVIGATION RADAR AND THE PASTEL RADAR WARNER. THE MODERNIZED CABIN HAS HEADS-UP AND LCD COLOR DISPLAYS. IT CAN LAUNCH KAB-500KL LASER HOMING (SAL) AND KAB-500KR TV GUIDED BOMBS.

SU-25KM/SKORPION: ISRAELI/GEORGIAN UPGRADE, WITH A CHOICE AMONG WESTERN AVIONICS.

SWEDISH MULTI-ROLE ATTACK AIRCRAFT AJ37 / VIGGEN

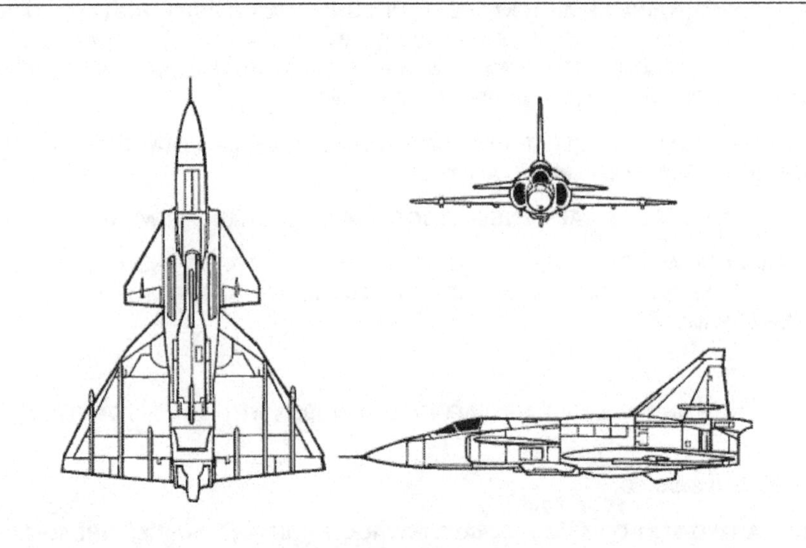

SYSTEM	SPECIFICATIONS	SYSTEM	SPECIFICATIONS
Weapon & Ammunition Types:		**Speed (km/h):**	
30 mm Oerlikon KCA automatic cannon (JA37) (rnds):	150	**Maximum (at altitude):**	2,135, Mach 2
Other Loading Options:		**Maximum (sea level):**	1,469, Mach 1.2
AJ37:	7 – 9	**Max "G" Force (g):**	+7 g
RB24 or RB74 Sidewinder:	7 – 9	**Ceiling (m):**	18,300
RB28 Falcon AAM:	7 – 9	**Vertical Climb Rate (m/s):**	203
RB75 Maverick AGM:	7 – 9	**Fuel (liters):**	
75mm 19-round rocket pods:	4	**Internal:**	5,700
135mm 6-round rocket pods:	4	**Range (km):**	
30mm Aden gun pod and drop tanks:	Yes	**With Aux Fuel:**	2000
JA37:	Yes	**Ferry:**	2250
RB74 Sidewinder AAM:	6	**Combat Radius (km):**	
2 RB 71 Skyflash AAM:	2	**Hi-lo-hi:**	>1000
SYSTEM		**Lo-lo-lo:**	> 500
Alternative Designations:		**Takeoff Run/Landing Roll (m):**	400/500
Date of Introduction:	1971	**Dimensions (m):**	
Proliferation:	Sweden	**Length:**	16.3
		Wingspan:	10.6

Description:		Height:	5.6
Crew:	1	Standard Payload (kg):	
Engines 14,750 lbs. thrust Svenska Flygmotor RM8A turbofan, 25,970 lbs. thrust with afterburner:	1	External:	6,000
Weight (kg):		Hardpoints pylons:	7 – 9
Maximum Gross:	20,500	Survivability/Countermeasures::	
Normal Takeoff:	16,000	0-75 km/hr. ejection seat.	Yes
Empty:	12,250	ECM system:	Yes
		Chaff dispenser	Yes
		Deception jammer:	Yes

NOTES

THE BASIC PLATFORM WAS THE AJ37 ATTACK AIRCRAFT, FOLLOWED BY THE S37 RECONNAISSANCE VERSIONS AND THE JA37 FIGHTER. THE NEW AIRCRAFT HAD A NOVEL AND ADVANCED AERODYNAMIC CONFIGURATION TO MEET THE SHOT TAKE-OFF/LANDING AND OTHER PERFORMANCE REQUIREMENTS: A FIXED FOREPLANE WITH FLAPS WAS MOUNTED AHEAD OF AND SLIGHTLY ABOVE THE MAIN DELTA WING. A TOTAL OF 329 AIRCRAFT WERE BUILT IN ATTACK, TRAINER, TWO RECONNAISSANCE VERSIONS AND THE MORE POWERFUL FIGHTER VARIANT THAT INCLUDED NEW AVIONICS, NEW AIR-TO-AIR MISSILES AND EUROPE'S FIRST PULSE-DOPPLER RADAR.

APPEARANCE:

WINGS: LOW-MOUNTED, DELTA-SHAPED FROM BODY MIDSECTION TO THE EXHAUST. SMALL, CLIPPED DELTA WINGS FORWARD OF MAIN WINGS AND HIGH-MOUNTED ON BODY. ENGINES: ONE TURBOFAN IN THE BODY. FUSELAGE: SHORT AND WIDE WITH A POINTED SOLID NOSE TAIL: NO TAIL FLATS. LARGE, UNEQUALLY TAPERED FIN WITH A SMALL, CLIPPED TIP. TAIL: NO TAIL FLATS. LARGE, UNEQUALLY TAPERED FIN WITH A SMALL, CLIPPED TIP.

VARIANTS

AJ37: ALL-WEATHER ATTACK AIRCRAFT WITH INTERCEPT CAPABILITY.

AJS37: VIGGENS REFITTED FOR MULTI-ROLE SERVICE WITH UPGRADED CENTRAL COMPUTER AND ESM/ECM PYLON JAMMING POD DEVELOPED FOR THE JAS 39.

JA37: AIR SUPERIORITY FIGHTER WITH STRIKE CAPABILITY; UPRATED RM8B ENGINE AND AVIONICS.

SF37: ARMED PHOTO RECONNAISSANCE VERSION. EXTENSIVE IR AND ESM FIT INCLUDING RWR AND ELINT DATA RECORDERS.

SH37: MARITIME RECONNAISSANCE/STRIKE VERSION HAS 2 SIDEWINDER AAM ON OUTER WING PYLONS.

SK37: TWO-SEAT TRAINER VERSION.

BRITISH/GERMAN/ITALIAN/SPANISH MULTI-ROLE AIRCRAFT EF-2000 EUROFIGHTER

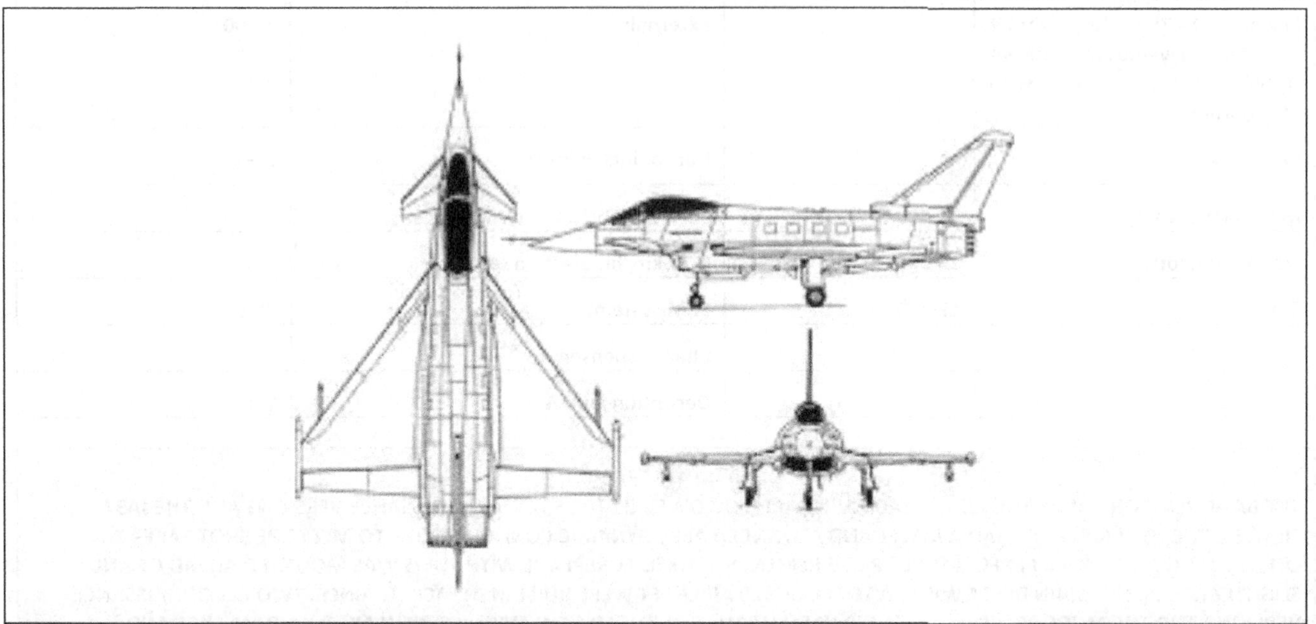

SYSTEM	SPECIFICATIONS	SYSTEM	SPECIFICATIONS
Weapon & Ammunition Types:		Combat Radius (km):	
Mauser BK 27-mm revolver cannon (rnds):	150	Ground attack, lo-lo-lo:	601
Air Superiority Packages:	Yes	Ground attack, hi-lo-hi:	1.389
BVRAAM:	6	Air defense with 3 hr. CAP:	185
ASRAAM:	6	AD with 10-min loiter:	1,389
Fuel tanks:		Takeoff Run (m):	300-700
1,500 (L):	2	Dimensions (m):	
1,000 (L):	1	Length:	16.0
Air Interdiction Package:		Wingspan:	11.0
Storm Shadow:	2	Height:	5.3
AMRAAM:	4	Standard Payload (kg):	
ASRAAM:	2	External:	6,500
Alarm	2	Hardpoints:	13
Fuel tanks:		Fuselage:	5
1,500 (L):	2	Wing (ea.):	4
1,000 (L):	1	Combat Radius (km):	

Suppression of Enemy Air Defense:	Yes	Ground attack, lo-lo-lo:	601
Alarm:	6	Ground attack, hi-lo-hi:	1.389
AMRAAMs:	4	Air defense with 3 hr. CAP:	185
ASRAAM:	4	AD with 10-min loiter:	1,389
Fuel tanks:		Takeoff Run (m):	300-700
1,000 (L):	1	Dimensions (m):	
Close Air Support Package:		Length:	16.0
Brimstone:	18	Wingspan:	11.0
AMRAAMs:	4	Height:	5.3
ASRAAM:	4	Standard Payload (kg):	
Fuel tanks:		External:	6,500
1,000 (L):	1	Hardpoints:	13
Maritime Attack Package:		Fuselage;	5
Penguin:	6	Wing (ea.):	4
AMRAAM:	4	Survivability/Countermeasures:	
ASRAAM:	2	Martin-Baker zero/zero ejection seat:	Yes
Fuel tanks:		DAAS (defensive aids sub-system) with electronic countermeasures/ support measures system (ECM/ ESM):	Yes
1,500 (L):	2	Front and rear missile warning:	Yes
1,000 (L):	1	Supersonic capable towed decoy system:	Yes
SYSTEM		Laser warning receivers:	Yes
Alternative Designations:	Typhoon	Chaff and flare dispensing system:	Yes
Date of Introduction:	2005	ARMAMENT:	
Proliferation:	5 countries (Britain, Greece, Germany, Italy, Spain)	Internal Mauser BK 27-mm revolver cannon:	Yes
Engines 13,500 lbs. thrust Eurojet EJ turbofans, 20,250 with afterburner:	2	AVIONICS/SENSOR/OPTICS:	
Weight (kg):		Helmet Mounted Symbology System (HMS):	Yes
Maximum Takeoff:	23,000	Heads-up-display:	
Normal Takeoff:		Flight reference data:	Yes
Empty:	9,750	Weapons Aiming and Cueing:	Yes

Speed (km/h):		FLIR Imaging:	Yes
Maximum (at altitude):	2,130, Mach 2.0	Head Down Display:	
Max "G" Force (g):	+9/-3 g	Tactical Situation:	Yes
Vertical Climb Rate (m/s):		System Status;	Yes
Fuel (liters):		Map Display:	Yes
Internal:		Multimode X-Pulse Doppler Radar:	Yes
External:	4,000	Infrared Search and Track System (IRST):	Yes
		Night / Weather Capabilities:	Yes

NOTES

EUROFIGHTER IS A SINGLE-SEAT, TWIN-ENGINE, AGILE COMBAT AIRCRAFT WHICH WILL BE USED IN THE AIR-TO-AIR, AIR-TO-GROUND, AND TACTICAL RECONNAISSANCE ROLES. THE DESIGN OF THE EUROFIGHTER IS OPTIMIZED FOR AIR DOMINANCE PERFORMANCE WITH HIGH INSTANTANEOUS AND SUSTAINED TURN RATES, AND SPECIFIC EXCESS POWER. SPECIAL EMPHASIS HAS BEEN PLACED ON LOW WING LOADING, HIGH THRUST TO WEIGHT RATIO, EXCELLENT ALL ROUND VISION AND CAREFREE HANDLING. THE USE OF STEALTH TECHNOLOGY IS INCORPORATED THROUGHOUT THE AIRCRAFT'S BASIC DESIGN.

APPEARANCE:

WINGS: CONSTANT LEADING EDGE SWEPT DELTA, WITH ALL-MOVING CANARD FOREPLANES PLACED AHEAD AND ABOVE THE MAIN WING. ENGINES: TWO TURBOFAN ENGINES FED BY A BROAD, ANGULAR GROUP UNDER THE FUSELAGE. FUSELAGE: CONVENTIONAL SEMI-MONOCOQUE WITH HEAVY BLENDING. TAIL: TALL SWEPT SINGLE FIN HAS AN INSET RUDDER. NO FLATS

VARIANTS

VARIANTS

TWO-SEAT OPERATIONAL CONVERSION TRAINER: RETAINS FULL COMBAT CAPABILITY. SECOND SEAT FITTED IN PLACE OF ONE FUSELAGE FUEL TANK, CANOPY LENGTHENED AND DORSAL LINE EXTENDED AFT TO BASE OF TAIL.

TYPHOON: ORIGINALLY, THIS WAS THE NAME FOR THE EXPORT VARIANTS, BUT IT IS LIKELY TO BE APPLIED TO ALL AIRCRAFT WITH APPROPRIATE SPELLING CHANGES.

NAVAL VARIANT: VERSION PROPOSED AS A POSSIBLE COMPETITOR TO THE JOINT STRIKE FIGHTER FOR OPERATIONS OFF FUTURE BRITISH CARRIERS.

INTERDICTOR VARIANT: LONG-RANGE, DEEP-STRIKE VERSION, CAPABLE OF SURGICAL STRIKE AGAINST GROUND TARGETS USING STAND-OFF PRECISION GUIDED MISSILES THAT COULD BE FITTED WITH CONFORMAL FUEL TANKS FOR INCREASED RANGE.

AMERICAN FIGHTER-BOMBER AIRCRAFT F-4/PHANTOM

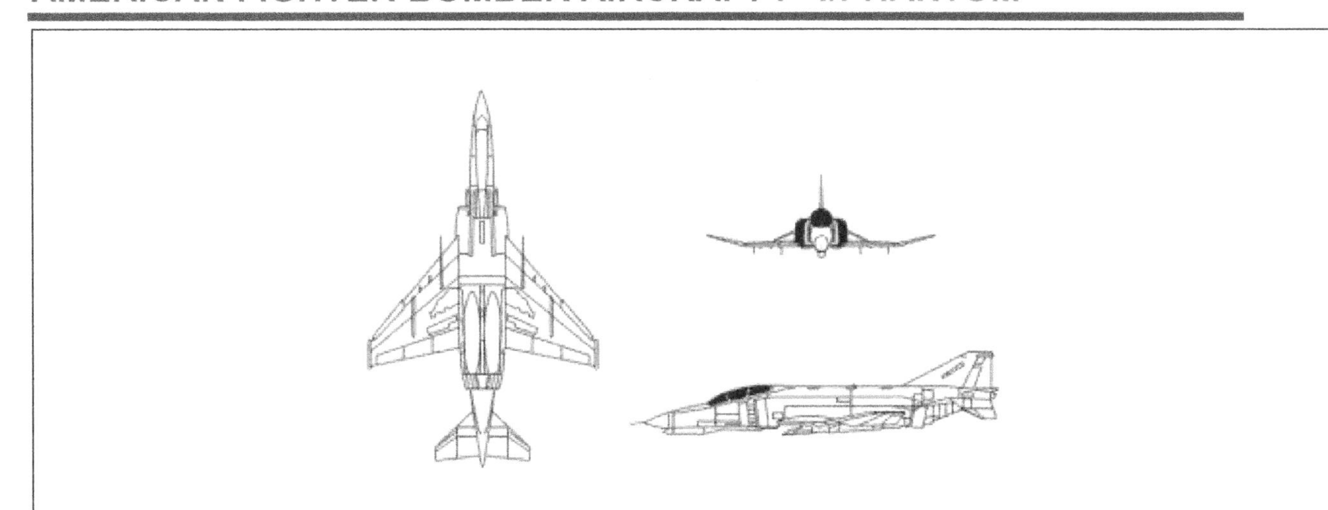

SYSTEM	SPECIFICATIONS	SYSTEM	SPECIFICATIONS
Weapon & Ammunition Types:		Combat Radius (km):	
Mauser BK 27-mm revolver cannon (rnds):	150	Ground attack, lo-lo-lo:	601
Air Superiority Packages:	Yes	Ground attack, hi-lo-hi:	1.389
BVRAAM:	6	Air defense with 3 hr. CAP:	185
ASRAAM:	6	AD with 10-min loiter:	1,389
Fuel tanks:		Takeoff Run (m):	300-700
1,500 (L):	2	Dimensions (m):	
1,000 (L):	1	Length:	16.0
Air Interdiction Package:		Wingspan:	11.0
Storm Shadow:	2	Height:	5.3
AMRAAM:	4	Standard Payload (kg):	
ASRAAM:	2	External:	6,500
Alarm	2	Hardpoints:	13 (5 fuselage, 4 ea. wing)
Fuel tanks:		Combat Radius (km):	
1,500 (L):	2	Ground attack, lo-lo-lo:	601
1,000 (L):	1	Ground attack, hi-lo-hi:	1.389
Suppression of Enemy Air Defense:	Yes	Air defense with 3 hr. CAP:	185
Alarm:	6	AD with 10-min loiter:	1,389

AMRAAMs:	4	Takeoff Run (m):	300-700
ASRAAM:	4	Dimensions (m):	
Fuel tanks:		Length:	16.0
1,000 (L):	1	Wingspan:	11.0
Close Air Support Package:		Height:	5.3
Brimstone:	18	Standard Payload (kg):	
AMRAAMs:	4	External:	6,500
ASRAAM:	4	Hardpoints:	13
Fuel tanks:		Fuselage;	5
1,000 (L):	1	Wing (ea.):	4
Maritime Attack Package:		Survivability/Countermeasures:	
Penguin:	6	Martin-Baker zero/zero ejection seat:	Yes
AMRAAM:	4	DAAS (defensive aids sub-system) with electronic countermeasures/ support measures system (ECM/ ESM):	Yes
ASRAAM:	2	Front and rear missile warning:	Yes
Fuel tanks:		Supersonic capable towed decoy system:	Yes
1,500 (L):	2	Laser warning receivers:	Yes
1,000 (L):	1	Chaff and flare dispensing system:	Yes
SYSTEM		ARMAMENT:	
Alternative Designations:	Typhoon	Internal Mauser BK 27-mm revolver cannon:	Yes
Date of Introduction:	2005	AVIONICS/SENSOR/OPTICS:	
Proliferation:	5 countries (Britain, Greece, Germany, Italy, Spain)	Helmet Mounted Symbology System (HMS):	Yes
Engines 13,500 lbs. thrust Eurojet EJ turbofans, 20,250 with afterburner:	2	Heads-up-display:	
Weight (kg):		Flight reference data:	Yes
Maximum Takeoff:	23,000	Weapons Aiming and Cueing:	Yes
Normal Takeoff:		FLIR Imaging:	Yes
Empty:	9,750	Head Down Display:	
Speed (km/h):		Tactical Situation:	Yes
Maximum (at altitude):	2,130, Mach 2.0	System Status;	Yes
Max "G" Force (g):	+9/-3 g	Map Display:	Yes

Vertical Climb Rate (m/s):		Multimode X-Pulse Doppler Radar:	Yes
Fuel (liters):		Infrared Search and Track System (IRST):	Yes
Internal:		Night / Weather Capabilities:	Yes
External:	4,000		

NOTES

F-4S ARE NO LONGER IN SERVICE IN THE U.S. MILITARY. THE QF-4 TARGET DRONE REMAINS IN US SERVICE. SEVERAL HUNDRED F-4S REMAIN IN SERVICE WITH GERMAN, JAPANESE, SOUTH KOREA, ISRAELI, GREEK, AND TURKISH AIR FORCES, WITH SEVERAL UPGRADE PROGRAMS UNDERWAY IN SEVERAL COUNTRIES. PLANNED AS AN ATTACK AIRCRAFT WITH FOUR 20 MM GUNS, IT WAS QUICKLY CHANGED INTO A VERY ADVANCED GUNLESS ALL-WEATHER INTERCEPTOR WITH ADVANCED RADAR AND MISSILE ARMAMENT. THE AIRCRAFT FLEW EVERY TRADITIONAL MILITARY MISSION: AIR SUPERIORITY, CLOSE AIR SUPPORT, INTERCEPTION, AIR DEFENSE, SUPPRESSION, LONG-RANGE STRIKE, FLEET DEFENSE, ATTACK, AND RECONNAISSANCE.

APPEARANCE: WINGS: SWEPT DELTA, LEADING EDGE HAVING GREATER SWEEP THAN THE TRAILING EDGES. ENGINES: TWO AFTERBURNING TURBOJETS HOUSED SIDE-BY-SIDE IN THE FUSELAGE. FUSELAGE: TUBULAR WITH POINTED NOSE AND TAPERED ENGINE HOUSING ON EACH SIDE. TAIL: SHORT, SHARPLY SWEPT FIN AND RUDDER.

VARIANTS

F-4B: FIRST PRODUCTION VARIANT FOR U.S. NAVY AND MARINE CORPS.

F-4C: FIRST PRODUCTION VARIANT FOR U.S. AIR FORCE.

F-4D: SIMILAR TO F-4C WITH IMPROVED RADAR, INS, GUN SIGHT AND WEAPONS RELEASE COMPUTER.

F-4E: IMPROVED AIR FORCE VERSION WITH NEW RADAR, SIX-BARREL CANNON, ADDED FUEL AND NEW ENGINE.

F-4EJ KAI: JAPANESE UPDATE PROGRAM, INCLUDED PULSE-DOPPLER RADAR, HUD, INS, AND RWR.

ISRAELI F-4E WILD WEASEL: F-4E CONFIGURED TO FIRE THE AGM-78B STANDARD ARM MISSILE.

F-4F: SIMILAR TO F-4E FOR GERMAN AIR FORCE. INTRODUCED LEADING-EDGE MANEUVERING SLATS.

F-4G WILD WEASEL: ATTACK/ELECTRONIC WARFARE (EW) VERSION OF THE F-4E FOR ANTI-RADAR ROLE.

F-4J: NAVY F-4B UPGRADE OF RADAR, FIRE CONTROL SYSTEM, ENGINE AND DROOPING AILERONS.

F-4K/FG1: ROYAL NAVY VERSION OF F-4J FOR CARRIER OPERATIONS.

F-4M/FGR2:ROYAL AIR FORCE VERSION OF THE F-4K

F-4N: UPGRADED F-4B WITH IMPROVED WEAPONS CONTROL SYSTEM AS WELLS STRUCTURAL STRENGTHENING.

F-4S: REBUILT F-4JS, BUT WITH OUTER LEADING-EDGE MANEUVERING SLATS.

RF-4: RECONNAISSANCE VARIANT.

KORNAS 2000/SUPER PHANTOM (SLEDGEHAMMER 2000): ISRAELI-DEVELOPED UPGRADE TO EXTEND SERVICE LIFE INTO THE 21ST CENTURY AND SERVE AS THE BASE OF THE IAF'S AIR-TO-GROUND CAPABILITY.

ISRAELI F-4E SUPER PHANTOM/PHANTOM 2000: KORNAS 2000 VARIANT FITTED WITH NEW TURBOFAN ENGINES. REDUCED TAKE-OFF DISTANCE, INCREASED RATE OF CLIMB, AND INCREASED LOW-LEVEL SPEED.

SWEDEN MULTI-ROLE FIGHTER AIRCRAFT JAS39/GRIPEN

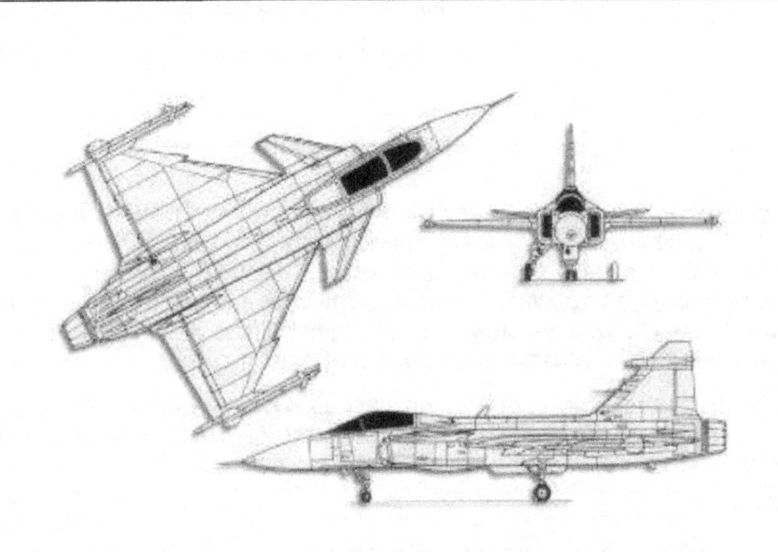

SYSTEM	SPECIFICATIONS	SYSTEM	SPECIFICATIONS
Weapon & Ammunition Types:		Range (km):	
Mauser BK 27-mm revolver cannon (rnds):	120	Combat Radius:	800
AIM-9 Sidewinder on the wingtips:	2	Ferry:	3,000
AIM-120 AMRAAM:	4	Takeoff Run/Landing Roll (m):	800/800
AGM-65A/B Maverick:	4	Dimensions (m):	
Saab RBS15F anti-shipping missile:	2	Length:	
Dasa DWS39 munitions dispenser or KEPD150 pods:	2	A/C:	14.1
Bofors rocket pods:	4	B/D:	14.8
Conventional bombs:	4	Wingspan over tip rails:	8.4
Description:		Height:	4.5
Crew:		Standard Payload (kg):	
JAS 39A/C	1	External:	3,600
JAS 39B/D	2	Hardpoints:	7
Engines 12,140 lbs. thrust Volvo Aero RM12 or 18,200 lbs. thrust with afterburner:	1	Wings:	4

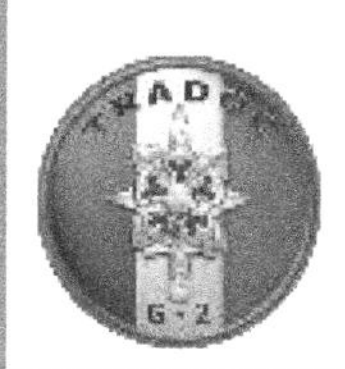

Weight (kg):		Centerline:	1
Takeoff:		Wingtip Rails:	2
A/C:	12,500	Survivability/Countermeasures:	
B/D:	14,000	Martin-Baker zero/zero ejection seat:	Yes
Empty:		IFF and an integrated EW system that provides radar warning:	Yes
A/C:	6.500	Electronic support measures:	Yes
B/D:	7,100	Decoy system:	Yes
Speed (km/h):		Chaff and flare dispensing system:	Yes
Maximum (at altitude):	2,150, Mach 1.8+	ARMAMENT:	
Max "G" Force (g):	+9/-3	Mauser BK 27-mm revolver cannon:	Yes
Ceiling (m):	16,000	AVIONICS/SENSOR/OPTICS:	
Fuel (liters):		Long-range multi-purpose pulse Doppler radar:	Yes
Internal:		Air-to-air operating mode:	
A/C:	3,008	Night / Weather Capabilities:	Yes
B/D:	2,852		
External:	3,300		

NOTES

THE JAS 39 GRIPEN IS A FOURTH GENERATION, MULTI-ROLE COMBAT AIRCRAFT. THE GRIPEN IS THE FIRST SWEDISH AIRCRAFT THAT CAN BE USED FOR INTERCEPTION, GROUND-ATTACK, AND RECONNAISSANCE (HENCE THE SWEDISH ABBREVIATION JAS – FIGHTER (J), ATTACK (A), AND RECONNAISSANCE (R)) AND IT IS NOW SUCCESSIVELY REPLACING THE DRAKEN AND THE VIGGEN .THE JAS 39 IS PART OF A SYSTEM THAT FIGHTS THE "INFORMATION WAR" IN WHICH AIRCRAFT RECEIVE AND CONVEY INFORMATION THROUGH AN AIR-TO-AIR TACTICAL INFORMATION DATA LINK SYSTEM (TIDLS).

APPEARANCE:

WINGS: MULTI-SPARRED DELTA. LARGE, SWEPT, ALL-MOVING FOREPLANE CANARDS MOUNTED ON ENGINE INTAKE SHOULDERS. ENGINES: TURBOFAN WITH INTAKE BOXES ON BOTH SIDES OF FUSELAGE. FUSELAGE: TAIL: LEADING EDGE SWEPT FIN WITH UPRIGHT INSET RUDDER.

VARIANTS

JAS 39A: ORIGINAL SINGLE-SEAT VERSION SUPPLIED TO THE SWEDISH AIR FORCE.

JAS 39B: DESIGN-STUDY CONTRACT FOR TRAINER/RECONNAISSANCE VARIANT AWARDED TO JAS IN 1989; FUSELAGE PLUG INSERTED TO MAKE ROOM FOR SECOND SEAT.

JAS 39C/D: NATO-COMPATIBLE EXPORT VARIANT EQUIPPED WITH OBOGS, FLIR, NVG-COMPATIBLE COCKPIT, LASER-DESIGNATOR POD, HMD. HIGHER GROSS TAKEOFF WEIGHT. THE 39D IS THE TWO-SEAT EQUIVALENT.

ISRAELI MULTI-ROLE FIGHTER KFIR (LION CUB)

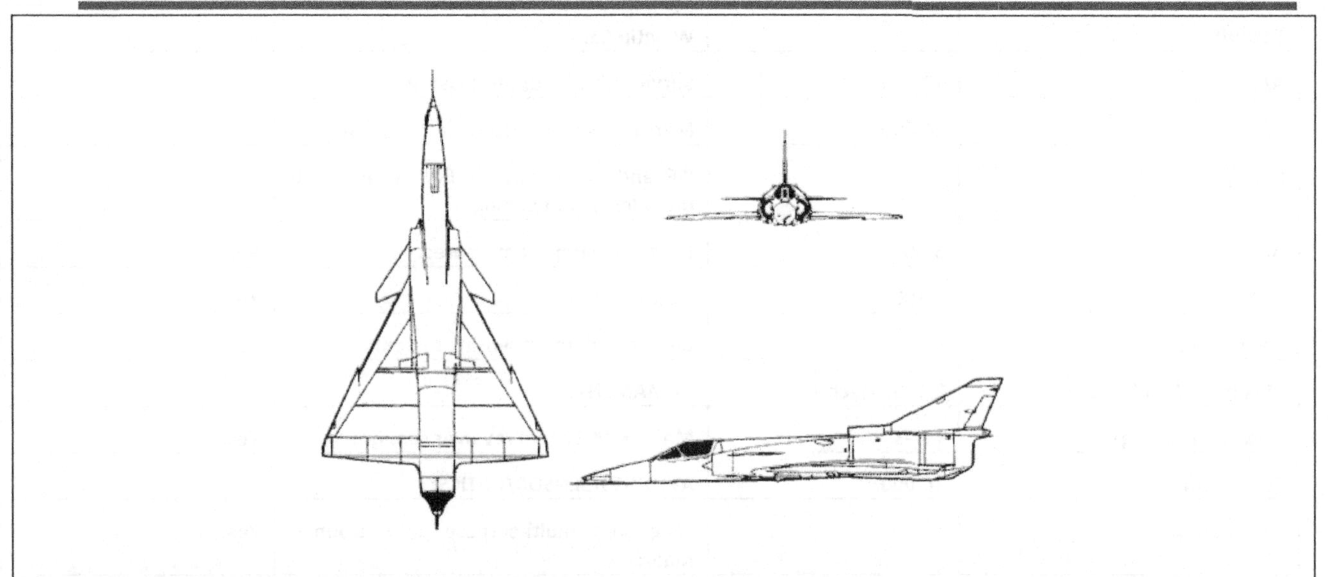

SYSTEM	SPECIFICATIONS	SYSTEM	SPECIFICATIONS
Weapon & Ammunition Types:		**Range (km):**	
Internal 30-mm cannons (x2):	140	**Ferry:**	
Other Loading Options:	Yes	**C2:**	2,991
Python/Shafrir/AIM-9:	2	**C7:**	3,232
AGM-45 Shrike ARM:	2	**Combat Radius (km):**	
AGM-65 Maverick:	1	**Intercept Mission:**	
1,500 (L):	2	**C2:**	347
1,000 (L):	1	**C7:**	776
Bombs:		**Combat Air Patrol:**	
GBU-15 glide bomb:	1	**C2:**	699
227 kg:	6	**C7:**	882
363 kg or 454 kg:	2	**Ground Attack:**	
1,361 kg:	1	**C2:**	768
Fuel tanks External:	3	**C7:**	1,186
(L):	4,700	**Takeoff Run/Landing Roll (m):**	
SYSTEM		**Max Load:**	
Alternative Designations:	C2; C7	**C2:**	1,455
Date of Introduction:	1975	**C7:**	1,555
Proliferation:	6 countries	**Dimensions (m):**	

Description:		Length:	15.7
Crew:	1	Wingspan:	8.2
Engines:		Height:	4.6
17,750 lbs. thrust General Electric J79-GE-1JE Turbojet (C2):	1	Standard Payload (kg):	
18,750 lbs. thrust (C7):	1	External:	
Weight (kg):		C2:	4,277
Maximum Takeoff:	16,500	C7:	5,775
Normal Takeoff:		Hardpoints including missiles:	
Empty:	7,285	C2:	7
Speed (km/h):		C7:	9
Max (at altitude):	2,440,Mach 2.3	Survivability/Countermeasures:	
Max (sea level):	1,389,Mach 1.1	Cockpit pressurized, heated, and air conditioned:	Yes
Takeoff/Landing Speed:	220	Martin-baker zero/zero ejection seats:	Yes
Max "G" Force (g):	+7.5 g	In-flight refueling:	Yes
Ceiling (m):	17,680	IFF, ECM pods:	Yes
Vertical Climb Rate (m/s):	233	Radar warning receiver:	Yes
Fuel (liters):		Chaff and Flares:	yes
Internal:	3,243	ARMAMENT:	
External:	4,700	2 internal 30-mm DEFA 552 cannons:	Yes

NOTES

OVER 230 AIRCRAFT WERE IN MILITARY SERVICE WITH ISRAEL AND SEVERAL OTHER NATIONS, BUT MOST OF THE ISRAELI KFIRS ARE NOW IN STORAGE.

APPEARANCE: WINGS: LOW-MOUNTED, DELTA-SHAPED WITH A SAW TOOTH IN THE LEADING EDGE. ENGINES: ONE TURBOJET. FUSELAGE: TUBE SHAPED WITH A LONG, SOLID, POINTED NOSE. TAIL: NO TAIL FLATS. FIN IS SWEPT-BACK AND TAPERED WITH A STEP IN THE LEADING EDGE.

VARIANTS

KIFR C1: INITIAL PRODUCTION MODEL, LEASED TO THE US NAVY AND MARINE CORPS AND RE-DESIGNATED F-21A.

F-21A: SLIGHTLY MODIFIED C1, USED TO TRAIN US COMBAT PILOTS IN ADVERSARY TACTICS. USED BY US NAVY FROM 1985 TO MAY 1988. USED BY USMC FROM 1987 TO SEP 1989.

KFIR C2: REVISED AIRFRAME WITH FOREPLANES AND NOSE STRAKE ADDED.

KFIR C7: UPGRADED VERSION WITH NEW AVIONICS: WDNS-391 WEAPONS CONTROL SYSTEM WITH STORES MANAGEMENT DISPLAY, UPRATED GE J79-1JE ENGINE AND GREATER INTERNAL FUEL CAPACITY.

KFIR-2000: DESIGNED AS A COST-EFFECTIVE MULTIROLE AIRCRAFT.

RUSSIAN MULTI-ROLE FIGHTER AIRCRAFT MIG-21/FISHBED

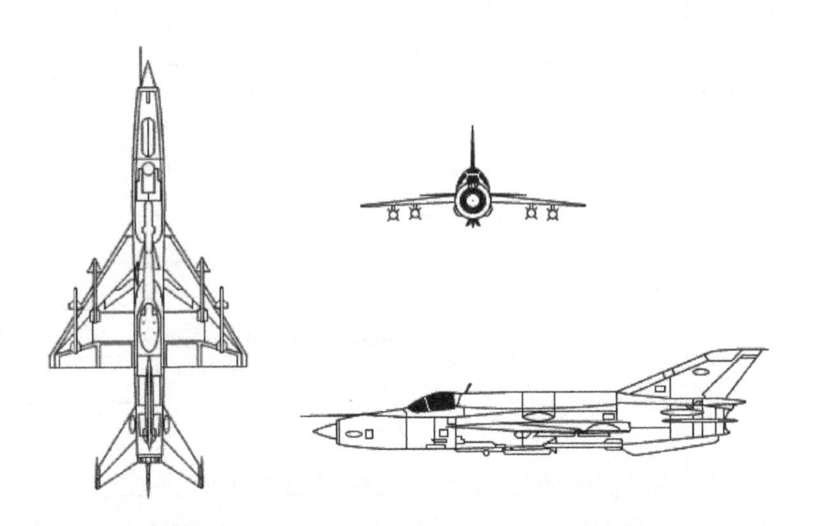

SYSTEM	SPECIFICATIONS	SYSTEM	SPECIFICATIONS
Weapon & Ammunition Types		Fuel (liters):	
23-mm Gsh-23 2-barrel cannon (rnds):	200	Internal:	2,650
Other Loading Options:	2	External (x2):	1,470
AA-8 Aphid:			
AA-2C or D Atoll:	2 - 4	Range (km):	
Gun Pods:	2	Ferry:	2,100
Unguided bombs:		High Alt w/internal fuel and 2 AAM:	1,000
Rockets:	4	Low Alt w/internal fuel and 2 AAM:	560
SYSTEM		Takeoff Run/Landing Roll with drag chute (m):	900/650
Alternative Designations:	J-7 (Chinese)	Dimensions (m):	
Date of Introduction:	1958	Length:	
Proliferation:	≥40 countries	w/out probe:	14.5
Description:		w/probe:	15.8
Crew:	1	Wingspan:	7.2
Engines:		Height:	4.5
12,675 lbs. thrust w/afterburner Tumansky R-11F-300 (MiG-21):	1	Standard Payload 4 under wing pylons (kg):	1,200

14,550 lbs. thrust Wopen-13 turbofan (J-8):	1	Survivability/Countermeasures:	
Weight (kg):		Pressurized cockpit with ejection seat:	Yes
Normal Takeoff:	8,825	Radar warning receiver:	Yes
Speed (km/h):		Chaff and flares:	Yes
Max (at altitude):	2,175 (Mach 2.05)	ARMAMENT:	
Max (sea level):	1,300 (Mach 1.05)	NR-30 guns in the forward fuselage (early models) (ea.):	2
Landing Speed:	270	GSh-23 23-mm cannons and 200 rounds (Fishbed-D and later models):	2
Max "G" Force (g):	+8.5 g	AVIONICS/SENSOR/OPTICS:	
Ceiling (m):	18,000	Spin Scan or Jay Bird airborne interception radar and a gyro-stabilized gun sight:	Yes
Vertical Climb Rate (m/s):	225	Night/Weather Capabilities:;	Limited

NOTES

A PRINCIPAL WEAKNESS OF THE MIG-21 DESIGN IS THE REARWARD SHIFT OF THE CENTER OF GRAVITY AS THE FUSELAGE TANKS ARE EMPTIED. A FULL ONE-THIRD OF THE FUEL ON BOARD CANNOT BE USED FOR THIS REASON. THE SAME LIMITATION EFFECTIVELY REDUCES MACH 2 FLIGHT TIME TO PERFUNCTORY HIGH-SPEED TESTS. THE MIG-21 IS A SIMPLE, RELIABLE AIRCRAFT WITH HONEST FLYING CHARACTERISTICS. IT IS ALSO CONSIDERED TO BE A COMPETENT DOG-FIGHTER AGAINST MOST WESTERN AIRCRAFT. INDIA HAS SUFFERED AN ALMOST INCREDIBLE STRING OF MIG-21 CRASHES SINCE 1998, INCLUDING SEVERAL NOTABLE INCIDENTS THAT HAVE KILLED PEOPLE ON THE GROUND. FROM JANUARY 1998 TO DECEMBER 2002 THERE WERE OVER 50 MIG-21 CRASHES, INCLUDING THREE THAT KILLED A TOTAL OF 13 PEOPLE ON THE GROUND. ANALYSTS ARE DEBATING IF THE AGE OF THE AIRCRAFT IS AN ISSUE OR IF THERE ARE SERIOUS ERRORS IN PILOT TRAINING. THE MIG-21 IS A SHORT-RANGE DAY FIGHTER-INTERCEPTOR WITH LIMITED POSSIBILITIES IN ADVERSE WEATHER CONDITIONS

APPEARANCE: WINGS: MID-MOUNT, DELTA, SQUARED TIPS. ENGINES: ONE TURBOFAN IN FUSELAGE. FUSELAGE: LONG AND TUBULAR, WITH BLUNT NOSE AND BUBBLE CANOPY. TAIL: SWEPT-BACK, TAPERED WITH SQUARE TIP. FLATS ARE MID-MOUNTED ON THE BODY, SWEPT-BACK, AND TAPERED WITH SQUARE TIPS.

VARIANTS

MIG-21 FISHBED-C, D, AND F VARIANTS ARE FIGHTERS. LATER RUSSIAN VARIANTS ARE MULTI-ROLE FIGHTERS, EXCEPT H (RECON).

MIG-21F FISHBED-C: FIRST PRODUCTION VARIANT WITH RD-11 ENGINE. 1 X NR-30 30-MM CANNON.

MIG-21PF FISHBED-D: INTERCEPTOR WITH ENLARGED INTAKE THAT BECAME STANDARD. SPIN SCAN RADAR. PITOT TUBE RELOCATED TO TOP OF INTAKE.

MIG-21PF FISHBED-E: PRINCIPAL PF PRODUCTION VERSION. GP-9 23-MM GUN PACK. PROVISION FOR ROCKET-ASSISTED TAKE-OFF, GROUND (RATOG).

MIG-21FL FISHBED-E: EXPORT VARIANT OF PF WITHOUT RATOG. FITTED WITH SPIN SCAN RADAR.

MIG-21 FISHBED-G: DERIVATIVE USED TO TEST LIFT AND CRUISE ENGINE VERTICAL TAKE-OFF AND LANDING (VSTOL) DESIGN. ALTHOUGH NOT PRODUCED, CONFIGURATION LATER REAPPEARED IN YAK-38 FORGER NAVAL VSTOL AIRCRAFT.

MIG-21R FISHBED-H: RECON VERSION WITH ELECTRONIC INTELLIGENCE EQUIPMENT IN BELLY PACKS, FOR DAY/NIGHT PHOTOGRAPHIC, LASER, IR OR TV SENSORS.

MIG-21PFMA FISHBED-J: TWO ADDITIONAL WING PYLONS. JAY BIRD RADAR CAPABLE OF GUIDING SEMI-ACTIVE RADAR HOMING ADVANCED ATOLL AAM.

MIG-21MF FISHBED-J: UPDATED PFMA USING 14,550-LB STATIC THRUST TUMANSKY R-13-300 ENGINE. WING STRESSED FOR LOW-LEVEL FLIGHT PERMITTING MACH 1.06 AT LOW ALTITUDE.

MIG-21M FISHBED-J: EXPORT VERSION OF MIG-21 PFMA WITH TUMANSKY R-11F2S-300 ENGINE. BUILT IN INDIA FROM 1973 TO 1981.

MIG-21SMB FISHBED-K: SIMILAR TO MIG-21MF, WITH EXTENSION OF DEEP DORSAL SPINE FOR FUEL TANK AND AERODYNAMIC SHAPING, ECM FAIRINGS ON WING TIP.

MIG-21 BIS FISHBED-L: THIRD GENERATION MIG-21, SIMPLER CONSTRUCTION, LONGER FATIGUE LIFE, GREATER FUEL CAPACITY. IT HAS IMPROVED COMPUTER-BASED FIRE CONTROL.

MIG-21 BIS FISHBED-N: SIMILAR TO FISHBED-L, BUT WITH 16,535-LB STATIC THRUST TUMANSKY R-25 ENGINE.

MIG-21-93 FISHBED-N: MIDLIFE UPGRADE PACKAGE BASED ON THE MIG-21 BIS. THE LATEST VERSION WAS ALSO DEVELOPED FOR UPGRADE OF OLDER MIG-21S, WITH UPGRADED FIRE CONTROL AND THE COHERENT PULSE-DOPPLER KOPYO RADAR, (PERMITTING USE OF RADAR-GUIDED AND OTHER PRECISION MUNITIONS). MISSILES AVAILABLE INCLUDE: AA-12 ADDER, AA-11 ARCHER, AA-10 ALAMO, AS-10, AS-12, AND AS-17.IT CAN ALSO DELIVER KAB-500R AND KAB-500L GUIDED BOMBS. A FACTORY UPGRADED AND EXPORTABLE VERSION IS OFFERED.

MIG-21 BISON. INDIAN LICENSED UPGRADE FOR THEIR MIG-21S TO THE MIG-21-93 STANDARD, BEGUN IN THE EARLY 2000S. THIS PROGRAM IS PROBABLY ENDED, WITH A RECENT REPORT THAT INDIA WILL SCRAP ITS FLEET OF MIG-21S, AND REPLACE THEM WITH NEWER RUSSIAN AIRCRAFT.

MIG-21-2000: ISRAEL AIRCRAFT INDUSTRIES (IAI) UPGRADE. CAPABLE OF USING RUSSIAN STANDARD ARMAMENT AND THE RAFAEL PYTHON 4 AAM

MIG-21 LANCER: ROMANIA'S AEROSTAR AND ISRAEL'S ELBIT JOINTLY DESIGNED THIS UPGRADE PROGRAM FOR 110 ROMANIAN AIR FORCE MIG-21S: 25 AIR DEFENSE, 75 GROUND-ATTACK AND 10 TWO-SEAT TRAINERS.

MIG-21U MONGOL-A: TRAINER VERSION WITH TWO-SEATS AND WITH WEAPONS REMOVED.

MIG-21US MONGOL-B: A MODIFIED VERSION WITH NO DORSAL FIN AND BROADER VERTICAL TAIL SURFACES. SIMILAR TO MONGOL-A, WITH SPS FLAP-BLOWING AND RETRACTABLE INSTRUCTOR PERISCOPE.

MIG-21UM MONGOL-B: TRAINER WITH R-13-300 ENGINE. SIMILAR TO MIG-21F.

J-8:CHINESE AIRCRAFT IS LOOSELY BASED ON MiG-21and MiG-23 features.

Russian Multi-role Fighter Aircraft MiG-23/MiG-27/FLOGGER

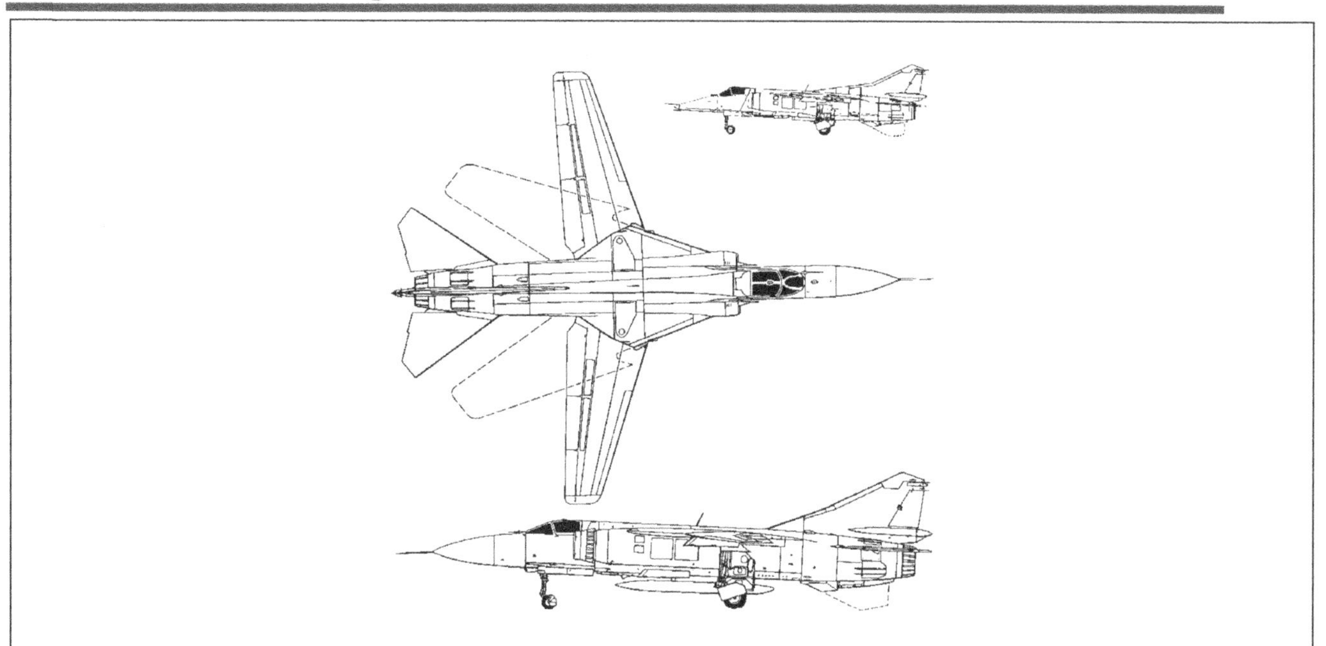

SYSTEM	SPECIFICATIONS	SYSTEM	SPECIFICATIONS
Weapon & Ammunition Types		Fuel (liters):	
23-mm Gsh-23L-twin gun (rnds):	200	Internal:	
23-mm Gsh-6-23 Gatling gun	260	MiG-23:	4,250
Other Loading Options:	2	MiG-27:	5,400
AA-7 APEX (K-23R/T):	2	External 800 liter tanks:	5
AA-8/APHID AAM launchers:	2	Range (km):	
AS-7/KERRY ASM:	4	Max Load:	1,500
AS-10/KAREN ASM:	4	With Aux Fuel:	2,500
AS-12/KEGLER ASM:	4	Combat Radius:	1,150
AS-14/KEDGE ASM:	4	Takeoff Run/Landing Roll (m):	
Rockets:	4	Prepared Surface:	
240-mm S-24 rockets:	1	MiG-23:	500/750
80-mm S-8 rkt pods:	20	MiG-27:	950/1,300
57-mm S-5 rkt pods:	32	Dimensions (m):	
Unguided Bombs:		Length:	
50 kg:		MiG-23:	16.8
100 kg:		MiG-27:	17.1
200 kg:		Wingspan:	

1,000 kg):		Extended:	14.0
SYSTEM		Swept;	7.8
Alternative Designations:	MiG-27, Bahadur, or Valiant (Indian variant)	Height:	
Date of Introduction:	1972	MiG-23:	4.8
Proliferation:	>23 countries	MiG-27:	5.0
Description:		Standard Payload (kg):	
Crew:	1	External:	
Landing Speed:	270	MiG-23:	3,000
Max "G" Force (g):	+8.5	MiG-27:	4,000
Ceiling (m):	18,000	Hardpoints:	
Vertical Climb Rate (m/s):	225	MiG-23:	6
Engines:		MiG-27:	7
28,660-shp Soyuz/Kachaturov R-35-300 turbojet, afterburner (MiG-23):	1	Survivability/Countermeasures:	
25,335-shp R-29B-300 turbojet, afterburner (MiG-27):	1	Pressurized cockpit with zero/130 ejection seat:	Yes
Weight (kg):		Infrared and radar jammer:	Yes
Max Gross:		Radar warning receiver:	Yes
MiG-23:	17,800	Decoy:	Yes
MiG-27:	20,700	Chaff and flares:	Yes
Normal Takeoff:		Armored Cockpit (MiG-27):	Yes
MiG-23:	14,840	ARMAMENT:	
MiG-27:	18,900	Gsh-6-23	
Empty:		23-mm twin gun, Gsh-23L:	
MiG-23:	10,200	Range (practical) (m):	2,500
MiG-27:	11,908	Elevation/Traverse (rigidly mounted):	None
Speed (km/h):		Ammo Type:	
Max (at altitude):		HEFI:	Yes
MiG-23:	Mach 2.35	Rate of Fire (rpm):	9,000
MiG-27:	Mach 1.7	23-mm 6x barrel Gatling gun, Gsh-6-23:	
Max (sea level):	Mach 1.2	Range (m) (practical):	2,500
Takeoff/Landing Speed:	315/270	Elevation/Traverse (rigidly mounted):	None
Max "G" Force (g):		Ammo Type:	HEFI

MiG-23:	+8.5	Rate of Fire (rpm):	9,000
MiG-27:	+7.0	AVIONICS/SENSOR/OPTICS;	
Ceiling (m):		Acquisition and tracking radar (MiG-23):	Yes
Service (clean):	18,600	IR Sensor (MiG-23):	Yes
With External Stores:	INA	Doppler Nav System (MiG-23):	Yes
Vertical Climb Rate (m/s):	240	Laser rangefinder/designator (MiG-23B and MiG-27):	Yes
		TV sighting system (MiG-23B and MiG-27):	Yes
		Acquisition and tracking radar (MiG-23B and MiG-27):	Yes
		MiG-23 (Night and day only):	Yes
		MiG-27:	Yes

NOTES

INSET LINE-DRAWING SHOWS NOSE AND INTAKE DIFFERENCES OF THE MIG-27.THIS DIFFERENCE ALLOWS FOR A LASER RANGEFINDER/TARGET DESIGNATOR. THE SWEEP WING IS CAPABLE OF THREE ANGLES: 16, 45, AND 72 DEGREES. THE VENTRAL FIN ON THE BOTTOM REAR OF THE FUSELAGE FOLDS FOR TAKEOFF AND LANDING. UP TO FIVE EXTERNAL FUEL TANKS CAN BE CARRIED ON THE MIG-23, AND FOUR ON THE MIG-27, BUT THE MIG-27 CAN ALSO BE FITTED FOR AERIAL REFUELING. AVAILABLE MUNITIONS ARE SHOWN ABOVE; NOT ALL MAY BE EMPLOYED AT ONE TIME. MISSION DICTATES WEAPONS CONFIGURATION. EXTERNAL STORES ARE MOUNTED ON UNDERWING AND UNDERBODY HARDPOINTS. EACH WING HAS ONE POINT, TWO POINTS ARE UNDER THE INTAKES ALONG THE FUSELAGE, AND THE CENTER FUSELAGE ATTACHMENT POINT GIVES FIVE TOTAL STATIONS. THE MIG-27 THEN ADDS TWO MORE BOMB RACKS UNDER THE WINGS FOR A TOTAL OF SEVEN STATIONS.

APPEARANCE: WINGS: MID-MOUNT, DELTA, SQUARED TIPS. ENGINES: ONE TURBOFAN IN FUSELAGE. FUSELAGE: LONG AND TUBULAR, WITH BLUNT NOSE AND BUBBLE CANOPY. TAIL: SWEPT-BACK, TAPERED WITH SQUARE TIP. FLATS ARE MID-MOUNTED ON THE BODY, SWEPT-BACK, AND TAPERED WITH SQUARE TIPS.

APPEARANCE: WINGS: HIGH-MOUNT, VARIABLE, TAPERED. ENGINES: ONE IN FUSELAGE. FUSELAGE: LONG AND TUBULAR, WITH BOX-LIKE INTAKES AND LARGE, SWEPT BELLY-FIN. TAIL: SWEPT-BACK, TAPERED WITH ANGULAR TIP, SWEPT, TAPERED FLATS MOUNTED ON FUSELAGE

VARIANTS

MIG-23M/FLOGGER B:FIRST PRODUCTION VERSION AS STANDARD INTERCEPTOR, - PULSE DOPPLER RADAR, IMPROVED ENGINE, IRST,
AA-7, ETC.

MIG-23U/-23UM/-23UB/FLOGGER C: A TANDEM SEAT COMBAT AND TRAINER VARIANT.

MIG-23MS/FLOGGER E: EXPORT BUILT TO B STANDARD. MIG-23MF DOWNSPEC VERSION

MIG-23B/FLOGGER F: INTERIM GROUND ATTACK VARIANT WITH AL-21 TURBOJET ENGINE, NO RADAR, AND TAPERED NOSE. THE MIG-23BN VARIANT RETURNED TO THE R-35-300 ENGINE.

MIG-23ML/FLOGGER G:LIGHTWEIGHT VERSION WITH IMPROVED ENGINE AND AVIONICS

MIG-23P/FLOGGER G: FIGHTER VARIANT SIMILAR TO FLOGGER B, BUT WITH DIGITAL AUTOPILOT FOR GROUND CONTROL.

MIG-23BK/-23BM/FLOGGER-H: GROUND ATTACK VERSIONS WITH THE UPRATED ENGINE, AND AVIONICS PODS BORROWED FROM THE MIG-27.

MIG-23MLD/FLOGGER K: UPGRADED MULTI-ROLE FIGHTER WITH IMPROVED AERODYNAMICS, LATEST MISSILE, AND OTHER IMPROVEMENTS. THIS IS CONSIDERED THE BEST CURRENT PRODUCTION UPGRADE AVAILABLE.

MIG-27K/FLOGGER D: GROUND-ATTACK VARIANT WITH INTERNAL GSH-6-23 23-MM GUN. APPEARANCE DIFFERS BY TAPERED NOSE.
MIG-27D/-27M/FLOGGER J: APPEARANCE DIFFERS BY A LONG DOWNWARD-SLOPING, POINTED NOSE. AIRCRAFT HAS A TV/LASER DESIGNATOR. CAN BE FITTED WITH A THREE-CAMERA RECON POD.
MIG-27L: EXPORT VERSIONS BUILT BY HINDUSTAN AERONAUTICS IN INDIA.
J-8: CHINESE AIRCRAFT IS LOOSELY BASED ON MIG-21AND MIG-23 FEATURES.

RUSSIAN MULTI-ROLE FIGHTER AIRCRAFT MIG-29/FULCRUM

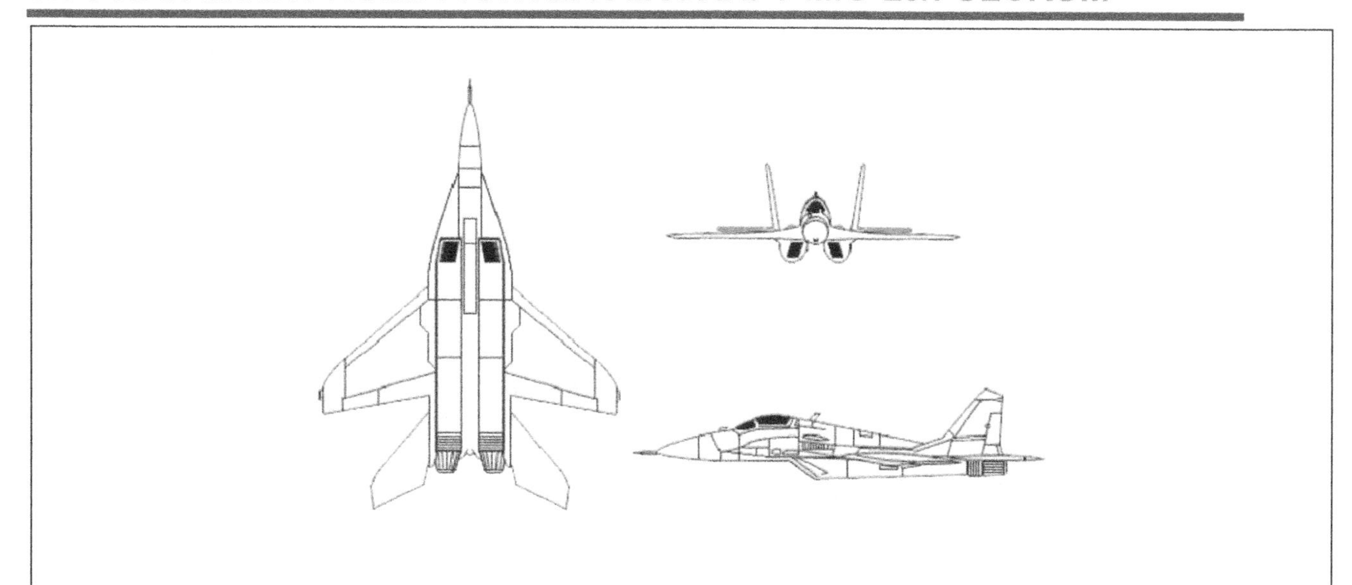

SYSTEM	SPECIFICATIONS	SYSTEM	SPECIFICATIONS
Weapon & Ammunition Types		Ceiling (m):	
30-mm Gsh-30-1 cannon (rnds):	150	Service (clean):	18,000
Other Loading Options:		With External Stores:	17,500
AA-8 APHID AAM :	6	Vertical Climb Rate (m/s):	330
AA-10 ALAMO AAM:	4	Fuel (liters):	
AA-11 ARCHER AAM:	4	Internal:	4,300
AA-12 ADDER AAM;	4	External:	4,150
AS-14 KEDGE:	2	Range (km) (3 drop tanks):	
AS-17 KRYPTON:	2	Maximum:	1,500
Bombs:	4	Low altitude (on internal fuel):	710
250 kg Bombs:	8	Ferry (3 external tanks):	2,900
500 kg Bombs:	4	Takeoff Run/Landing Roll (m):	
ZB-500 (Napalm tanks):	4	Prepared Surface:	550/900
KMGU-2 (sub munition dispensers):	4	Afterburner/Drag Chute:	250/660
		Dimensions (m):	
Rockets:		Length:	17.3
130 mm and 240 mm rockets:	4	Wingspan:	11.4
B-8M1 (20 x 80 mm) rocket pack:	4	Height:	4.8

Fuel:		Standard Payload (kg):	
3 External Tanks;:	4150	External FULCRUM-C/D and MiG-29SMT:	3,000 - 4,000
SYSTEM		Hardpoints (wing pylons):	6
Alternative Designations:		Survivability/Countermeasures:	
Date of Introduction:	1983	Zero/zero ejection seat:	Yes
Proliferation:	≥ 25 country	Radar warning receiver:	Yes
Description:		Radar jammer:	Yes
Crew:	1	Chaff and flares:	Yes
Engines:		ARMAMENT:	
Kimov/Sakisov RD-33 Turbofans (18,300 lbs.) each:	2	30-mm cannon in the left wing root:	1
14,550 lbs. thrust Wopen-13 turbofan (J-8):	1	AVIONICS/SENSOR/OPTICS:	
Weight (kg):		Coherent pulse-Doppler look-down/ shoot-down radar range 9km):	Yes
Max Gross:	22,000	Search range (km):	70
Normal Takeoff:	16,800	Tracking range (km0;	35
Empty:	10,900	Targets tracked;	10
Speed (km/h):		Targets engaged;	2
Max (at altitude):	2,400	Heads-Up-Display (HUD):	Yes
Max (sea level):	1,500	infrared search and track system (IRST):	Yes
Takeoff/Landing Speed:	240	Night/Weather Capabilities:;	Yes
Max "G" Force (g):	+9.0		

NOTES

APPEARANCE:

WINGS: SWEPT-BACK AND TAPERED WITH SQUARE TIPS. ENGINES: TWIN JETS MOUNTED LOW AND TO THE SIDES OF THE FUSELAGE. DIAGONAL-SHAPED AIR INTAKES. FUSELAGE: LONG, THIN, SLENDER BODY WITH LONG POINTED DROOPING NOSE. TAILFINS HAVE SHARPLY TAPERED LEADING EDGES, CANTED OUTWARD WITH ANGULAR CUTOFF TIPS. FLATS ARE HIGH-MOUNTED ON THE FUSELAGE, MOVABLE, SWEPT-BACK, AND TAPERED WITH A NEGATIVE SLANT.

VARIANTS

MIG-29/FULCRUM A: SINGLE SEAT TACTICAL FIGHTER DESIGNED TO OPERATE UNDER GROUND CONTROL.

MIG-29UB/FULCRUM B: OPERATIONAL CONVERSION TRAINER; TWO-SEAT CONFIGURATION. AIR-DEFENSE ROLE.

MIG-29S/FULCRUM C: PRODUCTION MULTI-ROLE VARIANT FITTED WITH DORSAL HUMP HOUSING UPGRADED AVIONICS, AND UPRATED FLIGHT-CONTROL SYSTEM WITH SOME AERODYNAMIC TWEAKING. PRINCIPAL UPGRADE WAS NO-19 FIRE CONTROL RADAR, WHICH CAN ENGAGE TWO TARGETS SIMULTANEOUSLY.

MIG-29SD: FULCRUM AN EXPORT UPGRADE VERSION OF MIG-29 TO FULCRUM C STANDARD.

MIG-29SM: CURRENT PRODUCTION UPGRADE WITH ASM CAPABILITY.

MIG-29K/FULCRUM D: A CARRIER BORNE VERSION OF THE FULCRUM.

BAAZ (FALCON): NAME GIVEN TO THE MIG-29 INDIAN AIR FORCE, WHICH BEGAN OPERATING THE AIRCRAFT IN 1987.

MIG-30: PROPOSED GROUND-ATTACK VARIANT OFFERED TO INDIA IN 1991 AS SUBSTITUTE FOR THE LIGHT COMBAT AIRCRAFT (LCA).

MIG-29SMT: ADVANCED MULTI-ROLE DESIGN, WITH CAPABILITY FOR IMPROVED ASMS, SUCH AS AS-14 AND AS-17.

FRENCH MULTI-ROLE FIGHTER AIRCRAFT MIRAGE 2000

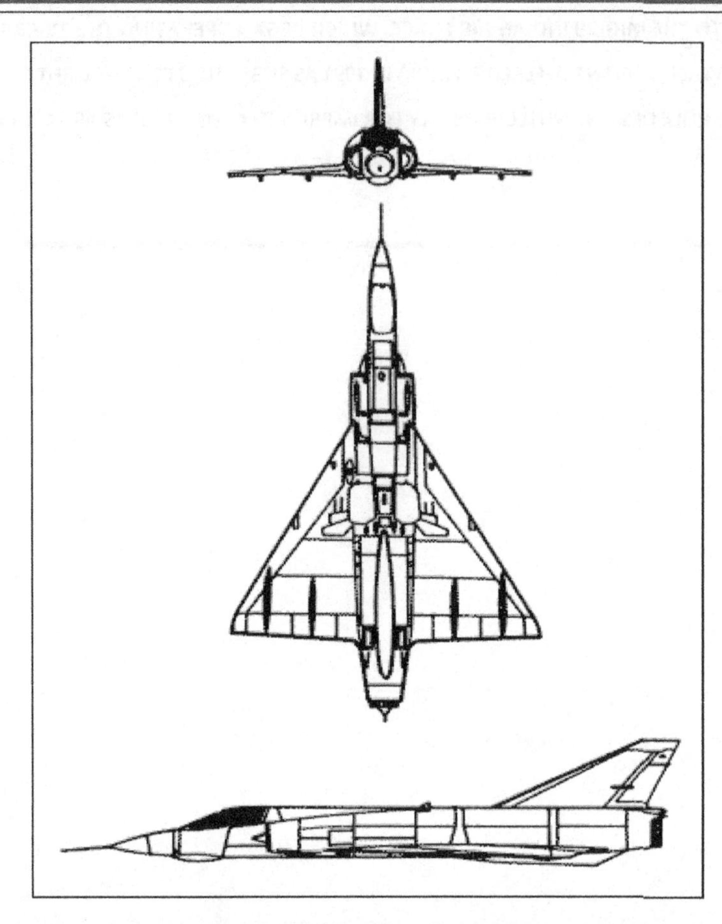

SYSTEM	SPECIFICATIONS	SYSTEM	SPECIFICATIONS
Weapon & Ammunition Types		Engines:	
Two 30-mm DFEA 554 guns(C/E/-5)	125	14,462 lbs. thrust SNECMA M53-P2 Turbofan:	1
Other Loading Options		21,385 lbs. thrust with afterburner:	1
AAMs:		Weight (kg):	
R550 Magic 2:	2 - 4	Maximum Takeoff:	
AIM-9 Sidewinder:	2 - 4	C:	17,000
Super 530:	2 - 4	Empty:	
MICA (2000-5):	4 - 6	C/E/-5:	7,500
AGMs:		B/N/D/S:	7,616
AS30L:	2	Speed (km/h):	

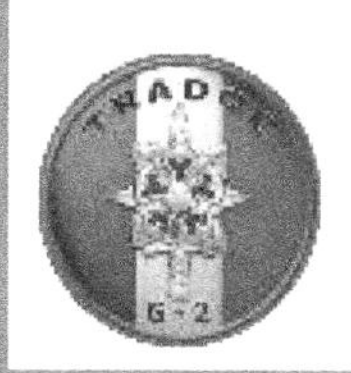

BGL laser-guided rocket/gun pods:	1 - 2	Maximum (at altitude):	2,630, Mach 2.2
18-round 68 mm rocket pods:	4	Maximum (sea level):	Mach 1.2
100 mm rocket packs:	2	Max "G" Force (g):	+9
CC630 twin 30 mm cannon pack:	1	Ceiling (m):	18,000
BOMBS:		Vertical Climb Rate (m/s):	285
35 kg BAP100 anti-runway:	18	Fuel (liters):	
250 kg conventional:	18	Internal:	3,978
200 kg Durandal anti-runway:	18	External:	4,700
Belouga cluster:	5 - 6	Range (km):	
400 kg BM400 modular:	5 - 6	Maximum Load:	2,960
1,000 kg BGL laser-guided:	1 - 2	With Aux Fuel (3 tanks):	3,600
Anti-radar:		Combat Radius:	900
Armat:	2	Dimensions (m):	
Anti-ship:		Length:	14.4
AM39 Exocet:	2	Wingspan:	9.2
Nuclear:		Height:	5.2
ASMP cruise missile (2000N):	1	Maximum Payload (kg):	6,300
Pods:		Hardpoints:	9
Recce/Offensive or intelligence ECM:	1	Under fuselage:	5
FLIR navigation:	1	Under each wing:	2
Fuel:		Survivability/Countermeasures:	
3 External fuel tanks (liters):	4,700	Martin-Baker zero/zero ejection seats:	Yes
SYSTEM		Canopy covered in gold film to reduce radar signature:	Yes
Alternative Designations:		ARMAMENT:	Yes
Date of Introduction:		30-mm DFEA 554 guns (C/E/-5):	2
C:	1983	AVIONICS/SENSOR/OPTICS:	
D:	1993	Pulse Doppler radar:	Yes
Proliferation:	8 countries	Look-down-shoot-down capacity:	Yes
Description:		Fly-by-wire:	Yes
Crew:		Automatic pilot:	Yes
B/C/D (Pilot):	1	Inertial guidance systems:	2

B/C/N (Pilot and Nav/Weapons officer):	2	Terrain following radar:	Yes
		Digital map:	Yes
		Integrated GPS:	Yes
		LASER designation pod with thermal camera:	Yes

NOTES

APPEARANCE: WINGS: LOW-MOUNTED DELTA, CLIPPED TIPS. ENGINES: TURBOFAN IN THE FUSELAGE. FUSELAGE: TUBE-SHAPED WITH A POINTED NOSE AND BUBBLE CANOPY. TAIL: TALL, SWEPT-BACK AND TAPERED WITH A CLIPPED TIP. THERE ARE NO TAIL FLATS.

VARIANTS
MIRAGE 2000B: TWO-SEAT, COMBAT-CAPABLE TRAINER VERSION. LACKS INTERNAL GUNS.
MIRAGE 2000C: INITIAL PRODUCTION SINGLE-SEAT VERSION.
MIRAGE 2000N: TWO-SEAT, NUCLEAR-CAPABLE FIGHTER/BOMBER VERSION IN FRENCH SERVICE ONLY. NO INTERNAL GUN. MOVING MAP DISPLAY, 60 M PENETRATION ALTITUDE.
MIRAGE 2000D: TWO-SEAT, CONVENTIONALLY ARMED VARIANT OF THE 2000N FOR LOW-LEVEL AND NIGHT-TIME STRIKE MISSION; SOME STEALTH MEASURES APPLIED INCLUDING GOLD-FILM COATING ON THE CANOPY AND CAMOUFLAGE.
MIRAGE 2000-5: CONVENTIONAL MULTI-MODE FIGHTER OFFERED FOR EXPORT.22, 050 LBS. THRUST SNECMA M53-P20 ENGINE OFFERED AS AN ALTERNATIVE.
MIRAGE 2000R: RECONNAISSANCE VERSION OF 2000C. FITTED WITH CAMERA PODS, ELECTRONIC INTELLIGENCE AND ECM EQUIPMENT

FRENCH MULTI-PURPOSE FIGHTER AIRCRAFT MIRAGE F1

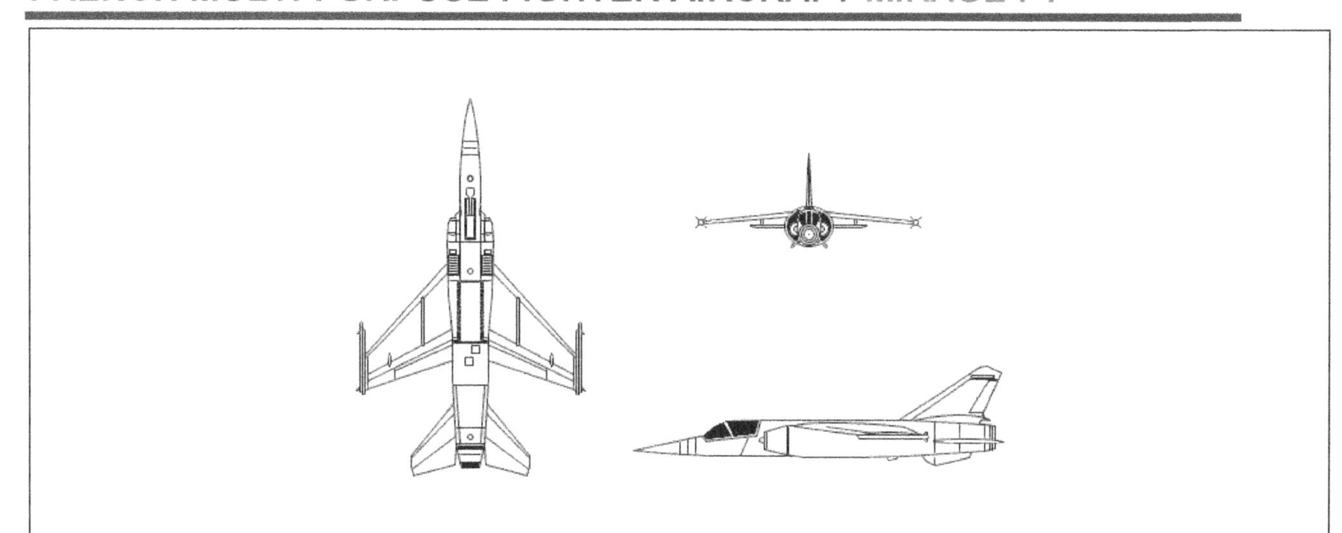

SYSTEM	SPECIFICATIONS	SYSTEM	SPECIFICATIONS
Weapon & Ammunition Types		Maximum (at altitude):	2,334, Mach 2.2
Two integral 30-mm DFEA Cannons:	135	Maximum (sea level):	1,471, Mach 1.2
Other Loading Options		Ceiling (m):	20,000
AAMs:		Vertical Climb Rate (m/s):	213
Super R530 AAM:	2	Fuel (liters):	
Armat ARM:	1	Internal:	4,200
AM 39 Exocet anti-ship missile:	1 - 2	External:	4,460
AS30L AGM:	1 - 2	Cruise:	2,170
30-mm DEFA gun pods:	2	Range (km):	
BOMBS:		Cruise:	2,170
Laser guided:		Ferry:	3,300
400 kg conventional:		Dimensions (m):	
1000 kg conventional:		Length:	15.3
ATLIS laser designation pod:		Wingspan:	8.4
250kg/BAP 100/BAT-100:	14	Height :	4.5
200 kg Durandal anti-runway:		Standard Payload (kg):	
Belouga cluster:		External:	6,300
Rockets:		Hardpoints:	5

68 mm rocket pods:	18	Centerline:	1
Anti-radar:		Each Wing:	2
R550 Magic or AIM-9 Sidewinder AAM:	2	Survivability/Countermeasures:	
SYSTEM		In-flight refueling:	Yes
Alternative Designations:		Martin-Baker zero/zero ejection seats:	Yes
Date of Introduction:	1974	IFF:	Yes
Proliferation:	≥ 11 countries	Infrared jammer:	Yes
Description:		Radar Warning Receiver:	Yes
Crew:	1	Electronic Countermeasures:	Yes
Engines:		AVIONICS/SENSOR/OPTICS:	
11,023 lbs. thrust SNECMA Atar 9K-50 turbojet:	1	Cyrano IVM radar (air-to-air, air-to-ground):	Yes
15,873 lbs. thrust with afterburner:	1	Inertial navigation system:	Yes
Weight (kg):		Panoramic camera:	Yes
Maximum Takeoff:	16,200	Vertical camera:	Yes
Normal Takeoff:	10,900	IR thermographic captor:	Yes
Empty:	7,400	Night/Weather Capabilities:	
Speed (km/h):		Interceptor:	Yes
		Fighter-Bomber:	Yes
		Dedicated Reconnaissance:	Yes

NOTES

THE MIRAGE F1 IS A MULTI-PURPOSE ATTACK/FIGHTER AIRCRAFT OF CONSIDERABLE VERSATILITY. IT CAN BE EMPLOYED IN THE INTERCEPT, GROUND ATTACK, RECONNAISSANCE, TRAINING, ELECTRONIC WARFARE, AND ELECTRONIC INTELLIGENCE ROLES. THE FRENCH AIR FORCE ORDERED THE MIRAGE F1 FOR ITS INTERCEPTOR SQUADRONS, AND THE FIRST F1S ENTERED SERVICE IN 1973.THE F1 PROVED A VERY POPULAR EXPORT, WITH OVER 500 OF THEM SOLD ABROAD IN THE FIRST 10 YEARS OF PRODUCTION. MORE THAN 700 MIRAGE F1'S HAVE BEEN SOLD TO SOME 11COUNTRIES.

APPEARANCE: WINGS: HIGH-MOUNTED, SWEPT-BACK, AND TAPERED. ENGINES: ONE TURBOJET BURIED IN THE AFT FUSELAGE. FUSELAGE: LONG, SLENDER, POINTED NOSE AND BLUNT TAIL. TAIL: SWEPT-BACK AND TAPERED FIN WITH A BLUNT TIP. FLATS ARE MID-MOUNTED ON THE FUSELAGE, SWEPT-BACK, AND TAPERED WITH BLUNT TIPS

VARIANTS

F1-C: FIRST PRODUCTION VERSION FOR SERVICE WITH FRANCE AND FOR EXPORT. AVIONICS ORIENTATED TOWARD AIR-TO-AIR INTERCEPTION.

F1-A:INITIAL PRODUCTION GROUND ATTACK VERSION WITH SMALL ADIA 2 TARGET-RANGING RADAR, RETRACTABLE REFUELING PROBE, GROUND ATTACK SYSTEM AVIONICS SUITE.

F1-B: TWO-SEAT COMBAT-CAPABLE TRAINER VERSION OF F1-C. INTEGRAL CANNON REMOVED.

F1-D: TWO-SEAT COMBAT-CAPABLE TRAINER VERSION OF THE F1-E.

F1-E: EXPORT VERSION WITH STRETCHED FUSELAGE AND IMPROVED AVIONICS.

F1-R (F1-CR-200): RECONNAISSANCE/ELINT/EW VERSION. FITTED WITH GROUND MAPPING AND OTHER LOW-ALTITUDE MODES RADAR, DIGITAL NAVIGATION/ ATTACK COMPUTER, HEADS-UP-DISPLAY, INERTIAL NAVIGATION SYSTEM, AND AIR DATA COMPUTER.
F1-CT: CANADIAN AIR FORCE REPLACEMENT FOR THE OLDER MIRAGE III AND SOME JAGUAR AIRCRAFT. USED AS STRIKE AIRCRAFT.
MIRAGE F1/M53: FITTED WITH 18,740 LBS. THRUST SNECMA M53, ENGINE LATER ADOPTED FOR MIRAGE 2000.
C-14: SPANISH DESIGNATION FOR MIRAGE F1.

CHINESE MULTI-ROLE FIGHTER AIRCRAFT Q-5/FANTAN

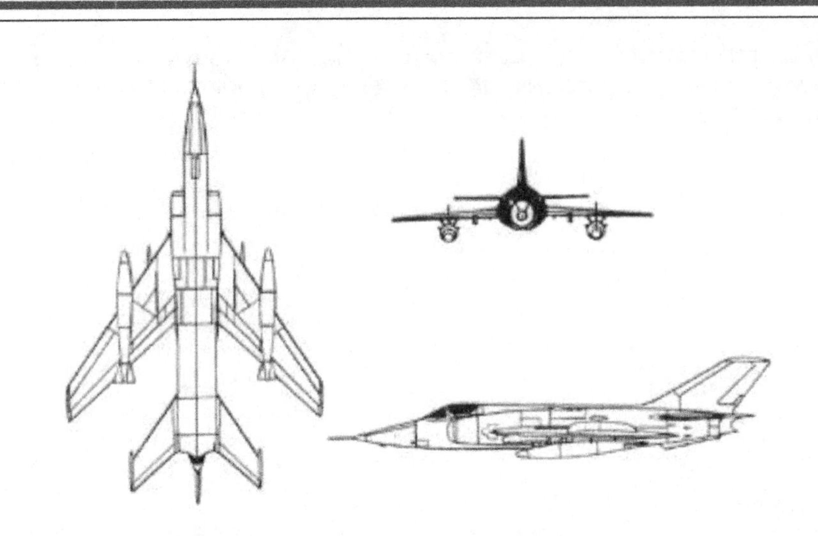

SYSTEM	SPECIFICATIONS	SYSTEM	SPECIFICATIONS
Weapon & Ammunition Types		Maximum (sea level):	1,120
2 x Norinco 23-2K 23mm cannon (rnds):	200	Max "G" Force (g):	+7.5
Other Loading Options:	2	(Max armament):	+5.0
Bombs:		Ceiling (m):	
225 kg;	6	Service (clean):	15,900
250 kg;	6	Vertical Climb Rate (m/s):	148
340 kg	2	Fuel (liters):	
Duranal anti-runway	6	Internal:	3,648
BL755 cluster	2	External:	1,520
Rocket pods:	4	Range (km):	
8-round 57mm/68mm:	2	Maximum Load1,816	
7-round 90mm:	2	Combat Radius:	550
130mm rockets:	4	Takeoff/Landing Roll (m):	1250/804
Missiles:		Dimensions (m):	
PL-2/PL-2B/PL-7 anti-air:	2	Length:	15.7
AIM-9 Sidewinder anti-air:	2	Wingspan:	9.7
Matra R550 Magic anti-air:	2	Height (gear extended):	4.5
CSS-N-4 Sardine anti-ship;		Standard Payload (kg):	
ECM Pods:		External:	2,000

Fuel:		Hardpoints:	10
2 External Fuel Tanks (liters ea.):	760	On fuselage:	4
SYSTEM		Under each wing:	3
Alternative Designations:	A-5 export version	Survivability/Countermeasures:	
Date of Introduction:	1970	Pressurized and air conditioned armored cockpit with one-piece jettisonable canopy:	Yes
Proliferation:	≥5 countries	Zero/250 to 850 km/h ejection seat:	Yes
Description:		ECM pod and RWR:	Yes
Crew:	1 (pilot)	ARMAMENT:	
Engines:		Norinco 23-2K 23-mm cannons, one per wing root:	2
5,400 lbs. thrust Wopen-6 turbojets 7,165 lbs. thrust w/afterburner:	2	AVIONICS/SENSOR/OPTICS:	
5,400 lbs. thrust Wopen-6A turbojets 8,930 lbs. thrust w/afterburner (Exports):	2	IFF:	Yes
Weight (kg):		VHF transponder;	Yes
Maximum Gross:	12,000	Radio Compass:	Yes
Empty:	6,636	Low-altitude radio altimeter:	Yes
Speed (km/h):		Horizon gyro:	Yes
Maximum (at altitude):	1,340	Optical sight :	Yes

NOTES

THE Q-5 IS A SINGLE-SEAT, TWIN-ENGINE SUPERSONIC FIGHTER DEVELOPED BY THE NANCHANG AIRCRAFT COMPANY OF CHINA. IT OFFERS ENHANCED COMBAT PERFORMANCE PARTICULARLY AT LOW AND SUPER-LOW ALTITUDE. IT IS USED MAINLY TO ASSIST GROUND TROOPS IN ATTACKING CONCENTRATED TARGETS ON LAND, KEY TRANSPORTATION POINTS AND SHIPS NEAR THE COAST. IT CAN ALSO INTERCEPT AND FIGHT ENEMY AIRCRAFT.

APPEARANCE: WINGS: MID-MOUNTED, SWEPT BACK, AND TAPERED WITH BLUNT TIPS AND WING FENCES. ENGINES: TWO TURBOJETS IN THE FUSELAGE WITH SEMICIRCULAR AIR INTAKES AND TWO EXHAUSTS. FUSELAGE: THICK, FLATTENED, WITH AN UPWARD TAPER TO THE REAR SECTION. TAIL: FLATS ARE HIGH-MOUNTED ON THE BODY, SWEPT-BACK, AND TAPERED WITH SQUARE TIPS. SHARPLY SWEPT-BACK TAIL FIN HAS A BLUNT TIP.

VARIANTS

Q-5: FIRST VERSION WITH INTERNAL BOMB BAY. THE Q-PREFIX INDICATES THE AIRCRAFT IS IN THE CHINESE MILITARY SERVICE. THE A-PREFIX DENOTES THE EXPORT VERSION

Q-5 I/A-5A: BECAME THE STANDARD CONFIGURATION. ENTERED PRODUCTION IN LATE 1970S. INTERNAL WEAPONS BAY CONVERTED INTO ADDITIONAL FUEL TANKS, TWO FUSELAGE PYLONS AND OUTER WING PYLONS ADDED. FORTY EXPORTED TO NORTH KOREA.

Q-5IA: INTRODUCED KEY REFINEMENTS, INCLUDING THE ADDITION OF TWO UNDER WING PYLONS, BETTER SELF-PROTECTION AND IMPROVED OPTICAL SIGHTS.

Q-5 II/Q-5B/A-5B: NEARLY IDENTICAL TO THE Q-5IA. INCLUDES RWR. MAY ALSO HAVE RANGING RADAR AND ALR-1 LASER TO WORK WITH PRECISION-GUIDED BOMBS. HUD, MISSION COMPUTER AND ECM ALSO INCLUDED.
Q-5-III/A-5C: MAJOR EXPORT VERSION. A SOMEWHAT LONGER AND WIDER Q-5 II. INCLUDES IMPROVED AVIONICS, MARTIN BAKER PKD10 ZERO-ZERO EJECTION SEAT

FRENCH MULTI-ROLE FIGHTER AIRCRAFT RAFALE

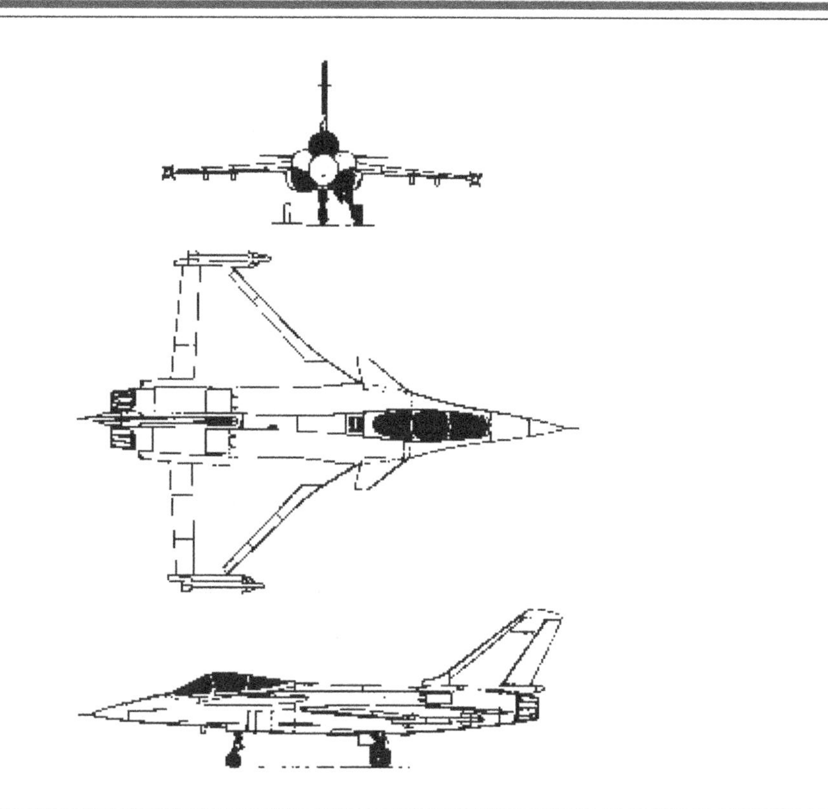

SYSTEM	SPECIFICATIONS	SYSTEM	SPECIFICATIONS
Weapon & Ammunition Types		Proliferation:	Expected to be exported
30-mm DFEA 791B Cannons (rnds):	300	Description:	
Other Loading Options		Crew:	
Magic:	6	M/C:	1
Mica:	10	B:	2
Sidewinder:	6	Engines19, 955 lbs. thrust SNECMA M-88-3 turbofans with afterburner:	2
ASRAAM	6	Weight (kg):	
AMRAAM:	5	Maximum Gross:	24,500
Exocet:	4	Maximum Takeoff:	20,000
Penguin 3:	4	Empty:	
Harpoon:	4	M:	9670
AS30L:	4	B/C:	9,060
Apache:	3	Speed (km/h):	

Alarm:	5	High-Altitude:	2,125
Harm:	5	Low-level:	1,853
Maverick:	4	Maximum:	2,390
Bombs;		Max "G" Force (g):	+9/-3.6
1000 kg;	3	Ceiling (m):	16,765
400 kg;	5	Vertical Climb Rate (m/s):	305
GBU-12	5	Fuel (liters):	
GBU-10	3	Internal:	5,325
250 kg-Mk 82:	20	External:	6,000
400 kg-Mk 83:	10	Range (km):	
Belouga cluster:	10	Maximum Load:	2,110
Bap 100:	10	With Aux Fuel (3 tanks):	3.520
Bat 120:	10	Combat Radius:	1,882
Derandal:	10	Takeoff Run/Landing Roll (m):	400-1000/450
Fuel:		Dimensions (m):	
1,300 L:	3	Length:	115.3
1,700 L:	3	Wingspan:	10.9
2,000 L:	3	Height:	5.4
Pods:		Standard Payload (kg):	9,500
PDLCT TV and FLIR:	1	External:	9,500
ECM:		Hardpoints:	14
RECCE IR:	1	Rafale M:	13
SLAR:	1	Survivability/Countermeasures:	
HAROLD:	1	Martin-Baker zero/zero ejection seat:	Yes
Twin gun pod (600 rounds):	1	Canopy gold coated to reduce radar reflections:	Yes
SYSTEM		ARMAMENT:	
Alternative Designations:		DEFA 791B 30-mm cannon:	1
Date of Introduction:			
M:	2001		
B/C:	2006		

NOTES

RAFALE IS A TWIN-JET COMBAT AIRCRAFT CAPABLE OF CARRYING OUT A WIDE RANGE OF SHORT- AND LONG-RANGE MISSIONS INCLUDING GROUND AND SEA ATTACK, AIR DEFENSE AND AIR SUPERIORITY, RECONNAISSANCE, AND HIGH-ACCURACY STRIKE OR NUCLEAR STRIKE DETERRENCE.

APPEARANCE: WINGS: MID-MOUNTED DELTA. ENGINES: TWO TURBOFANS BURIED IN AFT FUSELAGE. FUSELAGE: CONVENTIONAL SEMI-MONOCOQUE WITH SOME BLENDING. TAILFIN HAS SHARPLY SWEPT LEADING EDGE AND SWEPT, INSET RUDDER. IN PLACE OF HORIZONTAL STABILIZERS AFT IT HAS RELATIVELY LARGE, SWEPT, ALL-MOVING CANARDS SHOULDER-MOUNTED ABOVE AND AHEAD OF THE WING LEADING EDGE.

VARIANTS

RAFALE B/C ACT: SINGLE-SEAT VARIANT INTENDED TO REPLACE SEPECAT JAGUAR. ONE VERSION WILL BE ARMED WITH ASMP AND REPLACE THE MIRAGE IV PENETRATING BOMBERS. ONE VERSION WILL BE FITTED WITH SNECMA M-88-2 ENGINES AND RDX RADAR.

RAFALE M ACM: CARRIER-CAPABLE STRIKE AIRCRAFT TO REPLACE F-8 CRUSADER AND SUPER ETENDARD.EMPTY WEIGHT WILL BE 750 KG HEAVIER THAN ACT. IT ALSO HAS A CARRIER-LANDING ARRESTOR HOOK AND ONE LESS HARDPOINT FOR WEAPONS.

RUSSIAN MULTI-ROLE AIRCRAFT SU-24/FENCER

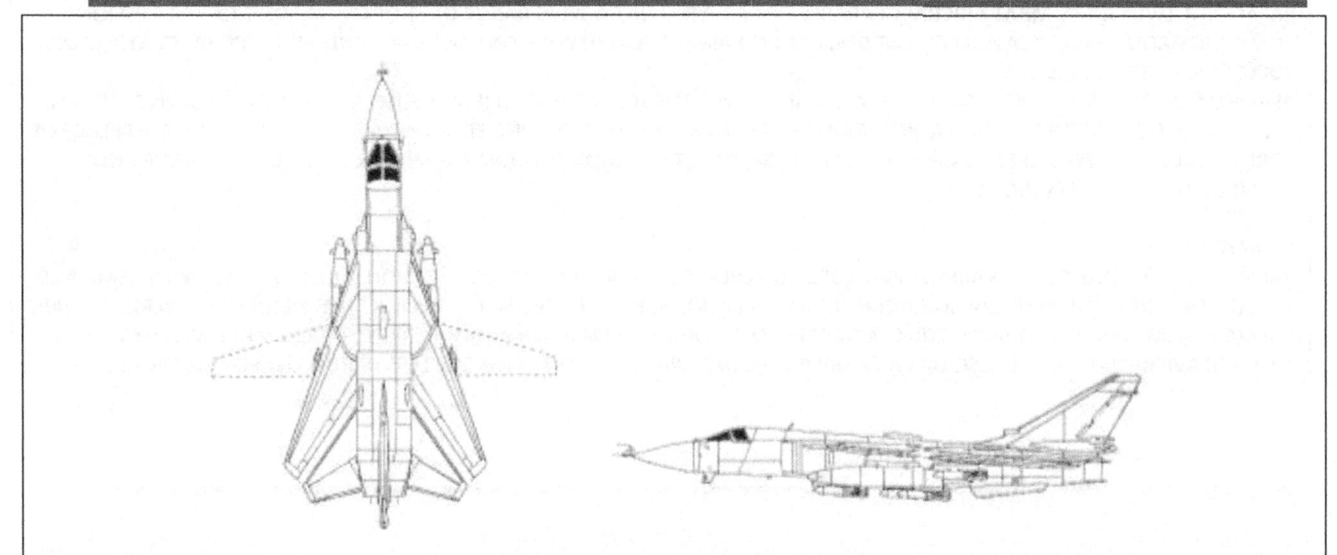

SYSTEM	SPECIFICATIONS	SYSTEM	SPECIFICATIONS
Weapon & Ammunition Types		With External Stores:	INA
23-mm 6x barrel Gsh-23 cannon (rnds):	250	Vertical Climb Rate (m/s):	150
Other Loading Options:	2	Fuel (liters):	
TN1000 or TN11200 nuclear weapons:		Internal:	11,760
100-kg FAB-100 bombs:	38	External:	8,000
TV or laser-guided bombs:	4	Range (km):	
		Maximum Load:	940
AS-7/KERRY ASM:	1	With Aux Fuel:	1,230
AS-10/KAREN ASM:		Combat Radius:	950
AS-12/KEGLER ASM:		Takeoff Run/Landing Roll (m):	
AS-13/KINGBOLT ASM:		Prepared Surface:	1,100-1,200/950
AS-14/KEDGE ASM:		Dimensions (m):	
AS-17/KRYPTON ASM:		Length:	24.6
S-25LD 266-mm precision rockets:		Wingspan:	
Gun pods:	3	Extended:	17.6
AA-8/APHID or AA-11 AAM:	2	Swept:	10.4
External fuel tanks (liters):	2,000 – 3,000	Height (gear extended):	6.2
SYSTEM		Standard Payload (kg):	

Alternative Designations:	See Variants	External:	8,000
Date of Introduction:	1975	Hardpoints underwing:	9
Proliferation:	≥ 11 countries	Survivability/Countermeasures:	
Description:		Pressurized cockpit with zero/zero ejection seats:	Yes
Crew (pilot, weapons operator):	2	Infrared and radar jammer:	Yes
Engines 17,200-shp Lyluka AL-21F-3A turbojet (24,700-shp with afterburner):	2	Radar and missile warning Receivers:	Yes
Weight (kg):		chaff and flares:	Yes
Maximum Gross:	39,700	ARMAMENT:	
Normal Takeoff:	35,910	23-mm 6x barrel Gatling gun, Gsh-6-23:	
Empty:	22,320	Range (m) (practical):	2,500
Speed (km/h):		Elevation/Traverse (rigid mount):	None
Maximum (at altitude):	2,320	Ammo Type:	HEFI
Maximum (sea level):	1,530	Rate of Fire (rpm):	9,000
Maximum Attack Speed:	1,200	AVIONICS/SENSOR/OPTICS:	
Cruise:	INA	Integrated navigation and	
Takeoff/Landing Speed:	INA	fire control radars:	Yes
Max "G" Force (g):	+6.5 g	Pulse-doppler terrain following radar coupled to autopilot:	Yes
Ceiling (m):		Laser/TV targeting and weapon guidance:	Yes
Service (clean):	17,500	Night/Weather Capabilities::	Yes

NOTES

THIS AIRCRAFT WAS THE FIRST DEVELOPED SPECIFICALLY FOR THE GROUND-ATTACK ROLE, BUT HAS BEEN ADAPTED FOR OTHERS. ITS VARIABLE SWEPT WING CAN BE SET AT 16, 45, OR 69 DEGREES. SOME AIRCRAFT ARE CAPABLE OF AERIAL REFUELING. ALL CAN CARRY UP TO THREE EXTERNAL FUEL TANKS FOR EXTENDED RANGE. THERE IS NO INTERNAL WEAPONS BAY. NOT ALL MUNITIONS MAY BE EMPLOYED AT ONE TIME. MISSION DICTATES WEAPONS CONFIGURATION. EXTERNAL STORES ARE MOUNTED ON UNDERWING HARDPOINTS. EACH WING HAS FOUR POINTS. THE CENTER FUSELAGE ATTACHMENT POINT GIVES NINE TOTAL STATIONS.

APPEARANCE: WINGS: HIGH-MOUNT, VARIABLE, TAPERED BACK. ENGINES: BOTH ALONG BODY, UNDER WINGS.

VARIANTS

SU-24M/-24MK/FENCER D: GROUND ATTACK VERSION AND EXPORT MODEL.
SU-24MK/FENCER D MODERNIZED: CURRENTLY MARKETED GROUND ATTACK VARIANT HAS UPGRADES SUCH ASILS-31 HEADS-UP DISPLAY COMPUTER GPS FCS, DIGITAL MAP DISPLAY, KS-418E RADAR JAMMER PODS, AND ACCESS TO RECENT MISSILES (E.G., AS-13, AS-17, S-25LD LASER DESIGNATED ROCKETS AND AA-11 AAM).

SU-24MR/FENCER E: RECONNAISSANCE VARIANT FOR MISSIONS TO 400 KM, WITH BKR-1SENSOR SUITE: A-100 SERIES AND AP-402M CAMERAS, AIST-MTV CAMERA, SHPIL-2M LASER RADAR SYSTEM, ZIMA IR CAMERA, AND SHTIK SIDE-LOOKING RADAR (24 KM RANGE, 5M ACCURACY).SYSTEM CAN OPERATE DAY OR NIGHT. THE BOK-2 ECM SYSTEM IS USED. OPTIONS INCLUDE EFIR-1M RADIATION DETECTION POD, KADR FILM DROP SYSTEM, AND TANGAZH ELINT POD. DATA OTHER THAN OPTICAL IS TRANSMITTED DIGITALLY. ANOTHER OPTION IS 2 XAA-8/APHID ASMS.

SU-24MP/FENCER F: ELECTRONIC WARFARE/ JAMMING/SIGINT VARIANT. BUKET SERIES JAMMERS ARE AKA SPS-22, -33, -44, OR -55.FASOL SERIES (SPS-5, -5M AND -5- 2X) RADAR NOISE JAMMERS ARE AVAILABLE. GERAN (SPS-161 OR GERAN F) IS A 2ND GENACTIVE JAMMER. GERAN/SPS-162 JAMS 6-12 GHZ, WITH 100 KW. ARMAMENT INCLUDES 23-MM GUN AND (OPTIONAL) 4 X AA-8 ASMS

RUSSIAN MULTI-ROLE FIGHTER AIRCRAFT SU-27/FLANKER-B AND VARIANTS

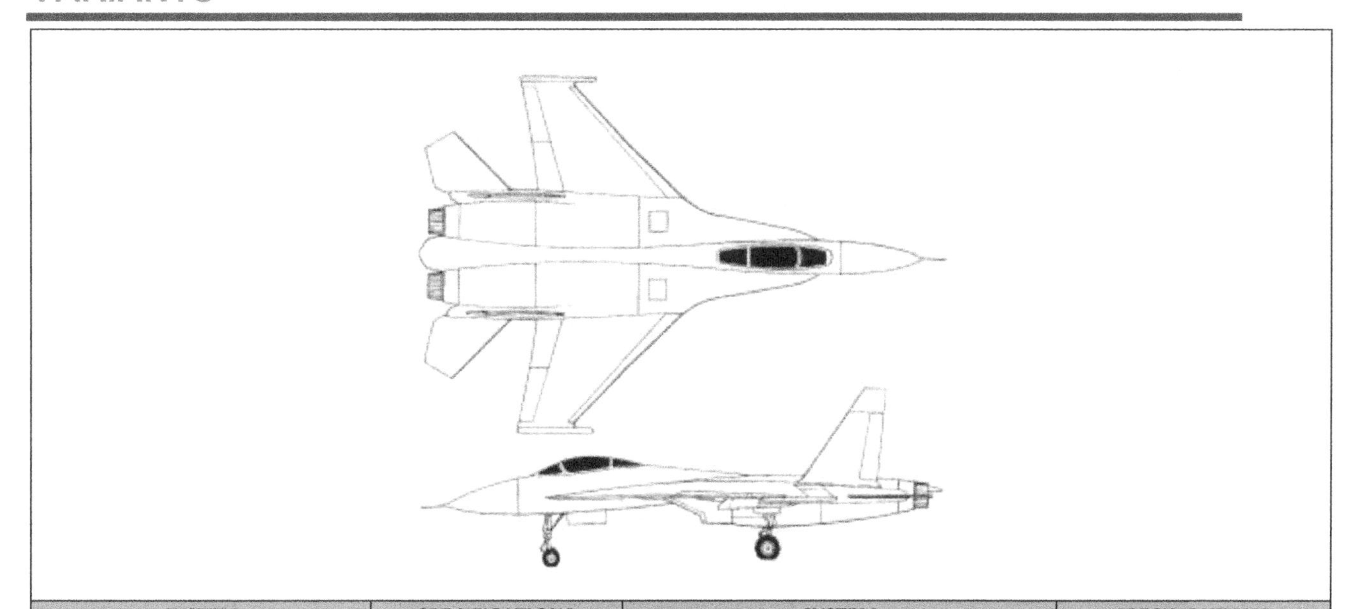

SYSTEM	SPECIFICATIONS	SYSTEM	SPECIFICATIONS
Weapon & Ammunition Types		Hardpoints:	
30-mm Gsh-30-1 cannon (rnds):	150	FLANKER-B:	10
Other Loading Options:	2	C:	12
AA-10A-D/ALAMOAAM	10	Takeoff Run/Landing Roll (m):	
AA-8/APHID AAM		Prepared Surface (variant dependent):	500 to 650/600
AA-9/AMOS AAM		Su-35:	720 / 1,200/1,200
AA-11/ARCHER AAM		Dimensions (m):	
AA-12 ADDER AAMs		Length:	21.9
AS-7/KERRY ASM:		Wingspan:	14.7
AS-10/KAREN ASM:	8	Height:	5.5
AS-12/KEGLER ASM:		Standard Payload (kg):	
AS-13/KINGBOLT ASM:		External:	6,000
AS-14/KEDGE ASM:	6	Survivability/Countermeasures:	
AS-17/KRYPTON ASM:		Zero/zero ejection seat:	Yes
AS-18/KAZOO ASM:	2	Infrared and radar jammer (SPS-171):	Yes
Gun pods:	3	Radar and missile warning receivers:	Yes
AA-8/APHID or AA-11 AAM:	2	Chaff and flares:	Yes
420-mm S-25 rockets (1 each):	4	ARMAMENT:	

80-mm S-8 rocket pod (20 ea.):	4	30-mm gun mounted in the right wing:	Yes
122-mm S-13 rocket pod (5 each):	4	30-mm gun, Gsh-30-1:	
250-kg, or 500-kg unguided and guided bombs:		Range (m) (practical):	4,000
SYSTEM		Elevation/Traverse (rigidly mounted):	None
Alternative Designations:	Chinese J-11	Ammo Type HEFI, APT, CC:	Yes
Date of Introduction:	1986	Rate of Fire (rpm):	1,500
Proliferation:	≥ 5 countries	AVIONICS/SENSOR/OPTICS:	
Description:	Variants in ( )	External:	8,000
Crew:	1	Hardpoints underwing:	9
Engines 27,557-shp Lyluka AL-31F turbojet with afterburner :	2	Survivability/Countermeasures:	
Weight (kg):		Pressurized cockpit with zero/zero ejection seats:	Yes
Max Gross (B/SM):	28,300/33,000	Infrared and radar jammer:	Yes
Normal Takeoff (B/SM):	23,000/23,700	Radar and missile warning Receivers:	Yes
Empty:	17,690	Chaff and flares:	Yes
Speed (km/h):		ARMAMENT:	
Max (at altitude):	Mach 2.35	23-mm 6x barrel Gatling gun, Gsh-6-23:	
Max (sea level):	Mach 1.1	Range (m) (practical):	2,500
Takeoff/Landing Speed:	250/231	Elevation/Traverse (rigid mount):	None
Max "G" Force (g)Control limited:	+9.0	Ammo Type HEFI:	Yes
Ceiling (m):		Rate of Fire (rpm):	9,000
Service (clean):	18,000	AVIONICS/SENSOR/OPTICS:	
With External Stores:	INA	Pulse-Doppler look-down/ shoot-down radar:	
Vertical Climb Rate (m/s):	305	Search range (km):	240
Fuel (liters):		Track range (km):	185
Internal (B/SM):	6,600/11,775	Multi Target Capability:	Yes
External:	no provision	IR Sensor:	Yes
Range (km):		Laser Designator:	Yes
Max Load:	3,790	Heads Up Display:	Yes
With Aux Fuel (B/SM):	4,390		
Combat Radius:	1,500		

NOTES

THE SU-27 IS PRIMARILY AN ALL-WEATHER INTERCEPTOR/FIGHTER AIRCRAFT USED FOR AIR DEFENSE. LATER VERSIONS ARE CAPABLE OF ALSO PERFORMING GROUND ATTACK MISSIONS. IT IS HIGHLY MANEUVERABLE BECAUSE OF A FLY-BY-WIRE CONTROL SYSTEM, WHICH AUTOMATICALLY RESTRICTS AIRCRAFT ANGLES OF ATTACK AND MAXIMUM G-LOADS DURING FLIGHT. EXTERNAL FUEL TANKS CAN BE CARRIED ON SOME VARIANTS, AND SOME ARE FITTED FOR AERIAL REFUELING, BUT THESE ARE GENERALLY NAVAL VERSIONS RATHER THAN AIR DEFENSE OR STRIKE VERSIONS. AVAILABLE MUNITIONS ARE SHOWN ABOVE; NOT ALL MAY BE EMPLOYED AT ONE TIME. MISSION DICTATES WEAPONS CONFIGURATION. EXTERNAL STORES ARE MOUNTED ON UNDERWING AND UNDERBODY HARDPOINTS. EACH WING HAS TWO POINTS AND AN ADDITIONAL RAIL ON THE WINGTIP. TWO POINTS ARE UNDER THE INTAKES ALONG THE FUSELAGE, AND TWO ARE CENTRALLY LOCATED UNDERNEATH THE FUSELAGE NEAR THE CENTERLINE AND BETWEEN THE INTAKES FOR A TOTAL OF TEN STATIONS.

APPEARANCE: WINGS: MID-MOUNT, SWEPT, SQUARE TIPS. ENGINES: TWO IN FUSELAGE, WITH SQUARE UNDERWING INTAKES. FUSELAGE: POINTED NOSE, RECTANGULAR FROM INTAKES TO TAIL. TAIL: TWIN TAPERED, SWEPT FINS, WITH MID-MOUNT, TAPERED, SWEPT FLATS

VARIANTS

SU-27/FLANKER B: PRODUCTION SINGLE-SEAT AIR SUPERIORITY FIGHTER USED IN RUSSIAN UNITS.

THERE ARE DOZENS OF UPGRADE PROGRAMS, MORE THAN A DOZEN FIELDED VARIANTS, AND SEVERAL DEVELOPED AIRCRAFT WITH DIFFERENT DESIGNATORS (SU-30, SU-34, SU-35, AND SU-37).

SU-27SK/-27P/FLANKER B: VARIANT EXPORTED TO CHINA WITH GROUND ATTACK CAPABILITY.J-11: CHINESE BUILT VERSION.

SU-27SM:MULTI-ROLE VERSION, WITH 12 HARDPOINTS, GREATER INTERNAL FUEL AND PAYLOAD CAPACITY, AND AIR REFUEL CAPABILITY.

SU-27UB/FLANKER C:TWO-SEAT MODEL (EXPORT -UBK), AS COMMAND AIRCRAFT, TRAINER AND INTERCEPTOR.JJ-11:CHINESE BUILT VERSION

SU-27K/FLANKER D: NAVAL VARIANT, READILY NOTICEABLE BY CANARDS FORWARD OF THE WINGS.

SU-27M/FLANKER E: MULTI-ROLE UPGRADE WITH HIGHER FINS, UPGRADED AVIONICS, ETC., DEVELOPED IN LATE 1980S.AN EXPORT VERSION CALLED SU-35 WAS MARKETED. IT HAD MORE POWERFUL 28,218-SHP LYLUKA AL-31FM ENGINES, THRUST-VECTORING NOZZLES FOR HIGHER GROSS WEIGHT AND GREATER RANGE. IT ALSO FEATURED BETTER RADAR AND TARGETING SYSTEMS FOR MULTIPLE ENGAGEMENTS. DIMENSIONS SLIGHTLY INCREASED, NOTICEABLE BY CANARDS FORWARD OF WINGS. FIELDING WAS MINIMAL, AND NONE WERE SOLD.SU-35UB WAS A TWO-SEATER UPGRADE VERSION.

SU-37/"SUPER FLANKER": SINGLE-SEAT MULTI-ROLE FIGHTER WITH THRUST VECTORING CAPABILITY AND SUFFICIENT MOBILITY FOR THE *KULBIT* PITCH-UP MANEUVER INTO A TIGHT 360 DEGREE SOMERSAULT, AS WELL AS IMPROVED LONG-RANGE WEAPONS AND FIRE CONTROL. EXPECTED FUTURE PRODUCTION VERSION IS SU-37MR. HOWEVER, AFTER THE ONE SU-27M CONVERSION TO SU-37 CRASHED DURING A FERRY FLIGHT, ALL WORK ON THE AIRCRAFT ENDED IN 2002.PRODUCTION IS UNLIKELY.

SU-27/SU-30 MAJOR/MINOR MODERNIZATION: UPGRADE PROGRAMS ARE BEING IMPLEMENTED TO BRING SU-27S UP TO SU-30 STANDARD, AND SOME SINGLE-SEAT UPGRADES TO THE STANDARD.

SU-30/FLANKER-F: PRODUCTION TWO-SEATER AIRCRAFT DEVELOPED FROM SU-27.

SU-34/FULLBACK: THIS 2-SEAT BOMBER VERSION HAS A SIDE-BY-SIDE COCKPIT, HIGH PAYLOAD FOR USE IN BOMBER MISSIONS AND MANEUVERABILITY SIMILAR TO FIGHTERS. EARLIER DESIGNATIONS INCLUDE: SU-27IB, SU-32, SU-32FN, AND SU-32MF.PRODUCTION AND EARLY FIELDING IS NOW UNDERWAY. THIS AIRCRAFT IS SCHEDULED TO GENERALLY REPLACE SU-24S IN RUSSIAN FORCES FOR THE STRIKE ROLE.

SU-35/SU-27BM: THIS NEW SINGLE-SEATER MULTI-ROLE FIGHTER IS DEVELOPED TO REPLACE SU-27M.THE 4+++ GENERATION PROTOTYPE FIRST FLEW IN 2008.IT INCLUDES A NEW AIRFRAME, WITH LARGER WINGS AND INTAKES, BUT NO CANARDS. IT HAS BIGGER ENGINES; NEW IRBIS-E PHASED-ARRAY RADAR, NEW IRST, AND 12 HARD POINTS FOR THE LATEST WEAPONS ARE INCLUDED. THE SU-35 EXPORT VERSION IS COMPLETELY DIFFERENT FROM THE PREVIOUS AIRCRAFT WITH THE SAME DESIGNATION. THE AIRCRAFT IS DUE TO BEGIN PRODUCTION

IN 2010, WITH FOCUS ON EXPORT CUSTOMERS. AN ATTRACTIVE FEATURE IS NO USE OF WESTERN TECHNOLOGY, WHICH IS VULNERABLE TO EXPLOITATION OR EXPORT RESTRICTIONS. THE RUSSIAN DOMESTIC VERSION IS SU-35S.

RUSSIAN MULTI-ROLE FIGHTER SU-30/FLANKER-F AND EXPORT SU-30MK SERIES

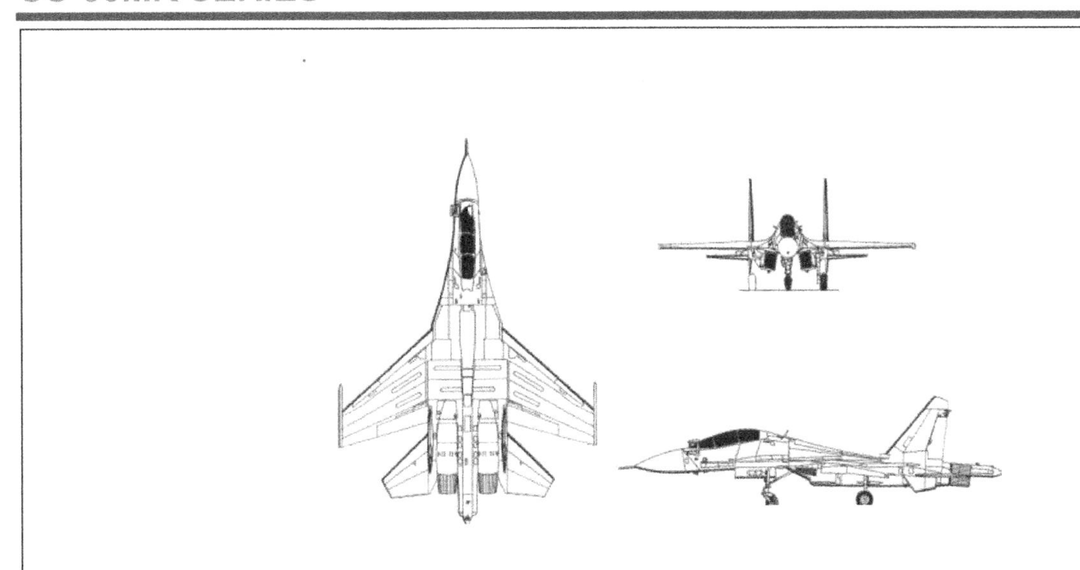

SYSTEM	SPECIFICATIONS	SYSTEM	SPECIFICATIONS
Weapon & Ammunition Types		Takeoff /Landing Roll (m):	550/670
30-mm Gsh-30-1 cannon (rnds):	150	Dimensions (m):	
Other Loading Options:	2	Length:	21.9
AA-10A-D/ALAMOAAM	6	Wingspan:	14.7
AA-11/ARCHER AAM	6	Height:	6.4
AA-12 ADDER AAMs	6	Standard Payload (kg):	
AS-14/KEDGE ASM:	6	External:	8,000
AS-17/KRYPTON ASM:	6	Hardpoints pylons:	12
AS-18/KAZOO ASM:	2	Survivability/Countermeasures:	
Gun pods:	3	Zero/zero ejection seat:	Yes
420-mm S-25 rockets (1 each):		Infrared and radar jammer:	Yes
80-mm S-8 rocket pod (20 ea.):		Radar and missile warning receivers:	Yes
122-mm S-13 rocket pod (5 each):		Chaff and flares:	Yes
250-kg, or 500-kg unguided and guided bombs:		Gaseous oxygen for 10 hours of flight:	Yes
KAB-500Kr Bombs:	6	ARMAMENT:	
KAB-1500Kr Bombs:	2	30-mm gun mounted in the right wing:	Yes

SYSTEM		30-mm gun, Gsh-30-1:	
Alternative Designations:	Su-27PU	Range (m):	(practical) 4,000
Date of Introduction:	1996	Elevation/Traverse:	None (rigidly mounted)
Proliferation:	China, India, Russia	Ammo Type:	HEFI, APT, CC
Description:	Variants in ( )	Rate of Fire (rpm):	1,500
Crew Su-30MK:	3	AVIONICS/SENSOR/OPTICS:	
Engines 16,755 lbs. thrust Saturn AL-31F turbofans, 27,558 lbs. thrust with afterburner:	2	Pulse-Doppler look-down/ shoot-down radar:	
Weight (kg):		Search range (km):	240
Maximum Takeoff:	38,000	Track range (km):	185
Normal Takeoff:	24,140	Multi Target Capability:	Yes
Empty:	17,900	IR Sensor:	Yes
Speed (km/h):		Laser Designator:	yes
Maximum (at altitude):	2,125, Mach 2.0	Heads Up Display:	Yes
Maximum (sea level):	1,350	AVIONICS/SENSOR/OPTICS:	
Max "G" Force (g):	+8	Pulse-Doppler look-down/ shoot-down radar:	
Ceiling (m):	17,500	Search range (km):	240
Vertical Climb Rate (m/s):		Track range (km):	185
Fuel (liters):		Multi Target Capability:	Yes
Internal:	9,400	IR Sensor:	Yes
Range (km):		Laser Designator:	yes
Unrefueled:	3,000	Heads Up Display:	Yes
One refueling:	5,200		
With Aux Fuel:	4,390 (SM)		
Combat Radius:	1,500		

NOTES

A SMALL NUMBER OF THE AIR SUPERIORITY FIGHTERS HAVE BEEN PRODUCED. THE GREATER EXPORT MARKET IS FOR MULTI-ROLE VERSIONS

APPEARANCE: WINGS: MID-MOUNT, SWEPT, SQUARE TIPS. ENGINES: TWO IN FUSELAGE, WITH SQUARE UNDERWING INTAKES. FUSELAGE: POINTED NOSE, HUMPED PROFILE AT THE COCKPIT AND TAPERED TO NEARLY FLAT AT THE ENGINES

VARIANTS

TWO-SEATER AIRCRAFT IS SIGNIFICANTLY UPGRADED AND DERIVED FROM SU-27 SINGLE-SEAT AIRCRAFT.

SU-30M: THE FIRST REAL MULTI-ROLE AIRCRAFT IN THE SU-27 FAMILY, WITH ALL NECESSARY SUB-SYSTEMS. THESE WERE CONVERTED INTO DEMONSTRATORS FOR EXPORTS.

SU-30MK: EXPORT SERIES VERSION. THE SU-30MK2 ANTI-SHIP UPGRADE VERSION HAS BEEN EXPORTED.
SU-30MKK/FLANKER-G: MULTI-ROLE UPGRADE UTILIZING AIR-TO-GROUND WEAPONS TO A MORE ADVANCED VERSION INCORPORATING NEW RADAR, CANARDS AND THRUST VECTORING.JJ-11: CHINESE LICENSE-BUILT VERSION.
SU-30MKI/FLANKER-H: VERSION OF THE SU-30MK MADE FOR INDIA. MOST WILL BE PRODUCED BY AN INDIAN FIRM. SOME WESTERN EQUIPMENT REPLACED MUCH OF THE RUSSIAN SYSTEMS. SU-30MKM: VERSION FOR USE BY MALAYSIA.

BRITISH/GERMAN MULTI-ROLE AIRCRAFT TORNADO IDS

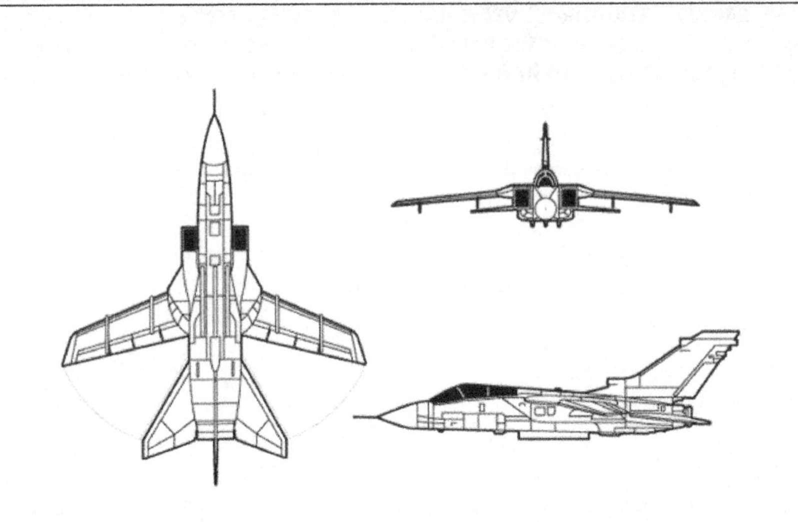

SYSTEM	SPECIFICATIONS	SYSTEM	SPECIFICATIONS
Weapon & Ammunition Types:		Ceiling (m):	+15,000
2 integral IWKA-Mauser 27-mm (rnds):	180	Fuel (liters):	
Other Loading Options lbs.:	≤ 9,000	Internal (RAF/RSAF):	6,393/5,842
Bombs:		External:	4,500
Air-to-air missiles:	≤ 8	Range (km):	
Anti-radar missiles:		Tactical Radius (hi-lo-hi profile with 2,629 kg) (km):	1,390
Anti-runway sub munition dispensers:		Ferry:	3,890
Stand-off weapons systems:		Takeoff Run/Landing Roll (m):	900/370
Air-to-surface missiles:		Dimensions (m):	
Brimstone ATGM:		Length:	16.7
Storm Shadow Cruise Missile:	1	Wingspan extended:	13.9
		Swept:	8.6
Sea Eagle Anti-Ship Missiles:	4	Height:	5.6
Raptor EO/IR Recon Pod:	1	Standard Payload (kg):	
Internal sensors:	3	External:	9,000
Paveway Laser-Guided Bombs:		Hardpoints:	7
		under fuselage	3

		under wing (ea.):	2
Flares:		Survivability/Countermeasures:	
EW equipment:		Martin-Baker MK-10A zero/zero ejection seat (2ea):	Yes
1500 L or 2250 L drop fuel tanks:		Radar Homing and Warning (RHAW):	Yes
SYSTEM		Active ECM pod:	Yes
Alternative Designations:		Chaff and flare dispensing system:	Yes
Date of Introduction:	1982	IFF:	Yes
Proliferation:	Germany, Great Britain, Italy, and Saudi Arabia	ARMAMENT:	
Description:		Internal Mauser 25-mm cannon (2ea):	Yes
Crew:	2 (pilot, weapons officer)	AVIONICS/SENSOR/OPTICS:	
Engines 9,000 lbs. thrust Turbo-Union RB199-34R turbofans, 16,000 lbs. thrust with afterburner:	2	Multi-mode, ground-mapping and terrain-following radar):	Yes
Weight (kg):		Digital Inertial Navigation System (INS):	Yes
Max Takeoff:		Doppler radar with Kalman filter:	Yes
Clean, full internal fuel:	20,411	Heads-up-display:	Yes
Full external load:	27,215	Laser Ranger and Marked Target Seeker (LRMTS):	Yes
Empty:	14,091	Night/Weather Capabilities:	
Speed (km/h):		All-weather close air support/battlefield interdiction	
Maximum (at altitude):	2,340, Mach 2.2	Interdiction/counter-air strike	
Max "G" Force (g):	+7.5	Naval strike and all-weather day and night reconnaissance capable	
External:	4,000		

NOTES

DESIGNED AND BUILT AS A COLLABORATIVE PROJECT IN THE UK, GERMANY, AND ITALY, THE TORNADO IS IN SERVICE WITH ALL THREE AIR FORCES AND THE GERMAN NAVY. TORNADO IS ALSO IN SERVICE IN SAUDI ARABIA AND OMAN. IT IS A TWIN-SEAT, TWIN-ENGINE, VARIABLE GEOMETRY AIRCRAFT AND IS SUPERSONIC AT ALL ALTITUDES.

APPEARANCE:

WINGS: HIGH-MOUNTED, VARIABLE-GEOMETRY, SWEPT-BACK, AND TAPERED WITH ANGULAR, BLUNT TIPS. ENGINES: TWO TURBOFANS INSIDE THE BODY. FUSELAGE: SOLID WITH A NEEDLE NOSE, THICKENS MIDSECTION AND TAPERS TOWARD THE TAIL.

TAIL: TALL, SWEPT-BACK, AND HAS A TAPERED FIN WITH A CURVED TIP AND A STEP IN THE LEADING EDGE. FLATS ARE LARGE, MID-MOUNTED ON THE BODY, SWEPT-BACK, AND TAPERED WITH BLUNT TIPS.

VARIANTS

TORNADO IDS: DESIGNATED GR1 IN RAF SERVICE. GROUND ATTACK/ INTERDICTION VERSION. SOME HAVE BEEN ADAPTED FOR THE ANTI-SHIPPING ROLE.

TAC-R TORNADO GR1A: RAF GR1S MODIFIED AS DEDICATED TACTICAL RECONNAISSANCE AIRCRAFT. FITTED WITH A MARCONI DEFENSIVE SYSTEMS EMITTER LOCATION SYSTEM. BOTH 27-MM CANNONS WERE REMOVED.

TORNADO GR1B: MODIFIED FOR MARITIME STRIKE MISSIONS WITH SEA EAGLE ANTI-SHIP MISSILES. RAF DISCARDED GR1B DESIGNATION IN JULY 2001.

TORNADO ADVAIR DEFENSE VARIANT

TORNADO ECR: ELECTRONIC COMBAT AND RECONNAISSANCE VARIANT FOR GERMAN AND ITALIAN SERVICE

RUSSIAN TRANSPORT AIRCRAFT AN-2/COLT

SYSTEM	SPECIFICATIONS	SYSTEM	SPECIFICATIONS
SYSTEM		Takeoff Run/Landing Roll (m):	
Alternative Designations:	INA	Prepared Surface:	150/170
Date of Introduction:	1948	Unprepared Surface:	200/185
Proliferation:	At least 32 countries	Max Load:	INA
Description:		Dimensions (m):	
Crew:	2 (pilots)	Length:	12.7
Engines 1,000-shp Shevetsov Ash-62 or PZL Kalisz Ash-621R 9-cylinder radial piston driving a four-bladed, variable-pitch propeller:	1	Wingspan:	18.2
Weight (kg):		Height:	4.0
Max Gross:	5,500	Cabin Dimensions (m):	
Normal Takeoff:	INA	Floor Length:	4.1
Empty:	3,450	Width:	1.6
Speed (km/h):		Height:	1.8
Max:	258	Standard Payload (kg):	
Min:	90	Internal:	1,500
Cruise:	185	Transports 12 troops or paratroops, or 6 litters.	
Takeoff/Landing Speed:	85	Survivability/Countermeasures:	
Max "G" Force (g):	-1.0 - +3.7	None	
Ceiling (m):		ARMAMENT:	

Service (clean):	4,400	12.7-mm machineguns:	Experimental
Vertical Climb Rate (m/s):	3.0	23-mm machineguns:	Experimental
Fuel (liters):		Unguided aerial rockets:	experimental
Internal:	1,200	AVIONICS/SENSOR/OPTICS	
External:	None	Flight avionics only.	Yes
Range (km):		Night/Weather Capabilities:	
Max Load:	900	The An-2 is capable of flight under day and instrument meteorological conditions.	

NOTES

THE WINGS AND ELEVATORS ARE FABRIC-COVERED, WHILE THE FUSELAGE IS METAL. THIS AIRCRAFT CAN OPERATE FROM UNIMPROVED AIRFIELDS, AND IS NOTED FOR SHORT TAKEOFF AND LANDING CAPABILITIES, AND RUGGEDNESS. ITS LOW ACOUSTIC SIGNATURE AND SLOWER SPEEDS ALLOW FOR STEALTHY OPERATION. CABIN CONTAINS TIP-UP SEATS, WHICH CAN BE EASILY FOLDED TO ALLOW SPACE FOR CARGO. SKIS OR PONTOONS CAN BE EMPLOYED ON THE MAIN LANDING GEAR STRUTS.

APPEARANCE: WINGS: BIPLANE AND RECTANGULAR-SHAPED WITH CURVED TIPS, ONE HIGH-MOUNT AND ONE LOW MOUNT (SHORTER), BRACED BY STRUTS. ENGINES: ONE MOUNTED IN NOSE. FUSELAGE: SHORT, THICK, WITH BLUNT NOSE. TAIL: TAPERED WITH ROUND TIP, RECTANGULAR, LOW-MOUNTED FLATS.

VARIANTS

THIS AIRCRAFT WAS ORIGINALLY BUILT IN RUSSIA. NOW IT IS PRODUCED IN CHINA AND POLAND.

AN-2D/-2TD: SPECIALLY MODIFIED FOR PARACHUTE TRAINING AND SPECIAL OPERATIONS.

AN-2P/-2T/-2TP: PASSENGER AND GENERAL TRANSPORT VARIANTS.

AN-2V/-2M/-4: SEAPLANE VARIANT WITH FLOATS IN PLACE OF MAIN LANDING GEAR.

AN-3: THIS VARIANT EMPLOYS AN UPGRADED 1,450-SHP GLUSHENKOV TVD-20 TURBOPROP ENGINE, AND A LARGER THREE-BLADED PROPELLER. THIS ALLOWS FOR AN INCREASED TAKEOFF WEIGHT OF 5,800 KG.

Y-5/C-5: CHINESE-BUILT VERSION AND CHINESE EXPORT NOMENCLATURE.

RUSSIAN CARGO/TRANSPORT AIRCRAFT AN-12/CUB

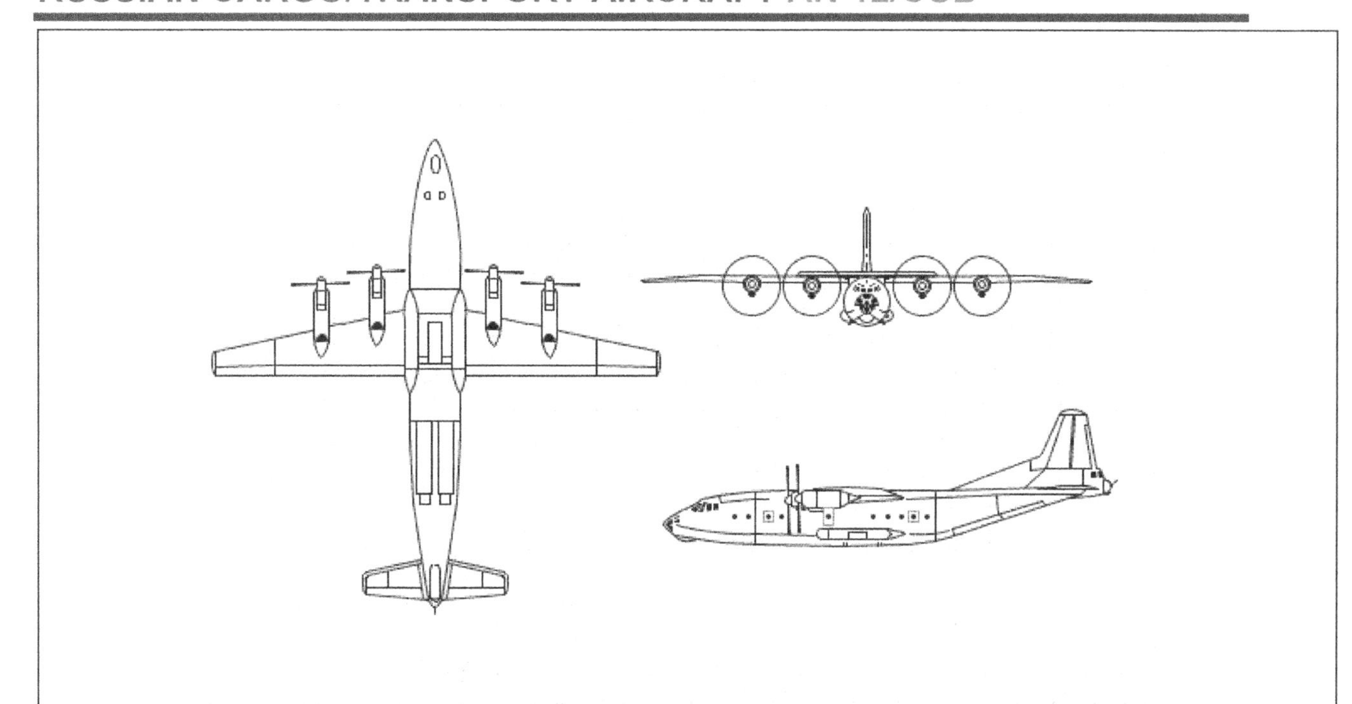

SYSTEM	SPECIFICATIONS	SYSTEM	SPECIFICATIONS
SYSTEM		Dimensions (m):	
Alternative Designations:		Length:	33.1
Date of Introduction:	1959	Wingspan:	38.0
Proliferation:	< 17 countries	Height:	10.6
Description:		Hatch Opening:	(m)
Crew (including tail gunner):	6	Length:	7.7
Engines 4,000-shp Ivchenko AI-20K with 4-blade reversible pitch propellers:	4	Width:	3.0
Weight (kg):		Cargo Hold (m):	
Max Gross:	61,000	Length:	13.5
Normal Takeoff:	55,100	Width:	3.5
Empty:	28,000	Height:	2.6
Speed (km/h):		Volume:	122.8 cu m
Max:	777	Standard Payload (kg):	
Min:	163	Internal:	

Cruise:		Troops:	90
Max 670		Paratroops:	60
Econ 580		Vehicles:	Yes
Landing Speed:	200	Weapons:	Yes
Ceiling (m):	10,200	Cargo:	Yes
Vertical Climb Rate (m/s):	10	Survivability/Countermeasures:	
Internal Fuel (liters):		Warning radar in the tail:	Yes
Normal:	13,900	ARMAMENT:	Yes
Maximum:	19,100	2 NR-23 23-mm cannons in tail turret:	Yes
Range (km):		AVIONICS/SENSOR/OPTICS:	
Max Load:	1,400	I-band ground mapping and precision location radar in chin radome.	Yes
10,000 kg Load:	3,600	Night/Weather Capabilities:	No
Max Fuel:	5,700		
Takeoff Run/Landing Roll (m):	700/500		

NOTES

THE AN-12 CUB IS A VERY WIDELY USED RUSSIAN CARGO AND PARATROOP AIRCRAFT, SIMILAR IN APPEARANCE, PAYLOAD AND ROLE TO THE C-130 HERCULES. IT IS A MILITARY VERSION OF THE AN-10. BEFORE THE COLLAPSE OF THE SOVIET UNION, THE CUB WAS THE PRINCIPAL MILITARY TRANSPORT AND WAS ADAPTED FOR THE ELECTRONIC INTELLIGENCE (ELINT) AND ELECTRONIC COUNTERMEASURES (ECM) ROLES BY THE SOVIET NAVY AND POSSIBLY SEVERAL OTHER COUNTRIES.

APPEARANCE: WINGS: HIGH WING, TAPERED LEADING EDGE, STRAIGHT TRAILING EDGES, AND BLUNT TIPS. ENGINES: 4 ENGINES IN THIN NACELLES EXTENDING FORWARD FROM THE UNDERSIDE OF THE WING. FUSELAGE: GLAZED ROUNDED NOSE; CONSTANT CROSS-SECTION CARGO HOLD; BROAD, FLAT BOTTOM TURNS UPWARD TO THE TAIL GUNNER'S POSITION. TAIL: SET HIGH ON AFT FUSELAGE WITH DOUBLE-TAPERED FIN AND FULL-HEIGHT RUDDER MOUNTED UP GUNNER'S POSITION. LARGE DORSAL FILLET SLOPES DOWN FROM FIN TO TOP OF FUSELAGE.

VARIANTS

CUB (AN-12BP): STANDARD TRANSPORT/CARGO VERSION; SEVERAL ELECTRONIC BLISTERS FITTED.

CUB-A: ELINT VERSION: BLADE AERIALS FITTED ON FRONT OF FUSELAGE, AFT OF FLIGHT DECK.

CUB-B: NAVAL ELINT VERSION. PALLETIZED PASSIVE RECEIVERS, FREQUENCY ANALYZERS, RECORDING EQUIPMENT AND ACCOMMODATION FOR EW PERSONNEL IN MAIN CARGO COMPARTMENT.

CUB-C: ECM VERSION. VENTRAL ANTENNA HOUSINGS, JAMMERS ON PALLETS, AND OTHER FEATURES INDICATE THE CAPABILITY OF ELINT COLLECTION.

CUB-D: UPGRADED CUB-C WITH ADDITIONAL ECM EQUIPMENT NAVAL ELECTRONIC WARFARE VERSION.

SHAANXI Y-8: CHINESE MANUFACTURED.

RUSSIAN TRANSPORT AIRCRAFT AN-26/CURL

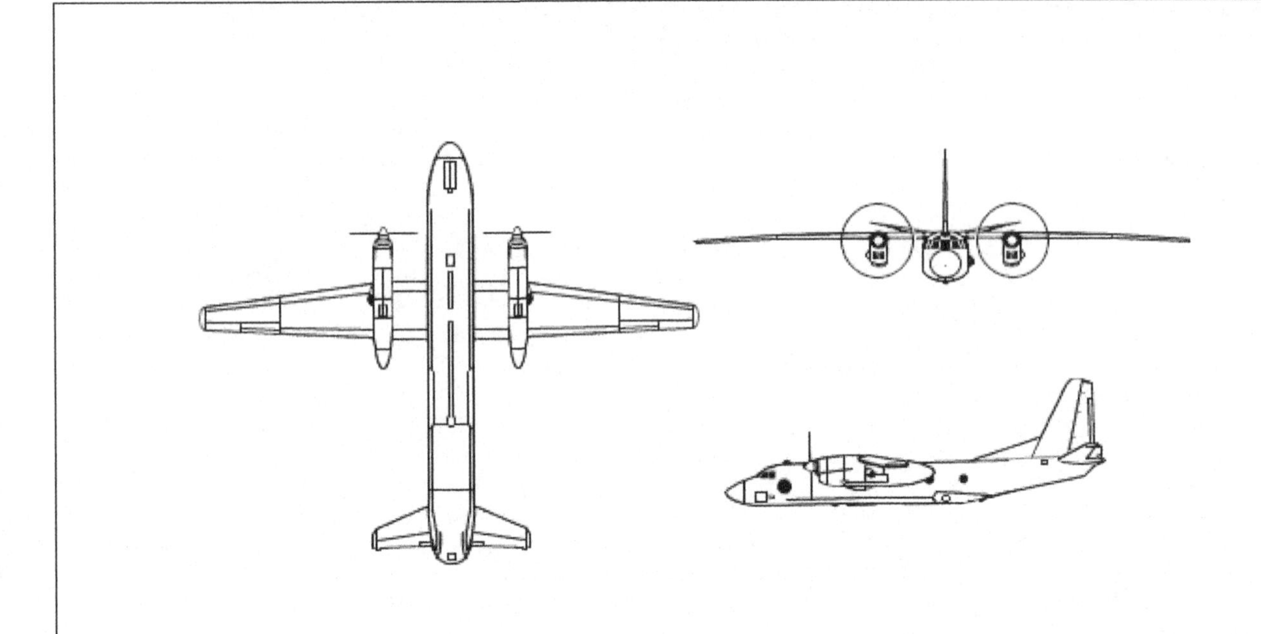

SYSTEM	SPECIFICATIONS	SYSTEM	SPECIFICATIONS
SYSTEM		Width:	2.4
Alternative Designations:		Height:	1.9
Date of Introduction:	1970	Standard Payload (kg):	
Proliferation:	> 28 countries	Internal:	
Description:		Normal:4,500	
Crew (pilot, copilot, navigator, flight engineer, radio operator):	5	Max:	5,500
Engines 2,820 ehp Ivchenko AI-24VT turboprops and 1 x 1,765 lbs. thrust RU 19A-300 turbojet for takeoff assist:	2	Transports:	
Weight (kg):		Seats in Pressurized Cargo bay:	38 - 40
Max Takeoff:	24,000	Litters with attendants:	24
Empty:	15,020	Survivability/Countermeasures:	
Speed (km/h):		Air-conditioned and pressurized cabin:	Yes

Max:	540	Emergency escape hatch in door immediately aft of flight deck:	Yes
Cruise:	440	Chaff/flare dispensers pylon-mounted:	Yes
Takeoff/Landing Speed:	200/190	Two ADF radio altimeter:	Yes
Ceiling (m):	7,500	Glide path receivers:	Yes
Vertical Climb Rate (m/s):	8	Glide slope receiver:	Yes
Fuel (liters):		Marker beacon receiver:	Yes
Internal:	7,050	Weather/navigation radar:	Yes
Range (km):		Directional gyro:	Yes
Max Payload:	1,100	Flight recorder:	Yes
Max Fuel:	2,550	Optional OPB-1R sight for pinpoint dropping of freight:	Yes
Takeoff Run/Landing Roll (m):	780/730	Medical equipment:	Yes
Dimensions (m):		Liquid heating system:	Yes
Length:	23.8	Night/Weather Capabilities:	No
Wingspan:	29.2		
Height:			
Length:	11.5		

NOTES

THE AN-26 CURL IS A WIDELY USED SHORT-HAUL CARGO/TRANSPORT. IT CAN BE MODIFIED TO PERFORM PARATROOP TRANSPORT, MEDICAL EVACUATION, OR PASSENGER TRANSPORTATION. THE AN-26 IS PRODUCED IN BOTH MILITARY AND CIVIL AIR VERSIONS WITH ESSENTIALLY THE SAME FEATURES.

APPEARANCE: WINGS: HIGH-MOUNTED, EQUALLY TAPERED FROM ENGINES TO THE BLUNT TIPS ENGINES: TWO TURBOPROPS MOUNTED IN PODS BENEATH THE WINGS, WHICH EXTEND BEYOND THE WINGS' LEADING AND TRAILING EDGES FUSELAGE: LONG AND SLENDER WITH AN UPSWEPT REAR SECTION AND A SOLID, ROUNDED NOSE, FEATURING A STEPPED COCKPIT. TAIL: BACK-TAPERED BLUNT TIPPED FIN. HIGH-MOUNTED BACK-TAPERED FLATS WITH BLUNT TIPS, AND HAVE A POSITIVE SLANT

VARIANTS

AN-26: ORIGINAL PRODUCTION VERSION.

AN-26B: UPGRADED VERSION WITH IMPROVED CARGO HANDLING EQUIPMENT.

Y7H/Y7H-500: CHINESE PRODUCTION VERSION

RUSSIAN TRANSPORT AIRCRAFT IL-18/COOT

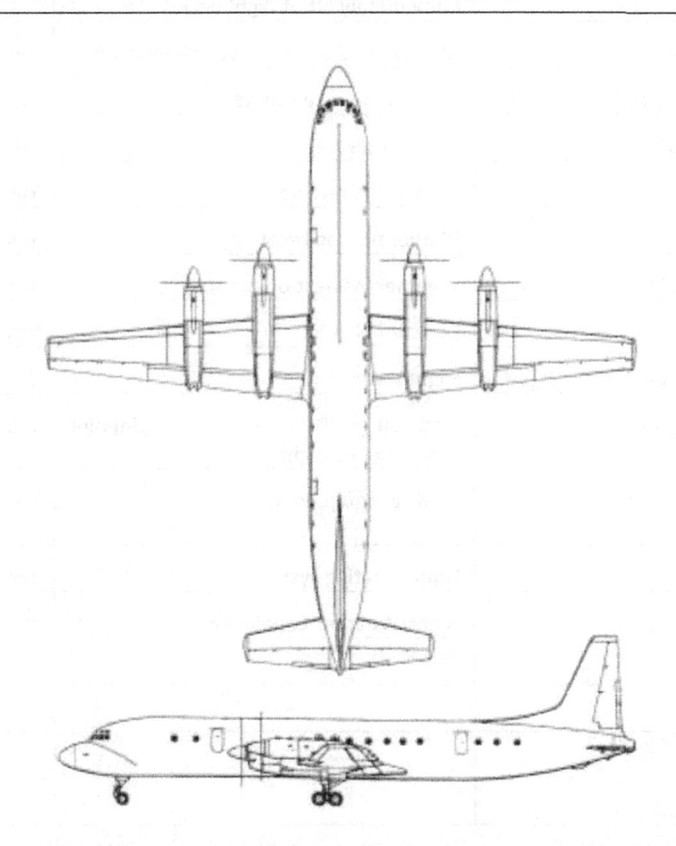

SYSTEM	SPECIFICATIONS	SYSTEM	SPECIFICATIONS
SYSTEM		**Fuel (liters):**	
Alternative Designations:	**Il-20, Il-22**	**Internal (D/E):**	**30,000 / 23,700**
Date of Introduction:	**1959**	**External: None**	
Proliferation:	**>5 countries**	**Range (km):**	
Description:		**Max Load (D/E):**	**4,000 / 3,200**
Crew (2x pilots, 1x navigator, 1x radio operator, 1x flight engineer):	**5**	**Normal Load (D/E):**	**6,500 /5,200**
Engines 250-shp Ivchenko AI-20M turboprop driving 4x four-bladed reversible-pitch propellers:	**4**	**Takeoff Run/Landing Roll (m):**	
Weight (kg):		**Prepared Surface (D/E):**	**1,300 / 850**
Max Gross (D/E):	**64,000 / 61,200**	**Unprepared Surface: INA**	
Empty(D/E):	**35,000 /34,610**	**Dimensions (m):**	

Speed (km/h):		Length:	35.9
Max:	675	Wingspan:	37.4
Min:	INA	Height:	10.2
Cruise:	625	Cabin Dimensions (m):	
Takeoff/Landing Speed:	INA	Floor Length:	24
Max “G” Force (g):	INA	Width:	3.2
Ceiling (m):		Height:	2
Service (clean):	10,000	Standard Payload (kg):	
Operating Altitude:	8,000-10,000	Internal:	13,500
Vertical Climb Rate (m/s):	INA	Troops:	122
Dimensions (m):		ELINT Operators:	20
Length:	23.8	Survivability/Countermeasures:	None
Wingspan:	29.2	ARMAMENT:	None
Height:	8.6	AVIONICS/SENSOR/OPTICS	
Cabin Dimensions (m):		Flight avionics:	Yes
Length:	11.5		

NOTES

APPEARANCE: WINGS: LOW-MOUNTED AND TAPERED WITH BLUNT TIPS. ENGINES: FOUR MOUNTED ON WINGS AND EXTENDING FORWARD. FUSELAGE: ROUND, CIGAR-SHAPED, TAPERED AT REAR WITH ROUNDED NOSE. TAIL: TAPERED WITH SQUARE TIP, FUSELAGE-MOUNTED, TAPERED FLATS

VARIANTS

THIS AIRCRAFT WAS ORIGINALLY DESIGNED AS A CIVILIAN TRANSPORT AIRCRAFT, BUT HAS BEEN ADAPTED FOR MILITARY USES.

IL-18D: HAS A CENTER FUEL TANK FOR LONGER FLIGHT DURATION AND EXTENDED RANGE.

IL-18E: VARIANT WITHOUT CENTER FUEL TANK.

IL-20/COOT A: UNARMED STRATEGIC ELECTRONIC INTELLIGENCE/ RECONNAISSANCE AND SURVEILLANCE AIRCRAFT. THE AIRFRAME IS ESSENTIALLY THE SAME AS THE IL-18D, BUT A CYLINDER CONTAINING A POSSIBLE SIDE-LOOKING AIRBORNE RADAR IS MOUNTED UNDER THE FUSELAGE FORWARD OF THE WING. SMALLER CONTAINERS ON THE FORWARD SIDES OF THE FUSELAGE HOUSES POSSIBLE CAMERAS AND SENSORS. MANY SMALL ANTENNAS ARE LOCATED UNDER THE FUSELAGE.

IL-20M: VERSION WITH A SIDE-LOOKING AIRBORNE RADAR (SLAR), A-87P LOROP CAMERAS, AND A ROMB 4 SIGINT SYSTEM.

IL-22M/COOT B: AN AIRBORNE COMMAND POST VARIANT OF THE IL-18D AIRFRAME.

COOT-C: LATER ELINT PLATFORM

RUSSIAN CARGO/TRANSPORT AIRCRAFT IL-76/CANDID

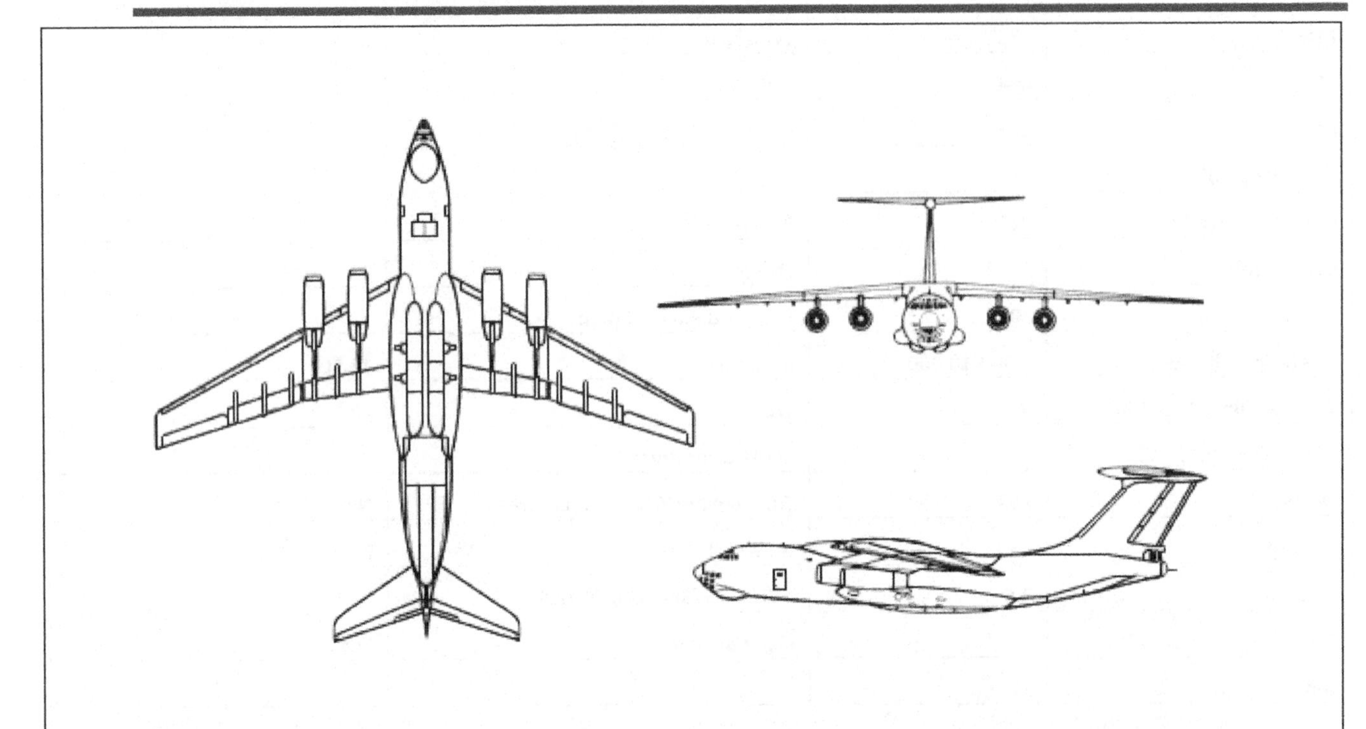

SYSTEM	SPECIFICATIONS	SYSTEM	SPECIFICATIONS
SYSTEM		**Dimensions (m):**	
Alternative Designations:		**Length:**	46.6
Date of Introduction:	1975	**Wingspan:**	50.5
Proliferation:	≥ 12 countries	**Height:**	14.8
Description:	(Il-76MD)	**Hatch:**	(m)
Crew (2x pilots, 1x navigator, 1x radio operator, 1x flight engineer):	5	**Height:**	3.4
Engines 26,455 lbs. thrust Rybinsk D-30KP II turbofan with thrust reversers:	4	**Width:**	3.5
Weight (kg):		**Cargo Hold (m):**	
Empty:	89,000	**Length to Ramp:**	20.0
Takeoff:		**Length with Ramp:**	24.5
General Max:	190,000	**Width:**	3.5
Allowable Max:	210,000	**Height:**	3.4
Unprepared Runway Max:	157,500	**Standard Payload (kg):**	

Speed (km/h):		Internal:	47,000
Max:	919	Troops:	140
Cruise:	780	Paratroops:	125
Ceiling (m):	10,500	Survivability/Countermeasures:	
Fuel (liters):		Entire aircraft pressurized:	Yes
Internal 12 tanks:	109,480	Crew emergency escape hatch forward of main entry door:	Yes
External:	None	Flares for illuminating landing area:	Yes
Range (km):		Radar warning receiver:	Yes
Length:	23.8	Electronic jammers:	Yes
Wingspan:	29.2	Chaff and flares:	Yes
Height:	8.6	ARMAMENT:	
Cabin Dimensions (m):		GSh-23L twin-barreled cannon in tail turret:	Yes
Length:	11.5	AVIONICS/SENSOR/OPTICS:	
Max Load:	3,800	Standard flight controls:	Yes
Normal Load:	4,760	Weather radar in nose:	Yes
Max Fuel:	7,800	navigation and ground mapping radar in radome:	Yes
Small Load (20,000) Payload kg:	7,300	Night/Weather Capabilities:	Yes
Takeoff Run/Landing Roll (m):	1,700/900-1,000		

NOTES

APPEARANCE: WINGS: HIGH-MOUNTED, SWEPT-BACK, AND TAPERED WITH BLUNT TIPS. TRAILING EDGE HAS A SLIGHT CRESCENT SHAPE. ENGINES: FOUR MOUNTED PYLONS UNDER AND EXTENDING FORWARD OF WINGS' LEADING EDGE. FUSELAGE: LONG, ROUND AND TAPERING TO THE REAR, ROUNDED NOSE WITH CHIN RADOME. TAIL: T-TAIL WITH CURVED LEADING EDGE AND INSET RUDDER. SWEPT TAIL PLANES MEET AT TOP OF THE TAIL.

VARIANTS

IL-76 CANDID-A: FIRST PRODUCTION MODEL.

IL-76M CANDID-B: ADDED REAR TURRET WITH TWO 23-MM NR-23 GUNS AND SMALL ECM FAIRINGS EACH SIDE OF NAVIGATOR'S WINDOWS.

IL-76MD CANDID-B: MILITARY VERSION OF IL-76T. T STANDS FOR TRANSPORT; D STANDS FOR LONG-RANGE.

IL-76MF: MILITARY VARIANT WITH STRETCHED FUSELAGE AND MORE POWERFUL ENGINE.

IL-76MF-100: A DERIVATIVE OF THE IL-76M WITH CFM56-5C TURBOFANS. RANGE INCREASED TO 7,000 KM WITH 40,000 KG LOAD.

IL-76PP: ELECTRONIC COUNTERMEASURES AIRCRAFT

IL-76PS/IL-84: SEARCH AND RESCUE CAPABLE OF 3-HOUR PATROL WITH RADIUS OF 3,000 KM.

IL-76VPK/IL-82 AIRBORNE COMMAND POST: VARIANT OF IL-76MD. FITTED WITH SPECIALIZED COMMUNICATIONS EQUIPMENT.

A-50/MAINSTAY: AWACS VERSION.

MIDAS: AERIAL TANKER VERSION.

CHINESE LIGHT BOMBER AIRCRAFT H-5 AND RUSSIAN/CZECH IL-28/BEAGLE

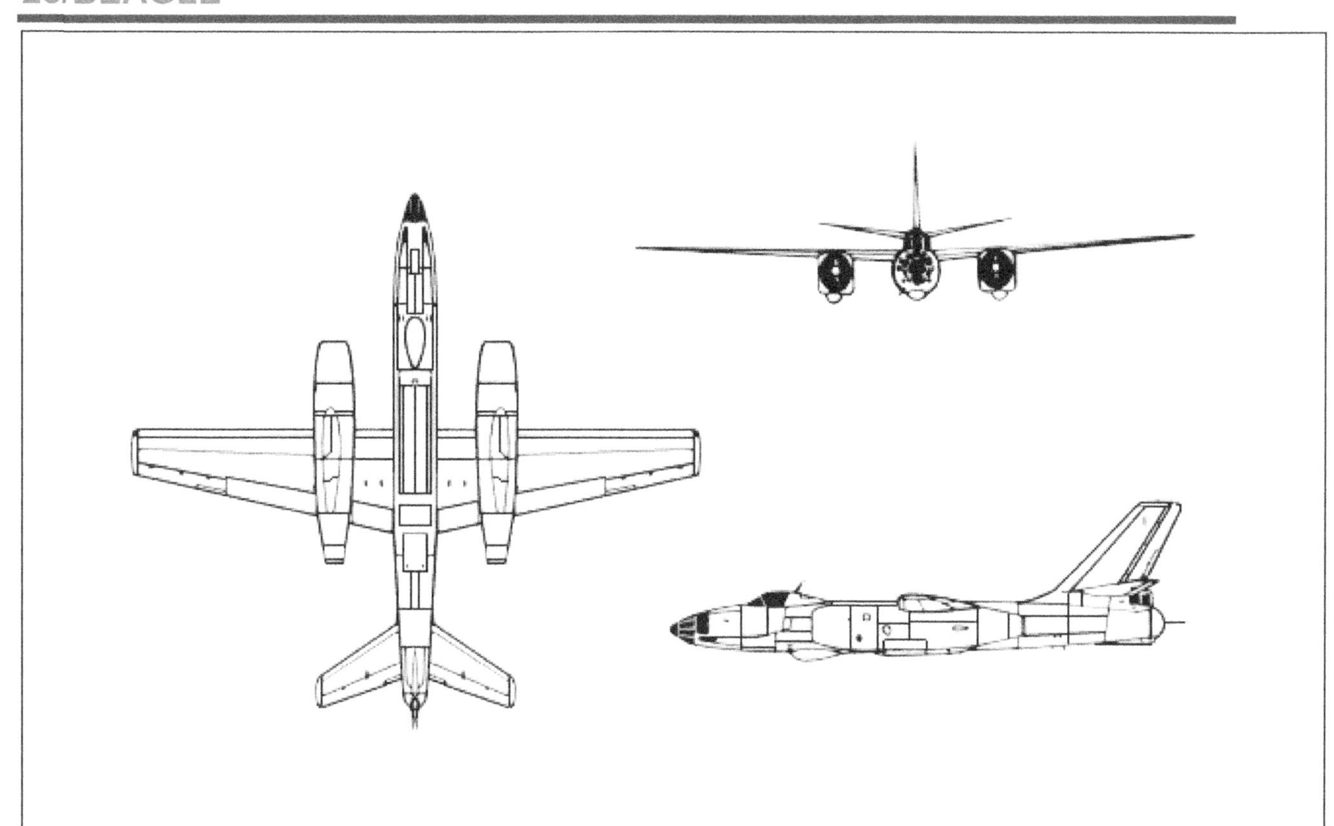

SYSTEM	SPECIFICATIONS	SYSTEM	SPECIFICATIONS
SYSTEM		Speed (km/h):	
Alternative Designations:	Hongzhaji-5 H-5 is a Chinese adaptation of the IL-28.	Max:	900
Date of Introduction:	1966, 1950 for Il-28	Cruise:	769
Proliferation:	> 24 countries	Ceiling (m):	12,500
Description:		Vertical Climb Rate (m/s):	15
Crew (pilot, navigator/bombardier, radio operator/gunner):	3	Fuel (liters):	7,908
Engines 5,952 lbs. thrust Wopen-5 turbojets:	2	Range (km) empty:	3,550
Weight (kg):		Combat Radius (w/max payload):	1,100
Max Takeoff:	21,200	Takeoff Run/Landing Roll (m):	980/930

Empty:	12,890	Dimensions (m):	
General Max:	190,000	Length:	17.6
Allowable Max:	210,000	Wingspan:	21.5
Unprepared Runway Max:	157,500	Height:	6.7
Speed (km/h):		Weapons load (kg):	
Max:	919	Max:	3,000
Cruise:	780	Normal:	1,000
Ceiling (m):	10,500	Survivability/Countermeasures::	
Fuel (liters):		Pilot and navigator ejection seats:	Yes
Internal 12 tanks:	109,480	Gunner/radio operator has escape hatch:	Yes
External:	None	ARMAMENT:	
Range (km):		23-mm NR-23 cannons :	4
Length:	23.8	2 fixed in nose (rnds):	100
Wingspan:	29.2	Tail position (rnds):	250
Height:	8.6	Bombs:	
Cabin Dimensions (m):		Bombs or torpedoes in internal weapons bay (kg):	3,000
Length:	11.5	500 kg bombs:	Option
Max Load:	3,800	53 VA torpedoes:	Option
Normal Load:	4,760	250 kg bombs:	Option
Max Fuel:	7,800	single 3,000 kg bomb:	Option
Small Load (20,000) Payload kg:	7,300	AVIONICS/SENSOR/OPTICS:	
Takeoff Run/Landing Roll (m):	1,700/900-1,000	Standard flight controls:	Yes
		Navigation and ground mapping radar in radome:	Yes
		Night/Weather Capabilities:	Yes

NOTES
THE TWIN-ENGINE LIGHT BOMBER IS ALSO USED AS A MARITIME STRIKE AND TRAINER AIRCRAFT.
APPEARANCE: WINGS: SHOULDER-MOUNTED WELL AFT ON FUSELAGE.
VARIANTS
CHINESE VARIANTS INCLUDE THE FOLLOWING.
H-5: BASIC BOMBER VERSION.
HJ-5: TRAINER VERSION.
HZ-5: RECONNAISSANCE VERSION.
RUSSIAN VARIANTS INCLUDE THE FOLLOWING.
IL-28R: PHOTO-RECONNAISSANCE VARIANT
IL-28RTR: RADAR RECON (ELINT) VARIANT
AN ECM VERSION WAS ALSO DEVELOPED.
IL-46: SAME DESIGN BUT TWICE AS LARGE.

CHINESE MEDIUM BOMBER H-6 (HONGZHAJI-6)

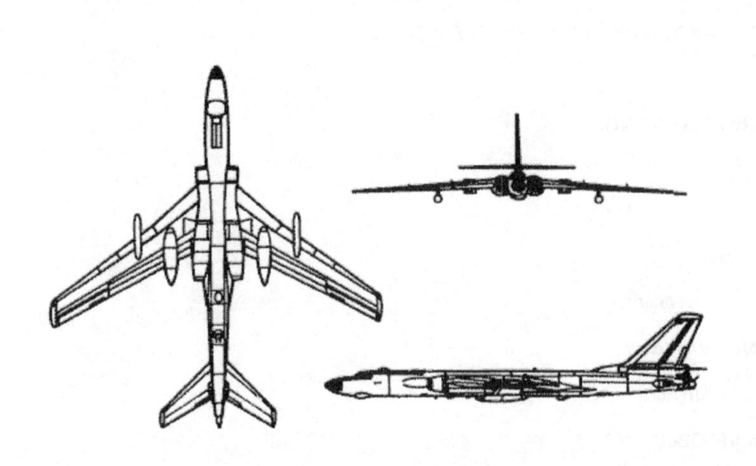

SYSTEM	SPECIFICATIONS	SYSTEM	SPECIFICATIONS
Weapon & Ammunition Types	Combat Load	Ceiling (m):	12,000
Type 23-1 30-mm Cannon:		Vertical Climb Rate (m/s):	19
Other Loading Options:		Fuel (kg):	
C502 ASMs externally (no internal):	2	Internal:	33,000
500kg Bombs:	12	External Underwing tanks (ea.):	2
1000kg Bombs internally:	6	Range (km):	
10 kt to 3 Mt (nuclear Bombs):	1 - 3	Maximum:	4,300
SYSTEM		Combat Radius:	1,800
Alternative Designations:	Hong-6, NOTES	Endurance hh:mm:	5:41
Date of Introduction:	1968	Dimensions (m):	
Proliferation:	Only China	Length:	34.8
Description:		Wingspan:	34.2
SYSTEM		Height (gear extended):	10.4
Alternative Designations:	Hong-6, NOTES	Internal Payload (kg):	
Date of Introduction:	1968	Normal:	3,000
Proliferation:	Only China	Maximum:	9,000
Description:		Survivability/Countermeasures:	

Crew:	2 pilots	Defensive electronic countermeasures system:	Yes
Navigator/bombardier:	1	Type 23-1 30-mm Cannon (ea.):	6
Tail gunner:	1	Twin-gun tail turret:	1
Observer positions in rear fuselage:	2	Twin remote controlled ventral/dorsal barbettes:	2
Engines 20,944 lbs. thrust Xian Wopen-8 turbojets:	2	AVIONICS/SENSOR/OPTICS:	
Weight (kg):		Automatic navigation system with Doppler and INS inputs:	Yes
Maximum Takeoff:	75,800	Offensive navigation/attack radar. RWR:	Yes
Empty:	38,530	Night/Weather Capabilities:	Poor
Maximum Clean Speed at 6,000 m:	992		
Max Cruise w/2 x C-601 ALCMs:	786		
Takeoff/Landing Speed:	302/233		
Max "G" Force (g):	+6.5		

NOTES

THE H-6 IS A CHINESE ADAPTATION OF THE FORMER SOVIET TU-16/BADGER MEDIUM BOMBER. IT IS USED FOR AIR-LAUNCHED CRUISE MISSILES AS WELL AS CONVENTIONAL AND NUCLEAR BOMB DELIVERY. IT CAN ALSO BE USED AS A NAVAL ANTI-SHIPPING STRIKE AIRCRAFT. IT HAS GONE THROUGH SEVERAL VARIANTS SINCE ITS INTRODUCTION IN THE 1950S. THE MOST CURRENT VERSION IS THE CHINESE NAVY'S H-6D IV.

APPEARANCE: WINGS: MID-MOUNTED, SWEPT-BACK, AND TAPERED WITH BLUNT TIPS. ENGINES: TWO TURBOJETS MOUNTED IN WING ROOTS, WHICH EXTEND BEYOND THE LEADING AND TRAILING EDGES OF THE WING ROOT. FUSELAGE: LONG, SLENDER AND BULGING WHERE ENGINES ARE MOUNTED AND TAPERED TO THE TAIL. TAIL: SWEPT-BACK, TAPERED FIN AND FLATS WITH BLUNT TIPS.

VARIANTS

H-6A I: PRODUCTION MODEL OF THE CHINESE REVERSE ENGINEERING OF THE TU-16 BADGER. EXPORT VERSION ARE DESIGNED B-6. NEARLY IDENTICAL TO THE ORIGINAL TU-16 BADGER, EXCEPT IT WAS POWERED BY XIAN WP8 TURBOJETS.

H-9A I/E: SECOND GENERATION OF THE H-6 BOMBER AND THE ONE USED BY THE CHINESE AIR FORCE. STARBOARD SIDE 23-MM NOSE CANNON WAS REMOVED AND IMPROVED ECM/ESM, BOMBING AND NAVIGATIONAL SYSTEMS WERE INSTALLED.

H-6B II, H-6C III: EQUIPPED WITH A DOPPLER RADAR, A NAVIGATION COMPUTER AND INERTIAL NAVIGATION EQUIPMENT.

H-6DU/H-6U TANKER: FIRST FLIGHT IN 1990. CARRIES TWO UNDERWING HOSE-AND-DROGUE PODS TO REFUEL TWO J-8DS SIMULTANEOUSLY.

H-6 ELECTRONIC WARFARE PLATFORM: MODELS HAVE BEEN SEEN. A LONG, CANOE-SHAPED RADOME ON THE LOWER FUSELAGE, AN EXTRA ANTENNA FAIRING ON THE TOP OF THE FUSELAGE AND A SOLID NOSECONE. COULD HOUSE A SIDE-LOOKING RADAR OR AIRCRAFT COULD SERVE IN AN ELINT OR OFFENSIVE ECM ROLE.

RUSSIAN LONG-RANGE BOMBER TU-22M3/BACKFIRE-C

SYSTEM	SPECIFICATIONS	SYSTEM	SPECIFICATIONS
Weapon & Ammunition Types		Maximum (sea level):	1,050, Mach 0.9
23-mm twin barrel gun (ea.):	1	Cruise:	800
Other Loading Options:		Takeoff/Landing Speed:	370/285
Missiles:	1 - 3	Max "G" Force (g):	+2.5
AS-4 Kitchen ASM:		Ceiling (m):	17,000
AS-17 Krypton ASM:		Fuel est. (liters):	16,500
AS-20 Kayak ASM:		Maximum Unrefueled Combat Radius (km):	4,000
AS-9 Kyle ARM:		Supersonic, hi-hi-hi, 12,000 kg weapons:	1,500 - 1,850
AS-16 Kickback short range attack:	6	Subsonic, lo-lo-lo, 12,000 kg weapons:	1,500 – 1,665
Bombs:	8	Subsonic, hi-hi-hi, max weapons:	2,200
3,000 kg:	2	Takeoff Run/Landing Roll (m):	2,000 – 2,100/1,200 – 1,300
1,500 kg:	8	Dimensions (m):	
500 kg:	42	Length:	42.4
250 kg:	69	Wingspan extended / swept:	34.3 / 23.4
100 kg:	69	Height:	10.8
Mines:		Standard Payload Max (kg):	24,000
1,500 kg:	8	External:	12,000

SYSTEM		Internal:	12,000
Alternative Designations:		ARMAMENT:	
Date of Introduction:	1974	23-mm 2x barrel NR-23 gun, in the tai:	Yes
Proliferation:	Russia and Ukraine	AVIONICS/SENSOR/OPTICS:	
Description:		Automatic high- and low-altitude preprogrammed flight control, with automatic approach:	Yes
Crew (pilot, copilot, navigator, defensive systems operator):	4	Secure SATCOM datalink receiver and comms:	Yes
Engines 50,000 lbs. thrust NK-25 turbofans:	2	Missile targeting and navigation radar:	Yes
Weight (kg):		Video camera to provide visual assistance for weapons aiming at high altitude:	Yes
Max Takeoff:	126,000	TV remote gun and bomb sights:	Yes
Empty:	49,500	PRS-3/Argon-2 ranging radar:	Yes
Speed (km/h):		PNA-D attack radar:	Yes
Maximum (at altitude):	2,327 Mach 2.05	Night/Weather Capabilities:	Good

NOTES

THE BACKFIRE IS A LONG-RANGE AIRCRAFT CAPABLE OF PERFORMING NUCLEAR AND CONVENTIONAL ATTACK, ANTI-SHIP, AND RECONNAISSANCE MISSIONS. ITS LOW-LEVEL PENETRATION FEATURES MAKE IT A MUCH MORE SURVIVABLE SYSTEM THAN ITS PREDECESSORS. CARRYING EITHER BOMBS OR AS-4/KITCHEN AIR-TO-SURFACE MISSILES, IT IS A VERSATILE STRIKE AIRCRAFT, BELIEVED TO BE INTENDED FOR THEATER ATTACK IN EUROPE AND ASIA, BUT ALSO POTENTIALLY CAPABLE OF MISSIONS AGAINST THE UNITED STATES. THE BACKFIRE CAN BE EQUIPPED WITH PROBES FOR IN-FLIGHT REFUELING, WHICH WOULD FURTHER INCREASE ITS RANGE AND FLEXIBILITY.

APPEARANCE: WINGS: LARGE FIXED GLOVE FOR VARIABLE-GEOMETRY SWEPT WINGS ENGINES: TURBOFANS FITTED SIDE-BY-SIDE IN THE AFT FUSELAGE FUSELAGE: CIRCULAR FORWARD OF THE WINGS, CENTER FUSELAGE FLANKED BY RECTANGULAR ENGINE INTAKES. TAIL: ALL SWEPT TAIL SURFACES, WITH LARGE DORSAL FIN.

VARIANTS

TU-22M2 BACKFIRE-B: THE INITIAL PRODUCTION MODEL. A REFUELING PROBE CAN BE FITTED, HOWEVER MOST HAVE BEEN REMOVED. DEVELOPED FOR THE LONG-RANGE STRATEGIC BOMBING ROLE.

TU-22M2YE BACKFIRE-B: THIS VARIANT HAS THE NEW NK-55 ENGINES AND ADVANCED FLIGHT CONTROL SYSTEM. FLIGHT CHARACTERISTICS WERE NOT IMPROVED.

TU-22M3 BACKFIRE-C: UPGRADES RESULTED IN NEW RADAR, ENGINE INTAKES, AND ENGINES. THE AIRCRAFT HAS AN IMPROVED WEAPONS CAPABILITY, INCREASING THE BOMB AND CRUISE MISSILE PAYLOADS.

TU-22MR: 1985 RECON VARIANT WITH SHOMPOL SLAR AND ELINT EQUIPMENT.

TU-22MP: IW VARIANT, CURRENTLY UNFIELDED

FURTHER UPGRADES ARE EXPECTED FOR DELIVERY OF ADDITIONAL PRECISION MUNITIONS.

RUSSIAN LONG-RANGE BOMBER AIRCRAFT TU-95/BEAR

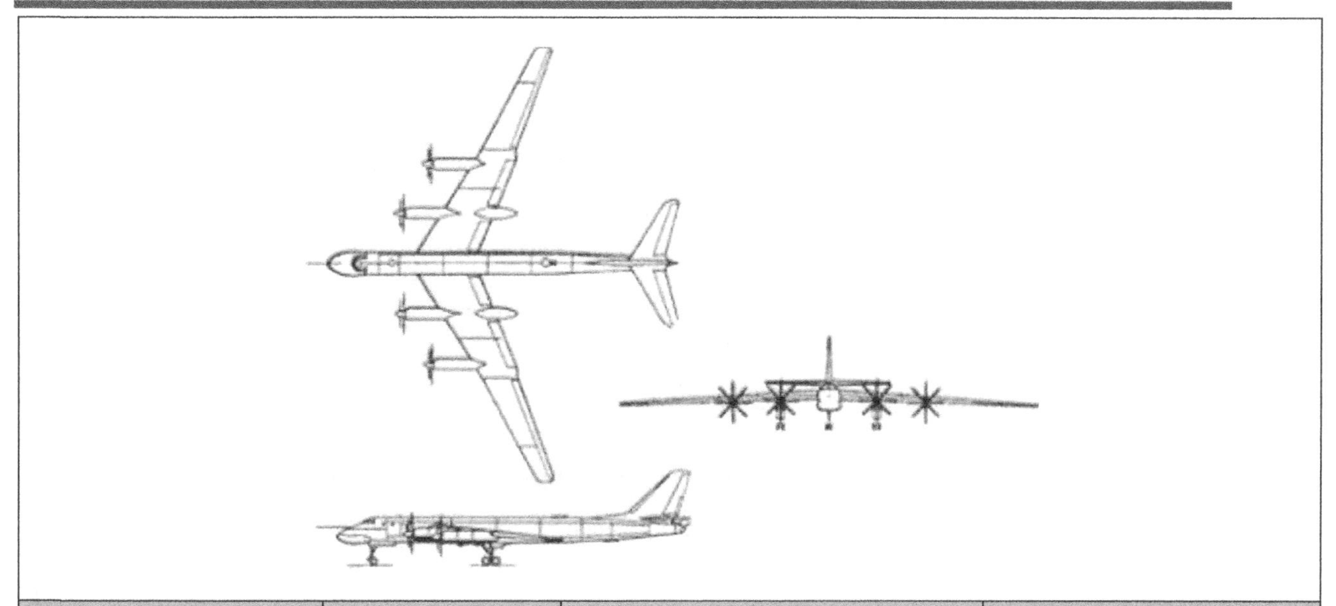

SYSTEM	SPECIFICATIONS	SYSTEM	SPECIFICATIONS
Weapon & Ammunition Types		Max load:	6,500
twin-barrel 23-mm GSh-23 in tail turret (ea.):	1 - 2	One Refueling:	14,100
Other Loading Options:		Takeoff Run (m):	2,450
Missiles:	1 - 3	Dimensions (m):	
AS-4 Kitchen ALCM:	2	Length:	49.1
AS-15 Kent ALCM:	10	Wingspan:	50.0
SYSTEM		Height:	13.3
Alternative Designations:		Internal Payload (kg):	
Date of Introduction:	1959	Normal:	9,000
Proliferation:	India	Maximum:	20,000
Description:		MKU-6 Rotary Launcher:	
Crew (pilot, copilot, navigator/weapons officer, defensive system officer, flight engineer, tail gunner)r):	7	AS-15 Kent missiles:	6
Engines 15,000 eshp Kuznetsove NK-12MP turboprops (max) 9,870 eshp (cruise):	4	Survivability/Countermeasures::	
Weight (kg):		Ejection seats:	No

Maximum Takeoff:	185,000	Crewmember Conveyor in flight deck floor:	Yes
Maximum In-flight:	187,000	Astrodome in roof:	Yes
Maximum Landing:	135,000	ECM pods:	Yes
Empty:	94,400	Infrared warning system:	Yes
Speed (km/h):		Gun fire control radar:	Yes
Maximum (at altitude):	830	Ground Bouncer ECM jamming system:	Yes
Maximum (sea level):	550	Radar warning receiver:	Yes
Cruise:	735	Chaff and flares:	Yes
Takeoff/Landing Speed:	300/275	ARMAMENT:	
Max "G" Force (g):	+2	twin-barrel 23-mm GSh-23 in tail turret::	1 - 2
Ceiling (m):	10,500	AVIONICS/SENSOR/OPTICS:	Yes
Fuel (liters) Internal:	95,000	Short range navigation system:	Yes
Range (km):		Navigation/ bombing radar:	Yes
No Refueling (normal load):	10,500	Weather radar:	Yes
		Terrain-following radar:	Yes
		IFF:	Yes
		Thermal anti-icing:	Yes
		Night/Weather Capabilities::	Yes

NOTES

THE BEAR IS A LONG-RANGE STRATEGIC BOMBER, WITH VARIANTS IN NAVAL SERVICE IN RECONNAISSANCE, ANTI-SUBMARINE WARFARE, AND COMMUNICATIONS RELAY ROLES. IT IS THE ONLY TURBOPROP-PROPELLED STRATEGIC BOMBER IN OPERATIONAL SERVICE IN THE WORLD AND IS HIGHLY REGARDED BY ITS CREWS.

APPEARANCE: WINGS: SWEPT, HIGH-MOUNTED MID FUSELAGE. ENGINES: FOUR 8-BLADE TURBOPROP ENGINES IN SEPARATE WING NACELLES. FUSELAGE: SLENDER, CIRCULAR-SECTION, SEMI-MONOCOQUE FUSELAGE. TAIL: SWEPT FIN, WITH DORSAL FILLET AND INSET RUDDER. SWEPT TAIL PLANES MOUNTED AT BASE OF FIN

VARIANTS

TU-95/TU-95M BEAR-A STRATEGIC BOMBER: BASIC PRODUCTION VERSION. TU-95M HAD MORE POWERFUL AND FUEL-EFFICIENT ENGINES.

TU-95V BEAR-A NUCLEAR BOMBER: ONE AIRCRAFT MADE TO CARRY LARGE HYDROGEN BOMBS. BOMB WEIGHED 27,500 KG AND HAD 58 MEGATONS YIELD.

TU-95K/TU-95KD BEAR-B MISSILE CARRIER: RADOME AND ADDITIONAL 23-MM GUN IN NOSE, UNDER FUSELAGE FITTINGS FOR LARGE CRUISE MISSILE, AND ELINT EQUIPMENT. TU-95KD RECEIVED AN AIR REFUELING SYSTEM.

TU-95KM BEAR-C MISSILE CARRIER/ RECONNAISSANCE: SIMILAR TO BEAR-B, BUT WITH TWO ELINT SYSTEMS AND CROWN DRUM RADAR AND BOX TAIL TAIL-WARNING RADAR.

TU-95RT BEAR-D ELINT RECONNAISSANCE: NAVAL RECONNAISSANCE AND TARGETING VARIANT. TU-95M BEAR-E PHOTO-RECONNAISSANCE: AIR FORCES PHOTO RECONNAISSANCE VERSION. TU-95MS/TU-95MS6/TU-95MS16 BEAR-H BOMBER: CURRENT MAIN SERVICE VERSION; WITH TOADSTOOL TERRAIN FOLLOWING AND CLAM PIPE GROUND MAPPING, TARGET ACQUISITION RADAR. TU-95MS6 WAS FIRST TO CARRY MISSILES IN AN INTERNAL ROTARY LAUNCHER. TU-95MS16 ADDS UNDER FUSELAGE AND UNDER-WING PYLONS TO CARRY MORE MISSILES.

RUSSIAN AIRBORNE WARNING AND CONTROL SYSTEM AIRCRAFT
A-50E/MAINSTAY

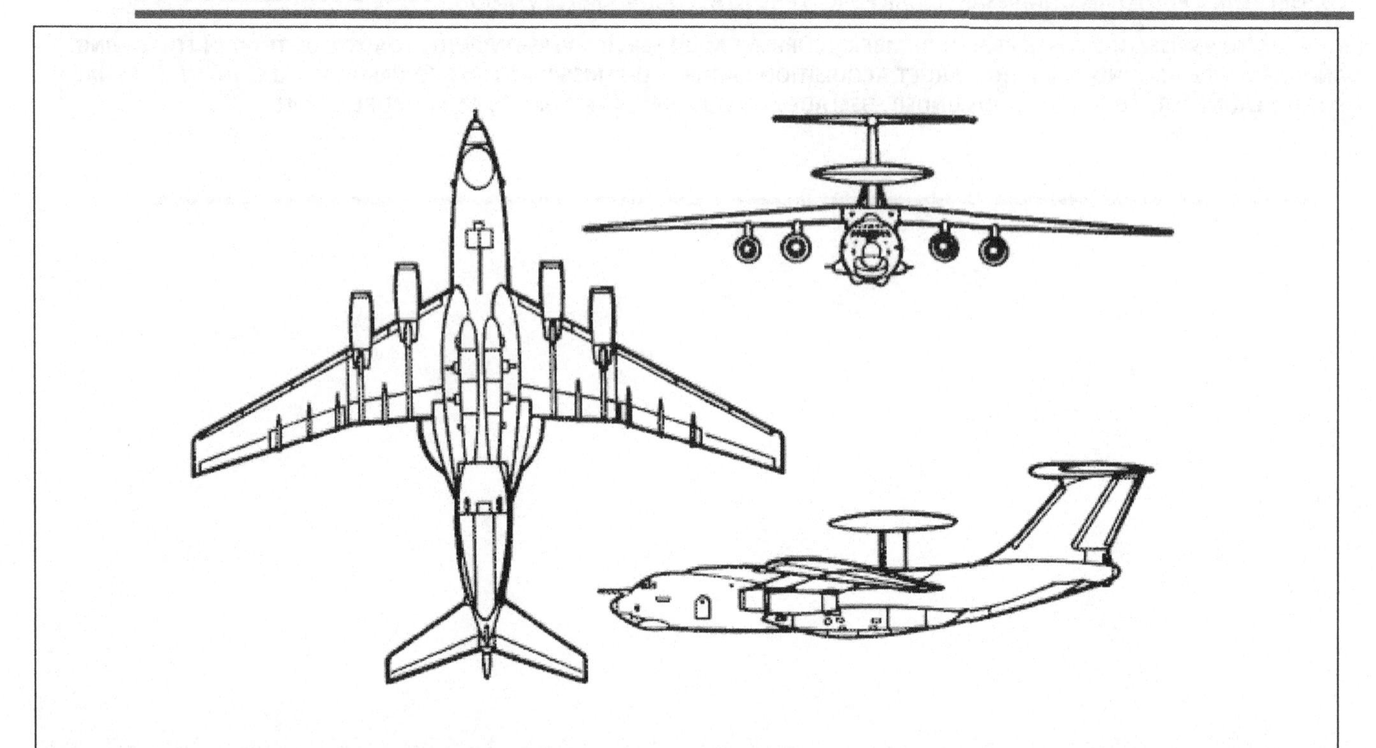

SYSTEM	SPECIFICATIONS	SYSTEM	SPECIFICATIONS
SYSTEM		Cargo Hold(m):	
Alternative Designations:	See Variants	Length to ramp:	20.0
Date of Introduction:	1987 original	Length including ramp:	24.5
Proliferation:	6	Width:	3.4
Description:		Height:	3.5
Crew:	5	Standard Payload (kg):	
Pilots:	2	Internal (M / MD):	40,000 / 48,000
Flight Crew:	3	Endurance with internal fuel and maximum payload hh:mm:	4:00
Mission Operators:	10	Survivability/Countermeasures:	
SYSTEM		IFF:	Yes
Engines 26,455 lbs. thrust Soloview D-30KP turbofans:	4	Wingtip countermeasures pod:	Yes
Weight (kg):		Flare pack each side of rear fuselage:	Yes
Max takeoff:	170,000 (Il-76M), 190,000 (Il-76MD)	IR warning receiver:	Yes

Empty:	61,000	ARMAMENT:	Yes
Speed (km/h):		TV remote gun and bomb sights:	Yes
Max:	850	PRS-3/Argon-2 ranging radar:	Yes
Cruise:	750 - 800	PNA-D attack radar:	Yes
Description:		Night/Weather Capabilities:	Good
Crew (pilot, copilot, navigator, defensive systems operator):	4	23-mm cannons fitted in a manned position at the base of the rudder (ea.):	2
Engines 50,000 lbs. thrust NK-25 turbofans:	2	AVIONICS/SENSOR/OPTICS	
Weight (kg):		Description (A-50E):	Yes
Max Takeoff:	126,000	Color CRT displays for radar observers:	Yes
Empty:	49,500	Satellite data link to ground stations:	Yes
Speed (km/h):		Weather radar in nose:	Yes
Maximum (at altitude):	2,327 Mach 2.05	Ground-mapping and navigation radar under nose:	Yes
Ceiling (m):	15,500	Signal detection radar:	50-500 MHZ
Fuel (liters):		Electronic Intel radar:	.5-18 GHZ
Internal:	81,830	Ground Target Detect Range:	Single Target:
Range (km):		Target Node:	250 km (tanks, etc.)
Max Payload:	5,000	Ship-size Target:	400 km
Max Fuel:	6,700	Air Target Tracking Range:	
Takeoff Run/Landing Roll (m):	850/450	Bombers:	650 km
Dimensions (m):		MiG- 21Target size:	230 km
Length:	46.6	Low-Flyers:	To radio horizon
Wingspan:	50.6		
Height:	14.8		

NOTES

MAINSTAY IS INTENDED TO DETECT AND IDENTIFY AIRBORNE OBJECTS, DETERMINE THEIR COORDINATES AND FLIGHT PATH DATA AND TRANSFER THE INFORMATION TO AIR DEFENSE CPS AND ACTS AS A CONTROL CENTER TO GUIDE FIGHTER-INTERCEPTORS. IT ALSO DETECTS GROUND AND SEA TARGETS AND GUIDES TACTICAL AIRCRAFT TO COMBAT AREAS TO ATTACK GROUND TARGETS AT LOW ALTITUDES. THE 10 MISSION OPERATORS CAN TRACK 50 TARGETS AND GUIDE INTERCEPTION OF 10 SIMULTANEOUSLY.

CAPABILITIES (A-50E): DETECTING AND TRACKING AIRCRAFT AND CRUISE MISSILES FLYING AT LOW ALTITUDE OVER LAND AND WATER, AND OF HELPING DIRECT FIGHTER OPERATIONS OVER COMBAT AREAS AS WELL AS ENHANCING AIR SURVEILLANCE AND DEFENSE.

APPEARANCE:

WINGS: HIGH-MOUNTED, SWEPT-BACK, AND TAPERED WITH BLUNT TIPS. TRAILING EDGE HAS A SLIGHT CRESCENT SHAPE. CHASSIS: IL-76/CANDID TRANSPORT. ENGINES: FOUR MOUNTED PYLONS UNDER AND EXTENDING FORWARD OF WINGS' LEADING EDGE. FUSELAGE: LONG, ROUND AND TAPERING TO THE REAR, ROUNDED NOSE WITH CHIN RADOME. LARGE ROTATING RADOME ABOVE THE FUSELAGE. TAIL: T-TAIL WITH CURVED LEADING EDGE AND INSET RUDDER. SWEPT TAIL PLANES MEET AT TOP OF THE TAIL.

VARIANTS

EARLIER VERSIONS INCLUDED A-50 WITH SHMEL RADAR. THE A-50U/MAINSTAY B HAS A SHMEL-M RADAR VARIANT.

A-50M: VARIANT HAS SHMEL-2 RADAR, RESISTS MOST CM, SIMILAR TO US AN/APY-1/-2.

A-50E: HAS BEEN EXPORTED. AN ISRAELI MODIFIED VARIANT WITH NEW ENGINES AND PHALCON RADAR WILL BE EXPORTED TO INDIA.

SWEDEN AIRBORNE ECM/EW POD, SAAB BOQ X-300 (ON JAS39/GRIPEN)

SYSTEM	SPECIFICATIONS	PERFORMANCE	SPECIFICATIONS
Alternative Designations:	None	Intern Range (km):	
Date of Introduction:	1997	Combat Radius:	800
Proliferation:	Sweden (Hungary and South Africa – planned)	Ferry:	3,000
Description:		Takeoff Run/Landing Roll (m):	800/800al:
Crew:	1 (pilot) (JAS 39A/C), 2 pilots (JAS 39B/D)	External:	3,300.
Appearance:		Dimensions (m):	
Wings:	Multi-sparred delta.	Length:	14.1 (A/C), 14.8 (B/D)
Engines:	Turbofan with intake boxes on both	Wingspan (m):	8.4
Sides of fuselage.		Height:	4.5
Tail:	Leading edge swept fin with upright inset rudder.	BOQ-X300 ECM/EW POD.	
Engines:	1 x 12,140 lbs. thrust Volvo Aero RM12, 18,200 lbs. thrust with afterburner	Alternative Designations:	None
Weight (kg):		DATE OF INTRODUCTION:	2012
Takeoff:	12,500 (A/C), 14,000 (B/D)	PROLIFERATION:	Sweden
Empty:	6.500 (A/C), 7,100 (B/D)	COUNTRY OF ORIGIN:	Sweden
Speed (km/h):		FREQ. BANDS:	S/C/X/Ku/K
Maximum (at altitude):	2,150, Mach 1.8+	FREQ. RANGE (MHz):	2-40,000
Max "G" Force (g):	+9/-3 g	RANGE:	INA
Ceiling (m):	16,000	POWER OUTPUT:	INA
Fuel (liters):		TYPE:	Airborne Electronic Countermeasures (ECM), radar jamming system.

NOTES
THE BOQ-X300 HIGH-PERFORMANCE JAMMING POD IS THE LATEST POD BEING DEVELOPED BY SAAB FOR THE GRIPEN FIGHTER. THE POD IS A MODULAR SYSTEM THAT INTEGRATES A SOPHISTICATED JAMMER, SUPPORTED BY A RWR AND ESM SYSTEM. AS AN OPTION, THE POD CAN BE CONFIGURED WITH A DUAL FIBER OPTIC TOWED DECOY TO PROVIDE EFFECTIVE COUNTERMEASURES AGAINST MONOPULSE THREAT. THE BOQ-X300 PROVIDES SELF-PROTECTION FOR HIGH VALUE ASSETS SUCH AS FIGHTER, ATTACK AND RECONNAISSANCE AIRCRAFT. THE POD IS DESIGNED TO SUPPRESS LEGACY THREATS, SURFACE BASED AS WELL AS AIRBORNE. A SECONDARY ROLE FOR THE BOQ-X300 IS TO PROVIDE JAMMING FOR TRAINING OF RADAR OPERATORS IN AIRBORNE AS WELL AS GROUND- OR SEA-BASED ENVIRONMENTS.

Worldwide Equipment Guide

Chapter 3: Unmanned Aerial Vehicles

Chapter 3: Unmanned Aerial Vehicles and Related Technologies

An aviation technology which has seen the greatest expansion of research, development, and fielding activity in recent years is the unmanned aerial vehicle (UAV). According to a 2015 research report by the Rand Corporation there are 960 UAVs being produced by 270 companies in 57 countries. They also claim that in the past two years the number of UAVs has gone up 40% with the number of UAV companies entering the market increasing by 20% and the number of countries involved up by 50%. Another market study from the Teal Group in 2014 estimated UAV spending will nearly double over the next decade from current worldwide UAV expenditures of $6.4 billion annually to $11.5 billion, totaling almost $91 billion in the next ten years.

Despite defense budget cutbacks, UAVs are projected to see steady growth as users continue to seek their versatility, robustness, and feasibility. Reasons for expanding the use of these systems are their capabilities to extend our vision and reach over any terrain, against any force, with fewer restrictions, dangers, and support requirements as opposed to manned systems. Since they are unmanned they can go into areas where risk to crews might hinder a mission. Uses for UAVs have also expanded beyond their initial RISTA mission, to include, security patrolling, delivery of information warfare (INFOWAR) systems (e.g., jammers), communications retransmission, attack, counter-air harassment of enemy aircraft, and remote materials delivery. Advances in lightweight materials, imagery systems, and navigation technologies, particularly commercial, have lowered costs and facilitated these changes.

This chapter provides characteristics of selected UAVs in use or readily available to the OPFOR. UAVs discussed are those likely to be encountered by U.S. forces in various environments and levels of conflict, or are representative of the range of systems fielded and available. The selection of UAVs is not intended to be all-inclusive.

UAVs come in various types, sizes, and levels of complexity, each having their own purpose and advantage in an operational environment. For example, fixed-wing, propeller-driven platforms excel in endurance and range. Jet-propelled UAVs trade endurance and maneuverability for speed. Rotary-wing UAVs can carry relatively large payloads, offer the best maneuverability, and trade higher initial cost for long-term reliability and reduced casualty rates.

A UAV is a system comprised of self-propulsion, maneuver capability, and guidance. Current UAV sizes range from large high-altitude long endurance (HALE) aircraft, to tactical mini-UAVs (MUAVs), which can now mount a stabilized gimbaled payload with multiple sensors. A rapidly expanding trend is the proliferation of MUAVs and micro-aerial vehicles (MAVs) for use at the lower tactical levels as well as civilian applications. The MAVs are normally hand launched, hard to detect, easy to use, and relatively inexpensive to purchase.

Among the most critical considerations for selecting UAVs are their operating range, operating altitude, and endurance (e.g. flight time). Tactical and operational systems must be reusable so the operating radius is critical. UAVs must at least range beyond the longest weapon range to provide warning time. Those not directly supporting weapons must have more range and time to observe larger areas. Usually,

fixed-wing systems are better suited for covering wide areas and rotary-wing for supporting tactical weapons and operating in defilade areas.

Diverse transport and launch configurations are available for UAVs. Israeli helicopters have carried Skylite A UAVs in ATGM racks, and launched them to survey areas where there may be some risk. The Skylite A has also be canister-mounted to fit on vehicles for launch at short halts, or launch from mortars. Another likely mini-UAV launch platform in the near-term (1-5 years) are airships (e.g. powered blimps and air defense aerostat balloons). Naval ships are using UAVs; and submarines have demonstrated their use while operating at periscope depth.

Several terms have recently been used to categorize UAVs and other unmanned aerial surveillance systems. However, the terms listed below should be understood to avoid confusion.

- The acronym, UAS, is currently used in some U.S. communities, with different meanings, but usually as unmanned aerial sensors, to emphasize the wide range of UAV designs available for U.S. force requirements, with a focus on RISTA applications.
- For some users, unmanned aerial sensors is an umbrella term which can include UAVs (vehicles both guided and self-propelled), as well as related technologies (e.g. unmanned aerial sensors other than UAVs). Thus, related technologies include remotely launched sensor munitions, with still cameras or video-cameras which sense and emit while in their trajectory. Another related technology are airships, such as balloons, with sensor pods mounted on them. The majority of airships are aerostats tethered to fixed sites or to vehicles, for long-term (days) or short-term (minutes) operations while others can be propelled. The above UAS terms are primarily used as sensors, but can be used in other roles such as air to surface attacks with guided missiles. Thus the term UAS is still misleading.
- Some organizations also use UAS to mean unmanned aerial systems, or unmanned aircraft systems. Selected sources have used one of these meanings as well as the one above the same paragraph, for the same system. Each meaning can exclude some aspect of the other or include one beyond the other. An aerial sensor may not be an aircraft, and an aerial system may have roles beyond that of a sensor. The OPFOR community should be wary of confusion between these two very different meanings for the same acronym.
- Because of the potential confusion with the acronym UAS, the OPFOR will avoid it. The WEG will use descriptions of specific technologies, such as UAVs, airships, etc., and generically precise categories like weapon-delivered aerial sensor munitions.

Questions and comments on data list in this chapter should be addressed to:

Mr. Patrick Madden

DSN: 552-7997 Commercial (913) 684-7997

E-mail address: patrick.m.madden16.ctr@mail.mil

Mini-UAVs and Micro-UAVs for Use in Military Forces

On the modern three-dimensional battlefield, military forces are developing missions for UAVs at all echelons and in many branches, for combat and supporting units. Tactical UAVs can be supplemented with lighter shorter-range UAVs at battalion and below. Air defense, anti-tank, artillery, theater missile, and other units with stationary facilities requiring security patrols can use these UAVs to execute the mission while reducing personnel and vehicle requirements.

System categories and descriptions can be vague and even contradictory. Producers, users, and publications use varied categorizations. For example, UAVs may be termed small UAV, short-range UAVs. International terms gaining the most use are mini-UAV (MUAV) and micro-aerial vehicles (MAVs). MUAVs are typically less than 25 kg and MAVs are typically less than 5 kg in weight. As UAVs have decreased in size, weight categorizations have also decreased.

Currently many MUAVs and the majority of MAVs are easily damaged. They must be low in cost and treated as disposable. A few, however, (e.g. rotary craft like the Russian Zala 421-12) offer stable flight control and designs with good survivability. The Zala 421-12 is used with security forces. Virtually all use electric motors for near silent operation at altitudes of 300 meters or less. Initial costs, repairs and maintenance are factors. They must be integrated into communications schemes and air space restrictions. Some training is required. Nevertheless, as in the commercial sector, the military sector has found a growing need for them. Paramilitary and special-purpose forces use these and other UAVs.

There is also growing interest in the development of MAVs. Key reasons for the interest include a widespread need for inexpensive aerial sensors to observe small areas rapidly. Commercial and scientific applications have resulted in an increase in development programs. Many are hand-size; but most conventional designs with front-mounted propeller have problems in control, wind stability, payload, range, and crash worthiness. Slightly larger sized hand-launched craft like the Zala 421-21, or MAVs close to the 5 kg limit offer better capability. Battery powered rotary-engine designs, especially multi-motors, have the most potential. The 6-rotor MAV 421-21 is stable with a 15 km range; GLONASS navigational feed, and notebook display.

Some Tier 1 forces have MUAVs in tactical battalions and companies. By the near term, forces will have MUAVs or MAVs in platoons. Squads and teams will carry MAVs or other aerial sensors (e.g., weapon-delivered sensors). By mid-term vehicles and dismounted squads and teams will have their own MAVs and small attack munitions will be fitted or optional. In addition to regular forces there is also a growing use by irregular forces to use these type of UAVs. Recent use by ISIS has shown that they use MAVs not only for surveillance but also for targeting and propaganda video footage. It is predicted that other irregular forces will use these inexpensive MAVs for similar purposes.

RUSSIAN MICRO UNMANNED AERIAL VEHICLE ZALA 421-08

SYSTEM	SPECIFICATIONS	SYSTEM	SPECIFICATIONS
Date of Introduction: 2007	2007	Wind speed at launch:	15 m/s
Proliferation:	At least one country	Recovery Method:	Parachute or Auto Return
Ground Crew:	2, backpack	System Composition:	
Engine: Electric	Electric	Number of UAVs	(2 X) UAVs
Propulsion:	2-blade propeller	Number of transport cases	(2x) transport cases
Weight empty (kg):	1.7	Video roll-stabilized cameras	(2x) Video cameras
Max Takeoff Weight (kg):	1.9	Infrared Camera (optional)	(1X) IR Camera
Max Payload (kg):	2.55	Catapult (optional)	1
Max Speed (level)(km/hr):	150	**PAYLOAD TYPES**	**SPECIFICATIONS**
Cruise Speed (km/hr):	65-130	Color Camera	10 (MPX) stabilized video
Maximum Ceiling (m):	3,600 Above Sea	Infrared Camera	Resolution Not Less Than 160x120
Minimum Ceiling (m):	15	Gas Detection Module	Chemical/Hazardous Emission
Operating:	3,600	**SYSTEM COMPONENTS**	**SPECIFICATIONS**
Endurance (min):	60	Ground Control Station (GCS):	
RPV Mode Range (km):	10	Transport Case	Ruggedized and Man Portable
Pre-programmed Mode Range (km):	40	Control Capability	UAV and payload controlled independently by GCS
Wing Span (cm):	81		
Length (fuselage) (cm):	42.5	GCS Power Supply:	120/220 V, 6 hr battery
Height (cm):	25	Setup Time (min):	5-10 min.
Launch Method:	Hand Launched, Self-powered		

RUSSIAN MICRO UNMANNED AERIAL VEHICLE ZALA 421-12

SYSTEM	SPECIFICATIONS	SYSTEM	SPECIFICATIONS
Alternative designations:	421-4M	Launch Method:	Elastic or pneumatic catapult
Date of introduction:	2001	Wind speed at launch:	10 m/s
Proliferation:	At least one country	Recovery Method:	Parachute (non-steerable)
Ground Crew:	Two, backpack	Dimensions (cm):	
Engine:	Electric, battery powered	Wing Span:	81
Propulsion:	2-blade propeller	Length (fuselage):	42.5
Max Takeoff Weight (kg):	4.8	Height:	25
Max Payload (kg):	1	**PAYLOAD TYPES**	**SPECIFICATIONS**
Max Speed (level)(km/hr):	130	Photo camera	10 Mpx
Cruise Speed (km/hr):	65-120	Color Video Camera	550 TVL
Maximum Ceiling (m):	3,600 Above Sea Level	Infrared Camera	160X120 (Optional)
Minimum Ceiling (m):	15	Hazard Gas Analysis Module	Hazardous Gas (Optional)
Operating Ceiling (m):	100-700	**SYSTEM COMPONENTS**	**SPECIFICATIONS**
Endurance (min):	130	Number of UAVs	(2 X) UAVs
RPV Mode Range (km):	25	Number of transport cases	(2x) transport cases
Pre-programmed Mode Range (km):	40	Ground Control Station (GCS)	(1x) man-pack, ruggedized
		GCS Power Supply:	120/220 V, 6 hour battery
		Setup Time (min):	5-10 min.
		Control Capability	UAV and payload controlled independently by GCS
		Video roll-stabilized cameras	(2x) Video cameras
		Infrared Camera (optional)	(1X) IR Camera
		Setup Time (min):	5-10 min.
		Elastic or Pneumatic Launcher:	(1x) (optional)

NOTES

PAYLOADS FIT IN THE STANDARD MOUNTING BLOCK THAT IS INTERCHANGEABLE WITH OTHER PAYLOADS AND IS A DUAL AXIS, GYRO-STABILIZED PAYLOAD. GLONASS/GPS SATELLITE NAVIGATION WITH CAPABILITY OF AUTONOMOUS OPERATION.

RUSSIAN MICRO UNMANNED AERIAL VEHICLE ZALA 421-21

SYSTEM	SPECIFICATIONS	PAYLOAD TYPES	SPECIFICATIONS
Alternative designations:	None	Photo camera	Single frame photo (color)
Date of introduction:	2010	Color Video Camera	INA
Proliferation:	At least one country	Infrared Camera (Optional)	INA
Ground Crew:	Two, backpack		All imagery is real-time transfer to ground control
Engine:	Electric, battery powered		
Propulsion:	6-two blade propeller, Vertical Take-Off and Landing (VTOL)	**SYSTEM COMPONENTS**	**SPECIFICATIONS**
		Number of UAVs	(2 X) UAVs
Max Payload Takeoff weight (kg):	.5	Number of transport cases	(2x) transport cases
Speed (km/h):		Ground Control Station (GCS)	(1x) hand-held, ruggedized
Maximum (level):	40	GCS Power Supply:	Independent 6 hour battery
Ceiling, Operational (m):	10-1,000	Setup Time (min):	5-10 min.
Endurance (min):	130	Control Capability	UAV and payload controlled independently by GCS
Range (km):	15	Roll-stabilized cameras	(2x) Video and photo camera
Launch Method:	Hand Launched	Infrared Camera (optional)	(1X) IR Camera
Wind speed at launch:	INA	Setup Time (min):	5-10 min.
Dimensions (cm):	Two fixed, skid landing gear		
Wing Span:	INA, See picture for est. scale		
Length (fuselage):	INA		
Height:	INA		
Launcher:	(1x) (optional)		

NOTES

PAYLOADS FIT IN THE STANDARD MOUNTING BLOCK THAT IS INTERCHANGEABLE WITH OTHER PAYLOADS. MULTI-SINGLE PHOTO/VIDEO/IR CAMERA SENSOR PAYLOAD IN DEVELOPMENT. 12/220 VOLT EXTERNAL CONNECTION FOR OPTIONAL POWER SOURCE FOR THE GCS.

ISRAELI MINI UNMANNED AERIAL VEHICLE SPYLITE/SKYLITE

SYSTEM	SPECIFICATIONS
Alternative designations:	Skylite
Date of introduction:	2003
Proliferation:	At least 4 countries
Engine:	Electric, battery powered
Propulsion:	two blade pusher propeller
Max Takeoff weight (kg):	6.3-8.0
Max Payload (kg):	1.3
Total System (kg):	39
Cruise Speed (km/h):	70-100
Max Altitude (ft):	36,000
Ceiling, Operational (m):	3000
Endurance (hr):	4
Range (km):	50
Wing Span (cm):	240
Length (cm):	110
Body Width (cm):	
Launch Method:	4-kg catapult launch
Recovery Method:	Combined parachute and inflatable bag

PAYLOAD TYPES	SPECIFICATIONS
Combined, triple axis, gimballed, gyro-stabilized sensor. Single sensor provides day/night optical zoom lens with auto-tracker, infrared dual field of lens with auto zoom, and laser pointer.	

PAYLOAD TYPES	SPECIFICATIONS
Survivability/countermeasures: Lightweight, composite structure. Small profile with low radar signature and very quiet engine. Excellent flight dynamics for use in all climates and severe weather, with winds of up to 35 knots, and gusts of up to 55.	

SYSTEM COMPONENTS	SPECIFICATIONS
Number of UAVs:	(2-3) UAVs
Number of transport cases	(2x) transport cases
Ground Control Station (GCS)	(1x) hand-held, ruggedized

VARIANTS	SPECIFICATIONS
SkyLite A: Briefly called SkyLark, the SkyLite A was a vehicle or shoulder canister launched UAV for use in tactical units.	
SkyLite B: Variant upgrade has improved cameras, larger wings, longer endurance, and 1.5-kg added weight. Launched by catapult.	

NOTES

THE CURRENT SPYLITE CONTINUES TO BE SUCCESSFULLY MARKETED GLOBALLY. THE MINI-SPYLITE SET A 2015 RECORD IN ITS CLASS FOR COMMUNICATING 120 KMS BETWEEN THE UAV AND ITS GROUND DATA TERMINAL.

ISRAELI MICRO UNMANNED AERIAL VEHICLE SKYLARK I, IV, LE

SYSTEM	SPECIFICATIONS	PAYLOAD TYPES	SPECIFICATIONS
Alternative designations:	None	Optical Camera:	Color CCD 10x zoom lens
Date of introduction:	2003	IR Camera:	Night FLIR
Proliferation:	10 Countries with deployment in Iraq and Afghanistan	User Image Capabilities:	All images can be overlaid on a downlinked integrated map
Ground Crew:	2, backpack (30-40 kg each)	Uplink:	Analog, encrypted, UHF
Engine:	Electric, battery powered	Downlink:	D/E-band telemetry/video
Propulsion:	Two blade propeller	**VARIANTS**	**SPECIFICATIONS**
Max Launch Weight (kg):	5.5	Skylark I LE (Long Endurance):	(column will only list changes)
Max Speed (level) (km/hr):	111	Weight (kg):	2.8
Cruising Speed (km/hr):	65	Engine:	Electric, battery powered
Endurance (min):	1 hr 30 min	Tail Wings:	Modified to adjust new engine
Max Ceiling, normal (m):	455	Wing Span (m):	2.9
Ceiling, Service (km):	5	Max Launch Weight (kg):	6.3
Radius of Operation (km):	10	Operational Ceiling (km):	4.9
Wing Span (m): 2.4	2.4	Radius of Operation (km):	15
Overall Length (m):	2.20	Endurance:	3 hrs
Flight Control:	GPS positioning, autonomous preprogrammed flight	**PAYLOAD TYPES**	**SPECIFICATIONS**
		Controp T-STAMP Sensor:	Miniature triple sensor:
Ground Control:	Handheld, Mini Ground Control Unit, color console	Optical:	Color CCD 10x zoom lens
		IR:	8-12 microns with x4 continuous optical zoom lens
Flight Control Method:	Sprectralink data link		
Launch Method:	Hand or bungee launched	Laser Pointer	high resolution panoramic scan mode
Recovery/Landing Method:	One button auto return, steep stall, inflatable cushion		
		VARIANTS	**SPECIFICATIONS**
		Skylark I LE Block II	(column will only list changes)
		Type of Engine:	hydrogen fuel cell propulsion
		Flight Endurance:	7 hours or greater

NOTES

ORIGINAL SKYLARK IV IS A SLIGHTLY IMPROVED VERSION (RUGGEDIZED AND GYRO-STABILIZED) OF SKYLARK 1. HOWEVER, IMPROVEMENTS IN COMBINED SENSORS AND EFFORTS TO CONVERT FROM ANALOG TO DIGITAL AVIONICS WILL CONTINUE TO IMPROVE THE SKYLARK IV. SKYLARK LE BLOCK 2 IN DEVELOPMENT AND TESTING. AWARDED FIVE YEAR US CONTRACT TO COMPETE FOR FUTURE SUAS SELECTION.

Weapon-Delivered Aerial Sensor Munitions

Several aerial imaging munitions have been developed for launch from weapon systems. They offer capability for real-time or near real-time overhead view of an enemy within or close to weapon range, even when the enemy may be concealed behind cover.

Weapon-delivered aerial sensor munitions were developed as back as the year 2000. However, they are not yet widely fielded, due to cost, difficulty of miniaturization, lack of portability, need for precise target location data, and lack of clear imagery. Advancements in image resolution, radio transmission and miniature servo-motor systems, now permit design of sensor and guided attack munitions for delivery by grenade launchers, mortars and rocket launchers. Linking the downloaded image or video to a digital transmission system can also permit it to be shared with other users. Because the sensor uses munition propulsion, it can reach the target area well before launch and employment of a UAV or MAV.

Several munitions are offered for under-barrel grenade launchers (UBGLs), and shoulder launchers users those grenades. The munitions offer overhead imagery for infantry squads and teams at lower cost than UAVs. Users can employ laptop or PDAs as terminals. Examples include the Israeli FireFly 40-mm UBGL round with a camera eye and parachute, to give a top-down view of features beyond line-of-sight 600 m away. The image footprint is approximately 1,200 m. Another, the Israeli Reconnaissance Rifle Grenade (RRG) is launched from a rifle barrel, provides 6-7 seconds of image, and also has 600 m range. The Singaporean S407/Soldier Parachute Aerial Reconnaissance Camera System (SPARCS) fits a 40-mm UBGL, with 300-600 m range (est.) and offers a real-time image to PDA or other display.

A Pakistani firm has developed the Firefly (not the same FireFly as above) hand-launched camera reconnaissance rocket. The pistol-styled launcher will direct a plastic rocket to a range of 800-1000 m in 8 sec, with a digital data link to a PDA. It is called a "mini-rocket UAV."

A few countries are developing mortar reconnaissance projectiles for 81 mm and 120 mm mortars. These are likely by the end of the near term (5 years). One developer predicts reconnaissance projectiles for 60 mm mortars. Prototypes and programs for 155-mm cannon fired reconnaissance projectiles are also underway and likely due by the mid-term (5-10 years).

One system developed in the 1990s is the Russian R-90 UAV rocket for launch by the 9A152 300-mm multiple rocket launcher. It is actually part weapon delivered sensor, part RISTA UAV, and part attack UAV. It reaches 70-90 km in less than a minute. On arrival the 42 kg UAV ejects, then loiters for 30 minutes to execute target confirmation, adjust MRL fires, and perform battle damage assessment afterward. As the UAV reaches the end of its flight time, it can acquire a remaining target for an impact kill. The attack option presages an increasing trend for UAVs and sensor projectiles - offering direct attack and munition launch options.

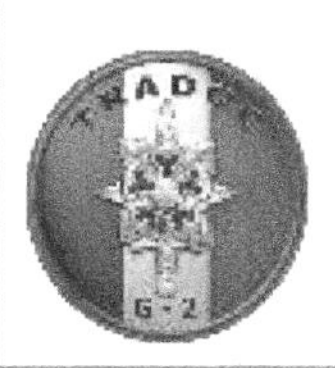

ISRAELI UNMANNED AERIAL VEHICLE SKYLARK II, III

SYSTEM	SPECIFICATIONS	PAYLOAD TYPES	SPECIFICATIONS
Alternative designations:	None	Optical Camera:	HD Color CCD 10x zoom lens
Date of introduction:	2006	IR Camera:	Night FLIR thermal imaging
Proliferation:	At least 3 Countries	Laser:	Marker
Ground Crew:	2		Range Finder (optional)
Engine (hp):	5.4 hp, Electric, battery	User Image Capabilities:	All images can be overlaid on a downlinked integrated map
Propulsion:	Two blade propeller	**VARIANTS**	**SPECIFICATIONS**
Max Launch Weight (kg):	65	Skylark III:	(column will only list changes)
Max Speed (level) (km/hr):	129	Alternative designations:	Skylark II LE
Cruising Speed (km/hr):	65	Weight (kg):	2.8
Endurance (hr):	5	Max Payload (kg)	10
Operating Altitude, (m):	150-1,525	Wings:	Curved wingtips
Ceiling, Service (m):	4,875	Wing Span (m):	4.8 mounted under fuselage
Radius of Operation (km):	60	Max Launch Weight (kg):	45
Wing Span (m): 2.4	6.5	Ceiling, Service (m):	4,600
Overall Length (m):	3.2	Max, Ceiling (km)	6.4
Ground Control:	Skylark dual station ground control, color console	Engine:	Electric, battery, rechargeable, rear mounted pusher engine
Launch Method:	Humvee class vehicle with mounted rail launcher, or can use optional rail launcher trailer	Endurance (hr):	6
		Launch Method:	Pneumatic launcher on Humvee class vehicle or launcher trailer
Recovery/Landing Method:	Parachute and airbag cushion	Recovery/Landing Method:	
		Radius of Operation (km):	100

NOTES

SKYLARK II AND III ARE DESIGNED TO SUPPORT BRIGADE AND DIVISIONS. THEY ARE BASED ON THE SKYLARK IV BUT MUCH LARGER. THE GUIDANCE SYSTEM IS A STARLINK AIR TERMINAL WITH SECURE DIGITAL DATA LINKS, REDUNDANT AVIONICS, AND AUTONOMOUS FLIGHT MODES. THEY ARE INTEROPERABLE WITH OTHER SKYLARKS WITH REAL-TIME VIDEO TRANSFER RATE OF 1.5 MB PER SECOND, ENCRYPTED UHF UPLINK, D/E BAND DOWNLINK. TWO SKYLARK III VEHICLES CAN ALSO BE ASSIGNED THE SAME MISSION WHILE SIMULTANEOUSLY USING A SHARED GROUND CONTROL STATION. FUSELAGE AND WINGS OF THE III MODEL HAS BEEN MODIFIED SIGNIFICANTLY TO INCREASE RANGE AND STABILITY. BOTH MODELS USE A TRIPLE SENSOR PAYLOAD THAT PROVIDES HIGH DEFINITION OPTICAL, THERMAL IMAGER, AND LASER CAPABILITIES.

FRENCH UNMANNED AERIAL VEHICLE FOX AT2

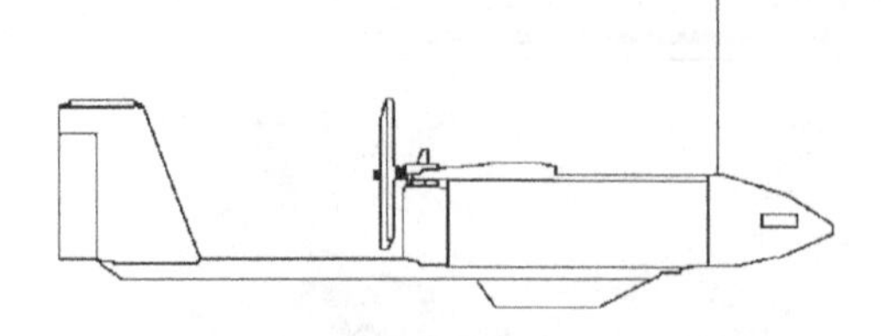

SYSTEM	SPECIFICATIONS	PAYLOAD TYPES	SPECIFICATIONS
Alternative designations:	None	Panoramic CCD color camera, low light television (with zoom), IR linescan CAMELIA camera, SAR camera, FLIR, multi-sensor gimball platform (IR and visible), etc.	
Date of introduction:	1988		
Proliferation:	France, the U.N., civilians		
Engine:	Limbach 22 hp L 275 E, 2 cylinder, 2 stroke, air-cooled		
Propulsion:	2 blade push propeller		
Take-off Weight (kg):	135		
Fuel and Payload (kg):	60		
Maximum (level)(km/h):	216	**VARIANTS**	**SPECIFICATIONS**
Cruising Speed (km/h):	145	Different versions are offered with varying ranges. The version selected for Tier 2 portrayl is the Fox AT2 (200), with 200km range.	
Endurance (hrs):	5		
Max Ceiling, normal (m):	3000		
Min Ceiling (m):	30		
RPV Mode Range (km):	50, 100, 150 (200 as an option)		
Programed Mode Range (km):	350		
Wing Span (m):	4		
Overall Length (m):	2.75		
Overall Height (m):	0.25		
Flight Control:			
Launch Method:	Hydraulic or sandow ramp		
Recovery Method:	Parachute		
Landing Method:	Airbag		

NOTES

THE FOX AT2 UAV IS ONE OF A FAMILY OF LOW-COST UAVS DESIGNED BY THE FRENCH FIRM CAC SYSTEMES. EACH UAV SYSTEM IS COMPOSED OF A TRANSPORT AND LAUNCHING SYSTEM, A GROUND CONTROL STATION (GCS) MOUNTED ON A 4X4 TRUCK FRAME, AND FOUR UAVS. THE FOX AT2 (LIKE THE FOX AT1) IS LAUNCHED FROM A MOBILE LAUNCHING CATAPULT (TRANSPORTATION AND LAUNCHING SYSTEM) THAT IS MOUNTED ON A TRAILER WITH TRANSPORTATION COMPARTMENTS FOR 4 UAVS. NORMALLY TWO OF THE FOUR UAVS ARE EQUIPPED WITH CCD CAMERAS FOR DAYTIME MISSIONS AND THE REMAINING TWO ARE FLIR EQUIPPED FOR NIGHTTIME MISSIONS. UPON MISSION COMPLETION THE UAV CAN BE RE-SERVICED AND AVAILABLE FOR ANOTHER MISSION IN LESS THAN 30 MINUTES. THE FOX AT2 IS CAPABLE OF CARRYING 30 KILOGRAMS OF VARIOUS PAYLOADS. ADDITIONALLY, TWO UNDER-WING PODS ALLOW FOR TWO LOADS TO BE CARRIED AND DROPPED. NORMALLY THE GCS CONSIST OF A CREW OF THREE PERSONNEL: PILOT, OBSERVER, AND A TECHNICIAN. HOWEVER, TWO PEOPLE CAN DEPLOY THE UAV SYSTEM AND HAVE IT AVAILABLE FOR OPERATION IN LESS THAN 20 MINUTES. THE GUIDANCE AND CONTROL CONSISTS OF AN UHF DATA LINK WITH FOUR PROPORTIONAL AND EIGHT NUMERIC CHANNELS, OF WHICH FOUR CONTROL THE AUTOPILOT. TELEMETRY IS THROUGH A 12-CHANNEL DATA LINK.

IRANIAN UNMANNED AERIAL VEHICLE MOHADJER 4

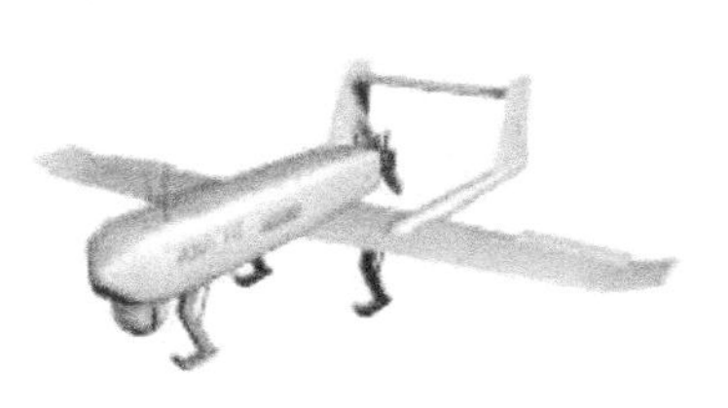

SYSTEM	SPECIFICATIONS
Alternative designations:	Mohajer, Hodhod, Shahin
Date of introduction:	2003
Proliferation:	at least 2 countries
Engine:	50 hp Limbach L550, four-cylinder, two-stroke engine
Propulsion:	
Max launch Weight (kg):	
Fuel and Payload (kg):	
Maximum (level)(km/h):	200
Cruising Speed (km/h):	
Endurance (hrs):	3
Max Ceiling, normal (m):	5400
Range (km):	150
Wing Span (m):	5.33
Overall Length (m):	3.74
Flight Control:	GPS, truck/trailer mounted system
Launch Method:	Rail launch from PL3 Pneumatic Catapult
Recovery Method:	Retractable landing skids on paved runway
Landing Method:	

PAYLOAD TYPES	SPECIFICATIONS
Gimbal optical freeze frame or forward, fixed color and monochrome cameras. Latest versions can use both types at the same time with multichannel imager. Later versions have IR camera and digital processor to downlink sensor imagery.	

VARIANTS	SPECIFICATIONS
Mohadjer 4B: Revealed during a 2014 exhibition. Changes include: wings mounted mid-way, new landing skids, wider wings, improved engine cooling, forward fuselage changed for improved aerodynamics and expanded payload bays.	
Armed Version: In 2014 a weaponized Mohadjer was shown with two QW-1 MANPADS. Primitive fire procedures. Operator uses the optics to line up the target and rely on the QW-1's IR system to acquire the target.	

NOTES

TWO MOUNTED STUB ANTENNAS SUPPORT HIGHER FREQUENCY DATALINKS FOR REAL-TIME MANUAL, SEMI OR AUTOMATIC CONTROL UNDER 10 GHZ.

CHINESE UNMANNED AERIAL VEHICLE (EW/ECM) ASN-207

SYSTEM	SPECIFICATIONS
Alternative designations:	D-4
Date of introduction:	2002
Proliferation:	At least 1 country
Engine:	51 hp, 4 cylinder, 2 stroke
Propulsion:	2 blade wooden push propeller
Max Launch Weight (kg):	222
Maximum (level)(km/h):	210
Cruising Speed (km/h):	150
Endurance (hrs):	6
Max Ceiling, normal (m):	5,000-6,000
Ceiling, Service (m):	100
Radius of Operation (km):	
Dimensions:	
Wing Span (m):	6
Overall Length (m):	3.8
Overall Height (m):	1.4
Flight Control:	
Launch Method:	Solid rocket booster on a zero length launcher
Recovery Method:	Parachute (non-steerable)
Landing Method:	2 spring loaded skids

PAYLOAD TYPES	SPECIFICATIONS
JN-1102 EW/ECM suite which can scan, intercept, analyze, monitor, and jam enemy ground to air communications at 20-500MHz. The system consists of a mounted intercept subsystem, mounted jamming subsystem, and a ground-based intercept and control subsystem.	

VARIANTS	SPECIFICATIONS
ASN-206:	
An older version of the ASN-207, with less control range smaller payload capacity and shorter endurance	

NOTES

THE UAV IS LAUNCHED FROM A ZERO-LENGTH LAUNCHER USING A SOLID ROCKET BOOSTER THAT IS JETTISONED AFTER TAKE-OFF.

RUSSIAN UNMANNED AERIAL VEHICLE SHMEL-1 AND PCHELA-1K

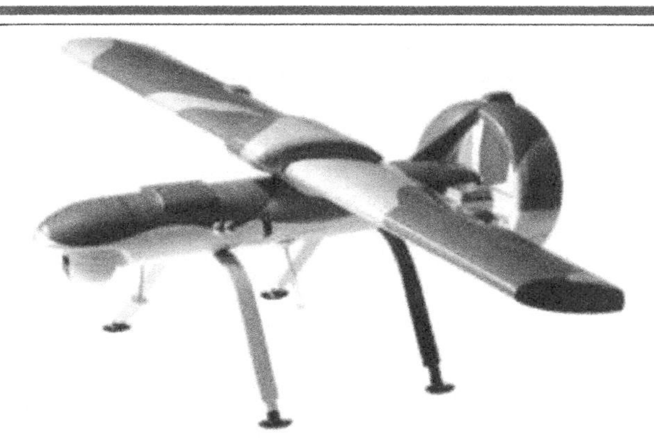

Pchela-1K modernized version of Shmel-1. Note the lack of turned down wingtips

SYSTEM	SPECIFICATIONS	PAYLOAD TYPES	SPECIFICATIONS
Alternative designations:	Bumblebee, Pchela-1, Malakhit	Video Camera, TV, IR linescan	
Date of introduction:	1991	TV Field of view (degrees):	3-30
Proliferation:	At least 6 countries	IR Linescan:	
System:	Launch vehicle, ground station, transporter/loader, tech support vehicle, and 3-10 UAVs	Length:	
		Resolution (milliradians):	3
Launch Vehicle:	BTR-D		
Engine:	32hp 2-cylinder 2-stroke gasoline		
Propulsion:			
Takeoff Weight (kg):	130	**VARIANTS**	**SPECIFICATIONS**
Payload Weight (kg):	70	Pchela-1k:	
Maximum (level) (km/h):	180	Upgrade design. It has 3.5 hrs endurance, 100km RPV-mode range, and 100-3,500 m altitude. Gyro-stabilized sensor ball has LLL TV, IR imaging for night, and earlier sensor options.	
Cruising Speed (km/h):			
Endurance (hrs):			
Max Ceiling, normal (m):	3000	Pchela-1T:	
Ceiling, Service (m):		System includes GAZ-66 truck launcher and varios Pchela-1 versions.	
Wing Span (m):	3.25	Pchela-2:	
Overall Length (m):	2.78	Developing upgrade with 62-hp engine, greater payload, and 100-km range.	
Overall Height (m):	1.10		
Flight Control:		Stroi-P:	
Launch Method:	Rocket-assisted catapult	Military UAV complex with Shmel-1 mounted on a tracked BTR-D launcher.	
Recovery Method:			
Landing Method:	4 Spring loaded landing legs		
		Modern complex, with Pchela-1K,-1T, or -1S launched from a GAZ-66 truck.	

NOTES

THE TRANSPORTER-LAUNCHER-CONTROLLER (TLC) HAS POSITIONS FOR TWO UAV OPERATORS. AUTOMATIC PRE-LAUNCH MONITORING, LAUNCH, FLIGHT CONTROL, AND DISPLAYING OF THE RECEIVED DATA IS CONDUCTED FROM THE TLC. THE DISPLAY IN THE TLC INDICATES AIRCRAFT POSITION OVERLAID ONTO THE TELEVISION IMAGE. GIVEN THE SYSTEM'S DIGITAL DOWNLINK, THE IR IMAGE COULD ALSO BE RECORDED ON MAGNETIC TAPE OR DISPLAYED ON A VIDEO MONITOR. HOWEVER, THE DATA IS ALMOST CERTAINLY RECORDED ON ELECTRONIC MEDIUM FOR PLAYBACK. THE DESCRIPTION OF THE SYSTEM MAY INDICATE A PROBLEM INVOLVING THE INABILITY OF THE OPERATOR TO TRANSLATE AIRCRAFT COORDINATES TO THOSE OF THE TARGETS BEING LOCATED. A LASER RANGEFINDER OR DESIGNATOR COULD EASILY ACCOMPLISH THIS, BUT SUCH A CAPABILITY IS

NOT INDICATED FOR THE SHMEL-1. THE CURRENT SYSTEM REQUIRES COORDINATE CONVERSION FROM MAP ASSOCIATION OR PHOTOGRAPHIC INTERPRETATION WITH A LASER CAPABILITY TO BE ADDED LATER.

THE AREA COVERAGE OF THE SENSOR PAYLOAD IS EXCELLENT. ANALYSIS INDICATES THAT THE CAMERA, AT AN ALTITUDE OF 1500 METERS AND A FIELD OF VIEW OF 30º, CAN IMAGE AN AREA OF APPROXIMATELY 500,000 M2 OR A CIRCLE WITH A RADIUS OF 400 METERS. THE IR LINESCAN AT THE SAME ALTITUDE WOULD SEE A STRIP APPROXIMATELY 5,100 METERS LONG AND 4.5 METERS WIDE. GROUND RESOLUTION WOULD DECREASE SIGNIFICANTLY AT THE ENDS OF THE SCAN. AT A NOMINAL SPEED OF 120 KM/H AND FLYING THE MAXIMUM ALTITUDE, THE AIRCRAFT COULD OBSERVE A MAXIMUM OF 192 KM2/HR WITH THE TELEVISION SYSTEM, OR 1,200 KM2/H WITH IR LINESCAN.

CIVILIAN VERSIONS INCLUDE FOREST, PIPELINE, AND COASTAL PATROL VERSIONS. MILITARY VERSIONS ARE OFTEN USED WITH ARTILLERY UNITS.

ISRAELI UNMANNED AERIAL VEHICLE HERMES 450

SYSTEM	SPECIFICATIONS	PAYLOAD TYPES	SPECIFICATIONS
Alternative designations:	450S	MOSP, high end: TV day/night, auto-tracker, auto-scan	
Date of introduction:	1997	FSP-1 mid-high end: FLIR with 3-FOV telescope	
Proliferation:	At least 8 countries	POP, low-mid-range: CCD Television day and/or night	
Ground Crew:	2	ESP-600C low end: Television, color, day only	
Engine:	70 hp gasoline UEL AR-80-1010 rotary engine	DSP-1: TV with recognition range of 10km and FLIR camera range of 3+km. Detection range is 25 km.	
Propulsion:	2 blade pusher propeller	Other options: MTI radar and SAR	
Take-off Weight (kg):	450-500	SURVIVABILITY/COUNTERMEASURES	
Payload (kg):	150	Light composite structure, low radar signature	
Maximum (level) (km/h):			
Cruising Speed (km/h):			
Endurance (hrs):			
Max Ceiling, normal (m):			
Ceiling, Service (km):	5.4	VARIANTS	SPECIFICATIONS
Radius of Operation (km):	200	Hermes 450S:	
Wing Span (m):	10.5	The original UAV had a weight of 450 kg, UEL 52-hp rotary engine, and flight duration of 20 hours.	
Overall Length (m):			
Overall Height (m):	2.36	Hermes 450 LE:	
Flight Control:	Ground control station vehicle	Has an improved engine, two payload bays, and two wing mounted fuel tanks with a longer duration of 30 hours. 13.2 gallons in two fuel tanks. It uses the DSP-1 sensor pod.	
Flight Control Method:	Preprogrammed/in-flight reprogram		
Launch Method:	Wheeled take-off	Hermes 450 Watchkeeper (WK)	
Recovery Method:	Conventional landing	Developed for the British Army. Fielded in 2014 to a British Artillery unit in Afghanistan. Has wing embedded into fuselage and ruggedized undercarriage for landing on semi-improved runways. Has the ability to mount underwing fuel tanks or can be weaponized using the underwing mounts.	
Landing Method:	3-wheeled, w/arrest cable		

NOTES

AN AVAILABLE OPTION IS DGPS AUTOMATIC TAKE-OFF AND LANDING. RECOMMEND THAT THIS OPTION BE PLAYED IN SIMULATIONS. AN ATTACK VERSION OF THE HERMES WITH MISSILES WAS ESTIMATED TO BE EMPLOYED IN THE SUDAN AGAINST IRANIAN TARGETS. THE MIKHOLIT, ISRAELI 10-KM VARIANT OF NIMROD LONG-RANGE MISSILE, IS ESTIMATED TO BE LAUNCH CAPABLE FROM THE HERMES. ANOTHER VERSION OF THE HERMES 450 WITH MISSILES HAS REPORTEDLY BEEN EMPLOYED AGAINST HAMAS AND HEZBOLLAH TARGETS. IT HAS ALSO BEEN REPORTED THAT RAFAEL SPIKE MISSILES HAVE BEEN INTEGRATED ONTO THE HERMES 450.

ISRAELI UNMANNED AERIAL VEHICLE HERMES 900

Source: Wikipedia/Author Permission: Tal Inbar

SYSTEM	SPECIFICATIONS
Alternative designations:	INA
Date of introduction:	May 2010
Proliferation:	At least 5 countries
Ground Crew:	
Engine:	105hp gasoline Rotax 914 turbocharged engine
Propulsion:	1 blade pusher propeller
Take-off Weight (kg):	970
Payload (kg):	300
Maximum (level)(km/h):	222
Cruising Speed (km/h):	130-175
Endurance (hrs):	36
Max Ceiling, normal (m):	30,000
Ceiling, Service (m):	
Radius of Operation (km):	
Wing Span (m):	15.3
Overall Length (m):	6.1
Overall Height (m):	2.36
Launch Method:	Wheeled take-off
Recovery Method:	Conventional landing
Landing Method:	
Flight Control Method:	Preprogrammed/in-flight reprogram; Secure redundant Line of Sight data link and redundant satellite communications beyond line of sight
Flight Control:	Ground control station vehicle; can control two Hermes simultaneously

PAYLOAD TYPES	SPECIFICATIONS
Gabbiano T200: long ranage surveillance radar, x band (8-12.5 GHz), 407km maritime range, MTI and SAr, >200 target Track-While Scan (TWS)	
DSP-1: TV with recognition range of 10km and FLIR camera range of 3+km. Detection range is 25 km.	
Tadiran Skyfix: COMINT DF and Elisra AES-210: ELINT	
Elop DCoMPASS (digital compact multi-purpose advanced stabilized system); stabilized turret incorporating thermal imager, color TV, dual-band laser designator rangefinder, inertial measurement unit and laser spot tracker	

SURVIVABILITY/COUNTERMEASURES
Light composite structure, low radar signature

VARIANTS	SPECIFICATIONS
An attack version of Hermes 900 may be possible. Each wing has two external hardpoints similar to the weaponized Hermes 450. The Mikholit, Israeli 10km variant of Nimrod long-range missile, is designed for launch from the Hermes 450.	

NOTES

AN AVAILABLE OPTION IS IATOL (INDEPENDENT AUTO TAKEOFF AND LANDING) SYSTEM FOR AUTOMATIC TAKE-OFF AND LANDING ON NON-INSTRUMENT RUNWAYS. RECOMMEND THAT THIS OPTION BE PLAYED IN SIMULATIONS.

FIRST USE BY THE ISRAELI AIR FORCE IN COMBAT OCCURRED IN JULY 2014 DURING OPERATION PROTECTIVE EDGE IN GAZA.

AUSTRIAN UNMANNED AERIAL VEHICLE CAMCOPTER S-100

SYSTEM	SPECIFICATIONS
Alternative designations:	Al-Saber
Date of introduction:	2006
Proliferation:	At least 4 countries
Engine (hp):	55 Diamond aviation engine
Propulsion:	2 blade rotary wing propeller
Take-off Weight (kg):	200
Payload (kg):	55+
Maximum (level)(km/h):	
Cruising Speed (km/h):	
Endurance (hrs):	
Max Ceiling, normal (m):	
Ceiling, Service (m):	
RPV mode (km):	130
Relay/Programmed (km):	130
Wing Span (m):	
Overall Length (m):	
Overall Height (m):	
Hover Capability:	Yes
Launch Method:	DGPS autonomous vertical launch from vehicle/ground base
Recovery Method:	
Landing Method:	
Flight Control Method:	Pre-programmed or in-flight re-program.
Flight Control:	Ground Control Station (GCS) inside vehicle Image processing: Real-time UAV video feed can also be routed to other subscribers.

PAYLOAD TYPES	SPECIFICATIONS
IAI/Elta POP-3000 gimbaled ball with TV and FLIR for night use, IAI/Tamam POP200 gimbaled ball with FLIR, 3km night acquisition range, (UAE version is projected with TV and high zoom for 20 km daytime acquisition)	
PicoSAR: Synthetic aperture radar for MTI surveillance and ground mapping	
Other options: Laser target designator (LTD), CBRN monitors, laser imaging radar (LIDAR), ground-penetrating radar (GPR), and signals intelligence sensors.	

SURVIVABILITY/COUNTERMEASURES
Light carbon fiber structure for low radar signature. It is very quiet, with narrow profile for low visual signature. It has auto-return and recovery mode for lost control signal. Inertial and GPS navigation: <1 meter accuracy.

VARIANTS	SPECIFICATIONS
An Unmanned Combat Aerial Vehicle (UCAV) attack version was developed and displayed in 2008, with 2 x Lightweight Multi-role Missiles (LMMs, see Vol 2, pg6-55). Missiles can engage light armored vehicles, aircraft, and other ground targets.	
The UCAV version could also mount guided rockets, machineguns, rockets, or automatic grenade launchers as needed for attack roles or self-protection. Small launchers for aerial rockets with homing devices could fit on the S-100 with a LTD for deep attack.	

NOTES

USED FOR VARIETY OF MILITARY ROLES, INCLUDING FIRE CONTROL AND OBSERVATION FOR FIRE AND STRIKE SYSTEMS, BORDER PATROLS, DE-MINING AND NAVAL SHIP-BASED ROLES. IN THE AIR DEFENSE ROLE, IT CAN BE USED FOR OBSERVATION OF LIKELY FLIGHT ROUTES, OR FOR HELICOPTER ATTACKS IN UCAV CONFIGURATION. A NOTED ROLE IS USING A LASER TARGET DESIGNATOR TO SELECT TARGETS AND DIRECT SEMI-ACTIVE LASER-HOMING MUNITIONS TO THE TARGET FOR A KILL. THE SYSTEM COULD ALSO CARRY A JAMMER, INCLUDING GPS JAMMING.

Unmanned Aerial Vehicles Used in Attack Missions

More modern forces are employing UAVs directly with fire support units. They offer responsive rapid fire observation with less risk to personnel and fewer terrestrial limitations to direct observation. Roles, capabilities, and configurations for integrated fires and strikes continue to expand. Range requirement for these tactical UAVs is 60+ km; and operational is 120+ km.

Abilities of UAVs to reconnoiter the battlefield, identify targets, give precise locations of targets, and provide fire correction depend on responsiveness stable viewing, and precision location. Improvements in GPS, stabilized sensor balls, and laser range-finders can now permit locations within 1-m accuracy, and stand-off viewing to 20+ km daytime and 3+ km at night. The image can be sent in real-time, and can be retransmitted with minimal delay. Some UAVs use SATCOM to extend the distance. Several forces use UAVs specifically designed for specific digital integrated fire and strike systems, for image and target location display at the battery or weapon monitor. The Russian Pchela-1K is designed for this target display with the 2S19M1. The South African Vulture UAV also directly links with the AS2000 fire control system.

Rotary-wing UAVs offer superior capabilities for fire support roles. Because they can hover, they can approach targets at nap-of-the-earth level (8 meters or level), between trees. They can also mount fairly hefty payloads of robust sensors (up to 55 kilograms for Camcopter S-100), in order to execute stand-off observation. Rotary aircraft generally offer better stability for precision viewing. All of these factors mean better all-weather capability with less risk of detection.

Other UAV missions include direct attack of fleeting targets. There are many programs to develop attack UAVs or convert UAVs for attack roles by mounting explosive warheads for an impact kill. The application goes back to WWII, with explosive-filled unmanned U.S. bombers directed by radio against German targets. UAV costs and limited fielding have limited use in attack roles. An exception is the Israeli Harpy attack UAV (see next page), specially designed as an attack UAV against high-value targets. This system can be called both a UAV and a cruise missile, as it can be piloted and/or programmed. The Russian R-90 UAV rocket is launched from 9A152 MRL, and has an attack option. Since MUAVs and MAVs have been fielded, their lower cost means that more attack versions will be likely. The Russian Pustelga MAV is noted to have an attack option. In the near term, weapon-launched sensor munitions will also have warheads and guidance for attack. UAVs, armed or not, can be used to harass and attack enemy RW aircraft. More attack UAVs or attack configurations will continue to increase world-wide. Russia, China, Iran, and a growing number of European countries already have or will have UAVs with attack configurations.

The U.S. has demonstrated another UAV design for direct attack by mounting ATGMs UAVs as unmanned combat aerial vehicles (UCAVs). UAV-based UCAVs operate similarly to larger aircraft-based UCAVs. They can fire guns or grenades or launch missiles against air and ground targets. Israel has also weaponized their UAVs with ATGMs such as the Hermes 450 in conflicts with Hezbollah and Hamas. The ATGMs were possibly Mikholits, a Nimrod variant designed for UCAVs.

Emerging attack UAVs/CAVs will compete with cruise missiles against deep-strike NLOS targets to 200+ km. Nevertheless, the most effective use of UAVs for attack remains in precision location and guidance. Best use is mounting a laser target designator to guide semi-active laser-homing munitions (from a UCAV

mount or delivered by artillery, tanks, aircraft, mortars, and ships) against targets otherwise inaccessible to ground-based designators.

ISRAELI UNMANNED AERIAL VEHICLE HARPY, CUTLASS

SYSTEM	SPECIFICATIONS	PAYLOAD TYPES	SPECIFICATIONS
Alternative designations:	See Notes	Optical Camera:	Electro-magnetic and optical
Date of introduction:	1988	Passive Radar Seeker:	Wide range of frequencies
Proliferation:	At least 5 countries	User Image Capabilities:	Receive images on possible targets via datalink
Ground Crew:	1-3 per truck launcher	Missile:	HE Fragmentation warhead
Engine (hp):	27.5, 2 cylinder, 2 stroke	Max Payload, Warhead (kg):	32
Propulsion:	Two blade pusher propeller	**VARIANTS**	**SPECIFICATIONS**
Weight (kg):		CUTLASS:	
Max Launch Weight (kg):	120	Weight (kg):	2.8
Speed (km/h):		Wing Span (m):	1.83
Maximum (level):	250	Max Launch Weight (kg):	125
Cruising Speed:	185	Max Ceiling (m):	4,575
Endurance (hrs):	6	Radius of Operation (km):	300
Max Ceiling, normal (m):	INA	Cruising Speed:	185
Ceiling, Service (m):	3,000	Endurance (hrs):	6
Radius of Operation (km):	500	Max Range (km):	1,000
Dimensions:		**PAYLOAD TYPES**	**SPECIFICATIONS**
Wing Span (m):	2	Missile:	HE Fragmentation warhead
Overall Length (m):	2.2	Max Payload, Warhead (lbs)	51
Overall Height (m):	.36	Data Link (direct LOS) (km):	150, 1,000 with GPS
Flight Control:	GPS positioning, autonomous preprogrammed flight	Infrared (IR)	Raytheon Seeker Head AIM-9X
Launch Method:	Booster rocket launched from truck launcher	Automatic Target Recognition and Classification	Raytheon Algorithms INA
Recovery/Landing Method:			
Launcher Trucks per Battery			
Missiles per Launcher:	18		
Total Missiles per Battery:	54		

NOTES
Harpy can be used as cruise missile in preprogramed mode. But it can also be considered a UCAV, which can be piloted or used without a pilot (e.g. programmed or homing attack mode). The Harpy and CUTLASS can also attack artillery counter-battery radars and ground surveillance radars. Harpy radar sensor can be remotely turned off to abort a target and continue searching. Day or night flight capability. CUTLASS stands for Combat Uninhabited Target Locate and Strike System. Built jointly with Raytheon. Primary difference from the HARPY is the CUTLASS is a semi-autonomous with a real time - data link to the ground control station to confirm target identity. It can also target ballistic missiles launchers and vehicles. Designed to be fired from naval ships.

Worldwide Equipment Guide

Chapter 4: Equipment Upgrades, Countermeasures, and Emerging Technology Trends

Chapter 4: Equipment Upgrades, Countermeasures, and Emerging Technology Trends

EQUIPMENT UPGRADES

Armed forces worldwide employ a mix of legacy systems and selected modern systems. In the current era characterized by constrained military budgets, the single most significant modernization trend impacting armed forces worldwide is upgrades to legacy systems. Other factors impacting this trend are:

- A need for armed forces to reduce force size, yet maintain overall force readiness for flexibility and adaptiveness
- Soaring costs for modern technologies, and major combat systems
- Personnel shortages and training challenges
- Availability of a wide variety of upgrade packages and programs for older as well as newer systems
- New subsystem components (lasers, GPS, imaging sensors, microcircuits, and propellants) which permit adaptation of new technologies to platforms, weapons, fire control systems, integrated C2, and munitions
- An explosion of consortia and local upgrade industries that have expanded worldwide and into countries only recently modernized or still in transition.

The upgrade trend is particularly notable concerning aerial and ground vehicles, weapons, sensors, and support equipment. From prototype, to low-rate initial production (LRIP), to adoption for serial production, minor and major improvements may be incorporated. Few major combat systems retain the original model configuration five or more years after the first production run. Often improvements in competing systems will force previously unplanned modifications.

Upgrades enable a military to employ technological niches to tailor its force against a specific enemy, or integrate niche upgrades into a comprehensive and well-planned modernization program. Because of the competitive export market and varying requirements from country to country, a vehicle may be in production simultaneously in many different configurations, with a dozen or more support vehicle variants concurrently filling other roles. In light of this trend, OPFOR equipment selected for portrayal in simulations and training need not be limited to the original production model of a system, but may also employ other versions that incorporate the armed force's strategic and modernization plans, along with likely constraints that would apply.

The adaptive OPFOR will introduce new combat systems and employ upgrades on existing systems to attain a force structure which supports its plans and doctrine. Because the legacy force mix and equipment were selected in accordance with past plans and options, upgrading an existing system will often present an attractive alternative to costly new acquisitions. A key consideration is the planned fielding date. For this document, the OPFOR planning time-frame is current to near-term. Thus, only upgrades currently available (or marketed, with production capability and fielding expected in the near term), are considered. Also, system costs and training and fielding constraints should be considered.

The following tables describe selected upgrades currently available for system modernization. These lists are not intended to be comprehensive. Rather, they are intended to highlight major trends in their respective genres. With armored combat vehicles, for instance, the focus is on upgrades in mobility, survivability, and lethality.

The category of survivability upgrades includes countermeasures (CM). Depending on their applicability and availability within the contemporary operating environment, the CM upgrades can apply not only to systems associated initially with specific branches (tanks, IFVs, and air defense guns), but over time to other systems that are vulnerable to similar threats. An example of this is the proliferation of smoke grenade launchers tailored for use with artillery and air defense vehicles.

Implementation of all upgrade options for any system is generally not feasible. Because of the complexity of major combat systems and the need for equipment subsystem integration and maintenance, most force developers will chose a mix of selected upgrades to older systems, augmented with limited purchases of new and modern systems. Please note that systems featured in this document may be the original production system or a variant of that system. On data sheets, the **VARIANTS** section describes other systems available for portrayal in training and simulations. Also, equipment upgrade options (such as night sights) and different munitions may be listed, which allow a user to consider superior or inferior variants. Within the document chapters, multiple systems are listed to provide a range of substitution options. Of course there are thousands of systems and upgrade options worldwide that could be considered for adoption by an innovative OPFOR.

OPFOR trainers have the prerogative to inject systems or upgrade packages not included in the OPFOR Worldwide Equipment Guide (WEG), in order to portray an adaptive, thinking OPFOR. In future WEG updates, we will expand the upgrade tables to include by-name descriptions of upgrade options and specific systems applications that have been noted elsewhere throughout the document. Our functional area analysts are available to assist OPFOR planners in selecting reasonable upgrade options that tailor system configurations to specific force portrayals. Questions and comments on tables and data contained in this chapter should be addressed to the respective POCs designated for corresponding individual chapters placed throughout all three volumes of the WEG.

Jim Bird

DSN: 552-7919/Commercial (913) 684-7919

e-mail address: james.r.bird.ctr@mail.mil

OPFOR AIR DEFENSE SYSTEMS UPGRADES

AIR DEFENSE GUN/GUN-MISSILE SYSTEM	MANPORTABLE AIR DEFENSE SYSTEM	SURFACE-TO-AIR MISSILE SYSTEM
Light AD vehicle: Combat support vehicle with light armor and TV, thermal sights, AD machine gun. Add encrypted voice and digital data capability, azimuth plotting board alert system and overhead launch turret or MANPADS. MMW radar for detection and possible fire control. High velocity laser beam rider MANPADS ranges to 7+ km, 0-5 km alt Armored AD vehicle: See IFV upgrades, e.g. Improved armor and suspension, 2-man turret. CM, e.g. multi-spectral smoke grenades, LWR Upgraded FCS: Cdr's independent viewer, 2-plane stabilized TV, auto-tracker, FLIR, multi-mode targeting (TV/radar, day/night modes). Dual-mode (TA/FC) low probability of intercept radar with longer range. Reduced radar mean-time to detect and system response time. Links to AD network, encrypted voice, digital data transmission capability, computer display GPS, and inertial land navigation, IFF. Improved multiple auto-cannons to 30 mm, with stabilized guns and fire-on-the-move capability. Improved rounds, e.g. electronic-fuzed HE, APFSDS-T, and frangible or canister rounds. Two-stage high-velocity laser beam-rider AD missiles with MMW radar, to out-range helicopter launch missile systems. Kinetic-energy missiles for use in AD role, and against ground vehicle targets. Altitude is 0-6 km. Range 0-8 km. Jam capability is 0.	Vehicle, ground platform, helicopter mounts, missiles in disposable launch tubes Early warning datalinks and alert display boards for mount on launcher. Upgraded IFF capabilities. FLIR night sight Improved missiles and seeker heads with better counter-countermeasure resistance. Proximity/PD fuze. Thrust-vectoring capability, all-aspect engagement capability, strap-on imaging infrared or thermal sights Improved larger warheads and blast/frag effects, base fuzing or propellant for increased blast Increased range to 7 + km. Improved aerodynamics, fuels, and materials, for increases in speed, reduced smoke signature, maneuverability, and accuracy ½ of missiles are high velocity ADAT missiles with laser beam rider KE missile and 3 LBR sub-missiles to 7 = km, 0 – 5 km altitude, and nil countermeasure vulnerability. Integrate AD defense with anti-helicopter mines	Improved vehicle or platform launcher for rapid emplacement/displacement CM e.g. multi-spectral smoke, LWR Upgraded FCS: 2-plane stabilized TV gunner sights, 1 – 2 gen FLIR, multiple target engagement capability, All-weather fire control, multi-mode targeting, with TV and radar, day and night. Improved EW and target acq radars, longer range, low probability of intercept, and signal processing in radars Reduced radar mean-time to detect, and system response time Links to AD network, encrypted voice, digital data transmission capability, computer display GPS and inertial land navigation, and graphic display battle management system, IFF Missiles with SACLOS, ACLOS radar, IR or multi-band terminal seekers, more lethal warheads, longer range, maneuverability with improved counter-countermeasure resistance Vertical missile launch

AIR DEFENSE GUN/GUN-MISSILE SYSTEM	MANPORTABLE AIR DEFENSE SYSTEM	SURFACE-TO-AIR MISSILE SYSTEM
UPGRADE PRIORITY Improved day/night optics and radar Light AD/MANPADS and MG Battalion AD fire support vehicle with HV launcher and MANPADS dismount teams Armored brigade AD vehicle with overhead turret and: High-velocity missiles 30-mm stabilized auto-cannon MMW TA radar Automated secure links to AD network	**UPGRADE PRIORITY** Improved sights and warning display boards Prox Fuze MANPADS with Strap-on II/FLIR, improved seekers, warheads, propulsion, wider FOV, IR CCM/Flare rejection capability Add ½ High velocity ADAT missiles MMW radar	**UPGRADE PRIORITY** Improved FCS with day/night optics and radars, and multi-target capability and modes Automated secure links, digital AD network Improved missiles and guidance CM protection from jamming and ARMs

OPFOR AERODYNAMIC SYSTEM UPGRADES

ROTARY-WING AND FIXED-WING AIRCRAFT	UNMANNED AERIAL VEHICLE (UAV)	THEATER MISSILES
Older airframes and utility helicopters can add upgraded sensors and weapons. Service life extension programs Western upgraded avionics, fire control computers, sights, and technology readily available to retrofit into existing older airframes Emerging belief in upgrade of existing platforms rather than developing new airframes, primarily due to financial constraints Two-seat conversions for adding weapons officers and multi-role use Development of quieter, more efficient main and tail rotor blades and more powerful engines to increase performance and load capacity Digital data-linking with ground systems and air defense networks Increased use of millimeter wave, FLIR, and NVG technologies to allow greater night/ weather weapons delivery and mission completion Self-protection jammers and IIR vs pyrotechnic IR seeker decoys Laser altimeters replace radar altimeters to reduce RF detectability Added weapon mounts to increase mission load capacity Improved weapons and munitions, including ATGMs, air-to-surface missiles, rockets and precision bombs Laser seekers and designators for missile/rocket/bomb conversion GPS course-corrected munitions (bombs and missiles)	Extend operational radius and endurance Reduce sensor-shooter timeline Continued development of micro-UAVs dwell time and image quality Advanced imagery fusion from multiple UAVs Real-time teaming between manned/unmanned fixed-wing aircraft Enhanced third-generation image intensifiers and second-generation thermal imagers may be available to limited countries. Multiple sensors will be employed on the same platform for enhanced target detection under all-weather conditions and may be linked to weapon delivery platforms. Integrated laser target designators for smart munitions in priority target areas Multiple sensors for chemical and biological agents will be employed on this platform and may be linked to comms platforms. Precision attack variants, such as anti-radiation UAVs for radar attack SATCOM stand-off navigation and sensor communications Miniaturization and reduced weight of munitions	Ballistic missile Improved launchers (swim capability, multiple missile capability, reduced signature) Reduced preparation time, emplace and displace times, shoot and scoot operation Launcher countermeasures: decoys, missile non-ballistic launch trajectory, smokeless solid fuel Autonomous operations or increased interval between launchers Missile countermeasures (e.g., non-ballistic trajectory, penetration aids, separating warhead, multiple maneuvering re-entry vehicles) Automated secure digital C2 network, linking with artillery, air, EW, and reconnaissance units Navigation system with GPS/inertial update, linked to automated net, and homing options Extended range missiles, some to 500+ km Multi-sensor or other improved homing, especially GPS with increased accuracy (10-50 m CEP) Advanced munitions (cluster munitions, FAE, jam, thermobaric munitions, biological, electro-magnetic pulse, anti-radiation missiles), larger payloads Cruise missiles (CMs) with pre-programmed multiple waypoints, and manned guidance option CM multi-seeker modes, including GPS, loiter and radar/IR homing, SAL-H, cluster PGM warheads

ROTARY-WING AND FIXED-WING AIRCRAFT	UNMANNED AERIAL VEHICLE (UAV)	THEATER MISSILES
UAV launch capability, permits them to precede aircraft or replace them in high threat areas **UPGRADE PRIORITY** MMW, FLIR, and NVG technologies GPS and SAL-H munitions Upgraded avionics Service life extension programs Conversion to multi-role systems	**UPGRADE PRIORITY** Extend operational radius and endurance Obtain improved EO capability Reduce sensor-shooter timeline Laser target designator integration	**UPGRADE PRIORITY** 10-50 CEP with GPS Improved smokeless solid fuel Separating warhead and larger payloads Decoys

COUNTERMEASURES

Countermeasures (CMs) are survivability measures that enhance the protection of assets and personnel by degrading enemy sensors and weapons effectiveness. These measures often fall under the rubric of the US Army term CCD (camouflage, concealment and deception) or within the OPFOR term C3D (camouflage, cover, concealment and deception). Decoys used by tactical units within branch operations are designed to aid survivability, and are considered to be countermeasures. Countermeasures can take the form of tactical CMs (also called reactive measures), or technical CMs. The various types of tactical CM change, alongside new unit tactics techniques and procedures (TTP), allowing adaptation to a given situation in compliance with prescribed rules of engagement. This document focuses on technical CMs. In certain specialized branches the development of new technical CMs is persistent and ongoing.

Modern forces will upgrade existing systems by augmenting them with selected force protection countermeasures. Many CMs noted are intended to protect combat vehicles from anti-armor sensors and weapons. Although the CMs shown below can be used to counter precision weapons, many were originally developed for use against conventional weapons. Priorities for countermeasures are dictated by the goals of survival, mission success, and maintaining effectiveness. The first CM priority is to avoid detection until you can control the events. Among goals for using countermeasures, mission success is the most important.

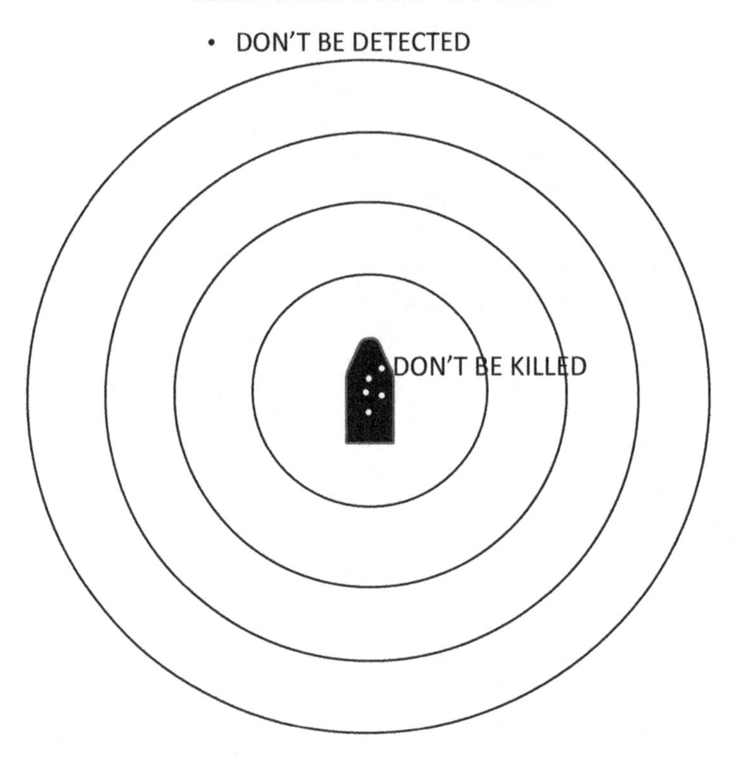
COUNTERMEASURE PRIORITIES
• DON'T BE DETECTED
DON'T BE KILLED

Survival ("Don't Be Killed") encompasses the following prerequisites in order of priority: operating system or network survival, vehicle survival, vehicle avoidance of major damage, crew survival, and vehicle avoidance of minor repair. A compatible suite of countermeasures may be limited to a more modest goal, to preserve a measure of effectiveness, even at the cost of system survival. Effectiveness in this context could be defined as: ability to successfully execute the immediate and subsequent missions, until system or subsystem failure interrupts this process. Effectiveness includes: crew effectiveness, crew fitness, mission success, operating system effectiveness, and vehicle/soldier readiness for employment.

Several factors must be considered when selecting countermeasures.

- Countermeasures should be fielded and mounted on systems with a holistic and rational approach to assure survivability. The rational developer will focus his countermeasures, assigning the highest priority to protection against the most likely and most lethal threats. However, evolution of threat capabilities over time, in combination with conflicting priorities, can prevent success of the current CM mix. Most CM are responses to specific perceived threats, and limited by prohibitive costs and budget constraints. With the modern reliance on precision weapons, military forces may seek out complex and expensive countermeasure "suites" to degrade their opponents' capabilities.
- Some countermeasures can diminish or neutralize the effects of numerous sensors and weapons. These CM can be categorized based on generic types of threats, such as artillery or ATGM. Others are more adversary technology-specific, and may only become available once the technology in question makes its appearance in the Operational Environment. Pressure from new and threatening technologies may compel designers to launch short-response programs to expedite the fielding of adaptive countermeasures.
- The R&D process has led to the development of counter-countermeasures, intended to negate the effects of CMs. However in a certain context these too fall under the rubric of CMs. To avoid confusion with semantics, the WEG regards all of them as countermeasures.
- When countermeasures are added to a vehicle or in close proximity to it, they must be mutually compatible and also compatible with other relevant subsystems. Accordingly issues like electromagnetic interference and self-blinding with smokes should also be taken into account.
- Although a variety of countermeasures are now marketed, many technical and financial factors can negate their advantages. Countermeasure development may be restricted due to limitations in resources, technology, and fielding capacity. All will vary by country and time frame. At times budget limitations may compel fielding of CM that fall short of desired standards. For instance, active protection systems can counter some weapons; but they are expensive, hazardous to soldiers, and ineffective against many weapons. Thus they may be unsuitable for application to many systems. OPFOR users should consult the appropriate WEG chapter POC if assistance is needed to adopt CMs against a specific system.
- Countermeasures will not replace the need for armor protection and sound tactics.

LETHALITY COMPONENT VERSUS COUNTERMEASURE RESPONSES

The intent of this table is to assist in selecting CM and understanding the categories used in upgrade schemes. Many of the more widely-fielded countermeasures are designed to degrade a variety of sensors and munitions, for minimal cost. Thus, countermeasure types may be replicated across several functions. Because new technologies are emerging rapidly, and systems are being applied throughout several CM modes, the placement of CMs can be somewhat arbitrary. CM uses against artillery, ATGMs, and mounted ground vehicle weapons systems will vary in type. The following list of CM can be used for artillery, air defense, antitank, armor, aircraft, theater missile, and other systems, depending on the platform, gun, sensor, and munition configuration of the system.

Capability to Be Degraded	Type of Countermeasure
Detection and location	Camouflage: nets, paints, fasteners for added natural materials Cover: entrenching blades, hole-blast device, underground facilities Concealment: screens, skirts, thermal engine covers, scrim, other signature reduction Deformers, engine exhaust diversion, other signature alteration measures Aerosols: smoke and flares, water spray systems Decoys, clutter, and acoustic countermeasures Counter-location measures: GPS jammers, laser and radar warning systems
C2/sensor-shooter links	See Information Warfare (IW) Chapter
Platform or weapon	Counterfire: directional warning systems, laser radars, for rapid response Directed energy weapons (DEW), such as high-energy lasers System prioritization for hard-kill, e.g., anti-helicopter mines (See Engineer Chapter)
Weapon sensors and fire control	CCD as noted above. Directed energy weapons, such as low-energy lasers (LEL) Electro-optical countermeasures (EOCMs)
Submunition dispensing/activation	Global positioning system (GPS) jammer Fuze (laser/IR/RF), RF barrage jammers, acoustic jammers
Precision munition and submunition sensors	CCD as noted above. False-target generator (visual, IR, RF/acoustic) Electromagnetic mine countermeasure system, to pre-detonate or confuse Fuze jammers (laser/IR/RF), RF barrage jammers, acoustic jammers
Munition/submunition in-flight, and its effects	Sensors to detect munitions: MMW radars, RF/IR/UV passive sensors Air watch and air defense/NBC warning net, to trigger alarm signal Active protection systems, for munition/submunition hard kill Cover, additional armor to reduce warhead effects
Other System Effects	Miscellaneous CM (See below)

COUNTERMEASURES AGAINST SENSORS

Type Countermeasure	Countermeasure	Example	Application
Camouflage	Camouflage nets, Camouflage paints, IR/radar/and laser-absorptive materials/paints Fasteners, belts for attaching natural materials	Russian MKS and MKT Salisbury screen rubber epoxy Chinese "grass mat" set	Variety of vehicles Variety of systems Uniforms and vehicles
Cover	Natural and manmade cover, civilian buildings Entrenching blade to dig in vehicles Hole-blast devices for troop positions, spider holes Underground facilities, bunkers, firing positions	Tree cover, garages, underpasses T-80U tank, BMP-3, IFV, 2S3 arty Hardened artillery sites, bunkers	TELs, vehicles, troops IFVs, tanks, SP arty Infantry, SOF Iraqi and NK sites
Concealment	Screens, overhead cover for infantry (conceal IR/visible signature) Canvas vehicle cover, to conceal weapons Thermal covers, vehicle screens Scrim, side skirts, and skirting around turret	Colebrand netting Cover on Chinese Type 90 MRL Kintex thermal blanket over engine French "Ecrim" track cover scrim	Infantry, weapon, sensor Truck-based weapons For combat vehicles Combat vehicles
Deformers/signature modification	"Wummels" (erectable umbrellas to change/conceal shape/edges) Exhaust deformers (redirect exhaust under/behind vehicle Engine and running gear signature modification (change sound) IR/radar deformers (in combination with RAM and RAP, etc.)	Barracuda RAPCAM/TOPCAM Russian exhaust deflectors Track pads, road wheel/exhaust change Cat-eyes, Luneberg lens	Vehicles, sites, weapons Combat vehicles Tracked, other vehicles Tracked, other vehicles
Aerosols	Visual suppression measures, smokes, WP rounds Multi-spectral smokes for IR and/or MMW bands Flares, chaff, WP, to create false targets, disrupt FLIR Toxic smokes (irritants to disrupt infantry and weapons crews) Water spray systems (to reduce thermal contrast)	Smoke generators, fog oil, S-4, RPO-D ZD-6 Smoke grenades (visual/IR) WP rounds, Galix 6 flare system Adamsite and CN in smoke mix Add-on kits for vehicles	Blinding, screening Vehicle protection Combat vehicles, arty Smoke generators Recon, C2, AD, arty

Type Countermeasure	Countermeasure	Example	Application
Decoys	Clutter (civilian/military vehicles, structures, burning equipment	Log site, truck park, tank farm, derricks	Artillery, combat vehicles
	Low to high-fidelity (multi-spectral) decoys	Shape International decoys	TBMs, SAMs, radars
	Radar/IR decoy supplements (to add to visual/fabricated decoys)	Corner reflectors, KFP-1-180 IR heater	Vehicle/site decoys
	Acoustic countermeasures (to deceive reconnaissance sensors)	Acoustic tape/speaker systems	Vehicles, sites
Counter-location measures	Degrade GPS by jamming to reduce precision location capability	Aviaconversia GS jammer	Infantry and others
	Jam radars/IR sensors	SPN-2 truck-borne jammer set	Tactical/operational area
	Laser, IR, and radar warning systems (to trigger move/CM)	Slovenian LIRD laser warner	Combat vehicles

COUNTERMEASURES AGAINST WEAPONS AND WEAPON SENSORS

Type Countermeasure	Countermeasure	Example	Application
Added protection (supplements to armor in reaction to specific capability	Armor supplements (ERA, screens, bar or box armor, sand bags) Armor skirts over road wheels Mine rollers, plows, and flails Vehicle belly armor, raised or redesigned belly design, skirt Vertical smoke grenade launchers (to counter PGM top attack)	Barracuda, SNPE ERA KMT-5, KMT-6	
EOCM	Use EOCMs such as IR jammer/IR searchlights to redirect ATGM	KBCM infrared CM system	Combat vehicles
False-target Generators	Acoustic jammers and directed acoustic countermeasures Laser false-target generator (against semi-active laser homing) Electromagnetic mine countermeasure system, counters, fuzes	In development, can be improvised In development	To distract acoustic seekers Combat vehicles
Jammers	Altimeter jammer (counters submunition dispersion altimeter) Fuze jammers (to spoof RF proximity fuzes on munitions) Incoherent infrared jamming (to jam IR fuzes on munitions) GPS jammers to confuse navigation and course correction systems	SPR-1 armored ECM vehicle	High priority sites, CPs, etc.
Active Countermeasures	Active protection systems, for munition hard kill High energy laser weapons to destroy munitions or sensors Low energy lasers to blind or dazzle Radio-frequency weapons to burn electronics and detonate munitions Directed MGs	Arena hard-kill system ZM-87 laser weapon VEMASID counter-mine system	Tanks, recon vehicles, IFVs AT, AD systems

Type Countermeasure	Countermeasure	Example	Application
Counterfire/Threat response warners	Directional warning system (locate laser/radar, to direct weapons) Employ sensors (RF/IR/UV –to detect munitions) Acoustic directional systems (to detect munitions) Laser radars (laser scanner to locate optics and direct weapons) Directed energy weapons (against optics) Anti-helicopter mines (against aircraft) Employ air watch/security, AD, NBC, nets to trigger alarm signal Dazzle grenades (temporarily blind personnel)	Pilar acoustic detection system Star-burst grenades	Infantry
Miscellaneous CM	Optical filters to degrade effect of battlefield lasers Pulse code/thermal CCM beacons on SACLOS ATGMs (to counter EOCM)	HOT-3 ATGM	

COUNTERMEASURES BY FUNCTIONAL AREA AND TYPE SYSTEM

Functional Area	System	Type Countermeasure	Countermeasure
Air Defense, Artillery, Radar units, Theater Missile units, Aviation, Headquarters	Command and communications vehicles, radars, missile launchers, aircraft (High value targets)	Camouflage	Camouflage paints, IR/radar/and laser-absorptive materials/paints
		Cover	Entrenching blade to dig-in vehicles Underground facilities, bunkers, firing positions
		Concealment	Canvas vehicle covers, to conceal weapons when not in use Thermal covers, vehicle screens
		Deformers/signature modification	Scrim, side skirts, and skirting around turret "Wummels" (erectable umbrellas to change/conceal shapes/edges Exhaust deformers (redirect exhaust under/behind vehicle Engine and running gear signature modification (mask sound) IR/radar deformers (in combination with RAM and RAP, etc.)
		Aerosols	Visual suppression measures, smokes, WP rounds Multi-spectral smoke grenades for IR and/or MMW bands, flares, chaff, WP, to create false targets, disrupt FLIR
		Counter-location measures	Degrade GPS by jamming to reduce precision location capability Jam radars/IR sensors Laser, IR, and radar warning systems (to trigger move/CM) Clutter (civilian/military vehicles, structures, burning equipment)
		Decoys	Low to high-fidelity (multi-spectral) decoys Radar/IR decoy supplements (to add to visual/fabricated decoys) Acoustic countermeasures (to deceive reconnaissance, sensors) Anti-helicopter mines (against aircraft)

Functional Area	System	Type Countermeasure	Countermeasure
		CM Operational Technologies	
			Beyond line-of-sight modes Non-ballistic launch modes Anti-radiation missiles Low energy lasers to blind/dazzle optics on designators/aircraft Encoded laser target designators to foil false target generators Radio-frequency weapons – burn electronics/detonate munitions High energy laser weapons to destroy munitions or sensors Laser false-target generator (against semi-active laser homing) Altimeter jammer (counters submunition dispersion altimeter) Fuse jammers (to spoof RF proximity fuzes on munitions) Incoherent infrared jamming (to jam IR fuzes on munitions) GPS jammers to confuse navigation and course correction systems Optical filters to degrade effect of battlefield lasers
Aircraft Units Reconnaissance UAVs Theater Missile Units	Helicopters Fixed-wing aircraft UAVs Attack UAVs Missiles	Camouflage Decoys Counter-location measures CM Operational Technologies	Camouflage paints, IR/radar/and laser-absorptive materials/paints Launcher decoys Flares, chaff, WP – decoy seekers, create false targets, disrupt FLIR Clutter (civilian/military vehicles, structures, burning equipment) Jam Radars Stealth materials and coatings GPS jammers to confuse navigation and course correction systems Jam IR sensors and seekers with laser/IR devices Fuze jammers (to spoof RF proximity fuzes on munitions)

Functional Area	System	Type Countermeasure	Countermeasure
			Radio-frequency weapons – burn electronics/detonate munitions Laser, IR, and radar warning systems (to trigger move/CM) Low energy lasers to blind or dazzle Optical filters to degrade effect of battlefield lasers Encoded CCM beacons on SACLOS ATGMs (to counter EOCM) Stand-off precision munitions (maneuvering) Beyond-line-of-sight and over-the-horizon modes Non-ballistic launch modes for missile launcher/missile survival Anti-radiation missiles to counter radars and aircraft Maneuvering re-entry vehicle (with warhead) for ballistic missiles
Information Warfare/ Deception Units	IW vehicles	Camouflage	Camouflage paints, IR/radar/and laser-absorptive materials/paints
		Cover	Natural and manmade cover, civilian buildings Underground facilities, bunkers, firing positions
		Deformers/signature modification	"Wummels" (erectable umbrellas to change/conceal shapes/edges) IR/radar deformers (in combination with RAM and RAP, etc.)
		Aerosols	Visual suppression measures, smokes, WP rounds Multi-spectral smoke grenades for IR and MMW bands; flares, chaff, WP, to create false targets, disrupt FLIR
		Counter-location measures	Degrade GPS by jamming to reduce precision location capability Jam radars/IR sensors Laser, IR, and radar warning systems (to trigger move/CM)
		Decoys	Clutter (civilian/military vehicles, structures, burning equipment)

Functional Area	System	Type Countermeasure	Countermeasure
All Units	Combat support vehicles (Light strike vehicles, Tactical utility vehicles, Motorcycles, ATVs, Armored CSVs, etc.), Trucks	Camouflage Cover Concealment Deformers/signature modification Aerosols Decoys CM Operational Technologies	Low to high-fidelity (multi-spectral) decoys Radar/IR decoy supplements (to add to visual/fabricated decoys) Acoustic countermeasures (to deceive reconnaissance, sensors Camouflage paints, IR/radar, and laser-absorptive materials/paints Fasteners, belts for attaching natural materials Natural and manmade cover, civilian buildings Underground facilities, bunkers, firing positions Armor supplements (ERA, screens, bar or box armor, sand bags) Thermal covers, vehicle screens Engine and running gear signature modification (change sound) IR/radar deformers (in combination with RAM and RAP, etc.) Multi-spectral smoke grenades for IR and/or MMW bands; Flares, chaff, WP to create false targets, disrupt FLIR Clutter (civilian/military vehicles, structures, burning equipment Air watch/security, AD, NBC, nets to trigger alarm signal Acoustic-directed counter-fire system

EMERGING TECHNOLOGY TRENDS

In order to provide a realistic OPFOR for use in Army training simulations, we must describe a spectrum of contemporary and legacy OPFOR forces that currently exist, as well as capabilities in emerging and future operational environments (OEs). Instead of trying to predict the future, this chapter notes known emerging adversary capabilities which can affect training.

The timelines for emerging OPFOR covered by this document are: 2015-2020 (Near Term) and 2021-2025 (Mid-Term). Time frames after 2025 would apply to "future" OPFOR that are considered beyond the scope of this WEG. Time categories were selected in part to facilitate building OPFOR systems and equipment well suited for Army training simulations. The timeframes are practical for use in focusing on and linking various technological trends. However, they also generally align with emerging force structures throughout the U.S. Army, as well as those known to be appearing among potential adversary nations with advanced technologies. These new technologies will pose a range of challenges to OPFOR planners and developers.

Within the designated Near Term and Mid-Term time frames, the mix of forces will continue to reflect tiered capabilities. The prevailing trend will most likely reflect an eclectic blend of forces that in large measure continue to rely heavily on legacy systems (see OPFOR tier tables). New OPFOR systems and an array of new technologies are bound to make their appearance between now and 2025. The most notable difference between the OPFOR force mix and U.S. forces is that the OPFOR will have a broader mix of older systems and a lower proportion of state-of-the-art systems. To compensate for disadvantages, OPFOR will rely more on adaptive applications, niche technologies, and selected proven upgrades to counter perceived capabilities of their adversaries. Force developers for OPFOR will retain expensive legacy systems, selectively adding affordable upgrades drawn from niche technologies. A judicious mix of equipment, strategic advantages, and sound OPFOR principles can enable even lesser (lower-tier) forces to challenge U.S. military force capabilities.

The OPFOR systems must represent reasonable responses to U.S. force developments. A rational thinking OPFOR would study force developments of their adversaries as well as methods used by the world's most technologically advanced militaries, then exploit and counter them. Thus equipment upgrades made by the world's major military powers will provide OPFOR with examples to follow in modifying their own equipment and tactics to deter, match, overmatch, or neutralize advantages enjoyed by their adversaries.

OPFOR TECHNOLOGIES AND EMERGING OPERATIONAL ENVIRONMENTS

As noted earlier in this chapter, the adaptive OPFOR will introduce new combat systems and employ upgrades on existing systems to attain a force structure which supports its plans and doctrine. Because a mix of legacy forces and equipment were selected earlier in conformance with past budgets, upgrades versus costly new acquisitions will always be an attractive option. A key consideration is the planned fielding date. To project OPFOR capabilities into the future, we should look at the technologies in various stages of research and development today, as well as those still in the conceptual stage, for applications in the future OPFOR time frame. Military engineering experience has demonstrated that the processes for identifying military requirements, and budgeting to fund research and development geared toward meeting future threats, can dramatically affect production timelines for equipment modernization programs. In addition, scientific discoveries and breakthroughs in the civilian sector have greatly accelerated the so-called "Revolution in Military Affairs," and have increased the capabilities for battlefield awareness, systems integration, and lethality. The table below shows OPFORs in emerging and future OEs, and offers some considerations relating to the impact and deployment of advanced technologies.

CONSIDERATIONS IN DETERMINING EMERGING OPFOR TECHNOLOGIES BY TIME-FRAME

OPFOR Considerations	Near-Term (2015-2020)	Mid-Term (2021-2025)
Challenging OPFOR	Emerging OPFOR	Objective OPFOR
Technology Source	Current marketed/fielded systems and subsystems	Recent major weapons, upgrade applications
Budget	Constrained but available for selected technologies	Improved, some major system acquisitions
Implications for OPFOR equipment	Many subsystem upgrades, BLOS weapons, remote sensors, countermeasures	More costly subsystems, recent major weapons, competitive in some areas
Implications for OPFOR tactics and organization; Implications for U.S.	COE tactics with contingency TTP updates. Slight subunit changes add BLOS and AT systems for integrated RISTA and strikes	Integrated RISTA with remotes. Strikes all echelons. Combined arms capacity within small units allows increased lethality and autonomy

The information revolution has also decreased the amount of time military system developers have to acquire a new technology, and either apply it to new systems or adapt it for use with older systems, before it presents a threat to friendly forces (see Equipment Upgrades section). The following technologies, and potential applications of those technologies, will influence R&D efforts as well as decisions related to future force modernization. They will, in turn, play a role in determining which OPFOR capabilities should be portrayed in future training environments.

TECHNOLOGIES AND APPLICATIONS FOR USE BY OPFOR: NEAR AND MID-TERM

TECHNOLOGY CATEGORY	TECHNOLOGY	TECHNOLOGY APPLICATION
Psychological Operations	Mood altering aerosols Reproductive terrorism Non-lethal technologies	Military and civilian targets, for short-term and long-term goals
Information Operations: Sensors	Higher-resolution multispectral satellite images New sensor frequencies for acquisition New sensor frequencies for operational security Use of light spectrum bandwidths (ultraviolet, etc.) Passive detection technologies and modes Auto-tracking for sensors and weapons Image processing and display integration Micro-sensors/imaging system miniaturization Unmanned surveillance, target acquisition/designation Multispectral integrated sensors and Multispectral integrated transmission modes Precision navigation (cm/mm three-dimension) Undersea awareness (sensors, activity) Underground awareness (sensors/mines)	High-intensity use of LITINT (Internet, periodicals, forums) Increased use of information from commercial, industrial, scientific, and military communities Increased adoption of dual-use technologies
Information Operations: Computers and Comms	Low-Probability-of-Intercept communications New power sources and storage technologies: Micro-power generation and Energy cells Advanced Human/Computer Interface Automatic Language Translators	New communities (Blogs, flash mobs, etc., to coordinate and safeguard comms) Secure encryption software New communications tools (Internet, social media and subscriber links)
Electronic Attack	Anti-Satellite weapons for RF, EMP, Hard kill Wide-area weapons (EMP, graphite bombs, etc.) EMP Precision (small area) weapons Computer Network Attack (worms, viruses, Trojan horses) Net-centric warfare (spoofing sensors, spoofing/intercepting data stream/spyware)	Attack electronic grid or nodes at critical times
Chemical/Biological/Radiological Attack	Dirty bombs Genetic/Genomic DNA tagging to assassinate	Agricultural attack (animal and plant stocks and supplies) Use of tagging to eliminate political leaders

TECHNOLOGY CATEGORY	TECHNOLOGY	TECHNOLOGY APPLICATION
	Genetic/Genomic/DNA targeting for Bio attack Designer Drugs/Organisms/Vectors Biologically based chemicals (Mycotoxins) Anti-materiel corrosive agents and organisms	
Physical Attack	Mini-cruise/ballistic missiles for precision, surgical strikes and widespread use Attack UAVs (land, sea, undersea UUVs) Micro-aerial vehicles-widespread use Swarming for coordinated attack Notebook-command semi-autonomous links Vehicle launch for NLOS attack/defense Multi-mode guidance systems: preprogrammed/guided/homing New types of warheads (wider area/different effects; tailorable warhead effects) Precision Munitions: Course-corrected/guided/homing; widespread – almost all weapons; Loiter/IFF Directed Energy Weapons (DEW): Blinding/high energy lasers; RF weapons against electronics; RF against people, vs. structures or systems; Directed acoustic weapons	
Sustainment, Protection	New battery/power cell technologies Neurological performance enhancers Better lightweight body armor Personal actuators, exoskeletons, anti-RF suits Active armor & protection systems Countermeasures to defeat rounds and sensors Counter-precision jammers, especially GNSS All-spectrum low-observable technologies Anti-corrosives Biometric prosthesis & cybernetics Robot-assisted dismounts/sensors/logistics Robotic weapons systems	Battlefield fabrication of spare parts Airborne/shipborne refineries Potable water processing systems Transportable power generation systems

PROJECTED OPFOR CAPABILITIES: NEAR-TERM AND MID-TERM

The next table provides projected OPFOR systems descriptions and capabilities that may confront US and coalition forces in future operating environments. Data for the first timeframe (2015-2020) reflects systems and subsystems already known to exist, and assumes their introduction to the emerging OPFOR adversary force. Timelines reflect capability tiers for systems which may already be fully fielded (not Interim Operational Capability or First Unit Equipped) in brigade and division-level units during the respective time frames indicated.

The systems projections are not comprehensive, and represent shifting forecasts. They may accordingly shift further as we approach the specified time frames. Because clarity diminishes as projections attempt to discern enemy capabilities beyond the turn of the present decade, current views on future trends become less specific for the out-years . Accordingly, the second column (Mid-Term 2021-2025) focuses more on technologies, and less on defined systems.

The columns can be treated as capability tiers for OPFOR operating within specified time frames. Please note: ***No force in the world possesses all systems at their most modern tier.*** The OPFOR, as with all military forces worldwide, is a mix of legacy and modern systems. Thus the emerging OPFOR force comprises a mix of COE time frame Tier 1-4 systems and newer systems. One would expect that some Near- or Mid-term adversaries with lower military technology capabilities could move up one or two capability tiers from (for instance) current COE capability Tier 4, to COE Tier 2. The most likely upgrade for emerging OPFOR used in most training simulations would be to move the OPFOR from COE Tier 2 to Tier 1, selectively adding some new systems that reflect emerging niche technologies.

We have previously stated that an OPFOR can portray a diverse force mix by separating brigades and divisions into different tiers. The OPFOR also has the option of incrementally adding higher tier systems to lower tier units, as selective upgrades. Because most of the systems shown below in the 2015-2020 column are currently fielded, an adversary might also incrementally upgrade COE Tier 1 or 2 units by adding fielded assets from 2015-2020 as described in that column. However, as time progresses through that period, we cannot be sure beforehand when and if all of those technologies will come online. Again, the tables are not predictive. The OPFOR force designer may choose a middle road between current Tier 1-4 and future systems; in many countries they are upgrading legacy and even recent systems to keep pace with state-of-the-art improvements. Thus they may look to subsystem upgrades discussed earlier in this chapter.

If a specialized system for a specific role is missing from the table below, continue to use the OPFOR system noted in Tiers 1-4. Please remember that these projections reflect "possible" technology applications for future systems. They incorporate currently marketed systems and emerging technologies and subsystems, and thus may be combined in innovative ways. The table below is not a product of the U.S. intelligence community, and is not an official U.S. Army forecast of future "threats". It is approved only for use in Army training applications and simulations.

Future OPFOR (2025 and after) is described in various media, but is generally FOUO or classified and is not included in the WEG.

PROJECTED OPFOR CAPABILITIES: NEAR – AND MID – TERM

RECONNAISSANCE, INTELLIGENCE, SURVEILLANCE, TARGET ACQUISITION

SYSTEM	NEAR-TERM OPFOR (2015-2020)	MID-TERM OPFOR (2021-2025)
Smart Dust	Rocket/UAV/aircraft emit signals for ½ hour that neutralize sensors	Scattered dust attaches to metal; acoustic/crush/seismic. Emits 1 hour
Acoustic sensor vehicle	Vehicle mounts microphones or dismount array, DFs/acquires aircraft, vehicles, or artillery. Rapid queuing and netted digital display. Range 10 km, accuracy 200m. Three-vehicle set can locate artillery to 30 km with 1-2% accuracy in 2-45 sec. DF/queuing rate 30 targets per minute.	Range extends to 20-30 km with 10 m accuracy. Micro-UAVs with microphones to supplement the network in difficult terrain. Tracks and engages multiple targets. Hybrid electric/diesel engine.
Ground or Vehicle Launched Mini-UAV and Micro UAV	Dual backpack system. Man-portable ground launcher, and laptop terminal. Vehicle-launch from rail or canisters. TV/FLIR. Range 35 km, 3-hour endurance.	IR auto-tracker. Laser designator. Cassette launcher for vehicles. Signal retransmission terminal. Dispenses micro-UAVs, UGSs mines.
Micro-UAV	Hand-launched 4-rotor, 4kg, 5 km/1 hr, GPS map/view on PDA/netbook. Atk grenade.	< 1kg for dismounted sqd/team, 2 km range. Add grenade for atk UAV.
Airborne (Heliborne) MTI Surveillance Radar	Range 200 km, endurance 4 hrs.	SAR mode added. Range to 400 km.
Commercial Satellite Imagery	Resolution 5 m for IR, SAR also available. < 2 days for request. Terminal on tactical utility vehicle at division. Can be netted to other tactical units.	Response time reduction (to , 6 hours). 1-m resolution.

ANTI-TANK

SYSTEM	NEAR-TERM OPFOR (2015-2020)	MID-TERM OPFOR (2021-2025)
Manpack Air Defense and Antitank (ADAT) Kinetic-Energy Missile Launcher (also listed in Air Defense)	Co/Bn substitute for ATGMs and AD. Targets helicopters and LAVs. Shoulder launched missile with 3 KE LBR submissiles 8 km, 0 m altitude. Submissiles have 25-mm sabot/HE warhead. Nil smoke. Mounted on robotic launcher (below). FLIR night sight.	Fits into 45-100-mm gun-tubes. Defeats all targets up to 135 mm KE. Range 8 km, time-in-flight 6 sec. Fused FLIR/II sight 10 km. Launch from enclosed spaces. Can mount on robotic ADAT launcher or ADAT Robot vehicle (below)
Robotic ADAT Launcher ADAT Robot Vehicle	Pintle mounted, shoulder/ground/ATV/vehicle launched. Robotic launcher-60 m link. Twin auto-tracker. Operator in cover/spider hole. MMW/IR absorbent screen and net for operator, launcher, and surrounding spall. CPS/ATS.	Masted 4-missile, hybrid drive. Self-entrenches, moves to launch point. Fused FLIR/II sight 10 km. Remote link to 10 km. Most AD and AT vehicles have 2 control stations, 2 robots. ATGM is SAB. CPS.
Attack UAV	Hit-to-kill system. Day/night 60+ km, up to 2 hours. GNSS/inertial	Cargo UAV 100 km dispenses IR/MMW/SAL DP (600-mm HEAT)

SYSTEM	NEAR-TERM OPFOR (2015-2020)	MID-TERM OPFOR (2021-2025)
	navigation, TV/FLIR, Frag-HE warhead. They include an anti-radiation variant.	submunitions, EMP munitions, SAL ATGMs – UAV LTD 30 km.
Attack UAV Launcher Vehicle	Hit-to-kill UAV launched from modular launcher, 18 UAVs. GPS/inertial nav, to 500 km. First version anti-radiation homing. Added TV guided and multi-seeker attack (hit-to-kill) UAV. Laser designator range 15 km. CPS/ATS.	Hybrid drive. Bus reusable UCAV with 4 ATGMs to 10 km, SAL-H bombs, or bus dispensing 16 terminally-homing submunitions (with MMW/IR seekers, or laser-homing DP submunitions). CPS. LTD.
Micro-Attack UAV	Hand or canister-launched UAV with TV and FLIR guidance to 10 km, 100-600 m altitude, with .25-.5 kg warhead.	Cassette/smoke grenade launcher launch for tactical vehicles. Recon and attack (top-attack) UAVs.
Mini-Attack UAV	Hand or vehicle canister-launched UAV with TV and FLIR guidance to 35 km, 100-600 m altitude, 1-4 kg warhead.	Cassette launcher launch for tactical vehicles. Recon and attack (DP with tandem 600 mm top-attack).

ENGINEER

SYSTEM	NEAR-TERM OPFOR (2015-2020)	MID-TERM OPFOR (2021-2025)
Scatterable Mines	Deliver by artillery, cruise missile, UAV, rotary or fixed-wing aircraft. Non-metallic case, undetectable fill, resistant to EMP and jammers, with self-destruct.	Advanced multi-sensor mines with wake-up and target discrimination. Prox fuze mines. Controlled mine-fields and intelligent mines.
Off-Route Mines (Side-Attack and Top-Attack)	Autonomous weapons that attack vehicles from the side as the vehicles pass. 125-mm Tandem HEAT (900+ mm). Target speed 30-60 km/h, range 150m, acoustic and infrared sensors.	Sensor-fuzed EFP 600mm KE top attack. Remote or sensor-activated (controller turn-on/off), 360-degree multi-sensor array. Hand/heli/UAV/arty/ATGL mortar emplace.
Controlled Mines and Minefields	AT/AP, machine emplace able. Armed, disarmed, detonated by RF command. Chemical fills and non-metallic cases are undetectable. With CM and shielding, can negate jammers/pre-detonating systems.	Control may be autonomous, based on sensor data and programmed into decision logic, or by operators monitoring with remote nets.
Smart Mines	Wide-area munitions (WAM) smart autonomous, GNSS, seismic/acoustic sensors. AT/AV top-attack, stand-off mine. Lethal radius of 100 m, 360 degrees. Hand-emplaced.	Can discriminate among targets. Reports data to monitor, evaluates target paths, built-in logic. Uses GNSS to arty/heli-emplace. Non-nuclear EMP or HPW options.

INFORMATION WARFARE

SYSTEM	NEAR-TERM OPFOR (2015-2020)	MID-TERM OPFOR (2021-2025)
Electronic Warfare Radio Intercept/DF/Jammer System, VHF	Intercepts, DF, tracks & jams FH; identifies 3 nets in non-orthogonal FH, simultaneously jams 3 fixed	Integrated intercept/DF/jam for HF/VHF/UHF

SYSTEM	NEAR-TERM OPFOR (2015-2020)	MID-TERM OPFOR (2021-2025)
	freq stations (Rotary/fixed wing/UAV capable)	
Radio Intercept DF/HF/VHF/UHF	Intercept freq range 0.1-1000 MHz. (Rotary/fixed wing/UAV capable)	Wider Freq coverage. SATCOM intercept. Fusion/cue wuth other RISTA for for target location/ID
Radio HF/VHF/UHF Jammer	One of three bandwidths; 1.5-30/20-90/100-400 MHz, intercept and jam. Power is 1000W. (Rotary/fixed wing/UAV capable)	Increased capability against advanced signal modulations. UAV abd mini-UAV Jammers
Portable Radar Jammer	Power 1100-2500W. Jams airborne SLAR 40-60 km, nav and terrain radars 30-50 km. Helicopter, manpack.	UAV and long range fixed wing jammers
High-Power Radar Jammer	Set of four trucks with 1250-2500 watt jammers at 8,000-10,000 MHz. Jams fire control radars at 30-150 km, and detects to 150 km.	UAV jammer and airship jammer. Hybrid electric/diesel drive.
Portable GPS jammer	4-25 W power, 200 km radius. Man-portable, vehicle & airborne GPS jammers, airship-mounted jammers.	Man-portable, vehicle & airborne (UAV) GPS jammers-increased range and power, and improvements in antenna design.
Missile and UAV-delivered EMP Munition	Cruise missiles and ballistic missile unitary warhead and submunition	Increased capability against advanced signal modulations
Cruise Missile Graphite Munitions and Aircraft "Blackout Bombs"	400-500 kg cluster bombs/warheads with graphite strands to short-out transmission stations and power grids	Rocket precision and UAV-delivered munitions

COMMAND AND CONTROL

SYSTEM	NEAR-TERM OPFOR (2015-2020)	MID-TERM OPFOR (2021-2025)
Radio, VHF/FM, Frequency-hopping	30-88 MHz, 100 hps, channels: 2,300; mix of analog and digital radios, tactical cellular/digital phone, all nets digitally encrypted. Burst trans. UAV Retrans	Digital radios, tactical cellular/digital phone, and satellite phones, all nets encrypted

DECEPTION AND COUNTERMEASURE SYSTEMS

SYSTEM	NEAR-TERM OPFOR (2015-2020)	MID-TERM OPFOR (2021-2025)
Air Defense System Decoy	Manufactured and improvised decoys used with decoy emitter. Covered by AD systems in air defense ambushes	Multispectral simulators of varied gun and missile systems mounted on robotic chassis
Air Defense System Decoy RF Emitter	Expendable RF remote emitters with signal to match specific nearby radars, to trigger aircraft self-protection jammers	Mounted on robotic chassis

ROTARY WING AIRCRAFT

SYSTEM	NEAR-TERM OPFOR (2015-2020)	MID-TERM OPFOR (FY 2021-2025)
Attack Helicopter	30-mm auto-cannnon, 8 NLOS FOG/IIR-homing ATGMs, range 8 km. Two pods semi-active laser homing (SAL-H) rockets 80mm (20x 8 km) or 122mm (5x 9 km). 2x LBR KE ADAT msl (warhead w/3 KE sub-missiles, 8 km range). Laser designator 15 km. UAVs to 30 km. 2nd gen FLIR auto-tracker. Radar and IR warners and jammers, chaff, flares	Tandem cockpit, coax rotor, 30-mm auto-cannon. 8 x RF/SAL-H ASMs to 40 km (28+kg HE=1300+mm), 2x SAL-H rocket pods (80mm or 122mm), 2 ADAT KE msl 8 km, and 2x MANPADs. 1/3 have ASM to 100 km. Fire control fused II/FLIR to 30 km, and MMW radar, link to ground LTD. Radar jammer. Atk and LTD UAVs to 30 km
Multi-role Medium Helicopter and Gunship	24 troops or 5000kg internal. Medium transport helicopter. Range 460 km. 30-mm auto-cannon, 8 FOG-M/IIR ATGMs to 8 km, 40 x 80 mm laser-homing rockets, 4 AAMs. ATGM launchers can launch mini-UAVs and more AAMs. Mine pod option. Day/night FLIR FCS	Fused FLIR/II to 15 km. 6x SAL-H ATGMs 18 km, 2 AAMs, 2 x 80/122-mm SAL-H rocket pods (20 or 5 ea). Laser designator to 15 km, and link to ground LTD. Aircraft survivability equipment (radar jammers and IR countermeasures).
Multi-role Helicopter and Gunship	12 troops (Load 400 kg internal, 1,600 external). Range 860 km. 23 mm cannon, 2 AAM, 4 SACLOS ATGMs to 13 km, TV/FLIR, day/night. Mine delivery pods	Launches 6x SAL-ATGM to 18 km, 28+kg HE warhead. 2 x AAM, Air-to-surface missile to 100 km. Pod w/7x SAL-H 90-mm rockets. Fused FLIR/II to 15 km. ASE
Light Helicopter and Gunship	3 troops (Load 750 kg internal, 700 external). Range 735 km. 20 mm cannon, 1 x 7.62mm MG, 6 SAL-H ATGMs to 13 km, 2 AAMs. FLIR night sight. Laser target designator. Mine pods	4xSAL-H ATGMs, 18 km range. Fused FLIR/II to 15 km
Helicopter and Fixed-Wing Aircraft Mine Delivery System	Light helicopter pod scatters 60-80 AT mines or 100-120 AP mines per sortie. Medium helicopter or FW aircraft scatters 100-140 AT mines or 200-220 AP mines per sortie.	Controllable and intelligent mines for aircraft delivery. Larger aircraft can hold multiple pods.

FIXED WING AIRCRAFT

SYSTEM	NEAR-TERM OPFOR (2015-2020)	MID-TERM OPFOR (2021-2025)
Intercept FW Aircraft	30-mm auto-gun, AAM, ASM, ARMs TV/laser guided bomb. 8 pylons Range 3,300 km. Max attack speed: Mach 4.	Stealth composite. ASE. Max G12+ All weather day/night. Unmanned option.
Multi-Role Aircraft	30-mm gun, AAM, ASM, ARM pods, guided, GNSS, sensor fuzed bombs, 14 hardpoints. Thrust vectoring. FLIR.	Improved weapons, munitions. Unmanned option. ASE all radars. Max G12+ All weather day/night.
Ground-Attack Aircraft	Twin 30-mm gun, 8 x laser ATGMs 16 km 32 kg HE, 40 SAL-H 80mm rockets, ASMs, SAL-H and GNSS sensor fuzed bombs, AA-10 and KE	Stealth composite design. ASE. Unmanned option. Max G12+ 80-mm/122-mmrockets SAL-H, SAL-H ASM (28+kg HE=1300+ mm, to 40

	HVM AAM. 10 hardpoints.Range 500+km. FLIR.	km, 2 gen FLIR, radar jammer, day/night.

OTHER MANNED AERIAL SYSTEMS

SYSTEM	NEAR-TERM OPFOR (2015-2020)	MID-TERM OPFOR (2021-2025)
High-altitude Precision Parachute and Ram-air Parachutes	High-altitude used with oxygen tanks. Ram-air parachute includes powered parachute with prop engine.	Increased range and portability. Reduced signature. Increased payload.
Ultra-light Aircraft	Two-seat craft with 7.62-mm MG, and radio. Folds for carry, 2 per trailer.	Rotary-winged, two-seat, MG, 1/trailer. Auto-gyro, more payload.

UNMANNED AERIAL VEHICLES

SYSTEM	NEAR-TERM OPFOR (2015-2020)	MID-TERM OPFOR (2021-2025)
UAV (Brigade) It may also be employed in other units (e.g. artillery, AT missile, and naval)	Rotary wing, TV/FLIR/auto-tracker, with LRF and LTD acquires targets to 15 km. Flies 180 km/6 hours, 220 km/hr, 2-5,500 m alt, 100kg payload. Can carry 2 AD/anti-armor missiles +MG for attack.	Range extends to 250 km. Increased payload. Attack version can carry 2 SAL-H ATGMs (12 km range) or 1+4 70-mmSAL-H rockets (7 km, defeats 200 mm).
UAV (Divisional)	Day/night recon to 250 km. GNSS/inertial nav, digital links, retrains. SLAR. SAR, IR scanner, TV, ELINT, ECM suite, jammer/mine dispensers. Laser designator 15 km.	Increased range, endurance. Diff GNSS. Composite materials, lower signature engine. SATCOM Retrans/relay links. Attack sub-munitions.
UAV (Operational)	Day/night recon to 400+km. GNSS/inertial nav with digital links. SLAR, SAR, TV, IR scanner, ELINT, ECM suite. Jammer option. Mine dispensers. Laser target designator 15 km. Retrans/relay.	Increased ranges, endurance. Diff GNSS. High altitude ceiling- 35 km option. Retrans/relay/SATCOM links. UAV attack sub-munitions. Laser target designators.
Unmanned Combat Aerial Vehicle (on Operational UAV platform)	Medium UAV with 4 ATGMs (range 10 km), laser-guided bombs. Laser designator 15 km. Mine dispensers. GNSS jammer, EW jammers. Range 400+ km.	Stealth composite design. ASE. Twin dispensers (pylons) with 16 terminally-homing sub-munitions, MMW/IR seekers. Range 500+ km.

THEATER MISSILES

SYSTEM	NEAR-TERM OPFOR (2015-2020)	MID-TERM OPFOR (2021-2025)
Short-Range Ballistic Missile and Cruise Missile Launcher	Twin launch autonomous vehicle (GNSS/inertial nav, self-emplace and launch). Range 450 km. Non-ballistic launch, separating GPS corrected reentry vehicle (RV) with decoys, CCD, 10-m accuracy. ICM, cluster, nukes. EMP warhead. EMP warhead. Some convert to 6-Cruise missile launch capability (500 km, 3-m accuracy, below radar). Vehicle decoys. Vehicle has visual/MMW/IR signature of a truck	Improved missile range (TBM 800 km, cruise 1,000) with 1-m accuracy. TBM has GNSS-corrected maneuvering RV. Warheads for both: terminal-homing sub-munitions, precision cluster munitions, EMP. Cruise missiles pre-program or enroute waypoint changes. Countermeasures include penaid jammers.
Medium-Range Ballistic Missile	Autonomous vehicle. Separating maneuvering warhead to 1300 km. GNSS 10-m CEP. Warheads: ICM, cluster, EMP, and nukes. Penaids include decoys, jammers. Truck visual/MMW/IR signature.	Range 2,300 m, 1-m CEP, Diff GNSS, terminal homing, separating warhead. Warheads include EMP, terminal-homing cluster munitions. Non-ballistic launch and trajectory.
Cruise Missile Cassette launcher vehicle	Off-road truck, GNSS for autonomous ops. 16/lchr. Range 470 km; preprogram GNSS inertial guidance, with in-course correction, 10 CEP. Munitions include cluster, chemical, thermobaric, DPICM, and scatterable mine sub-munition	Launcher fire direction. Supersonic missile Diff GNSS/inertial nav, 1-m CEP. Range 900km. EMP warhead option. Warheads include homing cluster munitions. Penetration aids-countermeasures.
Cruise Missile/AD Missile (Multi-role) Launcher Vehicle Category includes specialized cruise missiles, long-range ATGMs, and SAM systems to engage targets at 12+ km.	Truck with 24 launchers. Range 100 km. 28-kg Frag-HE warhead = 1300 mm. AT Preprogrammed GNSS/inertial nav phase. LTD veh range 25 km. Thermal camera to 10 km. Radar 40 km. Support UAV with LTD. FW/ship/anti-ship versions. Anti-heli RF guided MMW radar.	Penetration aids (countermeasures). IR Terminal-homing warhead or IR-homing sub-munitions can be used. MMW lock-on before/after launch.
Land-attack SAM system (secondary role for system)	The SAM system uses its EO sight and LRF (short/med range, strat "hittiles").	Range extends with SAM ranges. Passive operation with TV/FLIR.

AIR DEFENSE

SYSTEM	NEAR-TERM OPFOR (2015-2020)	MID-TERM OPFOR (2021-2025)
General Purpose and Air Defense Machinegun	12.7 mm low recoil for ground tripod. Chain gun light strike vehicle, ATV, motorcycle, etc., on pintle. TUV/LAV use RWS. Remotely operated ground or robot option. Frangible rd 2 km, sabot 2.5 km. RAM/RAP/IR camouflage/screens. TV/FLIR fire control. Lightweight MMW radar 5 km. Display link to AD azimuth warning net. Emplace 10 sec. RF/radar DF set. ATS control option.	Stabilized gun and sights. Remotely operated computer FCS with PDA/laptop. Fused II/FLIR 5 km. Frangible, sabot rds to 3 km. Laser dazzler blinds sights. Robot mount and micro-recon/heli atk UAVs. Some light/AD vehicles replace gun with 30-mm recoilless chain gun on RWS firing a HEAT round 4 km; add-on ADAT missile launcher.
Improvised Multi-role Man-portable Rocket Launcher (AD/Anti-armor)	4-tube 57-mm launcher with high-velocity dual-purpose rockets. EO day/night sight. Blast shield. Range 1,000 m. Penetration 300 mm, 10 m radius.	Prox fuze, 1,500 m range. Penetration 400 mm, 20 m radius.
Man-portable SAM launcher	6 km day/night range/ 0-3.5 km altitude all aircraft, velocity mach 2.6. Thermal night sight. Proximity fuze, frangible rod warhead (for 90% prob hit and kill). Approach/azimuth link to AD warning net. Twin launcher vehicle quick mount. Nil smoke. Mount on robotic AD/AT launcher. RF/radar DF set on helmet.	Warhead/lethal radius increases for air/ground targets. Improved seekers – cannot be decoyed by IR decoys/jammers. Fused II/FLIR 10 km. Launch from enclosed spaces. Laser dazzler. Optional AD/AT LBR KE warhead missile – 8 km. Mounted on AD/AT robot vehicle.
MANPADS Vehicle Conversion Kit (Lt, Stryker, vans, recon TUV, truck etc.)	Twin launcher and ADMG on improvised IR SAM vehicle. Day/night IR auto-track FCS, MMW radar. Display link AD net. RF/radar DF set to 25 km. Camouflaged.	Launcher replaced with 3-missile launcher: 2x ADAT KE SAMs, 1x IR SAMs. Total 6 missiles (3 & 3)
Manpack Air Defense and Antitank (ADAT) Kinetic-Energy Missile Launcher (also listed in Anti-tank)	At company/battalion, can replace ATGMs and SAMs. Targets helicopters and LAVs. Missile has 3 KE LBR darts (sub-missiles), 8 km, 0 m altitude. Camo screen. Dart is 25-mm sabot with HE sleeve. Nil smoke. Fits on robotic ADAT launcher. Helmet RF/radar DF.	Larger sabot kills all targets up to 200 mm (KE) armor. Range 8 km, time of flight 5 sec. Fused II/FLIR 10 km. Launched from enclosed spaces. Can mount on 3x remote launcher w/ IR auto-tracker, which fits on AD/AT robotic vehicle.
Towed/Portaged/Vehicle Mounted AA Short Range gun/missile system	2x23mm gun. MMW/IR camouflage/screen. Frangible round, range 3,000 m (17mm pen). Onboard radar/TV FC with ballistic computer, 5 km MMW radar, thermal night sight, auto-tracker, net azimuth warner. Twin MANPADS.	Replaced with twin 30mm recoilless chain gun. Frangible, sabot, AHEAD rnds to 4 km. TV/fused II/FLIR auto-tracker 10 km. MMW radar, twin MANPADS/ADAT KE missile (8 km) launcher. APU for self-relocating or robotic mount. Laser dazzler.

SYSTEM	NEAR-TERM OPFOR (2015-2020)	MID-TERM OPFOR (2021-2025)
	RF/radar DF set, 25 km. RWS on veh hull/turret. CPS/ATS.	
Air Defense System Decoys (visual decoy, decoy emitter)	See DECEPTION & COUNTERMEASURE SYSTEMS	
Brigade gun/missile turret for mount on tracked mech IFV, wheeled mech APC, truck (motorized) chassis	Twin 30-mm gun, APFSDS/frangible rds, 4 km. 30-mm buckshot rd for UAVs. Mounts 4x hyper-velocity LBR-guided SAMs to 8 km, 0m min altitude. Passive IR auto-tracker, FLIR, MMW RADAR. 2 per battalion. Track/launch on the move. Targets: air, LAVs, other ground. RF/radar DF set 25 km range. CPS/ATS.	Dual mode (LBR/radar guided) high velocity missile, 12 km, 0m min altitude. Auto-tracker (launch/fire on move). Phased array radars. Fused II/FLIR 19 km. Twin 30-mm recoilless chain gun with AHEAD type rds to 4 km. Micro recon/heli atk UAVs. TV/IR attack grenades.
Divisional gun/missile system on tracked mech IFV, wheeled mech APC, truck (motorized) chassis	Target tracking radar 24 km. TV/FLIR. 8x radar/EO FCS high velocity missiles to 18 km/12 at 0 m min altitude. Auto-track and IR or RF guided. 2 twin 30 mm guns to 4 km. 30-mm buckshot rd for UAVs. RF/radar DF. CPS/ATS.	Hybrid drive. Missile 18 km at 0 m, and kill LAVs. Fused II/FLIR auto-tracker, launch on move. Radar 80 km. Home on jam. Twin 30-mm recoilless chain gun, electronic fuzed air-burst rds to 4 km. Micro-recon/heli-atk UAVs. TV/IR atk grenades.
APC Air Defense/AT Vehicle in APC Bn (Company Command Vehicle, MANPADS Vehicle in Bn/Bde)	1-man turret on 8x8 chassis. 30 mm gun, 30-mm buckshot rd for UAVs. 100-X TV 2 gen FLIR. 2x LBR ATGM lchrs 6 km, 2x veh MANPADS lchrs. 2 dismount teams. 1xMANPADS lchr, 1xADAT KE lchr. Total 18 missiles. 12.7-mm MG. RF/radar DF to 25 km. CPS/ATS.	10x10 whld hybrid drive, box armor. 30-mm recoilless gun RWS. Add AHEAD-type 4 km, 2 veh launchers for 5 AD/AT KE LBR HV SAM 8 km. Anti-helicopter surveillance/atk micro-UAVs. Fused II FLIR 10 km. MMW radar. TV/IR atk grenades.
IFV, HIFV, or Tank ADAT Vehicle in Bn/Bde MANPADS	Vehicle on IFV, HIFV, or tank chassis with above features and weapons.	See AIR DEFENSE, APC ADAT above for weapons and upgrades.
Towed Medium Range AA gun/missile system	35mm revolving gun 1,000 rd/min. Rds: frangible, HE prox, electronic-fuzed. 4 SAMs/lchr, 45 km, 0 m min alt. Radar 45 km, 4 tgts. Resists all ECM. 2 gen FLIR auto-tracker 20 km. RF/radar DF 25 km. SAM includes active homing, home-on-jam. RAP/RAM/IR camo. CPS/ATS.	Hybrid-drive auxiliary power unit short moves. Improved FCS, radars phased array low probability of intercept acq to 80 km. Fused II/3rd gen FLIR auto-tracker to 35 km in day/night all-weather system. Track and engage 8 targets per radar.
Medium-range ground SAM system	Tracked lchr. Radar to 150 km. 4x radar-homing SAMs to 45 km, 0 m min altitude (4 targets at a time). Home on jam. Use as cruise missile – priority ground tgts to 15 km, water 25km. Fused 3rd gen FLIR auto-track. RF/radar DF. CPS/ATS.	Hybrid drive. Improved FCS with radars and EO, fused II/3rd gen FLIR day/night all-weather system to range 50 km. Radar range 200 km.
Strategic SAM System	Cross-country truck launchers, 1 x track-via-missile SAMs 400 km, at Mach 7. 1x ATBM/high maneuver missile to 200 km. Also 8 x "hittile" SAMs to 120 km. Modes are track-via-missile and ARM (home-on-jam). All missiles 0 m 50 50 km	Off-road trucks or tracked with hybrid drive. Most units, launchers have 2 big missiles+8 small "hittile" missiles ranging 200 km, altitude 0 m – 50 km. All missiles Mach 7. OTH radars operate on the move 600 km range. Targets include all

SYSTEM	NEAR-TERM OPFOR (2015-2020)	MID-TERM OPFOR (2021-2025)
	altitude vs stealth aircraft, UAVs, and SAMs. All strat/op missiles in IADS. Local IADS all AD. Battery autonomous option. Over-the-horizon (OTH) TA radar vehicle to 400 km. Mobile radar 350 km. Site CM, decoys.	IRBMs. Increased target handling capacity (100/battery in autonomous operations).
Operational-Strategic SAM System	Same as above on tracked chassis. Mobile FOs all batteries. AD radars on airships.	Same as above on tracked chassis
Anti-helicopter Mines (Remote and Precision Launch)	In blind zones force helos upward or deny helo hides and landing zones. Range 150m. Acoustic and IR fuse, acoustic wake-up, or cmd detonation. Directed fragmentation. Precision-launch mines use operator remote launch, proximity fuze for detonation. RF/radar DF.	Stand-alone multi-fuse systems. Remote actuated hand-emplaced mines with 360-degree multi-sensor array, pivoting/orienting launcher, 4-km IR-homing missile. Operator monitors targets and controls (turns on or off) sections, mines or net.
Helicopter Acoustic Detection System	Early warning of helicopters. Acoustic sensors to 10km, 200m CEP. IR sensors can also be linked to air defense net.	Range 20 km, 50 m CEP. Track and engage multiple targets. Digital link to AD net, AD unit, IADS.

MILITARY TECHNOLOGY TRENDS FOR VOLUME 2 SYSTEMS IN 2025

Year 2025 is a demarcation line for focusing on future military technologies. Even with the "Revolution in Military Affairs", most major technology developments are evolutionary, requiring one or more decades for full development. Most of the technologies noted below are in conceptual or early developmental stages, or fielded at this time. Many exist in limited military or commercial applications, and can be easily extrapolated to 2025 and the near future time frame. Throughout this period and beyond, military forces will see some legacy systems become obsolete, then either be replaced, or relegated to lesser roles with lower priorities. Most will be retained and updated several times. New technologies will emerge and be widely adopted, only to be overtaken by still more modern technologies that will drive OPFOR modernization. Additional technologies/adaptations not yet conceived will surface with little warning, be quickly adopted, and significantly impact OPFOR force structure.

SENSORS

- Multi-spectral immediate all-weather sensor transmission with real-time display
- Remote unmanned sensors, weapon-launch and robotic sensors and manned sensors
- Sensor nets integrated and netted from team to strategic and across functional areas

AIRCRAFT

- Continued but selective use of FW and rotary wing for stand-off weapons, sensors
- Aircraft critical for transport, minelaying, jamming, other support missions
- Light aircraft and UAVs adapted with multi-sensor pods for real-time fused intelligence and laser target designation

OTHER AERIAL SYSTEMS

- High-altitude UAVs, long-endurance UAVs, and UCAVs seamlessly integrated with other intelligence and support systems
- Recon/attack low-signature UAVs and UCAVs and stand-off munitions at all levels down to squads
- Ballistic missiles with non-ballistic trajectories, improved GNSS/homing re-entry vehicles, precision submunitions, EMP
- Shift to canister launchers of tactical cruise missiles with precision homing and piloted option, cluster warheads, EMP
- Airships and powered airships for long-endurance and long-range reconnaissance, and variety of other roles

- Increased use of ultra-lights and powered parachutes

AIR DEFENSE

- Integrated Air Defense System with day/night all-weather RISTA access for all AD units
- Improved gun rounds (AHEAD/guided sabot) and missiles (anti-radiation homing, jam-resistant)
- Autonomous operation with signature suppression, counter-SEAD radars and comms
- Shoulder-launch multi-role (ADAT) hypervelocity missiles/weapons immune to helicopter decoys and jammers,
- UAVs and airships for multi-role use includes air defense recon and helicopter attack
- Acquisition/destruction of stealth systems and aerial munitions and ground rockets to 500+ km

INFORMATION WARFARE

- Jammer rounds most weapons, electro-magnetic pulse rounds, weapons of mass effects
- UAVs, missiles and robots carry or deliver jammers/EMP/against point targets and for mass effects
- Multi-spectral decoys for most warfighting functions
- Computer network attack and data manipulation

ACCESS DENIAL

- Use of nuclear/bacteriological/chemical weapons to deny entry, access to areas or resources
- Use of media and public opinion for access denial
- Remotely delivered RF-controlled, smart and sensor-fuzed mines and IEDs defeat jamming

NON-LETHAL WEAPONS

- EMP/graphite/directed energy weapons to degrade power grid, information networks, and military systems
- Space-based data manipulation to deny adversary use of satellite systems
- Population control effects (acoustic devices, bio-chemical and genetic weapons, resources attack, dirty bomb)
- Anti-materiel agents and organisms (microbes, chemicals, dust, and nanotech)
- Countermeasures, tactical and technical, in all units to degrade enemy sensor and weapon effectiveness.

Jim Bird

DSN: 552-7919/Commercial (913) 684-7919

e-mail address: james.r.bird.ctr@mail.mil

Worldwide Equipment Guide

Chapter 5: Unconventional and SPF Aerial Systems

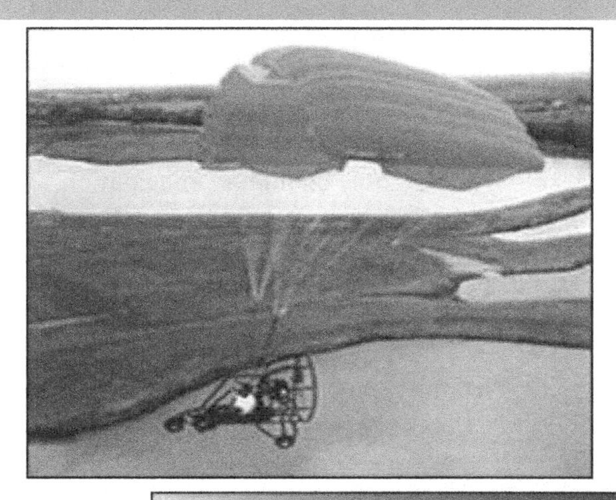

Distribution Statement: Approved for public release; distribution is unlimited.

Chapter 5: Unconventional and SPF Aerial Systems

Chapter 5 includes information on unconventional and SPF aerial systems.

The conflict spectrum in the Contemporary Operational Environment includes forces across the capability spectrum. They will use specially-designed military technologies, as well as improvised weapons and other systems. They will also employ all available assets for innovative applications.

That creativity will also extend into the vertical dimension. Increasingly, as modern forces are able to gain air superiority, adversaries will seek innovative ways to deny airspace, while operating in that airspace. They will increasingly turn to innovative and improvised systems. Aerial roles will include reconnaissance for ground forces and for air defense and air attack.

Improvised air and ground systems will also be used for air defense. Creativity in air defense includes decoy and camouflage arrangements. The threat from rotary-wing aircraft has led to responses such as obstacle systems in likely landing zones, use of mines, and improvised explosive devices (IEDs). New technologies such as unmanned aerial vehicles (UAVs) can be used in counter-helicopter roles. The list of improvised weapons available is limited only by human imagination.

Mrs. Jennifer Dunn

DSN: 552-7962 Commercial (913) 684-7962

E-mail address: jennifer.v.dunn.civ@mail.mil

Airships in Military Applications

Airships ("lighter-than-air" craft) have been used in warfare since the 1800s, when balloons offered elevated platforms for military observers. Airships are increasingly used in civilian venues and offer capabilities for military use. Primary roles are:

- Communication support
- Support to electronic warfare (EW) and artillery units
- Surveillance platforms
- Air defense support

With their low cost, low upkeep, commercial availability, and ability to stay aloft with minimal signature for substantial periods, they will offer more and wider uses for military forces.

Airships can be categorized as non-rigid, semi-rigid, and rigid. Non-rigid describes balloons and blimps. ***Balloons*** can be of various shapes but without internal structure except air pockets for shaping. Most are round. ***Blimps*** (see right) generally fit the characteristic shape. Blimps can orient better in wind than round craft. Airships which are moored to a winch on the ground or on a vehicle are also ***aerostats***. Semi-rigid airships have some struts or framing, but use inflation to fill part of the structure. Rigid airships have their overall structure supported with framing. Some aerostats, especially larger ones, are semi-rigid or rigid. ***Dirigibles*** are airships powered by electric or internal combustion engines, and are rigid or semi-rigid. Their max speed varies up to70 km/hr. ***Zeppelins*** are special-designed airships trademarked by a German company.

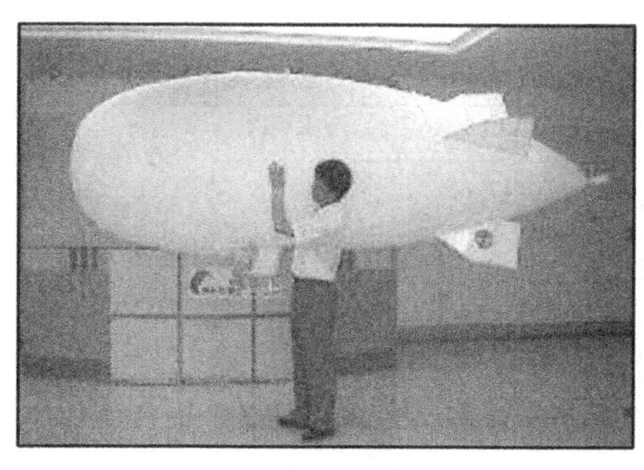

Airships come in various shapes and sizes. They are made of varied materials, mostly PVC or UV-treated nylon. Wind speed should not exceed 25-35 km/hr during flight. Although they can be filled with hydrogen, hot air, etc., the vast majority use helium. Helium can be produced by generators in ground stations or in trailers, compressed in tanks, and distributed to airship users. Helium tanks will sustain a small airship for days. Most airships can absorb several hits while remaining aloft. Most rips and bullet holes can be easily and quickly repaired. An electric hoist can be vehicle-mounted for stationary launch, frequent relocation, and re-launch.

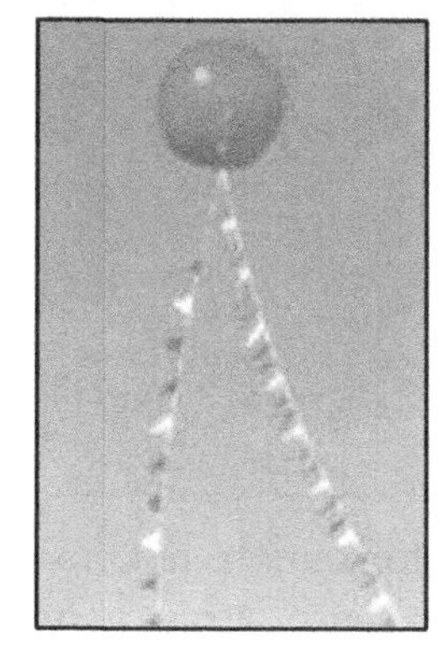

Support to Communications. Balloons can be used in a manner similar to ancient use of pennants and mirrors, to passively signal change in conditions or start an action, while avoiding intelligence and jamming systems. Commercial users often use balloons to trail streamers behind or stretched to the ground to draw attention and mark location of an activity. They can mark location of an LZ, flight corridor, or a registration point for navigation or fires.

Balloons can be used for rescue missions. The below helikites are offered for military uses. A jungle backpack includes aerostat, valve, helium bottle, line, handle, strobe light, bag, and instructions.

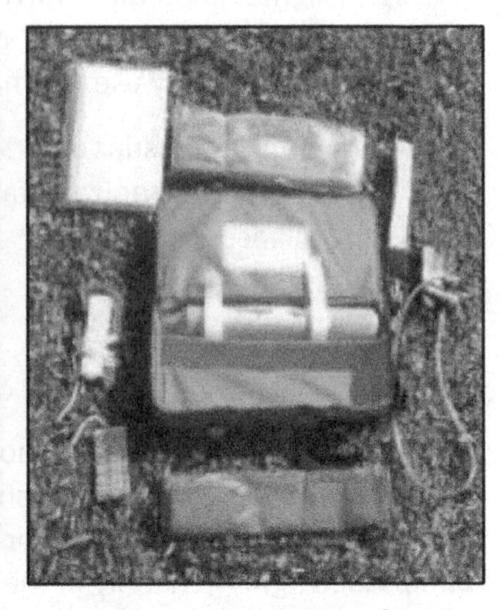

Some signal intelligence and communications units have the option of using aerostats to raise antennae for increased operating range. British Allsopp developed the Mobile Adhoc Radio Network (MANET), with three steerable Low Visibility Skyhook Helikites bearing ITT Spearnet radios to 65-m height. They demonstrated that an infantry radio, usually limited to 1 km range, can send video data (with a 15 kg helikite backpack) to a receiver 10 km away. The set can also be used to retransmit, or to control UAVs in almost any terrain. The company claims that antenna altitude could rise up 500 m.

Electronic warfare units can use aerostats to raise antennae on jammers and recon systems. A simple method would be to attach a jammer round on a cable. A GPS jammer could be mounted on a vehicle-based aerostat or on a dirigible moving within protected zones. Artillery units have long used weather balloons in meteorological units to supply data for calculating fire adjustments. Those units also have helium generators for supplying the gas.

The most widely-used role for airships is reconnaissance. In the U.S. Civil War, balloon gondolas were used by some military observers. Today some military and civilian forces use large aerostat balloons with cameras for border and aerial surveillance. Some sporting events use blimps and dirigibles to feed TV imagery for real-time broadcast. Survey, engineering, and land use organizations also use airship sensor products. The elevated view offers a long-range unobstructed field of view, and extended viewing duration. With the proliferation of small and medium-size commercial balloons, stabilized and gimbaled sensor mounts, and smaller high-resolution optical systems, use of improvised systems is expanding.

Technologies developed for commercial and recreational video-photography, and for remote military sensors and robot systems can be readily adapted to airships. Thus airship-mounted sensor arrays vary

from a simple camera or camcorder hung underneath to day/thermal video-camera or TV transmitting real-time to a palm pilot or laptop, or over a digital net. Gondolas can have a camera bar, stabilized mount, or even a gimbaled sensor ball with multiple sensors, laser-rangefinder (LRF), auto-track, and 60+ power digital/optical zoom. Navigation can include GPS location, ground-based location with a LRF, or inexpensive in-viewer display.

The easiest and most numerous applications would be to attach a camera or camcorder underneath. On page 7-7 is a demonstrated sensor set for RC aircraft. It can be mounted on aerostat balloons less than 1-m for quick over-the-hill surveillance. A separate cord can be attached to the camera or balloon to orient it in the desired direction.

Controllable Camera Mount

Mount on a Camera Bar

Gimbaled Ball

Manufacturers such as Inflateable4less offer small aerostat blimps (3-m, below) which can carry a camera. Range for an HF transmitter can limit distance to a ground station (2 km for a low-cost unit); but a hand-held display unit can operate from a vehicle.

Mini-zepp blimps come in sizes 6-13 m, for use as aerostats or as dirigibles. The dirigibles include 2 electric motors and a gas-powered motor. Options include a video head and HF transmission system. In event of a power failure, a cable drops to the ground for recovery.

The Skymedia Pro aerostat system is offered for $4,999. It includes:

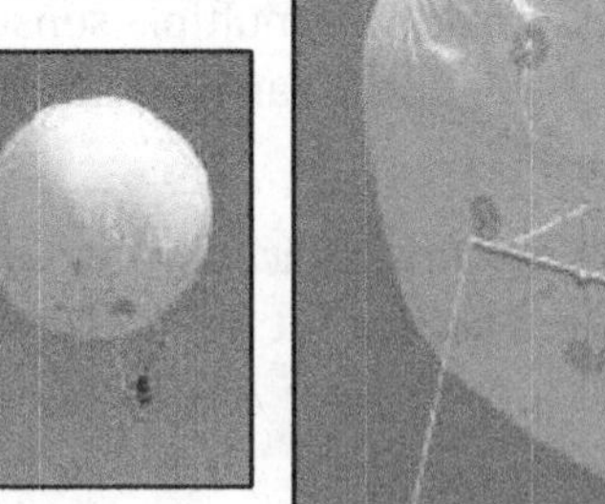

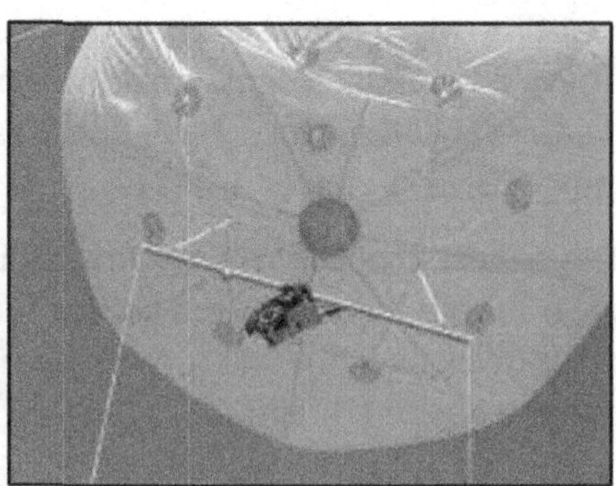

- 2.4-m urethane-coated ripstop nylon balloon
- Highly precise camera bar (210° tilt pan, 2 x 360° zoom shoot carbon fiber camera platform)
- HF transmitter on the platform (2 km range)
- A remote control unit, system integrated (HF receiver patch antenna LCD color display 13 cm) -- a suitcase with all necessary chargers, battery, etc.

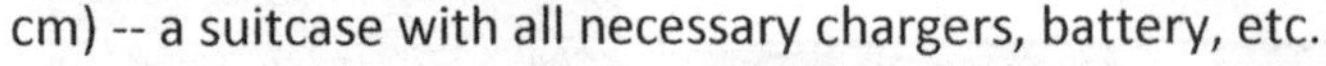

As airships become better-controlled and more stable, other sensors can be added to the payload. An airship could be used in tactical reconnaissance units to mount a small light-weight radar antenna, such as on the FARA-1E (Vol 1, pg 4-29). The Israeli Speed-A stabilized payload system with automated EO/thermal imager and laser rangefinder fits on lightweight airships.

Air defense units will use airships in above roles. Airship antennae can extend the range of tactical AD radio nets. Airship-mounted camera systems can detect helicopters flying at low altitudes (using forest canopy for cover) earlier than their ground-based counterparts. Airships could also raise a cordon of light-weight radar antennae over obscured approaches for early detection of helicopters and other threats.

Another air defense use can be resurrected from the World War II era using modern airships as barrage balloons. They can deny low-level airspace to enemy aircraft by:

- Forcing aircraft to fly at higher altitudes, thereby decreasing surprise and attack accuracy,
- Limiting direction of attack, permitting more economical use of AD assets, and
- Presenting definite mental and material hazards to pilots by cables and airships.

During WWII in 1944, the UK had 3,000 aerostats operating. During the Blitz, 102 aircraft struck cables (66 crashed or forced landings), and 261 V-1 rockets were downed. The blimps were 19 m long. Modern more compact airships offer more flexible options, with fast vehicle-mount winches, powered dirigibles, and lighter and stronger cables. Although modern aircraft have better sensors (such as thermal sights for night use), most airships have no thermal signature and can be camouflaged and concealed for rapid rise with minimal visual signature. Latest recorded catastrophic collision of an aircraft with aerostat cable was 2007 in the Florida Keys. The Iranians have demonstrated air mines, barrage balloons with explosive charges.

The tether cable and loose lines are the main threat to low- flying aircraft. Tether cables are next to impossible to detect in either day or night conditions, and can be steel, Kevlar, PBO or nylon. Type and length of tether material is determined by lift capacity of the balloon. Multiple loose lines and/or tethers may be suspended from the balloon. Short-notice balloon fields can be emplaced in 10-20 minutes, and raised or lowered with fast

winches in 1-5. Netting, buildings, and trees can be used to conceal inflated balloons between uses. Smaller (e.g., 1-m) inflated shaped balloons can be used in target shaping, altering appearance of buildings, vehicles, weapons, etc. They can also be raised as AD aerostats.

Although some balloons will use concealment, others will be clearly displayed to divert aircraft, or trigger a response and draw aircraft into air defense ambushes. Captured marker balloons can divert search and rescue aircraft into ambushes. Balloons can be used in deception as decoys to draw aircraft away from high-value targets.

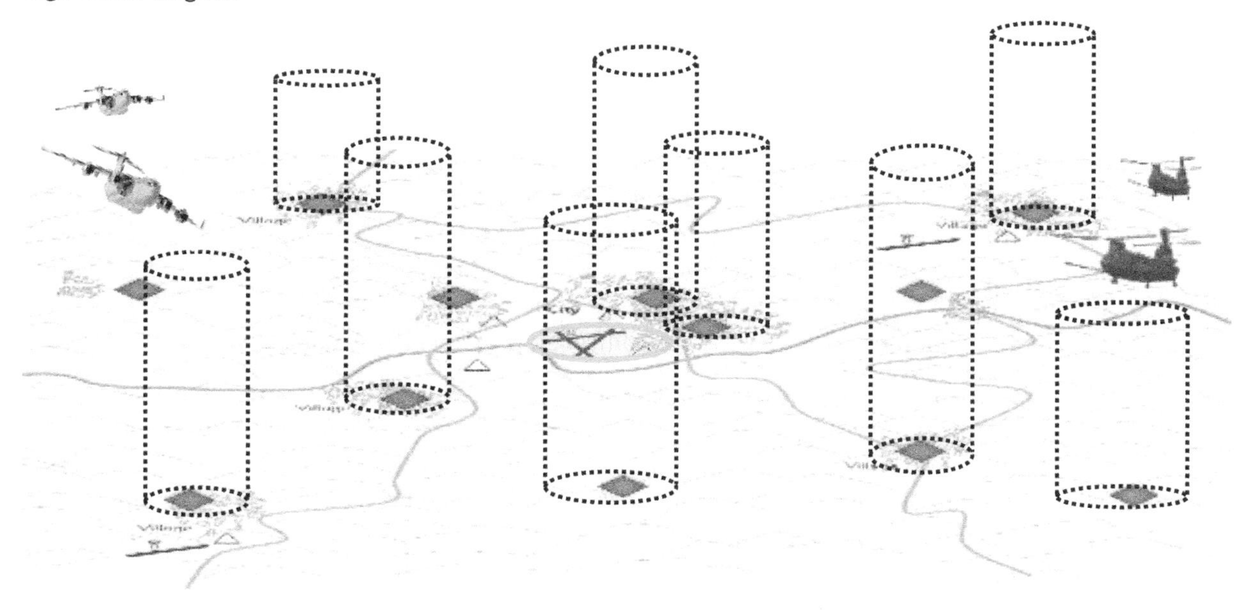

Two areas where airships are most effective in air defense are urban and complex terrain.

Remote-Controlled Aircraft and Micro-UAVs for Military Use

A wide variety of unmanned aerial vehicles are available in commercial and military sectors for use in military roles. However, cost can be a limiter for wide use. Some forces have turned to use of ***micro-UAVs***, in order to more widely distribute assets for close-in aerial surveillance. There is a burgeoning array of commercial and military options for these aerial systems. The term micro-UAV is open to wide variation, from palm size, to 1-2 meters. They can be almost as costly as mini-UAVs (up to $150,000 per set), or can cost only a fraction of that ($10,000 per set for a Russian Pestulga set). For even lower cost (and reduced capability), some forces turned to remote-controlled (RC) aircraft.

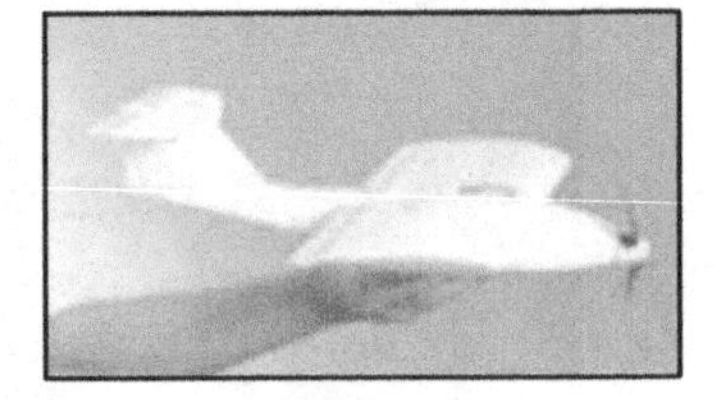

These aircraft can be used for several roles, including surveillance, electronic warfare, and attack. Some use gas engines; but others are battery-powered. Most are composed of wood, plastic, or composite materials, with almost no radar signature. With camouflage and a flying altitude of more than 100 meters, most have almost no acoustic, visual, or thermal signature, and would be very hard to shoot down with current weapons. The greatest threat to them is shotguns. The surveillance role is obvious, with range of 1 or 2 km and flight range varying from a 0.5 kilometer (RC aircraft) to 10-100 km for some micro-UAVs. Beyond surveillance, other roles include electronic warfare (mounting a pocket-size GPS jammer onboard), and attack (with onboard IED charges or grenades). Piloted aircraft do not like to fly where UAVs may operate. Thus micro-UAVs can be used in air defense to challenge/attack incoming aircraft. Micro-UAVs can fly harassing flights over military and civilian targets in a PSYOPS role. Low cost of the systems means that they can be used as reusable or disposable assets, with ample re-supply.

Hobbyists have been flying RC aircraft for decades. In the last decade, camera technology has advanced to the point that commercial applications for the technology have been used. They permit acquisition of affordable aerial views of buildings, wildlife areas, industrial sites, and terrain, which otherwise would require expensive use of aircraft. Military applications have been used. Tamil Tigers in Sri Lanka were found to have two aircraft with small cameras mounted inside.

A recently demonstrated RC aircraft conversion with video camera showed potential of this technology. The aircraft had a 20-km 900MHz telemetry link and 32-km flight path. Navigation data from GPS permitted precise aircraft location and image orientation. Sharp PDA for display and flight recording was used. The same imagery system could be used with airships.

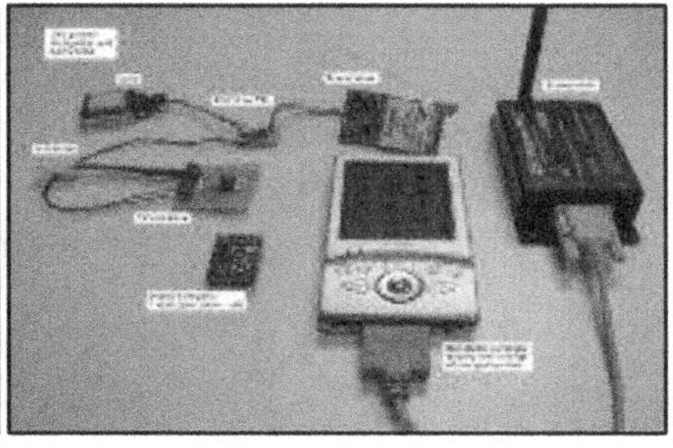

The Nokia N95 camera was displayed in an RC aircraft system described as "your personal Google Earth". It features the Multiplex EasyStar battery-powered RC aircraft with a 1.37-m wingspan, weighing 680 grams. With a GPS display unit and hand-held Optic 6 RC terminal, the system is ready to use in 10 minutes. Initial system assembly from kit, set-up, and training time is 2 hours. Pict'Earth software is used to download imagery.

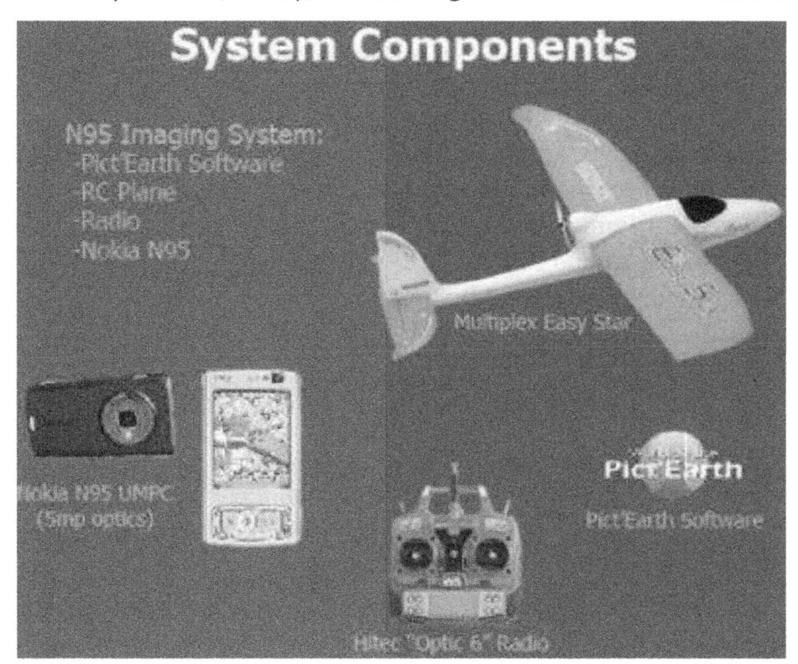

There are now clubs and internet forums for radio-controlled UAVs. More robust recreational aircraft are being marketed. An example is the E2 Electronic Surveillance Infrared UAV from Imaging1. The battery-powered craft (with pusher prop) is a flying wing configuration 1.85 m wide, weighing 2.7 kg. It can fly 3 hours (up to 160 km) and 1,500 m in altitude. Standard payload (up to 2.7 kg) is a CCD camera. It features autonomous take-off and landing. Thermal camera is optional. Cruise speed is 30 knots. With graphite construction, this craft offers durability for military and law enforcement applications.

A Russian micro-UAV is the Pustelga, which they call a "flying micro-vehicle" (FMV). The composite aircraft weighs less than 0.3 kg, and is hand launched. The whole system, with battery-powered UAV weighs less than 5 kg. It features a TV camera, laptop terminal, inertial/GPS navigation, digital map and azimuth display. With a skeletal frame, it has virtually no visual or acoustic signature. The "strike version" can mount a charge for attack missions.

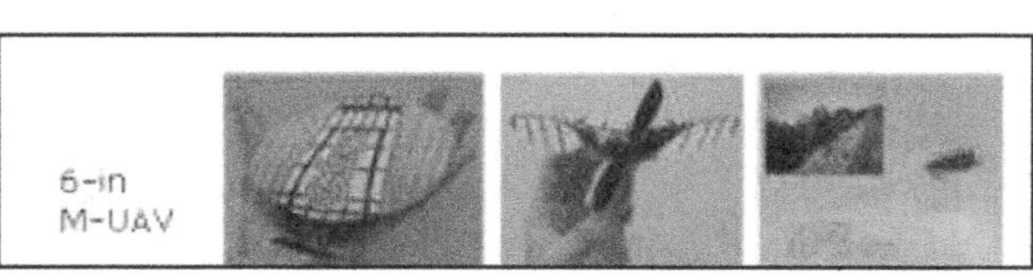

Other micro-UAV programs are underway. These will yield even smaller systems for military applications. Most MAVs are intended as disposable sensors, for hand or canister launch from ground units or vehicles. Attack

versions are being considered, with hit-to-kill attack profiles. Use of loiter and "swarm" behaviors have been demonstrated for MAV control.

Powered Parachutes, Paragliders, Hang Gliders, and Wingsuits

Often military, paramilitary, and insurgent forces will attempt to operate in areas where they do not control the skies. At key times forces will attempt to surreptitiously emplace teams behind enemy lines. To do so quickly may require the use of aircraft. But aircraft flying beyond unit can be detected, and perhaps engaged, endangering the mission and exposing inserted teams.

Aids for airborne insertion permit troops to more accurately land at the intended point and at the same time mask their landing location. Assets include the use of rotary-wing aircraft, and low-flying low-signature fixed-wing aircraft (see, An-2/Colt, pg 3-35). Terms ***parachute, paraglider,*** and ***hang glider***, are not standardized, and are used indiscriminately. They are sometimes classed as ultralight aircraft; but the link is random and due in part to sharing of facilities and the sky.

Parachuting has greatly advanced with development of the cruciform shaped (rectangular) steerable canopies, which can stay aloft longer and offer glide ability (3-4:1 glide angle) to veer from the aircraft flight path and land precisely at selected landing points. Their superior lift permits them to launch from heights, e.g., cliffs, bridges, or balloons. These parachutes can take off from the ground at lower speeds as well as descend at a slower rate than older round chutes with a soft landing, usually erect and without injury. With their drogue-type pilot chute to open the main chute, they can launch from a towing boat or vehicle. An unattached cart can bear the chutist in a tow launch. Without propulsion, parachutes lack the lift and glide ability to stay aloft for a prolonged period after launch. Chutes tethered to a towing system are called ***Parasails.***

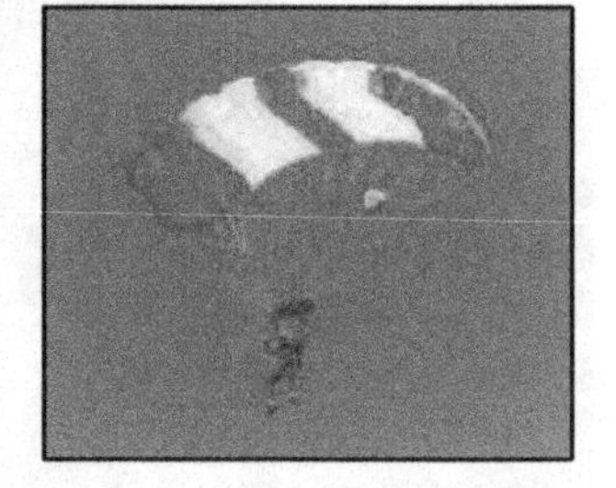

A spin-off technology is the ram-air parachute, also commonly known as a ***Paraglider.*** The airfoil design has two layers of fabric with an open front to catch air and inflate the semi-rigid structure. Like parachutes, paragliders use nylon, which is subject to UV ray degradation. Medium performance canopies are rectangular, whereas high-performance canopies are elliptical, weighing 55-139 kg. Some are triangular wing structures, with greater glide angles (5-6:1) to extend flight distance for longer range and stay aloft longer. For experienced users, the technology offers capabilities beyond those of parachutes. They are also more subject to mishap. Poor wind can limit performance. User mistakes, and wind turbulence can result in catastrophic results, such as spin or canopy collapse. Another phenomenon is "cloud suck", which can carry the chutist to 9 km or more, where temperatures can drop to -40 $^{\circ}$ Fahrenheit. A chutist can also carry a reserve parachute. In most cases, a collapsed paraglider will recover on its own in about 100 m. Glide speeds can mean faster landing speeds with paragliders.

An adjunct to parachuting or paragliding is ***powered parachuting*** or ***powered paragliding***. This can involve a backpack ***paramotor,*** which can propel and steer troops. Units for parachutists generally require 40 to 70 hp. ***Powered parachutes (PPC)*** convert parachutes into aerial vehicles. With them, troops can stay aloft for long periods and long distances. The paramotor is mounted on the chutist's back, and is surrounded by a cage. A user can launch from a stationary standing position, and land erect. Most use a gasoline engine, and weigh 20-37 kg. With easily assembled cages, the motors can be transported in the trunk of a car. A Chinese electric paramotor, the Yuneec ePAC, is in pre-production testing and will likely soon be marketed.

When linked with paragliders, paramotors transform them into ***powered paragliders (PG)*** to fly 100 kilometers on a tank of gas. Paramotors for paragliders need a power range of only 15 to 30 hp. The equipment can be set up in 15 minutes. Disassembly into 3-4 parts takes about 3 minutes. Flight speed is 32-40 km at 150-5,500 meters altitude. They generally cannot launch from standstill.

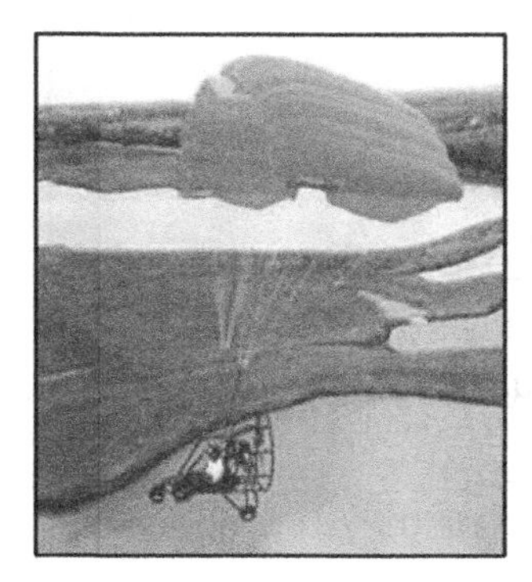

An efficient design for military units is to suspend a ***trike*** under the canopy and mount the paramotor onto it. Then the operator is freed to fly the craft; and can suspend combat gear to the frame. Some are erected at launch site, whereas others are solid welded structures. Trikes require larger parachutes than for parachutists or paragliders (discussed below). The chutes have 30 cells, compared to a normal design with 13. Wind and gust should not exceed 10-15 mph in flight. Paragliders and parachutes with trikes usually take off and land from paved surfaces; however, parachute versions have lower stall speeds, and can use unpaved areas as well. One example of a commercial trike is the Powrachute Sky Rascal. The one-seat craft weighs 105/117 kgs, with 40 or 52-hp engine, max payload of 136/159 kgs, and air speed of 67-90 km/hr. Typical trike specifications are as follows:

- Continuous flight capability: ~3hrs w/ 10gal fuel tank
- Take off distance: < 30 meters
- Flight speed: 40 – 111 km/hr
- Flight elevation: up to 5,500 m AGL (150-450 typical)
- Range: Approx. ~185 km round trip
- Glide Ratio: 4-5:1
- Cost: Single Seat $6000 - $10000 USD, Two Seat: $15000 - $20000 USD
- Payload: Up to 1,100 kg (varies by engine type, GVW, and canopy)
- System Assembly / Disassembly: ~10 min w/ 1 person
- Training: 5 – 7 days

Various other structures have been added to powered paragliders, including rubber inflatable boats (RIBs, Vol 3, pg 3-11), pontoons (right) , and wheeled cab designs. A new feature for PPC is Rapid Launch Amphibious Powered Parachute, a rectangular ram-air canopy with helium-filled chambers forming a balloon. The rigid canopy lifts even at standstill, permitting launch from stationary position. Various mounts are permitted, but the one displayed with Rapid launch is a catamaran boat.

Missions with these craft include reconnaissance, insertion, and delivery of critical materials. Trikes can also be used to launch parachutists. With night vision goggles, GPS, coordination with ground support, and nighttime illumination along flight routes, they can operate at night. Illuminated areas are safer for take-off and landing. For powered PPCs and PGs,

most of the time, altitude is low (less than 500 m) to reduce likelihood of detection. Flight time is about 2-2.5 hours between refills. With refills and ground support, the craft can fly hundreds of kms. A passenger on a trike could use a laptop or PDA controller to operate small UAVs to fly ahead or conduct area surveillance along the flight path. A paramotor FARP can be as simple as a pickup truck with communications and 5-gallon fuel cans at a pre-coordinated point. Powered parachutes and paragliders are similar to ultra-light aircraft in that reliability, operator errors, wind conditions, and landing/take-off conditions can cause accidents and injuries. However, because of their slower speed and superior lift, consequences of PPC and PG accidents are usually less severe than with ultra-light aircraft.

Powered parachutes and paragliders are an inexpensive alternative to UAVs, or they can be used in conjunction with them. Iran, India, Pakistan, China, Cuba, and Lebanese Hezbollah have all demonstrated either a PPC or PG capability. In 2002 Beijing's China Central Television showed members of Special Forces reconnaissance militia using trikes and a powered paraglider with a small rubber boat similar to a small Zodiac RIB. In 2014 Hamas had plans to use Paragliders as "flying suicide bombs". Authorities in Spain, India, and Norway have also uncovered plots to use paragliders as weapons or to transport fighters into restricted territory.

Hang gliders can be classified as paragliders. Some hang gliders use rectangular parachutes or paragliders, or paraglider wings to bear them when aloft. Higher performance hang gliders use erectable Dacron rigid wings or triangular structures, with bars underneath. The operator lies prone underneath. Hang gliders offer glide angles of up to 20:1, for long flight times and distances. The wing above can block the user's skyward view; so some use transparent material to expand viewing area. Many hang gliders use erectable struts, which can be disassembled and fit into a tube 6 m long, for vehicle mount. A few makers, such as Wills Wing and Finsterwald, offer structures which can fit into 2-meter tubes and inside of vehicles.

Triangular wing paragliders with paramotors are often included in the category of ultra-light aircraft, and operate with similar capability and vulnerabilities.

Another recreational development with possible application to military actions is the ***wingsuit*** (aka ***jumpsuit***). Developed for base jumping, the suit permits a user to glide to earth, and then pop a parachute for a safe landing. Obviously, there is risk associated with this arrangement, with flight speeds of 80-200 km/h and glide ratios of 2-3:1. Training is critical. With schools, clubs, and competitions, designs vary greatly. Brands include Phoenix-Fly, V3, and many others, plus experimental and privately made creations. The jet-powered Go Fast has demonstrated a safe landing without parachute. Wingsuits permit SOF to insert personnel with less visible signature and shorter vulnerability time than those on paragliders. Wingsuits can deploy from 2-man ultralights or trike-powered paragliders, enabling insertion personnel to exit the aircraft quickly. Military designs include the German Gryphon, which has been demonstrated and displayed at exhibitions. With rigid wings and jets, it is intended to offer 40 km range and payloads for military missions. In the Near Term, more composites and inflatable sections may add rigidity for stabilization. It is likely that military versions will offer safer and practical designs for tactical roles.

Ultralight Aircraft and Military Uses

Recreational use of ultralight aircraft has generated a myriad of activities and flying organizations worldwide. Their designs are much less regulated than conventional aircraft, which has led to thousands of makes and designs. They require much shorter and less developed airfields than other aircraft, with few organizational procedures, with primary focus on operational procedures to fly the aircraft.

Many operate on water, to ease dangers of takeoff and landing. In many cases, these are the only craft that can operate in some remote areas. At right is one of several craft operating in the Nepalese mountains. Ultralight aircraft are generally cheap to operate and operators can be trained in a matter of days. The craft can travel for thousands of kms, stopping only for refueling. A number of them can hold more than two persons as well as several hundred kilograms of cargo.

Key descriptors that set ultralights apart from other aircraft are that they are manned, are smaller than conventional aircraft, and are powered. The most common configurations are the following:

- Hang-glider type with a paramotor and seat,
- Smaller conventional wing-over-cab design, and
- Rotary-wing design.

Powered hang-glider type ultralights are easy to produce, maintain, and fly. They were an outgrowth of the expansion in recreational hang-gliding. Designs widely differ; but they usually use Dacron fabric, and a triangular wing design. Similar versions employ conventional

wings with swept angles. They are light and require less fuel than other designs. If the paramotor were to fail, the craft can glide to a landing.

Most ultralights have rigid structures; but many combine those structures with fabric wings and shock units. Many are fitted for water take-off and landings. The Italian Polaris FIB (left) has sold more than a thousand units in several models throughout Asia, Europe, and in the U.S. The FIB 2001 Flying Inflatable Boat is an upgraded design using a Lomac RIB hull and weighting 58 kg. It is fitted with a 48-hp Rotax 503 twin-cylinder 2-stroke engine selected for noise suppression. Other FIBs include the 503 (right), with a tandem overhead wing.

Conventional tandem wing-over-cab designs vary from finished craft with attractive designs, dashboard gauges, and shocked retractable landing gear, to Spartan frame structure. The Fotos Seamax is an example of the former. For military use, the craft are apt to be closer to the latter, but with additional features. Military craft are apt to have an open cockpit design with two seats, light weight, ample cargo capability for military gear, and ruggedized for long use and wear and tear of possible combat conditions in difficult weather and terrain. The craft should also be able to accommodate night missions. An example of this kind of craft is Quicksilver Sport 2S (see data sheet next page).

Ultralight helicopters are made mostly in the U.S., Russia, and European countries. They have been sold in other areas. Many are often referred to as gyrocopters and rotorcraft. Most are built from kits, and are 1-seater designs. The Russian K-10 (left) was an early craft used to support Naval icebreakers. An example of a more finished design is from the Italian firm Elisport. The Kompress (Angel CH-7) is a single-seat craft with a 65-hp Rotax 582 engine. It weighs 1,078 lbs, with 2.5 hrs endurance. There are a few 2-seaters available. Civilian and military roles for these rotorcraft include ambulance duty, surveillance, search and rescue, agricultural spraying, etc. Some military versions are equipped to fly unmanned.

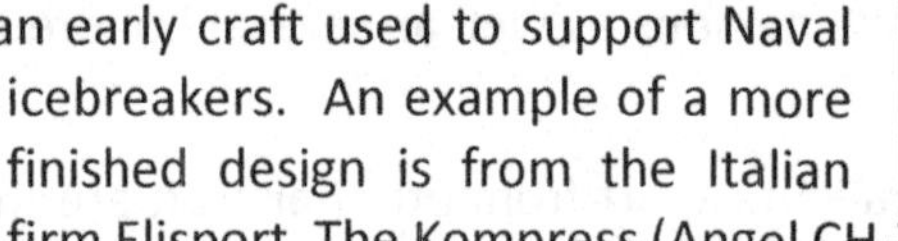

Ultralight aircraft vary widely in their reliability and capabilities. All are more subject to weather and terrain considerations than conventional craft. Recently a Hamas-operated ultralight craft broke up off the coast of Israeli in the Mediterranean Sea. Even well-designed craft are subject to adverse events. Nevertheless, these craft offer cost-effective aerial use by civilian and military organizations.

Today ultralight craft are employed in military operations. Most common military missions are insertion of special operating forces, reconnaissance, patrol and quick-reaction units, and delivery of materiel in difficult terrain. They generally have reduced signatures. They can fly low (below radars), and land in areas where conventional fixed-wing aircraft cannot land. Military versions of these craft are used in various countries, including India, Iran, and China. Iran produces ultralight aircraft in a variety of designs. The Iranian Saba Airline Company ultralight is offered for sport flying, short-haul freight, crop dusting, fire fighting, urban taxi service, police patrolling, as well as military roles. The Saba Company offers an unmanned version of its craft for military surveillance. Ultralights could also launch small UAVs, conduct jamming missions, retransmit signals, and attack targets. Craft useable for crop dusting could also deliver chemical agents.

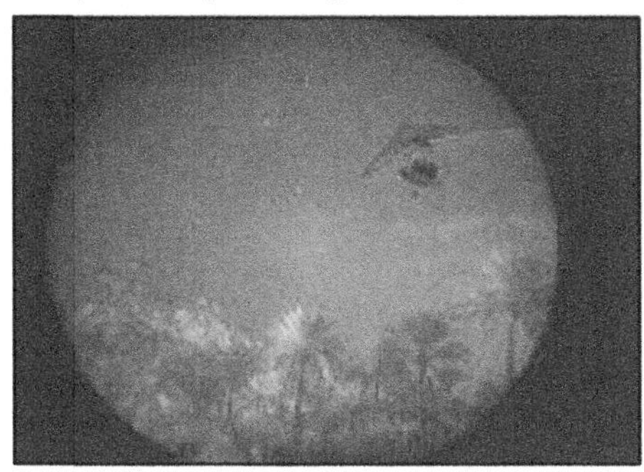

US ULTRALIGHT AIRCRAFT QUICKSILVER SPORT 2S

SYSTEM	SPECIFICATIONS	SYSTEM	SPECIFICATIONS
Min crew:	1	Minimum sink rate:	660 ft/min
Seats:	2	Required Distances:	Takeoff, ground roll – 240 ft
Blades:	Propeller - 68in x 36, Selected for less noise		50 ft obstacle – 660 ft
Engine:	Rotax 582, 2-stroke, 64 hp+		Landing with brake – 220 ft
	No. of Cylinders: 2	Design: Tapered stabilizer, tubular-braced tail	
	Displacement: 580.7cc	Double Surface wings	
	Dual CDI Electronic Ignition	Aluminum steerable nose wheel	
	Dual Carburetor Engine	Main wheel brakes	
Dimensions:	Length: 18ft 1/2 in	Conventional 3-axis controls	
	Height: 8 ft	Kit Assembly Time:	40-60 hours
	Wingspan: 31 ft	Breakdown for Transport:	Considered "quick"
	Wing area: 174.1 sq ft	VARIANTS	
Weight:	Empty: 430 lbs	An amphibious version of the Sport 2S is available. Similar modification with pylons could be made with most ultralights; but factors such as endurance and performance on takeoff and landing in water can vary.	
	Max takeoff: 996 lbs		
Fuel Capacity:	6 US GAL		
Speed:	Cruise: 70 mph	A Quicksilver cab aircraft offers 65 mph spray speed. It has 94 liter and 140 liter fiberglass spray tanks. Spray rate is 6 acres (2.5 hectares)/min, flying at 3.7-4.7 meters altitude. Spray mixtures can vary for different spray rates. Similar ultralite craft could be used in military roles for dispensing chemical agents.	
	Sea Level: 69 mph		
Rate of Climb:	500 ft/min		

NOTES
THERE ARE MANY SYSTEMS OF SIMILAR DESIGN, WITH DIFFERENT FEATURES AND PERFORMANCE LEVELS. THIS CRAFT HAS A REASONABLE CAPABILITY LEVEL TO EXPECT IN A KIT ULTRALIGHT. MOST ULTRALIGHTS CAN BE MODIFIED TO FIT SPECIFIC USES, SUCH AS ADDING CAGE FOR ADDING CARGO, MORE GAUGES (SUCH AS GPS FOR NAVIGATION), RADIO, AND EVEN MOUNTING A WEAPON PINTLE (FOR MG OR GRENADE LAUNCHER), OR WEAPON CRADLE FOR QUICK DEPLOYMENT. ACCESSORIES JUST AS NVGS COULD BE USED. NONE OF THESE MODELS ARE MARKETED BY THE MANUFACTURER FOR USE IN MILITARY ROLES; BUT THEY COULD BE USED FOR THEM.

Worldwide Equipment Guide

Chapter 6: Theater Missile Systems

Distribution Statement: Approved for public release; distribution is unlimited.

Chapter 6: Theatre Missiles

In an era of increased emphasis on lethality and protection against manned aerial forces, military forces world-wide are seeking to extend their deep-attack capabilities by means other than manned aircraft. Thus, new missile systems are being fielded. The trend among military forces for acquisition of theater missiles has expanded with the growth of regional rivalries and the strategy of using long-range strike capability to gain regional leverage. Theater missiles are generally categorized among two types - ballistic missiles (BMs) and cruise missiles (CMs). They are launched from ground launchers, aircraft, or naval vessels. These systems are designed for deep strike missions—beyond those of close battle assets. Where missiles are subordinate to the ground force commander, they will be used as another strike asset to support his plan. They may be used for purposes other than execution of conventional strike missions, such as delivery of mines, and information warfare missions.

Theater ballistic missiles (TBM) are an expanding threat to U.S. soldiers, allies, and interests in regions where military forces are deployed, such as South Korea, Japan, Iraq, or Afghanistan. The trend among military forces for acquisition of theater missiles has expanded along with the growth of regional rivalries and the strategy of using long-range strike capability to gain regional leverage. TBM provide the OPFOR commander the ability to strike a target(s) 3,000 km (1,864 mi) away with a nuclear warhead or with an array of conventional warheads.

The role of cruise missiles (CMs) has changed. Prior to the 1990s, fielded designs were generally limited to ***anti-ship missiles*** (WEG Naval Vol 3, Littoral Chapter). Improved in guidance systems, propulsion, warhead options, launch platforms, and affordable designs have vaulted CMs to the role of the first option for deep attack against point and small area targets.

New missile systems have been developed which do not fit in the BM or CM category. These are long-range missiles flying non-ballistic trajectories with a mix of pre-programmed phase and options for manned guidance, loitering in the target area, as well as separate homing by GPS, radar or passive RF seeker, and/or IR/MMW homing. These systems may also be categorized as non-line-of-sight antitank guided missiles (NLOS ATGMs), or as unmanned combat aerial vehicles (UCAVs). They can be launched from ground vehicle launchers, ships, and/or aircraft. Some are developed as anti-ship missiles. Most have high-explosive warheads for multi-role use; and are large enough to kill armored targets and bunkers. They will supplement lethal strikes against high-value targets, including moving targets.

Systems featured in this chapter are the more common systems, or represent the spectrum of missile systems which can threaten US Army forces or interests within an operational environment. Questions and comments on data in this specific update should be addressed to:

Mr. Rick Burns

DSN: 552-7922 Commercial (913) 684-7987

e-mail address: richard.b.burns4.ctr@mail.mil

Theater ballistic missiles

Theater ballistic missiles (TBMs) employ a high-atmosphere or exo-atmospheric ballistic trajectory to reach the target. Because of the high cost and limited numbers of these systems compared to artillery, they will be used against high-priority targets at critical phases of a conflict, or against political targets. Selected OPFOR forces with limited numbers of missiles may hold them in a separate missile unit at echelons above the supported ground force commander. The most critical component of a theater ballistic missile system, which defines its capabilities and limitations, is the missile. Unlike rockets, all missiles have guidance or homing for precision strikes. Missiles are generally classified according to their range—

- Short-range ballistic missile (SRBM), 0-1,000 km.
- Medium-range ballistic missile (MRBM), 1,001-3,000 km.
- Intermediate-range ballistic missile (IRBM), 3,001-5,500 km.

Numerous countries are adding technologies to extend range and improve accuracy of ballistic missile systems. Approaches for improve range include increased use of solid fuel, lengthening missiles for increased fuel and longer burn time, improving motors (in the propulsion section), using more efficient solid fuel motors, and employing smaller and lighter warheads. Key additions for precision are maneuvering re-entry vehicles (RVs), and GPS. Below is an example of a modern missile (Russian Tochka-U SRBM) and its major components.

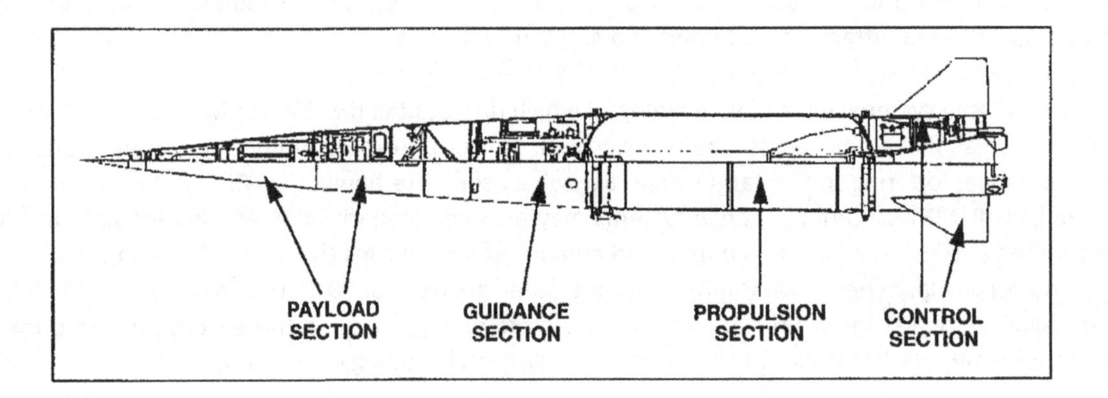

Mobility. These missiles employ a high-atmospheric or exo-atmospheric ballistic trajectory to reach the target. Most TBMs follow a set course that cannot be altered after the missile has burned its fuel. However some have the capability for non-ballistic trajectories and precision maneuver. Ballistic missiles have three categories of propellant for engines, which are liquid, hybrid, or solid, effect the distance a missile can travel and the CEP, or accuracy.

The majority of TBMs are able to launch from the ground, or naval assets. Missile ground launch platforms vary from fixed ground launchers, trailer launchers, mobile launch complexes (numerous vehicles) and transporter erector launcher (TELs). Fixed ground launchers may include hardened underground sites. Mobile ground launchers vary from older systems with simple modifications, to specialized vehicles designed for operation in all types of terrain. Newer launchers may incorporate improved mobility to reduce vulnerability to location by terrain analysis and intelligence preparation of the battlefield.

Lethality. Critical lethality considerations for TBMs include range, precision, munitions options, and responsiveness. The missile system is selected for a mission based on its ability to reach the target within targeting timelines, and its ability to deliver effective lethality on the target. Improved heavy multiple rocket launcher systems with course correction and increased-lethality warheads have replaced TBMs as preferred strike systems against selected deep targets. For instance, a Russian 9A52 MRL can deliver twelve 300-mm rockets 70-90 km with near-missile precision and minimal preparation time. However, a modern TBM can deliver twice the payload a farther distance with better precision against critical heavy targets.

The warhead (within the payload section) is the munition, the lethality mechanism which is selected for that strike mission and around which the system is designed. Many countries acquired ballistic missiles specifically to deliver weapons of mass destruction (WMD) against civilian targets such as urban centers. For such a mission, a less accurate system with a large payload capacity is sufficient for the mission. A substantial proportion of SRBM and some MRBM designs are copies or variants of the former-Soviet SCUD-B/SS-1c. Although these systems lack accuracy and responsiveness of some the newer systems, they can deliver large lethal payloads against fixed targets or targets whose limited mobility permits them to be stationary long enough for the TBMs' operational timelines.

Warhead developments include separating warheads, multiple warheads, maneuvering reentry vehicles (RVs), navigating and homing warheads, varied lethal and electronic warhead fills, warhead buses (e.g., submunitions), and warheads with countermeasures (penaids). Improved precision, in-flight targeting updates, warhead seekers, penaids, and other upgrades will further challenge theater missile defense assets to prevent strikes against priority targets.

Newer TBM designs with improved range, accuracy and operational considerations have been fielded. All missiles have some type of inertial guidance. Accuracy ranges 300 - 500m CEP for older systems, to less than 50m CEP for some advanced systems. These include several missiles with 10 m CEP. Some missiles add global navigation satellite systems (GNSS, e.g., GPS) for improved precision. Thus, older design systems can see immediate upgrades with that change. Further precision (5-9 m) is added with infrared (IR) or radiation-homing seekers.

Another critical consideration for effectiveness of TBMs is their responsiveness. Keys for timely delivery include target location, fire mission calculation and transmission, launcher and missile operational timelines. Therefore, modern missile system support equipment can include computerized fire control, location/navigation system (such as global positioning systems), as well as dependable secure communications. A key technology for increased TBM responsiveness is the use of solid fuel propellant, which removes the need for fueling a liquid fuel missile prior to launch. That step can increase preparation time at the firing point, and delay use or compel use when changing battlefield situation changes the mission. Solid fuel missiles are more consistent and reliable; and the modern trend is toward solid and away from liquid.

Operational timelines for missile crews of fixed launchers as well as mobile TELs are addressed in three phases: (1) time from leaving the hide to launch, (2) time from launch to leaving launch point, (3) and missile trans-loading time prior to next launch. These times are based on technology requirements as well as sound tactics. Steps in the launch sequence based on technology include surveying the launch site, launch coordination, emplacing the launcher,

preparing the launcher and missile for launch, initiating safety measures, and the launch. Post-launch sequence includes displacement of the launcher, and displacement of support equipment. Missile transloading is executed far from the launch site; therefore time includes travel time, service to the launcher, fueling liquid-fuel missiles for the next launch if the next launch is less than 24-48 hours, planning coordination, then movement time to the next launch area (but not to the launch point). Additional time is included in TBM operational time lines because of survivability tactics, as noted below.

The warhead (within the payload section) is the munition, the lethality mechanism, which is selected for that strike mission and around which the system is designed. A number of newer TBM designs with improved range, accuracy and operational considerations including maneuvering reentry vehicles (RVs) have been fielded. Modern warhead developments include nuclear and chemical warheads, separating warheads, and multiple warheads. TBM can also deliver a wide variety of conventional munitions. Some examples are HE, anti-radiation (ARM), fuel-air-explosive (FAE), DIPCM, ICM cluster munition, varied lethal and electronic warhead and EMP fills, warhead buses (varied submunitions), precision navigating and homing warheads (such as IR homing). Countermeasures, including separating and maneuvering warheads, penaids, and other technical measures will further challenge the capability of theater missile defense assets to prevent strikes against priority targets.

Survivability. Technologies for increased missile reliability include almost total conversion from liquid to solid fuel. Some missiles are canisterized to protect them prior to use and permit easier handling and loading. With increased use of GPS correction and computer digital loading of propulsion system commands, possibilities of misfire and guidance failure are greatly reduced.

The high lethality of the missiles and their launchers means that both are considered by their adversary to be high priority targets for defeat and destruction. Therefore, the OPFOR can be expected to employ a variety of tactical and technical countermeasures to protect them. Tactical countermeasures include: using the missile's long range to outrange most adversary systems, use of hides (such as hardened artillery sites and terrain near the launch point or at trans-loading points to reduce exposure time, high mobility (high speed or all-terrain chassis) to move rapidly and reduce exposure time, use of OPSEC and deception operations (decoys, launch

site emission control measures, movement in clutter, surge operations, etc.), and reduced launch sequence timelines (pre-surveyed site, pre-arranged communications, etc. These steps may sacrifice accuracy for reduced exposure time. More modern launchers will have a minimal preparation time between emplacement and execution of a fire mission.

Technical survivability measures for missiles include: improved coatings and camouflage patterns separating re-entry vehicles, non-ballistic trajectories (to foil trajectory prediction), cluster munitions, and penetration aides (such as jammers in warheads). Technical survivability measures for launchers include:

improved coatings and camouflage patterns and nets, high mobility (to expand useable launch areas), self-survey capability (to minimize emplace time), short displacement time (<5 min), rapid launch sequence, non-ballistic trajectories (to foil back-tracking for counter-battery fires), employment of high-fidelity decoys, and SATCOM encrypted digital burst communications. These measures are intended to degrade the enemy's detection, targeting, impact or effectiveness kill, and lethality effects.

Other Considerations. State-of-the-art TBMs can cost more than a million dollars each. If the systems are not accurate enough, or if the enemy has ABM capabilities, those TBMs may not have a high assurance of success, and may not be a factor in the OPFOR plan. Thus, budgetary, political, and military considerations affect TBM decisions. The OPFOR may limit its missile requirement to systems used to gain regional political leverage by targeting civilian targets. Given the budget limitations and systems costs impacting most military forces in recent years, the OPFOR will likely have a mix of older and newer systems and selected upgrades. They may also balance the mix of TBMs with other, less costly, long-range precision strike assets. These can include ***precision artillery rockets***, ***precision artillery missiles***, non-line-of-sight antitank guided missiles (NLOS ATGMs), ***unmanned combat aerial vehicles (UCAVs)***, and ***cruise missiles***. Cruise missiles (CM) are discussed in the section beginning at pg 5-11.

Conclusions. Updates to both launch platforms and missiles systems are allowing the threat to become increasingly mobile and accurate. The extended range of both missiles and their mobile platforms create a dangerous combination providing a potential adversary the ability to launch missiles and strike well beyond preconceived ranges. These assets are a critical component of deep strike mission planning for conventional forces. They are also used as an asymmetrical political tool for use in affecting strategic power calculus in peacetime international struggles.

RUSSIAN BALLISTIC AND CRUISE MISSILE LAUNCHERS ISKANDER-E, -M, AND -K

		Weapons & Ammunition Types	Typical combat load
		Missiles on launcher	2
SYSTEM	**SPECIFICATIONS**	Launcher Performance	
Alternative Designations	SS-26, SS-X-26 Iskander-M for Russian forces Iskander-E for export	Land Navigation	GNSS
Date of Introduction	1999	Missiles per Launcher	2
Proliferation	At least 1 country. 3 other countries are considering acquiring the system. Iskander-M is in Russian service.	Total Emplace-Launch-Displace Time (min)	15
ARMAMENT	**SPECIFICATIONS**	Time Between Launches (min)	1, for second missile
Transporter-Erector-Launcher		Reaction Time	1 min
Name	SPU 9P78E (MZKT-7930 variant)	Position Location	Gyroscopic inertial with GNSS updates
Crew	3	Missile	
Chassis	MAZ-7930 (8x8)	Name	Iskander-M/Iskander-E
Combat Weight (mt)	44.7 est based on chassis	Type	Single-stage, solid-fuel
Chassis Length Overall (m)	12.67	Launch Mode	Vertical launch
Height (m): TER down	3.02	Max Launch Range (km)	400/280*
Width Overall (m)	3.05	Min Launch Range (km)	50
Armor Protection	None	Length (m)	7.3
NBC Protection System	Yes	Diameter (mm)	920
Automotive Performance		Weight (kg)	3,800
Engine Type	Diesel, 500-hp		
Cruising Range (km)	1,100		
Max Road Speed (km/h)	70		
Max. Swim Speed:	N/A		
Fording Depths (m)	1.4		
Radio	INA		
Armor Protection	None		

NBC Protection System	Yes		

PRIMARY COMPONENTS

TRANSPORTER-ERECTOR-LAUNCHER (TEL) AND COMMAND VEHICLE: REAR SUPPORT INCLUDES A TRANSPORT AND LOADING VEHICLE (9T250E), MAINTENANCE VEHICLE, MOBILE TEST AND REPAIR STATION, DATA PREPARATION POST, AND LIFE SUPPORT VEHICLE. THE SYSTEM CAN ALSO BE LINKED INTO AN INTEGRATED FIRES COMMAND (IFC).

COUNTERMEASURES

OFF-ROAD MOBILITY TO CONCEALED LAUNCH POINT, AUTONOMOUS AND PASSIVE OPERATION AT LAUNCH POINT. MISSILE NON-BALLISTIC TRAJECTORY IN ASCENT CONCEALS VEHICLE/LAUNCH POINT LOCATION. MISSILE REENTRY VEHICLE HAS DECOYS, AND POSSIBLE FINAL-PHASE MANEUVER. WITH IR HOMING JAMMING IS INEFFECTIVE. FINAL PHASE IS MOST LIKELY NON-BALLISTIC PITCH-OVER INTO A DIVE.

GUIDANCE

INERTIAL, WITH OPTIONAL GNSS AND/OR OPTICAL/IR HOMING. ADDITIONAL COURSE CORRECTION USES THE RADAG RADAR CORRELATOR.

TRAJECTORY

BALLISTIC WITH NON-BALLISTIC BOOST PHASE FLY-OUT, AND POSSIBLE RE-ENTRY MANEUVER

ACCURACY (M)

5-7 WITH IR-HOMING; 10-20 WITHOUT

FIRE CONTROL COMPUTER

THE MODERN AUTOMATED FIRE CONTROL SYSTEM CAN BE USED AS THE BATTLE MANAGEMENT SYSTEM FOR A RECONNAISSANCE-STRIKE COMPLEX, OR "INTEGRATED FIRES COMMAND", IN CONCERT WITH ARTILLERY AND OTHER RECONNAISSANCE AND FIRES ASSETS.

FOR IR-HOMING MODE, COMPUTER LOADS TARGET IMAGE FROM A SATELLITE OR UAV INTO THE WARHEAD. THUS, EVEN WHEN THE GNSS OR SATELLITE IS JAMMED OR WEATHER CAUSES INTERFERENCE, THE REENTRY VEHICLE WILL FIND THE TARGET.

VARIANTS

EARLY TEL VARIANT (SPU 9P78) HAS ONE MISSILE. THE TELS CAN LAUNCH R-500 CRUISE MISSILES.

ISKANDER-E: EXPORT VARIANT TEL WITH SHORTER RANGE (280 KM). THIS MISSILE WAS DEVELOPED TO COMPLY WITH THE MISSILE TECHNOLOGY CONTROL REGIME, WHICH IS NO LONGER IN EFFECT.

ISKANDER-M: DOMESTIC TEL AND MISSILE WITH 400+-KM RANGE.

ISKANDER-K: CRUISE MISSILE ONLY TEL, WITH LAUNCHER ASSEMBLY ADAPTED TO MOUNT 6 X R-500 (3M14?) CRUISE MISSILES.

WARHEAD TYPE: HE, ARM, FAE, ICM CLUSTER MUNITION (10), ICM (54 SUBMUNITIONS), NUCLEAR, CHEMICAL, TACTICAL EARTH PENETRATOR

WARHEAD WEIGHT (KG): 700/480

OTHER MISSILES

R-500: CRUISE MISSILE RANGE IS INITIALLY 280 KM; BUT NEAR TERM RANGE IS 500 KM (EST). IT HAS GNSS PROGRAMMED FLIGHT PATH, <100 M ALTITUDE, MULTIPLE WAYPOINTS, IN-FLIGHT REPROGRAM ABILITY, A VELOCITY OF 250 M/S, AND <30-METER ACCURACY. TERMINAL GUIDANCE OPTIONS INCLUDE AN IR (CORRELATOR) OR ACTIVE RADAR HOMING. PRODUCTION WAS DUE 2009. MID-TERM UPGRADE COULD INCLUDE A SUBSTANTIAL RANGE EXTENSION.

NOTES

RANGE VARIES WITH DIFFERENT WARHEADS AND WARHEAD WEIGHTS. POTENTIAL RANGE WITH THE DESIGN IS 500 KM. FUTURE WARHEAD OPTIONS MAY INCLUDE BIOLOGICAL WARFARE AND NON-NUCLEAR EMP WARHEADS.

RUSSIAN THEATER BALLISTIC MISSILE TRANSPORTER-ERECTOR-LAUNCHER TOCHKA-U

		Weapons & Ammunition Types	Typical combat load
		Missiles on launcher	1
SYSTEM	**SPECIFICATIONS**	Launcher Performance	
Alternate Designations	System with 120 km was called the SS-21 Mod 2/9K79M (see VARIANTS). For Tier 2 use SS-21 Mod 3.	Land Navigation	GNSS for command vehicle
Date of Introduction	1989 for Tochka-U	Missiles per launcher	1
Proliferation	At least 11 countries all variants. At least 3 countries Tochka-U	Emplace-launch time (min)	16 from march
ARMAMENT	**SPECIFICATIONS**	Displace time (min)	1.5
Transporter-Erector-Launcher		Time between launches (min)	40
Name	9P129M-1	Position location system	Inertial with GNSS updates
Crew	3	Missile	
Chassis	BAZ-5921 (6x6)	Name	9M79-1F/SS-21 Mod 3
Combat Weight (mt)	18.3 loaded	Type	Single-stage, solid-fuel
Chassis Length Overall (m)	9.5	Launch Mode	Vertical launch
Height, TER down (m)	2.4	Max Launch Range (km)	120
Width Overall	2.8	Min Launch Range (km_	20
Automotive Performance		Length (m)	6.4
Engine Type	Diesel, 300-hp	Diameter (mm)	650
Cruising Range (km)	650	Weight (kg)	2,010
Max Road Speed (km)/h)	60	Warhead Weight	482 Frag-HE
Off-road Speed (km)/h)	30	Fuze	Laser proximity for Frag-H
Max Swim Speed (km)/h	8		
Radio	R-123, R-124 on TEL		
Armor Protection	None		
NBC Protection System	Yes		

PRIMARY COMPONENTS

BATTERY HAS 2 X TELS, 2 X 9T128-1 TRANSLOADERS, AND A C2 VEHICLE. REAR SUPPORT INCLUDES TEST VEHICLES, MISSILE TRANSPORTERS, AND MAINTENANCE VEHICLES. THE SYSTEM CAN ALSO BE LINKED INTO AN INTEGRATED FIRES COMMAND (IFC). A MET UNIT WITH END TRAY / RMS-RADAR AND RADIOSONDE BALLOONS PROVIDES UPDATED WEATHER REPORTS.

COUNTERMEASURES

OFF-ROAD MOVE TO CONCEALED LAUNCH POINT. LIKELY AUTONOMOUS AND PASSIVE OPERATION AT LAUNCH POINT. NON-BALLISTIC TRAJECTORY ON ASCENT CONCEALS VEHICLE LAUNCH POINT LOCATION. APU FOR MINIMUM IR/NOISE. ERECT-TO-LAUNCH TIME: 15 SEC.

GUIDANCE

INERTIAL, WITH IR HOMING FOR FRAG-HE. OTHER HOMING GUIDANCE FOR OTHER MUNITIONS.

TRAJECTORY

BALLISTIC WITH NON-BALLISTIC BOOST PHASE FLY-OUT, AND RE-ENTRY MANEUVER FOR HOMING MISSILES

ACCURACY (M)

5-10 IR-HOMING , OR PASSIVE RADAR HOMING 10 WITHOUT HOMING GUIDANCE.

FIRE CONTROL COMPUTER

AUTOMATED FIRE CONTROL SYSTEM CAN BE USED AS THE BATTLE MANAGEMENT SYSTEM FOR A RECONNAISSANCE-STRIKE COMPLEX (RSC), OR "INTEGRATED FIRES COMMAND" (IFC), IN CONCERT WITH ARTILLERY AND OTHER RECONNAISSANCE AND FIRES/STRIKE ASSETS.

FOR IR-HOMING MODE, COMPUTER LOADS TARGET IMAGE FROM A SATELLITE OR UAV INTO THE WARHEAD. THUS, EVEN WHEN THE GPS OR SATELLITE IS JAMMED OR WEATHER CAUSES INTERFERENCE, THE REENTRY VEHICLE WILL FIND THE TARGET.

VARIANTS

SS-21MOD 1/9K79M/TOCHKA: FIRST FIELDED SYSTEM IN 1976, WITH 70-KM RANGE, 150 M CEP.

SS-21 MOD 2: SYSTEM WITH THE 120-KM 9M79M-F FRAG-HE MISSILE. CEP IS 20-50 M.

TOCHKA-U/SS-21 MOD 3: IMPROVED SYSTEM (SEE PRIMARY COMPONENTS) WITH TEL, NAV, AND SURVEY SYSTEM AND NEW MISSILES. THEY INCLUDE

9M79-1F, THE TOCHKA-R, AND OTHERS (BELOW).

WARHEAD OPTIONS TYPE: FRAG-HE, CLUSTER MUNITION (50 APAM-SIZE SUBMUNITIONS). OTHER WARHEADS CLAIMED TO BE AVAILABLE ARE: FAE, ICM DPICM, NUCLEAR (10 KT AND 100 KT), EMP, AND CHEMICAL.

TOCHKA-R: MISSILE FOR SS-21 MOD 3 WITH ARM (ANTI-RADIATION HOMING MISSILE), WHICH LAUNCHES ON A NON-BALLISTIC TRAJECTORY, THEN TARGETS RADARS.

AN EXPORT MISSILE CAN SWITCH WARHEADS BETWEEN UNITARY FRAG-HE AND APAM CLUSTER. THERE ARE REPORTS OF TESTS WITH 2-MISSILE VERSIONS WITH 180-KM RANGE.

NOTES

SYSTEM ALSO REPRESENTS OTHER MODERN TBMS WHICH COULD THREATEN US ARMY FORCES. THIS IS THE TIER 2 SYSTEM FOR USE IN OPFOR PORTRAYAL IN ARMY TRAINING SIMULATIONS (SEE PG 1-5). IN LATER OPFOR TIME FRAMES, (NEAR TERM AND MID-TERM); THE TOCHKA-U IMPROVED WILL INCLUDE OTHER OPTION, SUCH AS BIOLOGICAL WARFARE AND NON-NUCLEAR EMP WARHEADS

IRANIAN THEATER BALLISTIC MISSILE MOBILE ERECTOR-LAUNCHER SHAHAB-3A AND -3B

		Weapons & Ammunition Types	Typical combat load
		Missiles on launcher	1
SYSTEM	**SPECIFICATIONS**	Launcher Performance	
Alternative Designations	INA	Land navigation	GNSS
Date of Introdution	INA	Missiles per launcher	1
Proliferation	Iran	Emplace-launch time (min)	60 (est)
Primary Components	INA	Displace time (min)	INA
ARMAMENT	**SPECIFICATIONS**	Time between launches	INA
Mobile Erector-Launcher		Position location system	
Name	INA	Missile	
Crew	3 (EST)	Name	Shahab-3A
Chassis	Based on No-dong 1type	Type	Single-stage liquid with separating re-entry vehicle (RV)
Combat Weight (mt)		Launch Mode	Vertical launc
Chassis Length Overall (m)		Max Launch Range (km)	1,300
Height, TER down (m):		Min Launch Range (km_	INA
Width Overall (m):		Length (m)	16.58
Automotive Performance		Diameter (mm)	1.38
Engine Type	V8, Diesel Engines	Weight (kg)	15,862-16,250
Cruising Range (km)	550 (est)	Guidance	Gyroscopic inertial
Speed (km/h)	Max. Road: 70 (est based off of No Dong) Off-road: UNK	Warhead Weight	760-1,158
Radio		Fuze	INA
Armor Protection	None	Accuracy (m)	190
NBC Protection System	None		

DESCRIPTION

LIKELY A HIGHLY MOBILE TRUCK (NFI) BUILT INDIGENOUSLY FOR THE SHAHAB 3 BASED OFF THE NO-DONG BALLISTIC MISSILE TEL.

FIRE CONTROL COMPUTER

INA

COUNTERMEASURES

OFF-ROAD MOVE TO CONCEALED LAUNCH POINT. THE WARHEAD ON A RE-ENTRY VEHICLE CAN MANEUVER SEPARATE FROM THE MISSILE BODY TO CHALLENGE INTERCEPT SYSTEMS. ERECT-TO-LAUNCH TIME: INA

VARIANTS

VARIANTS HAVE USED DIFFERENT TRUCKS AND TRAILER DESIGNS.

ORIGINAL SHAHAB-3: THE MISSILE AND WARHEAD RESEMBLED THE NODONG-1, WITH A 1,200 KG WARHEAD AND A RANGE OF 1,300 KM. ACCURACY IS SAID TO BE 190 M. WITH ADVENT OF THE NEW MISSILE DESIGN, IT IS NOW CALLED SHAHAB-3A.

SHAHAB-3B: THIS VERSION HAS A NEW DESIGN SEPARATING RV WITH 2,000 RANGE AND SMALLER 500-650 KG WARHEAD. ACCURACY IS SAID TO BE 190 M. IT MOUNTS ON A DIFFERENT MEL TRAILER.

SHAHAB-C AND D: REPORTS THESE ARE IN TESTING.

NO-DONG-A1: A NORTH KOREAN COUNTERPART VERSION OF THE SHAHAB-3B MISSILE.

WARHEAD OPTIONS TYPE: : NUCLEAR, HE, CHEMICAL, OR SUB-MUNITIONS

NOTES

THERE ARE REPORTS THAT PAKISTAN HAS A SIMILAR TECHNOLOGY SYSTEM.

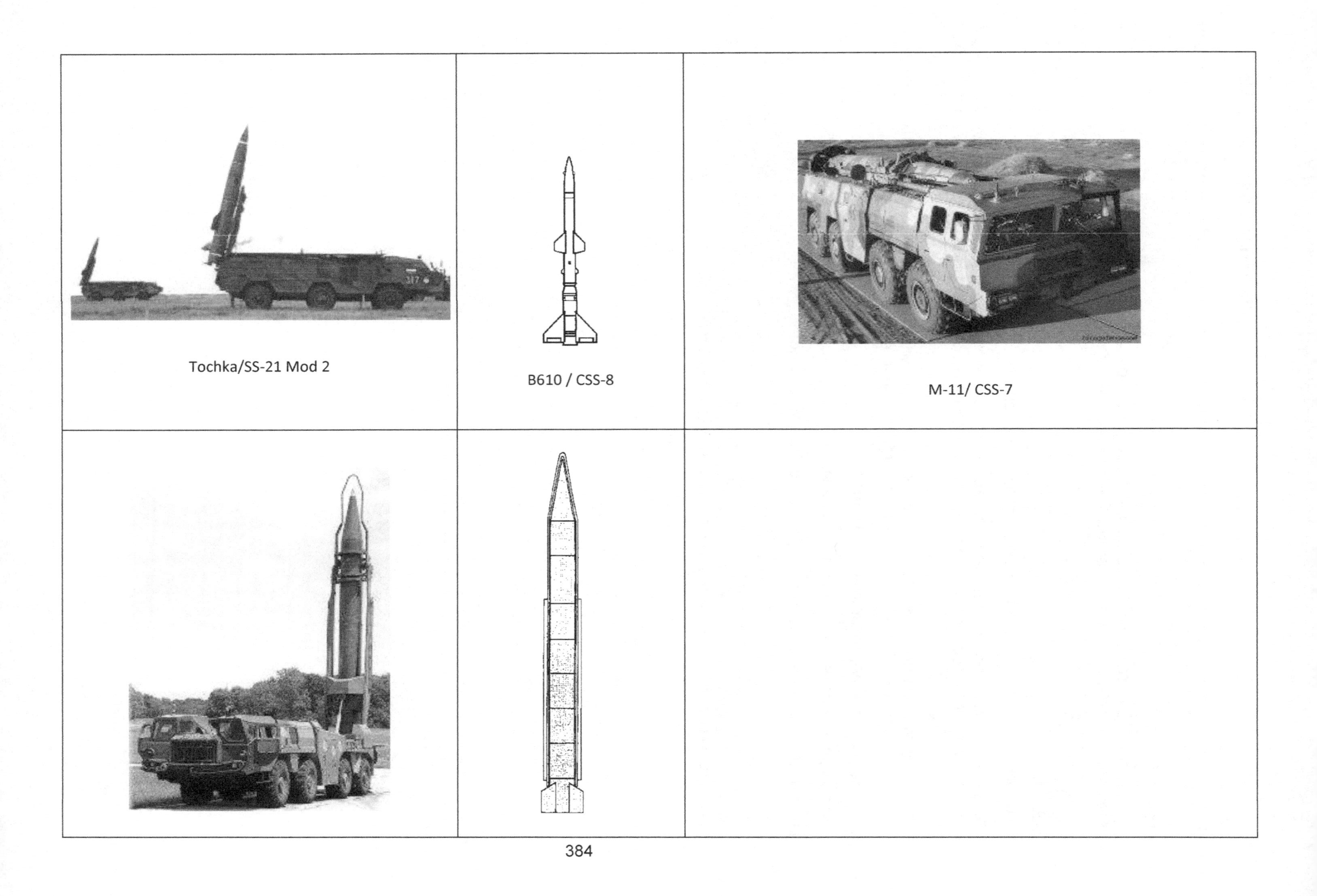

Tochka/SS-21 Mod 2

B610 / CSS-8

M-11/ CSS-7

SCUD-B / SS-1c	Nodong-1	 DF-3 / CSS-2

System Type	SRBM	SRBM	SRBM	SRBM	SRBM	SRBM	SRBM	MRBM	IRBM	Technologies & Trends
Name/ NATO Name Designator	Tochka-U SCARAB SS-21 Mod 3	M-7 B610 CSS-8	SCUD-B SS-1c	SCUD-B Mod 2 SS-1c Mod 2	M-11 DF-11 CSS-7	SCUD-C SS-1d	M-9 (export) DF-15 CSS-6	Nodong-1	DF-3 CSS-2	More SCUD variants
Producing Country	Russia	China	Russia North Korea	Russia	China	Russia North Korea	China	North Korea	China	Technology Transfer
Proliferation (countries)	At least 11 all variants	At least 2	At least 20	At least 1	At least 2	At least 5	At least 1	At least 1	At least 2	Increased proliferation
Type Launcher	TEL	TEL	Fixed, TEL	Fixed, TEL	TEL	Fixed, TEL	TEL	TEL	Fixed, Mobile complex	Mobile/decoy launchers
Propulsion	Single-stage Solid	Single-stage (est) Solid	Single-stage Liquid	Single-stage Liquid	Single-stage Solid	Single stage Liquid	Single-stage Solid	Single-stage Liquid	Single-stage Liquid	Non-ballistic trajectory
Range Min-Max (km)	20-120	50-150	50-300	300	50-300	500	200-600	170-1,300	1,500-3,000+	Increased range
Guidance	Inertial	Inertial	Inertial	Inertial IR homing	Inertial	Inertial	Inertial	Inertial	Inertial	Multi-sensor Homing
Accuracy (m) (Max Range)	5-10 IR-Hmg 15 without	150	1,000	50	300	<800	600	4,000	2,000-2,500	Improved Guidance
Payload (kg)	480	190	1,000	600	800	700	500-600	770	1,500-2,150	Separating multiple RVs
Warheads	HE, Chem, ARM, Nuc, IR Homing, APAM, ICM, EMP, DPICM	HE, Chem	HE, Chem, Nuc	Separating HE, Nuc	Separating HE, Nuc poss Chem	HE, Chem	Separating. HE, Nuc poss Chem Poss Fuel-Air	HE, Chem poss Nuc	HE, Nuc, or 3 separating reentry vehicles (RVs)	Cluster, Volumetric, Submunitions

							Submunitions			BW warheads ARM, EMP
Comments	TEL is amphibious 2 msls/TEL	Modified SA-2 SAM Tracked TEL	Technology widely used	Previously called SCUD-E Requires compatible IR imagery	Exported as M-11	SCUD-B variant Russia limited production	Mod 2 range 1000 km DF-15B CEP 150-500 m DF-15C CEP 35-50 m	SCUD-B variant ND-2 IRBM variant Poss export	Variants with varied warheads and ranges Towed launcher Lengthy prep time	Autonomous operation, Penaids*/ Counter-measures, Reduced prep /displace times

* Penaids - Penetration aids, such as RF jammer

Cruise Missiles

In the global arena many countries, including potential Threats to the U.S., are procuring cruise missiles (CM) as an inexpensive alternative to ballistic missiles and aircraft. CMs are economical and accurate delivery systems that can be used to deliver conventional, and nuclear, chemical and biological warheads. CM proliferation poses an increasing threat to U.S. National security interests. As the technology matures further, both State actors and non-state actors are becoming increasingly able to acquire cruise missile and effectively employ CM capabilities. The Hezbollah 2006 cruise missile attack on the INS Hanit illustrates the danger to units that are not technically prepared to meet this challenge.

Many older CMs are still used in less capable military forces. They fly a straight course to target with relatively slow speed (subsonic), are vulnerable to early detection, and can be shot down. Due to imprecision in guidance systems and the difficulty of flying long distance overland to ground targets, they are used as ***anti-ship missiles***. But in most forces they are being replaced by newer systems.

Cruise missiles (CM) are unmanned precision aerodynamic munitions with warheads propelled by rocket motors or jet engines, and designed to consistently fly a non-ballistic trajectory to the target. The diagram below illustrates the four main components of a basic cruise missile: (1) a propulsion system, (2) guidance and control system, (3) airframe, and (4) payload. CMs may have booster rockets which fall off after fuel is depleted. Then turbofan engine engages, the tail fins, and air inlet, and wings unfold. At the target the missile either dispenses its submunitions or impacts the target and is destroyed.

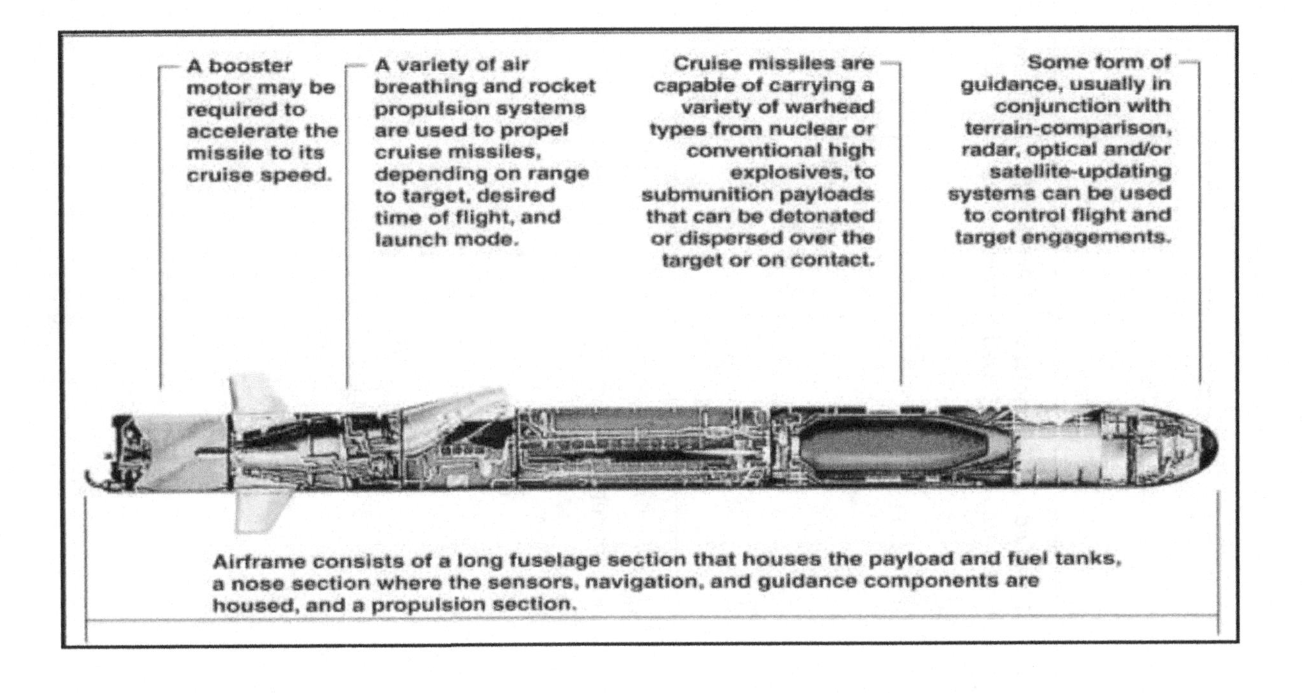

A booster motor may be required to accelerate the missile to its cruise speed.
A variety of air breathing and rocket propulsion systems are used to propel cruise missiles, depending on range to target, desired time of flight, and launch mode.
Cruise missiles are capable of carrying a variety of warhead types from nuclear or conventional high explosives, to submunition payloads that can be detonated or dispersed over the target or on contact.
Some form of guidance, usually in conjunction with terrain-comparison, radar, optical and/or satellite-updating systems can be used to control flight and target engagements.
Airframe consists of a long fuselage section that houses the payload and fuel tanks, a nose section where the sensors, navigation, and guidance components are housed, and a propulsion section.

The overall sophistication of CMs has increased greatly with technological advancements. This is especially true with regard to guidance systems in the era of more capable Global Navigation Satellite Systems (GNSS) like GPS, Russian GLONASS, Chinese Beidou and the European Galileo. These advanced guidance systems, in combination with autonomous onboard systems, have allowed CMs to become more accurate in acquiring targets. The basic CM guidance controls consist of one of four different systems (below) that direct the missile to its target. Most newer CMs use a combination of systems to provide redundancy and precision in a combat environment.

1. Inertial Guidance System (IGS) tracks acceleration via accelerometers from missile movement compared against a known first position, usually the launch position to determine current location.
2. Terrain Contour Matching (TERCOM) uses a radar or laser altitude system, and compares terrain features enroute to a pre- loaded 3-D map terrain database.
3. GNSS (e.g., GPS), uses satellites and an onboard receiver to verify the missile's position.
4. Digital Scene Matching Area Correlation (DSMAC) uses a camera and image correlator to identify the target (good versus moving targets).

The most effective mix is IGS on the airframe, with TERCOM and/or GNSS with multiple route waypoints. Upon arrival in the target area, the missile can loiter or home based on warhead identification of target DSMAC, GNSS, or radiation confirmation. Some CMs can change route and target assignment while enroute, to maximize their effectiveness.

Technology of CMs is changing; and their role is expanding. CMs are relatively mobile and easy to conceal. Even after launch the missiles can avoid detection by traveling at low altitude, under many radar horizon and use terrain masking until the CM reaches the target. The newer CMs present even greater challenges to aircraft and air defense assets by integrating stealth features that make them even less visible to radars and infrared sensors. CMs can take roundabout routes to engage their targets, and are usually programmed to circumvent known defenses and engage targets from gaps in radar and SAM coverage. Modern cruise missiles offer flexibility for different configurations, and for air, sea, and ground-launch. In the COE, ground-launched CMs (GLCMs) can fly to targets within artillery range to support artillery fires, or deep to attack high-value ground targets. A CM's size, alterable course, and unique low flight profile makes it a convenient system for dispensing chemical or biological agents, for jamming, and for designating targets with an LTD. Examples of applications include *Exocet* and *Apache*. Swedish Bofors, South African Denel, and German LFK offer similar systems.

CMs used against ground targets are referred to as ***land-attack cruise missiles*** (***LACMs***). They can be ground, ship, or air-launched. Precision guidance has permitted rapid growth of multi-role ***air-launched cruise missiles*** **(ALCMs)**, for use against various naval and ground targets. ALCMs for land-attack are included in WEG chapters on aircraft (9 and 10) or in later issues. Cruise missiles vary in size, range (25-2,500+ km), and warhead payload. Larger ones can actually be manned bomber aircraft loaded with ordnance and controlled by a remote pilot system. An innovative modern small CM is the Harpy (pg 5-13), which can launch 18 missiles from a truck "cassette launcher". The BrahMos (pg 5-14) is an example of an operational level supersonic GLCM system, with future applications on other platforms. Initial uses are against ships, as well as high value nodes, such as airfields, C4, and missile launch sites. BrahMos ALCM and ship-launched versions are due out soon.

ISRAELI MISSILE/ATTACK UNMANNED AERIAL VEHICLE HARPY, CUTLASS

SYSTEM	SPECIFICATIONS	PAYLOAD TYPES	SPECIFICATIONS
Alternative designations:	See Notes	Optical Camera:	Electro-magnetic and optical
Date of introduction:	1988	Passive Radar Seeker:	Wide range of frequencies
Proliferation:	At least 5 countries	User Image Capabilities:	Receive images on possible targets via datalink
Ground Crew:	1-3 per truck launcher	Missile:	HE Fragmentation warhead
Engine (hp):	27.5, 2 cylinder, 2 stroke	Max Payload, Warhead (kg):	32
Propulsion:	Two blade pusher propeller	**VARIANTS**	**SPECIFICATIONS**
Weight (kg):		CUTLASS:	
Max Launch Weight (kg):	120	Weight (kg):	2.8
Speed (km/h):		Wing Span (m):	1.83
Maximum (level):	250	Max Launch Weight (kg):	125
Cruising Speed:	185	Max Ceiling (m):	4,575
Endurance (hrs):	6	Radius of Operation (km):	300
Max Ceiling, normal (m):	INA	Cruising Speed:	185
Ceiling, Service (m):	3,000	Endurance (hrs):	6
Radius of Operation (km):	500	Max Range (km):	1,000
Dimensions:		**PAYLOAD TYPES**	**SPECIFICATIONS**
Wing Span (m):	2	Missile:	HE Fragmentation warhead
Overall Length (m):	2.2	Max Payload, Warhead (lbs)	51
Overall Height (m):	.36	Data Link (direct LOS) (km):	150, 1,000 with GPS
Flight Control:	GPS positioning, autonomous preprogrammed flight	Infrared (IR)	Raytheon Seeker Head AIM-9X
Launch Method:	Booster rocket launched from truck launcher	Automatic Target Recognition and Classification	Raytheon Algorithms INA
Recovery/Landing Method:	Non-recoverable		
Launcher Trucks per Battery	3 4X4 or 4x6 medium trucks		

Missiles per Launcher:	18		
Total Missiles per Battery:	54		

NOTES

HARPY CAN BE USED AS CRUISE MISSILE IN PREPROGRAMED MODE. BUT IT CAN ALSO BE CONSIDERED A UCAV, WHICH CAN BE PILOTED OR USED WITHOUT A PILOT (E.G. PROGRAMMED OR HOMING ATTACK MODE). THE HARPY AND CUTLASS CAN ALSO ATTACK ARTILLERY COUNTER-BATTERY RADARS AND GROUND SURVEILLANCE RADARS. HARPY RADAR SENSOR CAN BE REMOTELY TURNED OFF TO ABORT A TARGET AND CONTINUE SEARCHING. DAY OR NIGHT FLIGHT CAPABILITY. CUTLASS STANDS FOR COMBAT UNINHABITED TARGET LOCATE AND STRIKE SYSTEM. BUILT JOINTLY WITH RAYTHEON. PRIMARY DIFFERENCE FROM THE HARPY IS THE CUTLASS IS A SEMI-AUTONOMOUS WITH A REAL TIME - DATA LINK TO THE GROUND CONTROL STATION TO CONFIRM TARGET IDENTITY. IT CAN ALSO TARGET BALLISTIC MISSILES LAUNCHERS AND VEHICLES. DESIGNED TO BE FIRED FROM NAVAL SHIPS.

INDIAN/RUSSIAN SUPERSONIC CRUISE MISSILE BRAHMOS AND BRAHMOS II

		Weapons & Ammunition Types	Typical Combat Load
		Missiles on launcher	3
SYSTEM	**SPECIFICATIONS**	Missile	
Alternative Designations	PJ-10	Name	BrahMos
Date of Introduction	By 2006. First Army ground launch regiment was fielded in 2007.	Type	Two-stage, solid-propellant launch and kerosene ram-jet cruise
Proliferation	Developed and offered for export. Russian system is fielded in at least 1 country. Indian contract signed for $2 billion in missiles. Talks have been held with five other countries.	Launch Mode	Angular or vertical
ARMAMENT	**SPECIFICATIONS**	Max Launch Range (km)	290
Transporter-Erector-Launcher		Max Launch Range (km)	INA
Name	Tatra variant (NFI)	Max Altitude (m)	14,000
Crew	3 (est)	Min Altitude (m)	5-10
Chassis	12x12	Missile Speed	Mach 2.8-3.0
Description	It is described as a high-mobility truck (NFI) built indigenously for the MAL.	Length (m)	8.9
Radio	INA	Diameter (mm)	670
Armor Protection	None	Weight (kg)	3,000
NBC Protection System	None	Weight with Canister (kg)	4,500
Launcher Performance		Warhead Weight (kg)	Weight (kg): 250
Land Navigation	GNSS	Warhead Type	Shaped Charge anti-ship
Missiles per launcher	3	Other Warheads	BrahMos A weighs 300 kg. For ground targets, HE warhead is available.
Total emplace time (min)	5		

DESCRIPTION

PRIMARILY DEVELOPED AS AN ANTI-SHIP MISSILE. IT CAN BE USED AS A LAND-ATTACK CRUISE MISSILE (LACM). LAUNCHERS INCLUDE LAND-BASED TEL, AIRCRAFT AND SHIPS (E.G., DESTROYERS). IT CAN ALSO BE LAUNCHED FROM SUBMARINE, FIXED GROUND SITE OR PONTOON UNDERWATER SILO.

PRIMARY COMPONENTS

TRANSPORTER-ERECTOR-LAUNCHER (TEL) IS CALLED A MOBILE AUTONOMOUS LAUNCHER (MAL) LINKED INTO AN INTEGRATED FIRES COMMAND (IFC). THERE IS ALSO A MOBILE COMMAND POST (MCP) WITH IT. RELOAD MISSILES WILL BE LOADED AT A TRANSLOAD POINT FROM A TRANSLOADER VEHICLE (SEE ABOVE).

COUNTERMEASURES

MISSILE SHIFTS FROM RADAR TO INERTIAL AT THE END OF ITS HIGH APPROACH PHASE, USES TERRAIN DATA TO SHIFT TO THE LOW APPROACH, THEN AND USES RADAR FOR ITS COURSE CORRECTION. LOSS OF RADAR DUE TO JAMMING OR OTHER CAUSE STILL PERMITS INERTIAL GUIDANCE OFF ITS LATEST COURSE. HIGH SPEED AND LOW FLIGHT MODE WILL CHALLENGE ALMOST ALL DETECTION AND INTERCEPT RADAR AND WEAPON SYSTEMS.

GUIDANCE

INERTIAL, WITH GNSS MID-COURSE CORRECTION SENSOR WITH UP TO 20-KM ADJUSTMENT FROM A DISTANCE UP TO 50 KM OUT. TERMINAL HOMING RADAR CORRELATOR.

TRAJECTORY

NON-BALLISTIC. MOST LIKELY USE IS HI-LO PROFILE (HIGH, EARLY PHASE, LOW ON APPROACH TO TARGET).

ACCURACY

HOMES TO SHIP AND AIMS USING RADAR CORRELATION TO HIT CENTROID. ACCURACY VARIES BY SEEKER, WITH <20 M.

VARIANTS

THIS IS AN INDIAN-PRODUCED SYSTEM FROM A RUSSIAN-INDIAN JOINT VENTURE. IT IS A VARIANT OF RUSSIAN SS-N-26/YAKHONT, AKA 3M55 ONIKS. THE SUPERSONIC YAKHONT HAS BEEN EXPORTED.

THE RUSSIAN MISSILE HAS A RANGE OF 300 KM WITH HI-LO FLIGHT PROFILE. THE RUSSIANS EMPLOY THE YAKHONT IN -STRIKE COMPLEXES (RSCS - SIMILAR TO INTEGRATED FIRES COMMANDS).

BRAHMOS A: AERIAL LAUNCH VERSION. LAUNCH TESTS FROM SU-30MKI FIGHTERS ARE IMMINENT.

BRAHMOS ARMY VERSION: FEATURES INCLUDE TERRAIN FOLLOWING CAPABILITY. AN IR SEEKER WILL BE AVAILABLE FOR THE ARMY VERSION

BRAHMOS II: AIR-LAUNCHED HYPERSONIC CM IS APPROVED FOR FIELDING. EXPECTED SPEED IS MACH IS 5+.

A BRAHMOS SHIPBOARD LAUNCHER IS IN TESTING, AND IS DUE OUT SOON, AS IS A SUB LAUNCH VERSION.

NOTES

BRAHMOS 2 IS A CONCEPT FOR A FUTURE INDIAN HYPERSONIC CRUISE MISSILE WITH MACH 6-7 VELOCITY.

INDIAN/RUSSIAN SUPERSONIC CRUISE MISSILE BRAHMOS AND BRAHMOS II

		Weapons & Ammunition Types	Typical Combat Load
		Missiles on launcher	3
SYSTEM	**SPECIFICATIONS**	Missile	
Alternative Designations	PJ-10	Name	BrahMos
Date of Introduction	By 2006. First Army ground launch regiment was fielded in 2007.	Type	Two-stage, solid-propellant launch and kerosene ram-jet cruise
Proliferation	Developed and offered for export. Russian system is fielded in at least 1 country. Indian contract signed for $2 billion in missiles. Talks have been held with five other countries.	Launch Mode	Angular or vertical
ARMAMENT	**SPECIFICATIONS**	Max Launch Range (km)	290
Transporter-Erector-Launcher		Max Launch Range (km)	INA
Name	Tatra variant (NFI)	Max Altitude (m)	14,000
Crew	3 (est)	Min Altitude (m)	5-10
Chassis	12x12	Missile Speed	Mach 2.8-3.0
Description	It is described as a high-mobility truck (NFI) built indigenously for the MAL.	Length (m)	8.9
Radio	INA	Diameter (mm)	670
Armor Protection	None	Weight (kg)	3,000
NBC Protection System	None	Weight with Canister (kg)	4,500
Launcher Performance		Warhead Weight (kg)	Weight (kg): 250
Land Navigation	GNSS	Warhead Type	Shaped Charge anti-ship
Missiles per launcher	3	Other Warheads	BrahMos A weighs 300 kg. For ground targets, HE warhead is available.

Total emplace time (min)	5		

DESCRIPTION

PRIMARILY DEVELOPED AS AN ANTI-SHIP MISSILE. IT CAN BE USED AS A LAND-ATTACK CRUISE MISSILE (LACM). LAUNCHERS INCLUDE LAND-BASED TEL, AIRCRAFT AND SHIPS (E.G., DESTROYERS). IT CAN ALSO BE LAUNCHED FROM SUBMARINE, FIXED GROUND SITE OR PONTOON UNDERWATER SILO.

PRIMARY COMPONENTS

TRANSPORTER-ERECTOR-LAUNCHER (TEL) IS CALLED A MOBILE AUTONOMOUS LAUNCHER (MAL) LINKED INTO AN INTEGRATED FIRES COMMAND (IFC). THERE IS ALSO A MOBILE COMMAND POST (MCP) WITH IT. RELOAD MISSILES WILL BE LOADED AT A TRANSLOAD POINT FROM A TRANSLOADER VEHICLE (SEE ABOVE).

COUNTERMEASURES

MISSILE SHIFTS FROM RADAR TO INERTIAL AT THE END OF ITS HIGH APPROACH PHASE, USES TERRAIN DATA TO SHIFT TO THE LOW APPROACH, THEN AND USES RADAR FOR ITS COURSE CORRECTION. LOSS OF RADAR DUE TO JAMMING OR OTHER CAUSE STILL PERMITS INERTIAL GUIDANCE OFF ITS LATEST COURSE. HIGH SPEED AND LOW FLIGHT MODE WILL CHALLENGE ALMOST ALL DETECTION AND INTERCEPT RADAR AND WEAPON SYSTEMS.

GUIDANCE

INERTIAL, WITH GNSS MID-COURSE CORRECTION SENSOR WITH UP TO 20-KM ADJUSTMENT FROM A DISTANCE UP TO 50 KM OUT. TERMINAL HOMING RADAR CORRELATOR.

TRAJECTORY

NON-BALLISTIC. MOST LIKELY USE IS HI-LO PROFILE (HIGH, EARLY PHASE, LOW ON APPROACH TO TARGET).

ACCURACY

HOMES TO SHIP AND AIMS USING RADAR CORRELATION TO HIT CENTROID. ACCURACY VARIES BY SEEKER, WITH <20 M.

VARIANTS

THIS IS AN INDIAN-PRODUCED SYSTEM FROM A RUSSIAN-INDIAN JOINT VENTURE. IT IS A VARIANT OF RUSSIAN SS-N-26/YAKHONT, AKA 3M55 ONIKS. THE SUPERSONIC YAKHONT HAS BEEN EXPORTED.

THE RUSSIAN MISSILE HAS A RANGE OF 300 KM WITH HI-LO FLIGHT PROFILE. THE RUSSIANS EMPLOY THE YAKHONT IN RECONNAISSANCE-STRIKE COMPLEXES (RSCS - SIMILAR TO INTEGRATED FIRES COMMANDS).

BRAHMOS A: AERIAL LAUNCH VERSION. LAUNCH TESTS FROM SU-30MKI FIGHTERS ARE IMMINENT.

BRAHMOS ARMY VERSION: FEATURES INCLUDE TERRAIN FOLLOWING CAPABILITY. AN IR SEEKER WILL BE AVAILABLE FOR THE ARMY VERSION

BRAHMOS II: AIR-LAUNCHED HYPERSONIC CM IS APPROVED FOR FIELDING. EXPECTED SPEED IS MACH IS 5+.

A BRAHMOS SHIPBOARD LAUNCHER IS IN TESTING, AND IS DUE OUT SOON, AS IS A SUB LAUNCH VERSION.

NOTES

BRAHMOS 2 IS A CONCEPT FOR A FUTURE INDIAN HYPERSONIC CRUISE MISSILE WITH MACH 6-7 VELOCITY.

INDIAN/RUSSIAN SUPERSONIC CRUISE MISSILE BRAHMOS AND BRAHMOS II

		Weapons & Ammunition Types	Typical Combat Load
		Missiles on launcher	3
SYSTEM	**SPECIFICATIONS**	Missile	
Alternative Designations	PJ-10	Name	BrahMos
Date of Introduction	By 2006. First Army ground launch regiment was fielded in 2007.	Type	Two-stage, solid-propellant launch and kerosene ram-jet cruise
Proliferation	Developed and offered for export. Russian system is fielded in at least 1 country. Indian contract signed for $2 billion in missiles. Talks have been held with five other countries.	Launch Mode	Angular or vertical
ARMAMENT	**SPECIFICATIONS**	Max Launch Range (km)	290
Transporter-Erector-Launcher		Max Launch Range (km)	INA
Name	Tatra variant (NFI)	Max Altitude (m)	14,000
Crew	3 (est)	Min Altitude (m)	5-10
Chassis	12x12	Missile Speed	Mach 2.8-3.0
Description	It is described as a high-mobility truck (NFI) built indigenously for the MAL.	Length (m)	8.9
Radio	INA	Diameter (mm)	670
Armor Protection	None	Weight (kg)	3,000
NBC Protection System	None	Weight with Canister (kg)	4,500
Launcher Performance		Warhead Weight (kg)	Weight (kg): 250
Land Navigation	GNSS	Warhead Type	Shaped Charge anti-ship
Missiles per launcher	3	Other Warheads	BrahMos A weighs 300 kg. For ground targets, HE warhead is available.
Total emplace time (min)	5		

DESCRIPTION

PRIMARILY DEVELOPED AS AN ANTI-SHIP MISSILE. IT CAN BE USED AS A LAND-ATTACK CRUISE MISSILE (LACM). LAUNCHERS INCLUDE LAND-BASED TEL, AIRCRAFT AND SHIPS (E.G., DESTROYERS). IT CAN ALSO BE LAUNCHED FROM SUBMARINE, FIXED GROUND SITE OR PONTOON UNDERWATER SILO.

PRIMARY COMPONENTS

TRANSPORTER-ERECTOR-LAUNCHER (TEL) IS CALLED A MOBILE AUTONOMOUS LAUNCHER (MAL) LINKED INTO AN INTEGRATED FIRES COMMAND (IFC). THERE IS ALSO A MOBILE COMMAND POST (MCP) WITH IT. RELOAD MISSILES WILL BE LOADED AT A TRANSLOAD POINT FROM A TRANSLOADER VEHICLE (SEE ABOVE).

COUNTERMEASURES

MISSILE SHIFTS FROM RADAR TO INERTIAL AT THE END OF ITS HIGH APPROACH PHASE, USES TERRAIN DATA TO SHIFT TO THE LOW APPROACH, THEN AND USES RADAR FOR ITS COURSE CORRECTION. LOSS OF RADAR DUE TO JAMMING OR OTHER CAUSE STILL PERMITS INERTIAL GUIDANCE OFF ITS LATEST COURSE. HIGH SPEED AND LOW FLIGHT MODE WILL CHALLENGE ALMOST ALL DETECTION AND INTERCEPT RADAR AND WEAPON SYSTEMS.

GUIDANCE

INERTIAL, WITH GNSS MID-COURSE CORRECTION SENSOR WITH UP TO 20-KM ADJUSTMENT FROM A DISTANCE UP TO 50 KM OUT. TERMINAL HOMING RADAR CORRELATOR.

TRAJECTORY

NON-BALLISTIC. MOST LIKELY USE IS HI-LO PROFILE (HIGH, EARLY PHASE, LOW ON APPROACH TO TARGET).

ACCURACY

HOMES TO SHIP AND AIMS USING RADAR CORRELATION TO HIT CENTROID. ACCURACY VARIES BY SEEKER, WITH <20 M.

VARIANTS

THIS IS AN INDIAN-PRODUCED SYSTEM FROM A RUSSIAN-INDIAN JOINT VENTURE. IT IS A VARIANT OF RUSSIAN SS-N-26/YAKHONT, AKA 3M55 ONIKS. THE SUPERSONIC YAKHONT HAS BEEN EXPORTED.

THE RUSSIAN MISSILE HAS A RANGE OF 300 KM WITH HI-LO FLIGHT PROFILE. THE RUSSIANS EMPLOY THE YAKHONT IN RECONNAISSANCE-STRIKE COMPLEXES (RSCS - SIMILAR TO INTEGRATED FIRES COMMANDS).

BRAHMOS A: AERIAL LAUNCH VERSION. LAUNCH TESTS FROM SU-30MKI FIGHTERS ARE IMMINENT.

BRAHMOS ARMY VERSION: FEATURES INCLUDE TERRAIN FOLLOWING CAPABILITY. AN IR SEEKER WILL BE AVAILABLE FOR THE ARMY VERSION

BRAHMOS II: AIR-LAUNCHED HYPERSONIC CM IS APPROVED FOR FIELDING. EXPECTED SPEED IS MACH IS 5+.

A BRAHMOS SHIPBOARD LAUNCHER IS IN TESTING, AND IS DUE OUT SOON, AS IS A SUB LAUNCH VERSION.

NOTES

BRAHMOS 2 IS A CONCEPT FOR A FUTURE INDIAN HYPERSONIC CRUISE MISSILE WITH MACH 6-7 VELOCITY.

INDIAN/RUSSIAN SUPERSONIC CRUISE MISSILE BRAHMOS AND BRAHMOS II

		Weapons & Ammunition Types	Typical Combat Load
		Missiles on launcher	3
SYSTEM	**SPECIFICATIONS**	Missile	
Alternative Designations	PJ-10	Name	BrahMos
Date of Introduction	By 2006. First Army ground launch regiment was fielded in 2007.	Type	Two-stage, solid-propellant launch and kerosene ram-jet cruise
Proliferation	Developed and offered for export. Russian system is fielded in at least 1 country. Indian contract signed for $2 billion in missiles. Talks have been held with five other countries.	Launch Mode	Angular or vertical
ARMAMENT	**SPECIFICATIONS**	Max Launch Range (km)	290
Transporter-Erector-Launcher		Max Launch Range (km)	INA
Name	Tatra variant (NFI)	Max Altitude (m)	14,000
Crew	3 (est)	Min Altitude (m)	5-10
Chassis	12x12	Missile Speed	Mach 2.8-3.0
Description	It is described as a high-mobility truck (NFI) built indigenously for the MAL.	Length (m)	8.9
Radio	INA	Diameter (mm)	670
Armor Protection	None	Weight (kg)	3,000
NBC Protection System	None	Weight with Canister (kg)	4,500
Launcher Performance		Warhead Weight (kg)	Weight (kg): 250
Land Navigation	GNSS	Warhead Type	Shaped Charge anti-ship
Missiles per launcher	3	Other Warheads	BrahMos A weighs 300 kg. For ground targets, HE warhead is available.
Total emplace time (min)	5		

DESCRIPTION

PRIMARILY DEVELOPED AS AN ANTI-SHIP MISSILE. IT CAN BE USED AS A LAND-ATTACK CRUISE MISSILE (LACM). LAUNCHERS INCLUDE LAND-BASED TEL, AIRCRAFT AND SHIPS (E.G., DESTROYERS). IT CAN ALSO BE LAUNCHED FROM SUBMARINE, FIXED GROUND SITE OR PONTOON UNDERWATER SILO.

PRIMARY COMPONENTS

TRANSPORTER-ERECTOR-LAUNCHER (TEL) IS CALLED A MOBILE AUTONOMOUS LAUNCHER (MAL) LINKED INTO AN INTEGRATED FIRES COMMAND (IFC). THERE IS ALSO A MOBILE COMMAND POST (MCP) WITH IT. RELOAD MISSILES WILL BE LOADED AT A TRANSLOAD POINT FROM A TRANSLOADER VEHICLE (SEE ABOVE).

COUNTERMEASURES

MISSILE SHIFTS FROM RADAR TO INERTIAL AT THE END OF ITS HIGH APPROACH PHASE, USES TERRAIN DATA TO SHIFT TO THE LOW APPROACH, THEN AND USES RADAR FOR ITS COURSE CORRECTION. LOSS OF RADAR DUE TO JAMMING OR OTHER CAUSE STILL PERMITS INERTIAL GUIDANCE OFF ITS LATEST COURSE. HIGH SPEED AND LOW FLIGHT MODE WILL CHALLENGE ALMOST ALL DETECTION AND INTERCEPT RADAR AND WEAPON SYSTEMS.

GUIDANCE

INERTIAL, WITH GNSS MID-COURSE CORRECTION SENSOR WITH UP TO 20-KM ADJUSTMENT FROM A DISTANCE UP TO 50 KM OUT. TERMINAL HOMING RADAR CORRELATOR.

TRAJECTORY

NON-BALLISTIC. MOST LIKELY USE IS HI-LO PROFILE (HIGH, EARLY PHASE, LOW ON APPROACH TO TARGET).

ACCURACY

HOMES TO SHIP AND AIMS USING RADAR CORRELATION TO HIT CENTROID. ACCURACY VARIES BY SEEKER, WITH <20 M.

VARIANTS

THIS IS AN INDIAN-PRODUCED SYSTEM FROM A RUSSIAN-INDIAN JOINT VENTURE. IT IS A VARIANT OF RUSSIAN SS-N-26/YAKHONT, AKA 3M55 ONIKS. THE SUPERSONIC YAKHONT HAS BEEN EXPORTED.

THE RUSSIAN MISSILE HAS A RANGE OF 300 KM WITH HI-LO FLIGHT PROFILE. THE RUSSIANS EMPLOY THE YAKHONT IN RECONNAISSANCE-STRIKE COMPLEXES (RSCS - SIMILAR TO INTEGRATED FIRES COMMANDS).

BRAHMOS A: AERIAL LAUNCH VERSION. LAUNCH TESTS FROM SU-30MKI FIGHTERS ARE IMMINENT.

BRAHMOS ARMY VERSION: FEATURES INCLUDE TERRAIN FOLLOWING CAPABILITY. AN IR SEEKER WILL BE AVAILABLE FOR THE ARMY VERSION

BRAHMOS II: AIR-LAUNCHED HYPERSONIC CM IS APPROVED FOR FIELDING. EXPECTED SPEED IS MACH IS 5+.

A BRAHMOS SHIPBOARD LAUNCHER IS IN TESTING, AND IS DUE OUT SOON, AS IS A SUB LAUNCH VERSION.

NOTES

BRAHMOS 2 IS A CONCEPT FOR A FUTURE INDIAN HYPERSONIC CRUISE MISSILE WITH MACH 6-7 VELOCITY.

ISRAELI LYNX ROCKET/MISSILE LAUNCHER WITH EXTRA AND DELILAH MISSILES

Weapons & Ammunition Types	Typical Combat Load
Rocket/ Missile Modules	2
Grad-type Rocket	40
LAR/AccuLAR Rocket	26
EXTRA Missile	8
Delilah Cruise Missile	2
Mixed Loads on Modules	½ each module

SYSTEM	SPECIFICATIONS	AMMUNITION	SPECIFICATIONS
Alternative Designations	Lynx is both the launcher module which can fit on various mounts and the Israeli launcher vehicle name.	Launcher Performance	
Date of Introduction	By 2007. Delilah cruise missile used in combat in 2006.	Land Navigation	GPS/inertial
Proliferation	At least 3 countries. Two others are testing versions of the system and adaptations of rockets and/or missiles. Others are looking at adopting TCS to their MRLs	Missiles per launcher	See the Loads above. They can use separate loads on the 2 modules (or launch pod containers, LPCs).
ARMAMENT	**SPECIFICATIONS**	Total emplace time (min)	5
Transporter-Erector-Launcher		Reload Time (min)	20
Name	Mercedes 3341	**AMMUNITION**	**SPECIFICATION**
Crew	3	Name	LAR-160 Rocket
Chassis	6x6	Type	Composite solid-propellant
Range	500 km (est)	Max Launch Range (km)	45
Radio		Min Launch Range (km)	10
Protection	Armor Protection: None. The LAROM and perhaps other variants are armored. NBC Protection System: INA	Rocket Speed (m/s)	1,022
Armor Protection	None. The LAROM and perhaps other variants are armored	Length (m)	3.48
NBC Protection System	INA	Diameter (mm)	160
		Weight (kg)	110
		Warhead Options	Frag-HE/PD or DPICM with time-fuze dispense.

DESCRIPTION

BECAUSE THE LAUNCHER CAN LAUNCH A VARIETY OF ROCKETS (122 MM OF VARIOUS, 160 MM ISRAELI LAR, WITH OR WITHOUT TCS), AND EITHER EXTRA OR DELILAH-GL MISSILES, IT IS LIKELY THAT THE PRIMARY MUNITION MIX WILL DEPEND ON ORGANIZATION LEVEL OF THE LAUNCHER. IF IT IS AT TACTICAL LEVEL, IT IS LIKELY TO BE USED PRIMARILY TO LAUNCH ROCKETS, WITH A FEW MAYBE DESIGNATED FOR EXTRA MISSILES. THOSE LAUNCHERS AT THE OPERATIONAL/STRATEGIC LEVEL ARE MORE LIKELY TO LAUNCH MISSILES, AND PERHAPS ACCULAR (LAR-160 WITH TCS) ROCKETS.

PRIMARY COMPONENTS

TRANSPORTER-ERECTOR-LAUNCHER (TEL) AND MOBILE COMMAND POST (MCP) VAN. RELOAD MODULES WILL BE TRANSLOADED AT A TL POINT FROM A TRANSLOADER TRUCK WITH FOUR MODULES, TO SERVICE TWO LAUNCHERS.

OTHER AMMUNITION

GRADLAR: ISRAELI UPGRADE PACKAGE WITH IMPROVED FCS CONVERTS MRLS FOR MODULES OF 122-MM GRAD ROCKETS AND 21-45 KM RANGE. ANY TYPE OF GRAD 122-MM ROCKET CAN BE USED.

LAR-160 OR LAR: 160-MM ROCKET (13 PER MODULE) WITH A 45-KM RANGE. THE WARHEAD IS A CANISTER; TO CARRY FRAG-HE, SUB-MUNITIONS, OR ANY 155-MM ROUND.

GUIDED ROCKETS AND MISSILES ON LYNX AND OTHER MRLS/TELS CAN USE THE TRAJECTORY CORRECTION SYSTEM (TCS). TCS CAN CONTROL >12 ROCKETS/MISSILES EQUIPPED FOR INERTIAL/GPS GUIDANCE, VS 12 SEPARATE TARGETS. ACCURACY IS 10 M. INDIA TESTED TCS ON THE PINAKA MRL, AND USES IT IN THE RECENTLY TESTED PRAHAAR SRBM.

ACCULAR ROCKET: A GPS FUZED VARIANT OF LAR-160, WITH 14-40 KM RANGE AND 10 M CEP). AT LEAST 4 COUNTRIES USE THESE ROCKETS.

EXTRA (EXTENDED RANGE ARTILLERY): THE 300MM BALLISTIC MISSILE (4/LAUNCH MODULE) RANGES 150 KM WITH A 10-M CEP. IT HAS A 120-KG PAYLOAD, AND FLIES A BALLISTIC TRAJECTORY, CORRECTED WITH GPS. VARIOUS WARHEADS ARE OFFERED.

DELILAH: THIS CRUISE MISSILE HAS A LENGTH OF 3.2 M, WEIGHING 230 KG. IT CRUISES AT MACH 0.3-0.7, AND 8,600 M ALTITUDE. IT CAN BE LAUNCHED FROM SHIPS, AIRCRAFT, AND THE LYNX GROUND LAUNCHER (GL) TO 250 KM, WITH PROGRAMMABLE GUIDANCE, AND MULTIPLE WAYPOINTS. DELILAH-GL HAS LAUNCH ASSIST. AIR, SHIP, AND HELICOPTER VERSIONS ARE OFFERED. THE MISSILE USES GPS HOMING, OR CAN LOITER AND USE A CCD/FLIR SEEKER TO HOME TO TARGET.

VARIANTS

LYNX: IS BOTH A VEHICLE, AND A LAUNCHER TO FIT ON VEHICLES. GROUND LAUNCHERS INCLUDE TRACKED ARMORED VEHICLES AND 8X8 TRUCKS. ISRAEL MARKETS THE LYNX 6X6 TRUCK (ABOVE). BUT THE LAUNCHER FITS ON OTHER USER-PREFERRED CHASSIS. OTHER USER COUNTRIES HAVE LICENSES FOR THE CONVERSION. MANY OF THE CUSTOMERS HAVE SUBSTANTIAL SUPPLIES OF 122-MM ROCKETS.

AZERBAIJAN LYNX: INDIGENOUS MRL/MISSILE TEL WITH LYNX LAUNCHER ON 8X8 KAMAZ-6350 TRUCK. WITH AUTONOMOUS FCS, IT LAUNCHES 122/ 160 MM ROCKETS, OR EXTRA BALLISTIC MISSILES

NAIZA: KAZAKH IMPORT/PRODUCTION MRL WITH LYNX FOR LAR-160 ON KAMAZ TRUCK.

LAROM: ROMANIAN 2-MODULE MRL CAN LAUNCH 122-MM GRAD OR LAR-160 ROCKETS

NOTES

THE LAR-160 ROCKET OFFERS A LETHAL EFFECTS AREA PER ROCKET OF 31,400 M2. WITH TCS (E.G., ACCULAR), ROCKETS PERFORM A PITCH-OVER FOR TOP ATTACK AND AN OPTIMIZED CIRCULAR PATTERN FOR FRAG-HE WARHEAD EFFECTS OR SUB-MUNITIONS. THUS, ACCULAR ROCKETS SHOULD HAVE EVEN GREATER LETHAL EFFECTS.

Other Options for Land-Attack

The overall decline in military budgets is likely to restrict the number of high-technology cruise missiles for land-attack to strategic and operational-strategic systems. For operational level, newer and lower-cost technologies such as semi-active laser-homing (SAL-H) and fiber-optic guidance (FOG), coupled with preprogrammed inertial/GNSS navigation, offer more precision long-range strike systems for forces with somewhat constricted budgets. Examples are *Nimrod* and *Hermes*. These systems are extensions of ATGM technologies, but with fire control mechanisms which resemble those of precision-guided artillery (see Vol 1, pgs 6-72 to 75). An example of a bridge system is the Israeli Nimrod 3 (SAL-H), which is listed with the NLOS ATGMs; but its range (55+km) places it in the same range band as precision guided artillery. Better-equipped forces (Tiers 1 and 2) have some AT units for long-range AT strikes, and perhaps in artillery units in the Integrated Fires Command (IFC), against high value targets. A Russian counterpart is Hermes SAL-H missile (initially 18 km) also listed with NLOS ATGMs. By Near Term it will range 100 km, for strikes against deeper high-value targets and guided by UAVs with laser target designators.

Another type of affordable technology cruise missile has emerged—the attack UAV. UAVs differ from cruise missiles in that an operator can guide the aerial vehicle, using its downloaded camera view and ground station controls. Most early ones used less precise pre-programmed inertial guidance, but with camera guidance for a precise hit-to-kill terminal phase. High UAV costs delayed fielding for these attack UAVs. However, the difference has become more discrete with GNSS-based route programming on the approach and return phases to reduce operator fatigue. Thus the UAV operator can focus his attention to the attack phase. Most attack UAVs (see pg 3-15) use less precise programmed guidance than CM (e.g., the Italian/former Iraqi Mirach 150), since they have camera guidance for a precise hit-to-kill terminal phase. As systems have become more robust, recent attack UAVs now offer precise GNSS, with capability for dozens of waypoints and capability for immediate changes, better-stabilized camera guidance, and IIR or MMW radar-homing for the terminal phase, similar to CMs. High UAV costs similar to CM may limit their fielding. Still, modern CM like Israeli Delilah offer programmable navigation and camera view guidance for the terminal loiter/attack phase, similar to most attack UAVs. More successful were ***anti-radiation missiles (ARMs),*** such as Harpy (pg 5-13), special-designed to destroy high-value radar targets.

New technologies and a continued requirement for unpiloted deep strike systems have accelerated R&D activity offered new attack systems. Smaller, more effective, and less costly systems are available. They can be separate weapons, canister/MRL launched, or dispensed from bus UAVs as munitions/submunitions. Some use GNSS phase, camera guidance, and IIR or MMW radar-homing terminal guidance, which will blur the lines between attack UAVs and CMs. Recently, UCAVs as ordnance delivery platforms have been fielded (such as Hermes 450 with Mikholit missiles, see pg 4-18). New longer-range NLOS ATGM systems (see Vol 1, pg 6-75) can also serve in the role of cruise missiles. These, attack UAVs and UCAVs will compete with CM for most battlefield targets to a range of 200 km.

The potential for adaptation of new technologies into attack UAVs or ***LACMs*** strains current paradigms for weapon system boundaries. Artillery rocket launchers can launch course-corrected (or maneuvering) rockets or missiles. The Russian *R-90* reconnaissance UAV demonstrates the viability of such a vehicle for future attack variants. Russian developers also have demonstrated a niche capability, claiming that SA-11 variant (Buk-M1 and Buk-M1-2) SAMs can be used to attack high-value ground and sea targets. Modern LACMs, as well as adaptive applications such as the ones noted, can bridge requirements of ATGMs, artillery, SAMs, and TBMs for OPFOR deep attack.

Selected Non-Ballistic Land-Attack Systems

System Name	Producing Country	Proliferation (countries)	Type Launcher	Propulsion	Range Min/Max (km)	Guidance	Accuracy (m)	Warhead Types	Payload (kg)	Comments
Nimrod Nimrod 3	Israel	At least 3	Tracked veh or TUV	Missile motor	0.8-26 0.8-55	Semi-active laser Inertial mid-course	Home to beam (1)	HEAT (800 mm)	15 kg warhead	Dive attack Requires laser designator
Mirach-150 UAV (poss)	Italy, Iraq	At least 5	Ground veh ramp	Turbojet	up to 470	Radio and pre-program	INA	HE est	INA	Attack version of recon UAV
Polyphem/ TRIFOM Polyphem-S (Naval) Triton, torpedo based	Consortium France Germany Italy	In final testing in 2002	Ship, MRL-type, Truck, TUV/ATV	Missile motor	60 TRIFOM 100 future Triton 15	Fiber-optic Infrared. Pre-program mid-course phase	Guide to target (1)	HEAT + Frag/HE	20-25 kg warhead	. ATGM version expected. Concept for remote launch canister and TV control link
Brahmos and Yakhont	India Russia	At least 2	Truck, ship, FW (due)	Ramjet	290	GPS/Inertial	<20	Frag-HE	250	Supersonic
R-90	Russia	1	MRL	Rocket	90	Camera	.5	HE		Adjust fire, BDA
Harpy	Israel	At least 5	Truck	Rocket	500	GPS, Radar	1	Frag-HE	18-22	
Delilah	Israel	At least 1	Truck, ship, FW	Turbojet	250	GPS/Inertial	1	HE	30	Waypoints, loiter

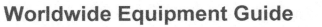

Worldwide Equipment Guide

Chapter 7: Air Defense Systems

Chapter 7: Air Defense

The increased effectiveness of aerial systems in modern warfare continues to drive a corresponding commitment for most forces to improving air defense forces, tactics, and technologies to counter them. Air defense (AD) is organized to address all capabilities of adversary aerial systems which can be used against a force. In addition, AD is integrated with other units (information warfare (INFOWAR), tactical units, ground reconnaissance, and aircraft units) to counter aerial threats. The AD plan means a force-wide strategy with active and passive all-arms counters, first to negate the effects of aerial systems, and second to destroy aerial systems when possible.

Air defense utilizes a variety of systems including: fixed-wing and rotary-wing aircraft, ballistic and cruise missiles, unmanned aerial vehicles (UAVs), unmanned aerial combat vehicles (UCAVs), air-delivered munitions (such as missiles, rockets, bombs, etc.), ground-launched rockets, and airships. For nearly a century, as developers of aerial systems developed new capabilities, AD developers responded with new tactics and technologies to counter them. In turn, aerial forces responded to the AD. Both sides of this antagonistic struggle continue fielding new technologies, counter-tactics, and countermeasures, even counter-countermeasures.

The AD forces are finding new ways to integrate those changes with more aggressive planning and organization. AD requires integration of separate functions: reconnaissance, target acquisition, C^4, and battle management, and target engagement – often with those assets separated by several kilometers. Assets for each can be vulnerable to physical attack, with links vulnerable to INFOWAR deception. Thus, AD forces continue updating systems and fielding new ones. As with aerial threats, AD is finding new missions and approaches for success.

Because of their ability to move and strike in any terrain and weather, aerial forces are generally the aggressors. Key capabilities for modern aerial assets must be addressed. The most challenging are traditional ones, but with new and greater technologies. They are:

- Suppression of enemy air defense (SEAD), for AD destruction, and INFOWAR attack (including jamming and cyber-attack).
- Surges, with multiple aircraft, multiple types of systems, and multi-aspect approaches,
- Strikes, with improved precision surveillance (satellites) and weapons (ballistic missiles),
- Stealth, in aircraft design, UAVs and UCAVs, and use of terrain flight profiles.

AD depends on efficient C^2 for responsive, integrated, and survivable counters to enemy aerodynamic weapons. Because increased threats from stealth, surges during air operations, aerial long-range weapons, and more forces are using improved C^2 to form integrated air defense systems (IADS). However, the increased challenges to air defense C^2 also require ability operate autonomously or in small units.

Key aspects of AD effectiveness against surges are: use of redundant overlapping systems with varied C^2 and RISTA nets, digitally linked and autonomous batteries, increased responsiveness, increased missile loads, and improved missiles for single missile kill per target. Modern battle management centers in IADS can de-conflict targets and maximize AD effects.

Sensors are a critical component of AD systems, since they perform surveillance and tracking functions against fleeting targets. Radars have dramatically improved, and receive the most attention among AD sensors. But increasingly, acquisition packages use multiple sensors, including acoustics, electro-optics, etc. In recent AD weapons, radars are integrated with passive sensors, such as optics, electro-optics, TV cameras, night vision sights, auto-trackers, and laser rangefinders. Throughout the force, air approach/attack warners are used, and may be linked with man-portable AD systems (MANPADS). Night sights are now common on weapons such as machineguns and MANPADS.

Weapons trends focus on guns and missiles, e.g., fitting both onto one chassis. Guns and missile launchers are increasingly more mobile and reliable under all conditions. They are becoming better integrated for responsive operation at AD brigade, in small units, and down to the single weapon. Most systems have onboard C^2 and passive electro-optical (EO) acquisition systems which permit them to operate precisely and autonomously, and slew quickly.

Improved long-range AD (LRAD) and medium-range AD (MRAD) surface to air missile (SAM) systems feature increased velocity and acceleration, high-G turn capability, and precision for use in ballistic missile defense (BMD). Short-range air defense (SHORAD) systems include the use of high velocity missiles (HVM), which can intercept high-speed anti-radiation missiles (HARMs). AD use of low probability of intercept (LPI) radars and signature reduction technologies challenge the ability of aggressors to locate and engage the systems.

Many SHORAD upgrades can counter low-flying helicopters using covert tactics and cruise missiles (CM). New technologies include laser and radio frequency weapons, and hypervelocity kinetic energy missiles. Modern MANPADS can be found in lower-tier forces. Improved missiles with proximity fuzes can fly lower to kill helicopters flying at nap-of-the-earth. New munitions such as frangible or electronically fuzed rounds increase gun lethality. Modular missile launchers and remote operated guns can transform vehicles or towed chassis into AD systems. MANPADS launchers can mount on vehicles with improved sensors and C^2 links for robust AD support. Upgrade sensors and weapons can rejuvenate older AD systems.

The greatest threat to AD is the use of stealth systems. New missile systems with multi-spectral nets and phased-array radars are being used to better detect stealth aircraft. Updated early warning radars and newer INFOWAR passive RF systems are being linked into IADS. AD aircraft, nets with substantial numbers of aerial observers, unattended sensors, and nets of modern infrared sensors are also used.

The priority for countering air threat applies force-wide. Most OPFOR weapons and sensors, including infantry and vehicle guns can engage helicopters and other AD targets. More weapons are multi-role or air defense/antitank (AD/AT). All machineguns can be used for AD. The OPFOR mixes legacy systems, improvised weapons, and recent equipment to improve AD across the AO. Modernization trends cover all aspects of the AD network, including SHORAD and long-range AD (LRAD).

Questions and comments on data listed in this chapter should be addressed to:

Mr. Rick Burns

DSN: 552-7962 Commercial (913) 684-7987

e-mail address: Richard.b.burns4.ctr@mail.mil

Air Defense Command and Control and RISTA in Training Simulations

Portrayal of combat systems capabilities in training simulations is never exact, and often may display serious limitations which hamper realism. Portrayal of air defense is particularly challenging because effective AD requires timely and effective integration of weapons, support assets, C^2, and skillful planning. Budget constraints, hardware, and other limitations can impact portrayals. The OPFOR is required to be ***reasonable, feasible, and plausible***. These priorities equally apply in OPFOR air defense systems portrayal. The following describes OPFOR air defense technologies and capabilities to be addressed in training simulations.

Responsive, efficient, effective, and survivable air defense requires effective C^2 in weapons units and the IADS. Flexible and integrated C^2 is particularly difficult to portray in simulations. These divergent priorities are in conflict. The AD system must link weapons with sensors. It begins with the individual air defense system, with the fire control system providing autonomous C^2. Increasingly, forces are providing autonomous capability for AD systems.

Many forces are producing mobile AD battle management centers. At the tactical level, they are in armored command vehicles (ACVs) for AD batteries and battalions. Tier 1 and 2 AD units have ACV/radar vehicles (e.g., Sborka). They can also be used in separate batteries plus link to the IADS. A modern ACV can receive, process, and pass a message in seconds (roughly 15, 4 for digital links), with parallel multi-function processing and multiple addressees (6-12). Older ACVs, e.g., PPRU, use analog voice and/or digital data links with longer processing/ transmission times. An IADS with analog C^2 is still an IADS, but may be a less responsive one. An IADS is physically dispersed for autonomous action, yet operationally integrated as required.

Air defense organizations balance capability with survivability by managing an array of sensors to provide full 360° coverage, surveillance in depth, with long-range assets supported by mobile reconnaissance assets and overlapping search sectors. The system requires: centralized linkage of various gun, missile, and gun/missile units, and coordination with AD aircraft units. Units will be relocated and re-assigned to prevent gaps in coverage. Airborne warning and command systems (AWACS), and other airborne air defense assets (aerial patrols, etc.) will be used. The IADS integrates AD nets and links them with other RISTA nets (air, ground recon, artillery, etc.) to fuse the battle picture, cue AD assets, and warn of approaching aircraft throughout the force. An IADS provides early warning (EW), assures that weapons resources are efficiently assigned to service all targets at the maximum possible stand-off, and reduces delay for vehicle halt and weapon response time. It also provides target acquisition (TA) data during jamming, avoids fratricide for aircraft operating in the area, and reduces redundant fires.

Missions are netted through the IADS with battery/battalion radars, command posts, longer-range radars for battle management at brigade and above, and various other sensors (acoustic, infrared, TV, visual, and other technologies). Modern EW units use long-range radars located behind the forward area to see for hundreds of miles, and use radar signal parameters to reduce jamming and terrain restrictions. These radars feed approach warnings throughout the net so that most AD systems can operate passively and not reveal their locations until the moment of engagement. They help facilitate AD ambushes by transmitting aircraft locations and allowing weapons radars to radiate only at the last minute when air targets are within range. Many SAM systems can use the IADS digital feed instead of their radars for passive operations.

The primary detection and acquisition system for an air defense unit is radar. Radars can more easily detect and track aircraft with less operator input than other sensors (e.g., EO sights). Radars are usually categorized by function and functions usually correlate to certain frequency bands. Older early warning (EW) radars generally operate in low frequency bands (A-E), for longer detection ranges. They can track targets and cue precision sensors to support an IADS.

Air Defense Radar Bands in the Electromagnetic Spectrum			
NATO Band	US Band	Low-End Freq (GHz)	Wavelength
A		0.0	
B		0.25	Decimetric
C		0.5	
D	L	1	
E	S	2	Centimetric
F	S	3	
G	C	4	
H		6	
I	X	8	
J	Ku	10	
K	K, Ka	20	Millimetric
L	L	40	
M		60	
X		8-12	

AD units employ a mix of radar systems operating at different frequencies in varied intervals with some radiating while others surveil passively. More mobile radar systems are being fielded with ability to quickly employ radars or operate radars while moving. Target acquisition (TA) radars are used to acquire aerial targets (and assign them to the fire control system for launch) often operate in I and J bands. Other bands offer precision and range while undetectable at scanned frequencies. Fire control (FC) radars (which track missiles and targets and direct weapons to target) often operate in H-J bands, but can operate in less detectible bands. Many more modern systems use dual-mode/multi-mode radars that can simultaneously perform EW, TA, FC or combination, with (automatic) target tracking during the engagement. For the OPFOR, unless air missions are scheduled, free-fire zones do not require IFF checks. Thus, most OPFOR sectors are free-fire zones and the OPFOR AD usually launches on first detection.

Radar performance is affected by technical factors such as: functional requirement (EW, TA or fire control), type (phased array vs continuous wave or pulse), operating paramaters (fan angle, power levels, operating time, frequency, etc), mount (stationary, mobile, missile mount on active homing missiles), target (radar cross section, countermeasures, speed, altitude, etc), and environment (curvature of the earth, terrain, weather, etc). Performance is also affected by tactical considerations of the target (aircraft dispersion, their use of stand-off weapons, etc), requirements for support systems, and survivability tactics for the radar (narrowing beam width, limited operation times, passive modes, frequent moves, etc.)

Increasingly, IADS also use passive sensor systems such as acoustic-triggered unattended ground sensors, remote-operated EO systems with auto-trackers, radio-frequency direction-finders, and sensors operating in other regions of the electro-magnetic spectrum. Acoustic sensors include acoustic arrays such as the HALO stationary microphone complex. They also include vehicle systems such as Israeli Helispot with microphones mounted onboard or dismountable. Russian sound-ranging systems (AZK-5, -7, etc.) can detect helicopters. Links from nearby units (recon, maneuver, artillery, etc.) can also supplement AD sensors.

An affordable low-technology response to air threats is AD observation posts (OPs). Forward OPs can support EW radars as well as AD OPs in tactical units. They can also include special purpose forces or civilian supporters near airfields or helicopter FARPs that can engage aircraft or notify AD units. Assets may include day/night observation systems, remote IR cameras, acoustic sensors (such as sound-ranging systems), anti-helicopter mines, and MANPADS. In Tiers 1 and 2 they will use laptop computer terminals and digital links to pass data. Sensors can include man-portable radars such as FARA-1. These OPs use goniometer-based laser range-finders, GPS, and radios for precise location and warning, and rapid reporting. In Tiers 1 and 2, MANPADS operators have azimuth warning systems which alert them day and night to approaching aircraft. In lower tier forces, radars can be supplemented with forward OPs (perhaps with binocs, compass and radio) to cover defilade areas and masked areas of approach. In the Near Term OPs will have micro-UAVs to detect and attack helicopters or chase them off.

An IADS does not limit autonomous fires, rather provides early warning and reduces delay for vehicle halt and weapon response time. Because the enemy will attack C^2 nodes and detected AD radars, most AD systems and subunits must be able to operate passively and autonomously with mobility and dispersion. It also provides target acquisition data for AD during jamming, avoids fratricide for aircraft operating in the area, and reduces redundant fires.

Most air defense systems have passive EO sights for use when radars cannot be used. They include TV day sights, infrared or thermal night sights, and target and missile trackers. Sights can have zoom capability with 24-50 + power, acquisition range equal to or greater than a radar, and minimum altitude down to the ground (0 meters). Range may be limited, however, by line-of-sight. Thus, EO range is comparable to a targeted aircraft's EO sensor acquisition range.

An IADS can operate as low as brigade level with AD working in concert with other units and other echelons. Even when a formal IADS is not established responsive and coordinated AD is possible. For instance FOs can notify AD weapons of enemy approach and direction. The FARA-1 radar can easily be mounted onto AD guns for day/night operation. Anti-helicopter mines can be used to cue AD ambushes. Innovations such as remote weapons and sensors and portable digital FCS are updating older AD weapons, permitting them to link to IADS. Battery ACVs such as Sborka feature EW/TA radars for RISTA and link to IADS.

Air Defense Systems and Domains

In modern warfare, the initial air operation is considered to be the critical component to success against modern enemy forces. That operation is expected to disrupt or destroy critical C3 nodes, exploit vulnerabilities in the air defense nets, and facilitate widespread aircraft and missile strikes against military targets. That operation would include stealth precision aircraft, missile strikes, and rotary wing aircraft flying low level deep strike missions. These would be generally conducted prior to entry of ground and naval forces in order to facilitate early entry safely. In modern forces using aggressive planning, the air defense plan will be designed in detail to counter each aspect of the air operation. Thus, the air defense operation must begin prior to the air operation to deny it success and insure integrity of the threatened forces and area. Air operation forces and air defense forces continue to see changes in plans, tactics, and equipment to counter the other's advantages, while operating within modern military budget constraints. A number of forces are choosing to reduce the size of costly fixed-wing aircraft, while increasing the sizes of theater missile and air defense forces, to deny adversaries air superiority. Trends noted on page 6-13 affect systems, fielding choices, and capabilities in all AD domains.

There are at least nine air defense plan domains with distinct missions, tasks, weapons, sensors, and phases in the air defense operation. Actions may require simultaneous effort in all nine domains. It is an all-arms effort involving more than just air defense forces. Range figures for these systems are general, variable, and changing with overlaps.

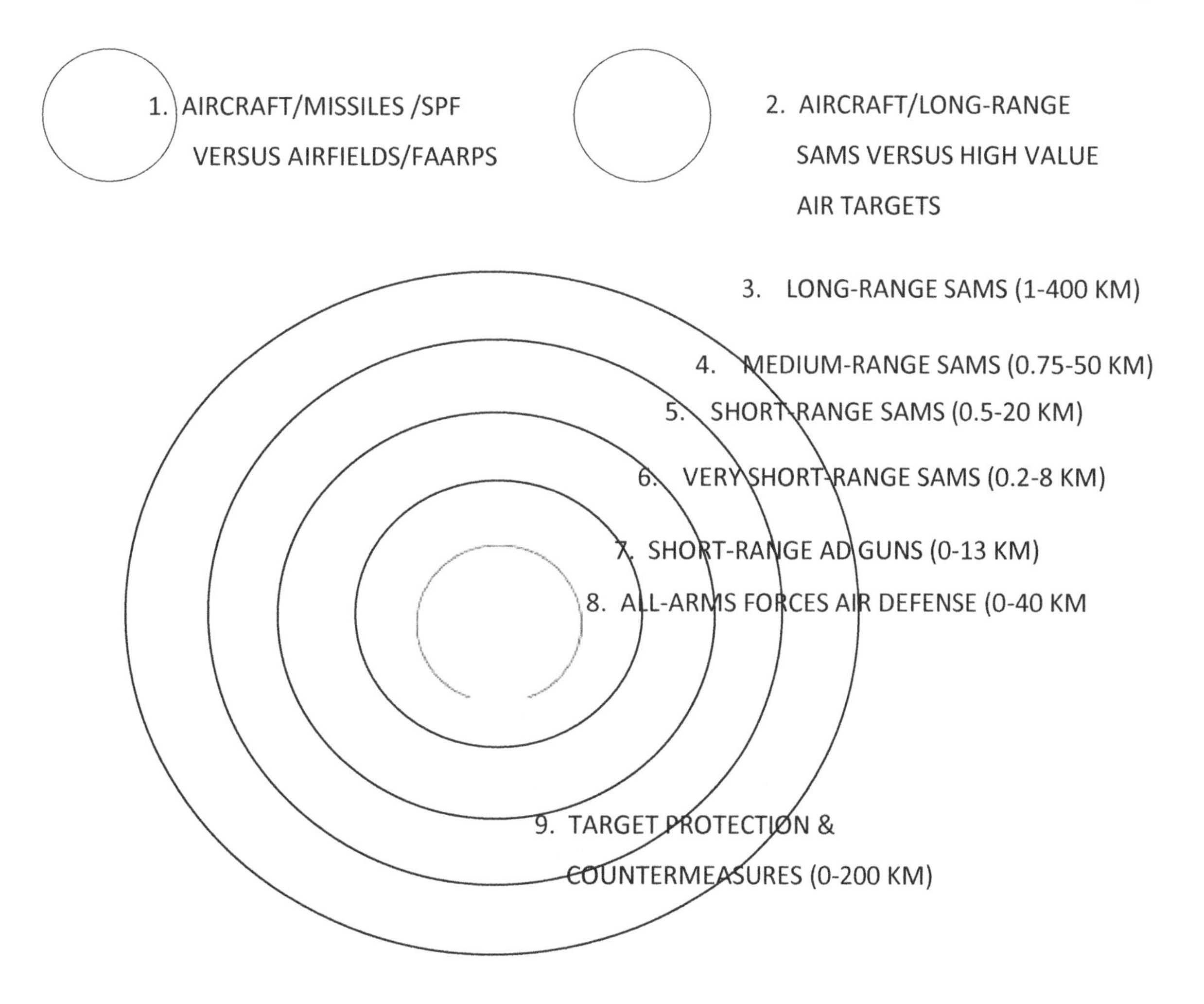

Domain #1. The plan calls for a combination of pre-emptive, reactive, and passive air defense measures being conducted simultaneously. Surveillance assets, especially forward observers, will be deployed around all potential adversary landing areas, including helicopter lighting points, to monitor activities. In the initial phase, and later as the adversary's aircraft reach forward locations, pre-emptive measures will be conducted by IW and deep-attack assets to degrade the adversary's air operations before they even reach defended airspace. Forces will attack airfields and helicopter lighting points and FAARPs with air-launched stand-off weapons, ballistic and cruise missiles, and special purpose forces. At critical phases of the operation, they will disrupt satellite systems and attack adversary long-range surveillance assets.

Domain # 2. Generally, AD air and ground forces will attempt to engage and disrupt enemy air activities as early and distant as possible to decrease the chance for enemy air success. Air intercept aircraft and long-range AD (LRAD) systems will attack reconnaissance aircraft, AWACS aircraft, SEAD aircraft, and bombers. Because of curvature of the earth limitations on SAMs, aircraft will operate at altitudes below the minimum altitudes of the SAMs, at ranges of 250 km, or more. Special nets of radars and passive Electronic Support (ES) systems will be created specifically for detection of stealth aircraft and cruise missiles, flying at lower altitudes.

The AD plan includes flexible prioritization of AD systems to deal with key events, such as enemy surges, ballistic missile and cruise missile strikes, and AD forces to survive air and SEAD operations, and ground forces attacks. The two main deployment priorities are site defense and area defense, and they activate as targets come into range. Forces hold out a portion of LRAD/MRAD launchers for site defense against ballistic missiles. Even when aircraft reach the range of MRAD systems, LRAD may service targets while MRAD SAMs conduct AD ambushes, monitor the IADS, and use passive electro-optical fire control systems (FCS).

Domain #3. Long-range SAMs (pg 6-79) include Russian and Chinese missile systems, e.g., SA-5, SA-10, SA-12, SA-20b, SA-21, SA-23, and HQ-11. These upgraded and new systems are networked with long-range early warning radars and electronic support measures (ESMs) to form the base for operational IADS. In the past, the size of the missiles limited them to selected roles, like counters to high priority aircraft (Domain #2), long range defense versus small formations (Domain #3), and anti-ballistic missile defense of high priority sites (Domain #9). However, several LRAD systems are being modified to fit canisters of "small missiles", to counter surges and all air targets in the other domains (#4-8) as well.

A wide variety of RISTA assets including forward observers, HUMINT assets and RISTA systems support AD operations. Early warning systems have lost appeal in certain AD circles, but are still useful. They operate in low bands outside of bandwidth of most radar detection systems, and have long detection ranges. Many are being modernized with multi-target precision tracking, and digital transmission and display systems. As aircraft approach the range of AD weapons, they are acquired by EW radars, which conduct IFF queries and feed intelligence to the Integrated Air Defense Systems (IADS). There is one overall IADS for the force. But other overlapping area and AD brigade IADS are used in case the central IADS is defeated by enemy SEAD. The IADS battle management center will select target acquisition radars to conduct surveillance and track targets, update the plan, and assign new targets. New phased-array TA radar and battle management systems have interface and networking features toform autonomous IADS, and autonomous firing units down to the battery level, and challenge SEAD and evasive aircraft tactics. As aircraft approach the targets, noted at 9, they have entered engagement zones of not one, but many types of AD systems and RISTA nets, each linked to the IADS and its RISTA nets. Thus, they are detected by multiple radar frequencies, ground observers, vehicle EO/IR acquisition systems, sound-ranging assets, and AD Infowar (IW) assets. Although the diagram depicts concentric circles with a single epicenter, defended forces are arrayed throughout the area; and multiple MRAD/SHORAD epicenters and assets overlap.

Domain #4. Most MRAD systems in Domain #4 (pg 6-71) are Russian, e.g., SA-3, SA-4, SA-6, SA-11, and Buk-M1-2. Some are highly mobile and can move with ground forces and challenge air surge capabilities of expected adversaries. Because of their high cost which approaches that of LRAD, most MRAD systems in used are older. But a number of users are updating them to approach modern capabilities, to counter short-range ballistic missiles, cruise missiles, stealth aircraft, and low-flying helicopters. Other countries are looking at the possibility of adapting LRAD systems to handle surge requirements and reduce the need to upgrade or produce MRAD systems. Other forces, e.g., Israelis (Spyder), Italians (Aspide 2000), and Indians (Akash) designed systems that in some cases are more affordable and mobile.

Domain #5. The short-range SAMs in Domain #5 include a wide array of systems produced and exported throughout the world. Leading producers include China, Russia, the US, and many European countries. Although the systems in this domain include semi-mobile towed systems, most are vehicle mounted and can be brought into action from the move in 0.5-5 minutes. Many can move with supported maneuver forces. Others cover critical assets which are likely air targets. Some are assigned to cover areas with defilade terrain and man-made features which could be approaches by aircraft flying contour or nap-of-the-earth (NOE) profiles. These systems have substantial missile inventories to respond during enemy surges. Many modern SAMs are configured as gun-missile systems (pg 6-34), to engage almost all aerial targets (including cruise missiles, UAVs, air-to-ground missiles, and helicopters flying NOE).

Domain #6. The very-short-range systems in Domain #6 also called Man-portable SAMs, or Man-portable Air Defense Systems (MANPADS). They can be dismounted; however, some vehicle mounted systems have been developed which use the same missiles. A wide variety of upgrades are expanding the lethality (and range) of these systems. Additionally, multi-role missile systems (such as Starstreak, pg 6-60) are being fielded.

Domain #7. AD gun systems are not as widely used as in the past. The primary reason for this is the limited effective range of most guns. Although gun range is noted as out to 13 km (KS-19M2 with radar, pg. 6-45), most AD guns are effective at ranges of 4 km or less. A lot of forces upgraded those guns by merging guns and missiles systems in gun-missile systems. Some forces still field new self-propelled AD guns. Substantial upgrades (pg 6-34 to 42) have increased effectiveness and utility of most weapons.

Domain #8. Modern forces proliferate weapons (especially machine guns) for self-defense, especially versus air threats (see pg 6-33). Ground and air responses to air threats include more medium guns, improved AD munitions, and more responsive missiles and FCS.

Domain #9. Developments in target protection and countermeasures include use of CCD technologies and tactics, as described on the next two pages, and in Chapter 9.

Aircraft Survivability and Air Defense Countermeasures

Modern forces focus much attention to protecting aircraft during air operations through a blend of tactical measures and technical capabilities which are collectively known as ***suppression of enemy air defense (SEAD)***. Separate SEAD aircraft and IW assets are engaged in locating AD assets, jamming AD C^2 and RISTA assets, and attacking systems in the AD network. Often SEAD aircraft will accompany FW and helicopters in carefully coordinated air missions. In addition, modern tactical aircraft and supporting aircraft can be equipped with aircraft survivability equipment (ASE) to countermeasure incoming AD missiles.

The OPFOR, like most forces in the world today, have developed technologies and tactics to counter ASE and SEAD. The first priority for AD effort is always force survivability. The OPFOR knows that SEAD usually facilitates other aircraft conducting missions; thus air protection measures are addressed in all units and at all levels. These include a network of air warning receivers to sound air alerts down to battalion level and below.

The most common and challenging air threat is from helicopters, because of their proliferation, their ability to use concealed approaches, and their ability to directly engage AD assets early in the air operation. Helicopters will use terrain and cover to mask their approach with terrain flight modes (low level, contour, or nap-of-the-earth - NOE). The OPFOR conducts an intelligence preparation of the battlefield (IPB) early on to determine routes and assign OPs, sensors and on-call AD weapons to cover areas which offer concealment. Air defense priorities are engaging all aerial targets primarily, and countering SEAD secondarily.

Selected Air Defense Tactics Used to Counter Air Attacks and SEAD Operations

Considerations	Examples
Protection and Countermeasures	Use concealment, mixing with civilian sites and traffic Use cover (dug-in positions, hardened facilities, urban structures) Disperse assets and use autonomous capabilities Relocate frequently Use protection envelope of friendly forces Deception operations for convoys, crossings, etc.
Tactics	AD conduct bounding overwatch during movement. Air defense ambush with passive mode (EO, radars turn just at launch) Direct attacks against AWACS, SEAD aircraft, airfields, and FARPs Engage SEAD/ASE aircraft from an aspect outside of the jamming arc Conduct beyond borders operations against air capabilities.
RISTA	Use intelligence preparation of the battlefield - approach routes, etc Passive radar and EO modes. Use IADS links for TA data.

	Emissions control measures Utilize civilians and insurgent links. Use lots of OPs linked to AD units, including forward-based SOF, etc. Employ non-AD sensors and units available to feed reports to IADS.
Command and Control	Mobile, redundant, concealed systems Comms OPSEC measures
Weapons	Engage aircraft, air-to-surface missiles, and ARMs beyond their range Prepare all weapons to respond to aircraft. All units conduct air watches with weapons at ready at all times

Airborne SEAD and SIGINT operations and technologies include radar acquisition systems, radar jamming assets, and anti-radiation missiles which can find and destroy radars, compel AD units to acquire more robust even longer range radar systems, and to more carefully manage radar assets. The OPFOR will use equipment and tactics to degrade SEAD effectiveness, deceive it, and attack SEAD directly. Some of those responses are listed below.

Selected Air Defense Technologies Used to Counter SEAD Operations

Technologies	Examples
Command and Control	IADS, directional comms, SATCOM, retransmission systems, etc. IADS links to artillery, recon, maneuver units, SOF, etc. Digital comms with reduced response time. Mobile, redundant, easily concealed systems
Radars	Low-frequency long-wave early warning radars (50-100 km setback) Low Probability of Intercept (LPI) radars (frequency, power control) Multiple-mode, multiple frequency, frequency-agile radars Phased array radars and guidance modes that negate jamming Counter stealth radars and passive sensors integrated for fast response Aerial radars on helicopters, UAVs, mobile airships, with retrans links Mobile radar systems for frequent moves, or operation on the move
Other Sensors	Sensors using passive modes (EO, IR, acoustic, other bands) Mobile goniometer based fire control sets with GPS and digital comms Remote sensors, unattended ground sensors, linked to AD nets. Remote IR and EO cameras (Sirene, ADAD), and on UAVs/airships
Weapons	SEAD-resistant missile guidance modes (semi-active radar homing, active radar homing, track-via-missile, laser beam rider, etc.)

	Home-on-jam missiles attack AWACS, SEAD, ASE (Aspide, SA-5) AD missiles can destroy ARMs and HARMs (Pantsir, SA-15b) Responsive autonomous or battery AD weapon systems (SA-11, 2S6) Passive guidance, e.g., IR-homing or EO FCS (Mistral, GDF-003)
Countermeasures	Encryption and secure comms modes Decoys: corner reflectors, multi-spectral, bridge mock-ups, etc, Electronic Warfare: SIGINT/ELINT, GPS/fuze jammers, deception

Most units operating in flight paths are subject to air attack and use active measures to respond to air threats. Dismounted infantry units will have AD OPs and will engage aircraft as required. Any AD weapon can alert its ACV and the IADS net of spotted aircraft. Of course the delays from transmitting reports through these links should be considering figuring response time (15 sec Tiers 1 and 2, 30 or lower for each message link from observer to AD weapon).

To counter the helicopter threat, a wide variety of tactical and combat support vehicles have MANPADs/MGs with AA sights to engage aircraft. Two of the greatest advantages for helicopters are weapons stand-off and ability to use terrain cover on approach. Many ground force and AD weapons can match the stand-off and inflict damage to force aircraft to disengage. When flying NOE (20-25 ft from the ground), a helicopter rotor is still 40 ft high. A helicopter terrain masking cannot easily engage targets or evade missiles, but can be targeted by ground weapons. Nearly all SAMs, small arms, and direct-fire crew weapons (ATGMs, ATGLs, AGLs, etc.) can engage it. ASE includes IR decoys which can be foiled by improved IR missile seekers, and RF jammers with dead zones (and limited effects against modern radars).

Engagement Factors and Data for Air Defense Simulations

No simulation can predict or reflect reality; but a well-designed air defense simulation can be robust enough and detailed enough to represent reality. Air defense engagements offer a difficult challenge for realistic portrayal in training simulations. A simulation might be expected to depict robust and responsive RISTA assets executing the acquisition stages (detection, classification, recognition, and identification) with early warning and target acquisition/ battlefield surveillance radars, C^2 processing (report posting on battle management nets, analysis, tracking, target assignment, and shooter assignment), target engagement (TA radar, location and tracking), missile launches, probability of hit (Ph) data, and probability of kill (Pk) by type of kill calculation. Degrading factors can be factored into calculations: e.g., target type, evasive tactics, battlefield environment constraints, AD systems limitations, and AD counter-tactics. In the real world, RISTA capabilities are affected by a variety of factors, which can affect capabilities calculations by system, by class of system, and in various ways. Here are key ones.

Selected Factors Which Affect Air Defense Functionalities

Technologies	Factors	Data Entry
Sensors for Acquisition, EW, And Fire Control	System range	km
	Target tracking range	km
	Night range (EO sensors)	km
	Range to target	km
	Radar down time	min
	Radar search sector (horizontal/vertical from mid-line)	degrees
	Radar altitude	km/m/ft
	Curvature of earth range limiter (based on sensor and target altitudes)	max km
	Terrain feature effects on line of sight (LOS) (limiter which interrupts LOS)	km
	Aircraft altitude	km/m/ft
	SEAD/aircraft ASE effects (sector of scan)	km x km
	Counter-SEAD capabilities (0 % degradation)	0 %
Command and Control (C^2)	Report time (x number of links)	min
	Report-processing time (x links)	min
	Authorization to fire	Yes/No
	IFF time	sec
	Target assignment time	sec
Weapons	Missile/gun effective range	Km
	Number of missiles/rounds per target	# x Ph
	Missile/gun minimum altitude	m, or band
	Weapon reaction time	sec
	Area of munition warhead effects (range and altitude)	m or Ph
	Aircraft ASE against missile seeker (degrader x Ph)	%
	Munition ASE CCM capability (0 % degradation)	%
	Ph against target types (RW, FW, ASM, UAV, TBM)	type/name
	Pk-Mobility, Firepower, Comms, Catastrophic, etc	type/name
	Munition approach/impact aspect vs target (if needed)	Ph
Target Effects	Target flight altitude, speed, range, etc	m, km, etc.
	Target countermeasures and counter-tactics	Ph factors

Many AD data adjustment factors are expressed in range or altitude, which can be used by the simulation to match AD system to target. Some of the factors (or degraders, such as LOS or ASE) can then adjust the capabilities. For time-based capabilities, degraders (such as report time) are critical considerations that can affect the likelihood of AD engagement within the time span of aircraft approach, while the aircraft is still outside of range for ordnance delivery.

Capabilities of AD weapons to engage, hit, and degrade aircraft physical viability and effectiveness is expressed in various data. These include range, altitude, time (noted above), and probabilities of hit and kill. Once target and shooter are within geospatial and time windows, with authorization to fire, the key data are probabilities of hit and kill.

Probability of hit can be affected by many factors (as noted in the table above). Sources vary widely in Ph data for the same systems. Often a range is listed, such as 40%-96%, without clear explanation of calculation criteria and with many detection variables rolled in the figures. Russian sources often state their figures as "single shot kill probability," combining hit and kill in one figure (Ph x Pk). The Ph figures noted in the WEG for missile systems are averages based on probabilities at all aspects, within operational ranges and altitudes, and against aircraft in noted classes. Different fixed-wing (FW) and rotary-wing (RW) aircraft will have different radar cross-sections, IR heat detection levels, and different Ph levels. Other aeronautical targets such as UAVs, cruise missiles, air-to-surface missiles, and theater ballistic missiles will have different Ph figures by type, system, and aspect. The simulations should use the Ph in the WEG as a single figure for the technical capability. Degraders such as factors noted in the above table could then be applied for use in the simulation. Often AD units will launch multiple missiles at a target. Two missiles will have greater Ph, possibly 2 x one Ph.

Developments in missile seekers, guidance, and gun ammunition technologies are greatly improving probabilities of hit for AD weapons. One of the most deadly AD missiles to threaten modern aircraft are anti-radiation missiles that home-in on an aircraft's ASE or SEAD radar jammer. Another modern AD missile capability is active radar homing missiles, which cannot be easily counter-measured. Both missile types have a higher P-hit. The Starstreak MANPADS system offers another new step in missile precision and countermeasure resistance with laser beam-rider guidance. Starstreak has a very high Ph against less maneuverable aircraft, especially helicopters conducting terrain flying such as NOE. Some modern AD guns now have rounds with proximity fuses, for higher Ph. Others rounds have an AHEAD-type fire control system (a laser range-finder-based computer sets electronic time fuzed rounds, for precision air bursts).

Probability of kill (given a hit) can require even greater variety of figures based on type, system, and aspect, and by munition type or specific munition. Because those Pk figures require laboratory-produced data based on precisely determined conditions, they will not be noted in this publication. However, a few concepts can be noted. A missile with a proximity fuze and large warhead will have a large lethal radius and a high probability of kill given a hit or detonation. For small missiles, partial kills have a greater probability than total (catastrophic) kills.

Dramatic improvements in AD weapon lethality are raising Pk figures. Increased use of HMX explosive has raised Pks. Frangible gun rounds fly like KE rounds, permitting better range and precision than HE rounds. But they shatter inside of the target, offering high explosive Pk figures similar to HE rounds. Some missiles (e.g., Pantsir and SA-18S) have frangible rods in their warheads. Others have multiple sabot penetrators and HE effects (3 x "darts" in Starstreak).

Air Defense Systems: Key Technology Trends

Aircraft upgrades and proliferation of other aerodynamic threats (cruise and ballistic missiles, air-delivered munitions, UAVs, etc.) have increased the aerial threat to military forces worldwide. Thus forces expanded their emphasis on all systems engagement of aerial threats or counters to those threats. Forces worldwide are fielding new air defense (AD) systems and upgrading legacy systems.

System Category	Technology Trend	Reference Vol/Page	System Example	Ref Page
Short-Range Air Defense (SHORAD) Systems	Missiles engage <1-20 km range, and 0-10,000 meters altitude	6-36	Pantsir-S1	6-65
	Radars integrated with passive electro-optical/thermal fire control	6-5	Crotale-NG	6-66
	High-velocity missiles engage aircraft, munitions, UAVs, and missiles	6-26	2S6M1	6-64
	Drop-in overhead turrets and remote weapons for AD vehicle systems	6-33	Strelets/Igla launcher Zu-23-2M1	6-60
	Guns and missiles integrated into gun-missile systems	6-34		6-50
	Many missiles, most guns defeat all countermeasures	6-10	Mistral 2	6-63
	New or upgrade robust shoulder-fired SAMs throughout the battlefield	6-26	SA-24/Igla-Super	6-60
Medium-Range Air Defense (MRAD) Systems	Missiles engage 1-45 km range, and 0-25,000 m altitude	6-72	Buk-M1-2	--
	Tracked or wheeled with increased mobility and responsiveness	6-72	Pechora-2M	6-28
	Some systems mix MRAD and SHORAD missiles for high surge rate	6-72	Spyder-MR	--
	New autonomous launchers melded in old units to add more FC radars	6-72	SA-6b with Buk-M1	6-43
Long-Range Air Defense (LRAD) Systems	Missiles engage 5-400 km range, and 0-50,000+ m altitude	6-80	SA-21/Triumf	6-19
	Vertical launch and increased velocity against ballistic missiles	6-80	SA-21b/Samoderzhets	6-13
	Anti-radiation/radar-homing missiles defeat SEAD/AWACS/JSTARS	6-80	FT-2000	6-14
	Launcher can add canisters of 1-120 km "small missiles" for surges	6-80	SA-20b/Favorit	6-11
C4ISR	Phased array/low probability of intercept (LPI) radars and more range	6-4	96L6E	6-80
	Radars mounted on RISTA/weapons to operate and rotate on the move	6-15	Sborka-M1-2	6-15
	Other RISTA sensors, e.g., forward observers, UGS, passive IR, IW	6-5	Orion	6-21

	Integrated air defense Systems (IADS) across echelons and branches	6-3	Giraffe AMB	6-16
	IADS, FOs, and radars for responsive AD even in high-jam areas	6-5, -9	SA-10/20	6-83
	Autonomous unit and systems capability in a jamming environment	6-5	2S6M1	6-64
	AD complexes to counter SEAD and stealth systems	6-20	Nebo-SVU	6-85
Multi-role Systems	Multi-role (AD/AT) missiles, gun, vehicles, for AD and AT, etc.	6-32	Starstreak	6-61
	IW and other Infowar add to RISTA and deceive/deny aircraft C4ISR	6-20	Orion	6-21
Other Systems for AD Use	All-arms AD weapons/munitions damage or defeat low-flying aircraft	6-27	12.7-mm/.50-cal MG	6-36
	Anti-helicopter mines or mines which can be used in the role	6-24	Helkir	6-31
	Airships in acquisition, jamming, or obstacle fields against helicopters	6-23	Helikite w/Speed-A	6-23
	Concealment or deception measures limit aircraft effectiveness	6-10	Barracuda RAPCAM	9-4,7

RUSSIAN MOBILE AIR DEFENSE RADAR VEHICLE LONG TRACK

SYSTEM	SPECIFICATIONS		
Alternative Designation	P -40. The name LONG TRACK is actually the radar.	Associated SAMs	SA-4/GANEF, SA-6/GAINFUL, SA-8/GECKO
Date of Introduction	IOC 1967	ADA Unit Level:	Employed at division and echelons above division The system is used in Tier 3 and 4 units.
Proliferation	More than 35 countries		
Chassis	A modified version of the AT-T heavy tracked transporter vehicle (426 U).	Other Radars	The radar system links to the IADS to provide analog warning and to pass analog data.
Engine	465-hp diesel	VARIANTS	SPECIFICATIONS
Weight (mt)	35		Polish Jawor (circa 1965) and Polish Farm Gate (Truck mounted).
Max Road Speed (km/h)	55		
RADAR	SPECIFICATIONS		
Antenna	Elliptical parabolic		
Auxiliary Power Unit	400 Hz gen and gasoline engine.		
Frequency Band	E-band (UHF)		
Frequency	2.6 GHz		
Sweep Rate (rpm)	12-15		
Display range (nm)	200		
Effective range (km)	167		
Tracking range (km)	150		
Effective altitude (km)	30		
Track targets on Move	No		
Emplacement time (min)	INA		

Displacement time (min)	INA		
Tracking range (km)	150		
Dead time (min)	0		
Max targets displayed	>8		

DESCRIPTION

TWENTY-FIVE FOOT HIGH SINGLE CONVENTIONAL PARABOLIC MESH REFLECTOR ANTENNA WITH MULTIPLE STACKED FEEDS THAT IS VEHICLE MOUNTED.

RADAR FUNCTIONS

SURVEILLANCE, TARGET ACQUISITION, AND EARLY WARNING. LONG TRACK WAS THE FIRST HIGHLY MOBILE EARLY WARNING RADAR. . THE ANTENNA IS FOLDED FOR TRANSPORT.

NOTES

BY COMPARING RESULTS AT VARIOUS FREQUENCIES, THE LONG TRACK CAN BE USED TO DETECT STEALTH AIRCRAFT.

RUSSIAN AIR DEFENSE ARMORED COMMAND VEHICLE SBORKA-M1 AND SBORKA-M1-2

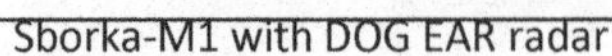
Sborka-M1 with DOG EAR radar

Sborka-M1 with DOG EAR radar

SYSTEM	SPECIFICATIONS	Communications Intercoms	2
Alternative Designations	9S80M-1, PPRU-M1. System is also called a "mobile aerial target reconnaissance and command post". Some sources incorrectly refer to DOG EAR radar as the name of the system.	Other Communications Links	7, including Integrated Air Defense System, brigade, and division for passive battle operations.
Date of introduction	Circa 1989, with –M1 upgrade by 2000	Vehicle can communicate on the move	Yes
Proliferation	At least 2 countries	Data formats	Graphic and digital data transmission and display
Crew	2 for vehicle	Onboard Generator	Yes
Troop Capacity	5-8 Command and staff workstations or modules	Whip antennae for mobile comms	2 HF whips, 3 VHF
Chassis	MT-LBu tracked vehicle, expanded variant of MT-LB	Other antennae	VHF discones masted, HF dipoles and 11-m mast
Combat Weight (mt)	16.1 est	Digital link to 1L15-1 MANPADS azimuth plotting board	Yes
Chassis Length Overall (m)	7.86	**RADAR**	**SPECIFICATIONS**
Height Overall (m)	2.72, with radar folded down	Name	DOG EAR
Width Overall (m)	2.97	Function	Target Acquisition
Automotive Performance		Frequency	F/G Band
Engine Type	240-hp Diesel	Range (km)	80 detection 35 tracking 500m and higher 22 tracking targets flying 25-499 m
Cruising Range (km)	500	Targets display and simultaneous tracking	63, 6 earlier version

Max Road Speed (km/h)	60	Target processing to assignment and track	1-step auto-track
Max Off-road Speed (km/h)	26	Scan rate (s)	2-5, 30 revolutions per minute
Average Cross Country Speed (km/h)	30	Data Transmission rate(s)	4
Max Swim Speed (km/h)	5-6	Scan coverage (°)	360 azimuth (rotating antenna) x 30 elevation
Fording Depth (m)	Amphibous	Antenna scan rate (rpm)	30
Emplace Time (min)	1-3	Antenna horizontal pattern width (°)	5.5 lower plane, 1.6 upper
Armor, Turret Front (mm)	15	Clutter suppression (dB) Operating time max Acquire on the move	30 or more 48 hrs, but usually use shorter on/off times Yes
NBC Protection System	Collective		
Smoke Equipment	Not Standard		
Target Missions Generated Simultaneously	1 or 2		
Target alert simultaneous rate	5-6		
Number of weapons with automatic control	6 separate weapons 12 with 2 per mission		
CP can operate autonomously/in network	Yes/Yes		
Number of sources which can generate targets	6 plus Sborka		
Encryption	Yes, E-24D		
Digital Navigation Monitor	Yes, GPS, Intertial and Map Display		
Automated Networks	Baget- 01-05 or -06 computer workstations		

COMMAND

CONFIGURATION FOR REAR COMPARTMENT HAS 1-4 OFFICER WORKSTATIONS, 3-6 COMMUNICATIONS/BATTLE STAFF CONSOLES, AND ONE RADAR OPERATOR CONSOLE (DEPENDING ON VEHICLE ROLE AND ECHELON).

COMMAND LEVEL

AD BATTERY AND BATTALION IN MECH AND TANK BRIGADES ASSOCIATED AD UNITS/SYSTEMS: SHORAD SYSTEMS (ZSU-23-4, 2S6, SA-9, SA-13, SA-15, MANPADS)

RADIOS, FREQUENCY, AND RANGE

3-6 X VHF WITH RANGE 30 KM (60 KM STATIONARY WITH MAST) 2 X HF WITH RANGE 50 KM (350 KM STATIONARY WITH MAST) NOTE: MAST AND DIPOLE ANTENNAE FOR LONGER RANGE OPTIONAL.

OTHER ASSETS

LINKS TO INTEGRATED AIR DEFENSE SYSTEM (IADS) FOR EARLY WARNING AND TARGET ACQUISITION DATA IN THE AIR DEFENSE NET. IT IS ALSO USED AS THE AD BATTERY CP FOR AD UNITS AT DIVISION AND BELOW.

VARIANTS

AN EARLIER VEHICLE VERSION WITH THE DOG EAR WAS PPRU-1/9S80/OVOD. PPRU-M1 HAS IMPROVED C3 AND TARGET PROCESSING FOR HIGHER TARGET VOLUME. IT SHARES THE MT-LBU CHASSIS WITH RANZHIR, MP-22 AND OTHER AD CP VEHICLES, BUT WITH DIFFERENT C3 EQUIPMENT AND THE ADDED RADAR.

SBORKA-M1-2/PPRU-M1-2/9S80M1-2: THE NEW VARIANT HAS SOLID- STATE RADAR, WHICH IS MORE COMPACT, AND FITS ON A HEAVIER MAST FOR OPERATION WHILE MOVING. THUS SET-UP AND DISPLACE TIMES ARE NEAR 0 SEC. THE RADAR IS SIMILAR TO THE TA RADAR ON THE LATEST VERSION OF PANTSIR.

NOTES

UNITS WITH TRACKED WEAPONS USE TRACKED CP VEHICLES (CPVS). WHEELED AD BATTERIES CAN USE THESE OR PU-12M6 OR PU-12M7 BRDM-2-BASED CP VEHICLES. TIER 1 OR 2 UNITS WITHOUT ONBOARD ACQUISITION RADARS ON WEAPON SYSTEMS USE SBORKA FOR THE RADAR. SBORKA'S RADAR CAN EXTEND THE RANGE FOR SYSTEMS WITH RADARS. SBORKA C3 OFFERS DIGITAL LINKS, IFF, IMPROVED BATTLE MANAGEMENT, AND REDUNDANT SUPPORT FOR MOST OF THE SYSTEMS. FOR INDEPENDENT OR AUTONOMOUS FORCE MISSIONS, THE VEHICLE CAN BE EQUIPPED WITH A SATCOM ANTENNA AND RADIOS TO EXTEND TRANSMISSION RANGE.

RUSSIAN AIR DEFENSE ARMORED COMMAND VEHICLE SBORKA-M1 AND SBORKA-M1-2

Sborka-M1 with DOG EAR radar

Sborka-M1 with DOG EAR radar

SYSTEM	SPECIFICATIONS	Communications Intercoms	2
Alternative Designations	9S80M-1, PPRU-M1. System is also called a "mobile aerial target reconnaissance and command post". Some sources incorrectly refer to DOG EAR radar as the name of the system.	Other Communications Links	7, including Integrated Air Defense System, brigade, and division for passive battle operations.
Date of introduction	Circa 1989, with –M1 upgrade by 2000	Vehicle can communicate on the move	Yes
Proliferation	At least 2 countries	Data formats	Graphic and digital data transmission and display
Crew	2 for vehicle	Onboard Generator	Yes
Troop Capacity	5-8 Command and staff workstations or modules	Whip antennae for mobile comms	2 HF whips, 3 VHF
Chassis	MT-LBu tracked vehicle, expanded variant of MT-LB	Other antennae	VHF discones masted, HF dipoles and 11-m mast
Combat Weight (mt)	16.1 est	Digital link to 1L15-1 MANPADS azimuth plotting board	Yes
Chassis Length Overall (m)	7.86	RADAR	SPECIFICATIONS
Height Overall (m)	2.72, with radar folded down	Name	DOG EAR
Width Overall (m)	2.97	Function	Target Acquisition
Automotive Performance		Frequency	F/G Band
Engine Type	240-hp Diesel	Range (km)	80 detection 35 tracking 500m and higher 22 tracking targets flying 25-499 m
Cruising Range (km)	500	Targets display and simultaneous tracking	63, 6 earlier version

Max Road Speed (km/h)	60	Target processing to assignment and track	1-step auto-track
Max Off-road Speed (km/h)	26	Scan rate (s)	2-5, 30 revolutions per minute
Average Cross Country Speed (km/h)	30	Data Transmission rate(s)	4
Max Swim Speed (km/h)	5-6	Scan coverage (°)	360 azimuth (rotating antenna) x 30 elevation
Fording Depth (m)	Amphibous	Antenna scan rate (rpm)	30
Emplace Time (min)	1-3	Antenna horizontal pattern width (°)	5.5 lower plane, 1.6 upper
Armor, Turret Front (mm)	15	Clutter suppression (dB) Operating time max Acquire on the move	30 or more 48 hrs, but usually use shorter on/off times Yes
NBC Protection System	Collective		
Smoke Equipment	Not Standard		
Target Missions Generated Simultaneously	1 or 2		
Target alert simultaneous rate	5-6		
Number of weapons with automatic control	6 separate weapons 12 with 2 per mission		
CP can operate autonomously/in network	Yes/Yes		
Number of sources which can generate targets	6 plus Sborka		
Encryption	Yes, E-24D		
Digital Navigation Monitor	Yes, GPS, Intertial and Map Display		
Automated Networks	Baget- 01-05 or -06 computer workstations		

COMMAND

CONFIGURATION FOR REAR COMPARTMENT HAS 1-4 OFFICER WORKSTATIONS, 3-6 COMMUNICATIONS/BATTLE STAFF CONSOLES, AND ONE RADAR OPERATOR CONSOLE (DEPENDING ON VEHICLE ROLE AND ECHELON).

COMMAND LEVEL

AD BATTERY AND BATTALION IN MECH AND TANK BRIGADES ASSOCIATED AD UNITS/SYSTEMS: SHORAD SYSTEMS (ZSU-23-4, 2S6, SA-9, SA-13, SA-15, MANPADS)

RADIOS, FREQUENCY, AND RANGE

3-6 X VHF WITH RANGE 30 KM (60 KM STATIONARY WITH MAST) 2 X HF WITH RANGE 50 KM (350 KM STATIONARY WITH MAST) NOTE: MAST AND DIPOLE ANTENNAE FOR LONGER RANGE OPTIONAL.

OTHER ASSETS

LINKS TO INTEGRATED AIR DEFENSE SYSTEM (IADS) FOR EARLY WARNING AND TARGET ACQUISITION DATA IN THE AIR DEFENSE NET. IT IS ALSO USED AS THE AD BATTERY CP FOR AD UNITS AT DIVISION AND BELOW.

VARIANTS

AN EARLIER VEHICLE VERSION WITH THE DOG EAR WAS PPRU-1/9S80/OVOD. PPRU-M1 HAS IMPROVED C3 AND TARGET PROCESSING FOR HIGHER TARGET VOLUME. IT SHARES THE MT-LBU CHASSIS WITH RANZHIR, MP-22 AND OTHER AD CP VEHICLES, BUT WITH DIFFERENT C3 EQUIPMENT AND THE ADDED RADAR.

SBORKA-M1-2/PPRU-M1-2/9S80M1-2: THE NEW VARIANT HAS SOLID- STATE RADAR, WHICH IS MORE COMPACT, AND FITS ON A HEAVIER MAST FOR OPERATION WHILE MOVING. THUS SET-UP AND DISPLACE TIMES ARE NEAR 0 SEC. THE RADAR IS SIMILAR TO THE TA RADAR ON THE LATEST VERSION OF PANTSIR.

NOTES

UNITS WITH TRACKED WEAPONS USE TRACKED CP VEHICLES (CPVS). WHEELED AD BATTERIES CAN USE THESE OR PU-12M6 OR PU-12M7 BRDM-2-BASED CP VEHICLES. TIER 1 OR 2 UNITS WITHOUT ONBOARD ACQUISITION RADARS ON WEAPON SYSTEMS USE SBORKA FOR THE RADAR. SBORKA'S RADAR CAN EXTEND THE RANGE FOR SYSTEMS WITH RADARS. SBORKA C3 OFFERS DIGITAL LINKS, IFF, IMPROVED BATTLE MANAGEMENT, AND REDUNDANT SUPPORT FOR MOST OF THE SYSTEMS. FOR INDEPENDENT OR AUTONOMOUS FORCE MISSIONS, THE VEHICLE CAN BE EQUIPPED WITH A SATCOM ANTENNA AND RADIOS TO EXTEND TRANSMISSION RANGE.

SWEDISH AIR DEFENSE RADAR/COMMAND VEHICLE GIRAFFE 50AT AND GIRAFFE AMB

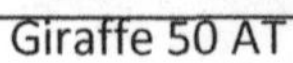

Giraffe 50 AT

Giraffe AMB

SYSTEM	SPECIFICATIONS	RADAR	SPECIFICATIONS
Alternative Designations	See Variants	Giraffe 50AT Specifications	
Date of Introduction	1992	Frequency Band	G-band, except for HARD (H/I-band)
Proliferation	Various configurations in at least 18 countries	Sweep Rate (rpm)	Antenna rotates 60 rpm
Crew	INA	Track Targets on Move	No
Weight (mt)	6.34, INA with arm	Effective Range (km)	50
Length (m)	6.9, INA with arm	Resolution 0.1 sq m target (km)	20-25
Width (m)	1.9	Effective Altitude (km)	10
Height (m)	2.4 for chassis, INA with arm	Low flying targets	up to 12 (in light of target resolution and aspect)
Engine Type	125-hp Mercedes Benz OM Diesel	Fire Units Controlled/Targets simultaneously handled	20
Cruising Range (km)	330	Track Targets on Move	No
Max. Road Speed (km/h	50		
Mobility	Off road mobility is very good on tracked chassis, off-road speed is slightly reduced due to arm.		
Fording Depth (m)	Amphibious; however, arm may affect it.		

DESCRIPTION

RADAR HAS A BROADBAND FULLY COHERENT TRAVELING-WAVE-TUBE (TWT) TRANSMITTER, AND A VERTICALLY POLARIZED PARABOLIC REFLECTOR ANTENNA LIFTED ON AN ELEVATING ARM. HYDRAULIC ELEVATING ARM HEIGHT IS 13M, 7M FOR GIRAFFE 50AT AND HARD.

GIRAFFE 50 AT CHASSIS (BV208)

THE MOST MOBILE SYSTEMS ARE GIRAFFE 50AT AND HARD, ON A SWEDISH HAGGLUNDS BV208 ALL TERRAIN TRACKED CARRIER, WITH AN ARTICULATED CHASSIS. IT IS A DIESEL-ENGINE VARIANT OF BV206.

ASSOCIATED AD SYSTEMS

RBS70, RBS90, RBS 23/BAMSE, STINGER, RAPIER, MISTRAL, AA GUNS, AND ANY OTHER AIR DEFENSE SYSTEMS WITH COMPATIBLE C2 NETWORKS. EMPLOYED TO SUPPORT SHORT-AND MEDIUM- RANGE FIRING UNITS, AD, AND COASTAL DEFENSE NETWORKS.

RADAR CAPABILITIES

FUNCTIONS: SURVEILLANCE, TARGET ACQUISITION AND EARLY WARNING. VEHICLE IS ALSO AD BATTLEFIELD MANAGEMENT CENTER FOR IADS.

FEATURES: RADAR IS DESIGNED TO OPERATE IN A GROUND CLUTTER AND ECM ENVIRONMENT. SIGNAL PROCESSOR USES DIGITAL MTI DOPPLER PROCESSING, WITH CURRENT ECCM, SUCH AS AUTOMATIC JUMPS TO AVOID JAMMED FREQUENCIES, AND EXTRACTS JAMMER BEARINGS FROM DISPLAY. RADAR HAS AUTOMATIC TARGET DETECTION AND TRACKING.

OTHER ASSETS

GIRAFFE RADARS LINK TO INTEGRATED AIR DEFENSE SYSTEM (IADS) FOR EARLY WARNING AND TARGET ACQUISITION IN THE AIR DEFENSE NET. THEY FUSE DATA FROM OTHER AD AND NON-AD UNITS, TO PERFORM BATTLE MANAGEMENT AT DIVISION AND BELOW. THEY ALSO PASS DATA TO OTHER UNITS AND IADS.

GIRAFFE AMB

SYSTEM HAS AN ISO MODULAR CONTAINER ON A 10-WHEEL CROSS-COUNTRY TRUCK, WITH A 3-D MONOPULSE PHASED ARRAY MULTI-BEAM RADAR ON A 12-M MAST. FREQUENCY IS 5.4-5.9 GHZ, WITH CAPABILITIES OF 100 KM RANGE, >20-KM ALTITUDE. LOW ANTENNA SIDELOBES AND FREQUENCY AGILITY OFFER OUTSTANDING JAM RESISTANCE. IT CAN TRACK OVER 100 TARGETS SIMULTANEOUSLY IN THE ONBOARD AD BATTLE MANAGEMENT CENTER. EMPLACE/DISPLACE TIMES ARE 10/3 MIN. SPLINTER AND NBC PROTECTION FOR THE CAB.

VARIANTS

GIRAFFE 50: SYSTEM FEATURED ABOVE, WITH REDUCED DETECTION AND REACTION TIME, AND BETTER CLUTTER RESISTANCE.

GIRAFFE (PS-70/R): ORIGINAL SYSTEM FOR USE WITH RBS70, WITH 40 KM SURVEILLANCE, 20 KM TARGET DESIGNATION RANGE.

GIRAFFE 40: TRUCK-MOUNTED SYSTEM FOR AA GUNS AND MANPADS SUPPORT NETS.

GIRAFFE 75 (PS-90): TRUCK-MOUNTED MEDIUM-RANGE SYSTEM, WHICH CAN CONTROL UP TO 20 FIRE UNITS.

COASTAL GIRAFFE: COASTAL DEFENSE VARIANT.

GIRAFFE AD: MEDIUM-RANGE VARIANT FOCUSED ON ECCM AND C^2.

GIRAFFE CS: SHORT-RANGE AND COAST DEFENSE VARIANT.

HARD (PS-91): SHORT-RANGE VARIANT ON BV-208 CHASSIS. THE H/I-BAND OPERATING FREQUENCIES PROVIDE LOW PROBABILITY OF INTERCEPT (LPI).

NOTES

THE AMB CAN BE MOUNTED IN VEHICLE CONFIGURATIONS, SUCH AS TRACKED VEHICLE, WHEELED APC, OR TRUCK, AND BE SHIP- MOUNTED. FIXED SITE VERSIONS ARE ALSO AVAILABLE. THE RADAR NET ALERTS MISSILE FIRERS, AND ASSIGNS SECTOR ON PLOTTING BOARDS WITHIN THE SIGHT UNITS FOR RBS-70 AND RBS-90 MANPADS.

Air Defense and Other Technology Counters to Unmanned Aerial Vehicles (UAVs)

UAVs are proliferating worldwide. These aircraft are used in various configurations and sizes and for an increasing variety of missions. Their size ranges from bomber size to palm-size micro-aerial vehicles. Missions include attack (attack UAVs and UCAVs), reconnaissance, fire support roles, C^2, INFOWAR, etc. Responses generally fit within the categories: C3D (pg. 9-1), information warfare (IW), and direct attack. Military tactical and technical responses can vary with the configurations and missions, and require an all-arms approach (see TC 7-100-2, Ch 11).

Forces will use C3D to counter a wide range of threats, including UAVs; but the proliferation of these aircraft throughout the area will require increased emphasis on C3D discipline. Measures include more use of IR/absorbent and vehicle conformant camouflage, screens for dismounted positions, and use of deformers, deception, and signature modification. Greater availability of responsive smoke and digging equipment will assist in rapid concealment.

INFOWAR assets can be used against UAVs. Intercept assets may be able to detect signals for UAV control and intercept the image display for their own RISTA. The US and Iran have demonstrated abilities to counter and crash UAVs. Even low-cost jam assets can jam UAV controls and GPS in critical areas at critical times, to neutralize/crash them or prompt auto-return to launch point. But jamming has its own vulnerabilities. Jammers generate a signature subject to detection and destruction, and must limit use time. Intermittent brief jamming can confuse and neutralize many UAVs, and challenge enemy counter-jamming capabilities.

The most likely and most widely available IW counter is the global navigation satellite system (GNSS) jammer. These can be miniaturized with low-power, significant range, and wide area effects. Stationary jammers can be detected and destroyed by direct attack but mobile jammers can fit on ground vehicles. They also can be mounted on UAVs flying prescribed routes with visual markers or on airships. They can also be linked with AD as a lure for air ambush. Although GPS jammers also jam their own forces, defenders and most adversary forces are generally less reliant on GPS precision than modern offensive-minded forces.

Most forces will prefer to destroy UAVs upon detection using direct action. Early detection is a critical factor. This task requires use of air watches and RISTA assets to surveil all approaches. Thus we see a trend in the proliferation of new, more flexible sensors for use on ground and vehicle mounts. They include aerial sensors, e.g., airships with radars and thermal/EO sensors (pg 6-18). They also include acoustic systems: sound-ranging sets, unattended ground sensors, and vehicle/tripod acoustic microphone counter-measure sets. Remote camera arrays offer 24/7 monitoring of large areas. Scores of lightweight remote weapon stations or EO sensor pods fit on vehicles or stands with 30-50+ magnification and fast slew. These can be linked to integrated AD nets, e.g., IADS (pgs 6-3 to 4), to cue other sensors and weapons, or to send warnings to possible units along the UAV flight path, using the attack alert systems and azimuth plotting board (pg 6-56). Lightweight, portable, and more responsive radars now fit ground and vehicle mounts for sector searches that include scanning the horizon for aircraft. Larger UAVs with signatures similar to FW aircraft and flying at higher altitudes will be treated like those targets. New stealthy designs in UAVs and unmanned combat aerial vehicles (UCAVs) will challenge conventional air defense radars. Thus, more forces will adopt recent IADS RISTA nets specifically designed to counter stealth aircraft (and their supporting radars).

The enemy will attack UAVs and support assets (i.e. launcher, ground station, and link assets) on encounter. Weapons for attack vary with UAV size. Conventional aircraft sized UAVs can be acquired and engaged by the same assets as their manned counterparts. Tactical UAVs generally feature smaller visual, thermal, and radar signatures. Reduced UAV thermal signature at night can challenge observation by systems other than air defense, and air defense systems without radars. However, most can be detected by modern radars and acoustics, some using high-resolution thermal sights. At range, missiles and rounds with proximity or AHEAD type fuzes (pg 6-39) can be used against these aircraft. More calibers of AD rounds will use these fuzes. Tactical UAVs which fly below 3,500 m altitude may be engaged by modern man-portable SAMs and guns. Vulnerability varies with design and flight profile. If a tactical UAV flies below 300 m altitude, it is vulnerable to nearly all weapons, including shoulder weapons. Rotary-wing UAVs are more likely to fly at a low altitude because of their low-speed control. Anti-helicopter mines can be command-detonated or sensor-fuzed to destroy low-flying UAVs.

UAVs which may present the greatest challenge to air defense are small UAVs of less than 25 kg (pg 4-3) -- mini-UAVs (MUAVs) and micro-aerial vehicles (MAVs). Battery power eliminates their acoustic and thermal signatures. Unless radars or other specialized AD sensors are used, there will not be timely detection to use most of the weapons in the UAV flight path. For MUAVs, small size almost eliminates radar signature beyond a few km. If they use a camouflage pattern and fly above 300 m, they are very difficult to see in daytime. However, due to limited camera range and wind patterns above tree lines, many will fly within 300 m of the ground. Machineguns can be somewhat effective. Rifle fire against them will be more difficult ("big sky – little bullet"). It is difficult to gauge range without ground level background as a gauge; therefore, a laser rangefinder is essential for aiming. A preferred weapon, found in some infantry units, is a shotgun with duck hunter loads. Automatic grenade launchers with precision optics and air-bursting munitions (Vol 1, pg 2-24) offer a counter to MUAVs. They have displayed AGLs fitted with bore-sighted FARA-1 man-portable radars (Vol 1, pg. 4-29) for near instantaneous cuing.

In the Near Term, as these MUAVs proliferate, forces will seek additional counters. A possible development will be proximity-fuzed grenades for 20-40 mm grenade launchers. About a dozen or so producers have developed shoulder-mount grenade launchers for these munitions, with a range of 500 - 1,000 m. Such a weapon with precision optics and a proximity-fuzed or ABM grenades would enable squads or weapons teams to respond quickly. Vehicle mount light remote weapon pods with multiple cameras for 360° monitor displays and rapid slew are likely.

Micro-aerial vehicles are less widely fielded. They vary from palm-size to hand-launched weighing 5 kg, with 0.67 m wingspan (see pg. 4-3). Many have small batteries short range (<5 km), close camera view (<300 m), and low altitude (often <300 m). With instability and high potential for crashing, many must be treated as disposable. Most are daytime only, but limited night capability is available. Good C3D practices such as camouflage and smoke can challenge them. Jammers can defeat them. Weapons in the above paragraph can defeat them, especially the larger MAVs. It is likely that forces will seek other weapon counters specifically against MAVs. But detection and rapid destruction will be a challenge. Most will detect targets before they are destroyed. Because MAVs are used by adversary low-level units or site security units, a target force must have assets and alert nets to quickly warn of their presence, and be ready to respond. Indeed, a weapon response may alert the adversary and accomplish the MAV's mission.

Air Defense Trends in Countering Low Observable (LO) and Stealth Aerial Systems

One of the greatest threats to AD is LO systems. For decades, aircraft used terrain flying (low altitude) flight profiles for stealth missions against AD. Modern forces also use LO and ***stealth*** (very low observable - VLO) aircraft in the early phase of air operations to neutralize or degrade target air defense capabilities and engage high value targets. Increasingly, stealthy unmanned aerial vehicles (UAVs) and unmanned combat aerial vehicles (UCAVs) are used to engage deep targets in cases where manned aircraft would not be employed. Increasing numbers of modern aircraft and UAVs employ stealth design principles by reducing or nearly eliminating their radar signatures. Thus, AD forces are finding new ways to detect, acquire, and engage them.

LO systems are designed or given technology upgrades to counter use of electronic signals, visual or thermal signature, noise, or change in those factors with stealthy modes of operation (refueling, etc) to avoid detection. Designs can reduce their radar cross section (RCS) in the head-on aspect against specific radar frequencies, but with some vulnerability for detection in other frequencies, and at other aspects. Most UAVs are inherently LO or stealthy. Aerial systems reduce their signal-to-noise ratio (SNR), hide in the radar clutter to reach their mission area, execute the mission, and escape AD response. Stealthy systems are costly designs especially for those roles. But even the stealthiest aircraft are not invisible all of the time or in all frequencies. AD must exploit their vulnerabilities by improving the SNR, improving acquisition means, and engaging more quickly upon acquisition.

Counter-stealth or counter-LO (CLO) systems and tactics have been in use since the 1980s. Some evolved from efforts to counter helicopters using terrain flying modes. Goals are improved responsiveness to counter limited warning time, better precision against protection countermeasures, use of integrated systems to improve responsiveness, expand links to neighboring units, better utilizing HUMINT capabilities, and reduce time out of action to reduce vulnerability. Methods include using overlapping AD assets to cover all approaches, using air observers to cover air avenues of approach, and integrating active and passive sensors directly with weapons to increase effectiveness. These and identification-friend-foe (IFF) systems are used to support autonomy and enable AD engagement by subunits and individual systems, even when the AD nets and IADS are degraded by SEAD and other methods. Counter-SEAD measures by AD units reduce the effects of SEAD, to maintain warning and data transmission systems. Digitization and GNSS map display in battle management centers are used at brigade, and all the way down to the AD vehicle or gun when possible. These provide faster planning and report updates for autonomous units. Blimps can be used in area defense (see pg 6-23). Space and airborne reconnaissance platforms supplement ground-based AD. Aircraft are tasked to intercept stealth systems long before they engage targets. Since most air stealth missions are conducted at night, improved night sensors (thermal sights, etc.) aid in AD counter-stealth effort.

Stealthy aircraft with high-speed anti-radiation missiles (HARMs) could attack AD units from stand-off range if AD lacks counters to HARMs. Some LRAD and MRAD units are increasing weapon loads on the TELs or adding launcher-loader vehicles to engage both aircraft and the precision ordnance which they carry. Increased missile loads can be added to LRAD systems like the Russian SA-20b (pg 6-86), SA-21, and recent South Korean KM-SAM system.

SA-20b with 4 "small missiles"

Most AD units use ***monostatic*** radars, which transmit and receive their own signal. The Russians have fielded a variety of complexes, which can transmit results in real time to battle management centers in LRAD radar units and IADS. Radars may operate in ***bistatic radar*** complexes, with one sending, and others receiving to operate passively and view targets from different aspects and detect stealth systems. There are also ***multistatic radars***, with multiple radars and frequencies in complexes, overlaying and comparing results. The need for multiple overlapping radar frequencies has led to new Russian radars of various frequencies in MRAD and LRAD units. Modern LRAD radars are often phased array systems (e.g., 30N6E2 for the SA-20b/Favorit) with 360° coverage, employing various search modes. They can be integrated, transmit results in real time to map displays, and are difficult to jam. Many AD radars common (centimeter) H-J bands; but others use less detectable or vulnerable bands. With real-time integration and analytic fusion, battle management centers can detect and track stealth systems. These complexes can be costly for most forces, unless costs are mitigated by updating and digitally integrating older radars. A number of countries offer similar radars for MRAD and LRAD systems, complexes, and IADS.

30N6E2 Radar Vehicle

The complexes can also include early warning (EW) radars which operate in lower frequency bands (A-C), updated with robust multiple target tracking and display, and secure responsive digital data links to integrate their results with IADS. Assets also include long-range passive electronic intelligence (ELINT) systems, to support the IADS with low probability of suppression. They scan wide areas over a wide band of frequencies including communications, guidance, and radar bandwidths (such as Russia's Vega 85V6-A/Orion 3-D complex with 400+ km range, 0.2-18 GHz frequency range, and 100-target handling capacity). The 4-vehicle complex has telescoping antennae on 6x6 off-road truck chassis to detect and track ground, sea, and air targets. Another example is Avtobaza, which is fielded and exported, reportedly sold to Iran and Syria. Other countries also produce/export ELINT systems (pg 6-21).

Avtobaza

Russian and other AD forces have developed *"counter-very low observable"* (CVLO) radar complexes, with meter-wave radars specially designed to find stealth aircraft. The Russian Nebo series from 1986 has seen several upgrades. The 1L13-1/Nebo-SV two-dimensional system was replaced by 55Zh6-1 and 1L13-3 automated 3-D versions. The 1L119/Nebo SVU 3-D active phased array system (pg 6-82) appeared in 2001, and links to modern systems, e.g., SA-20b. A new system is Nebo-M mobile radar complex (pg 6-

Nebo 1L13-1 (left) and 1L13 (right) radars

87), with frequency bands ranging from B through X band. Russia is now fielding this system. Nebo variants have been exported; and CVLO radars have been developed by other forces.

Aerial forces are expanding stealth capabilities with longer stand-off for aircraft (and missiles) and smaller UAVs, and UCAVs and UAVs with stealthy designs. Even in an era of reduced budgets, this sequence of improved aerial systems and AD counters will continue to drive the requirement for AD upgrades in acquisition, C^2, weapons, and tactics.

Electronic Intelligence (ELINT) Support to Air Defense

ELINT (or Electronic Support, e.g., ES) systems have been in AD forces for decades. They include specialized systems to specifically detect aircraft electronic emissions. General use ELINT systems to detect air, ground, and naval emitters can also be effective with AD forces.

These sensors offer key benefits, including: long ranges and the ability to operate passively and continuously (for days at a time). Thus, they are well suited as early warning assets particularly against aerial systems using radios, radars, or jammers. They can cue the IADS and use triangulation to locate approaching aircraft. Most systems use multiple stations and a control post, but an individual station could be data-linked with radars or other IADS sensors for location. ELINT systems are ineffective against stealth aircraft when the aircraft are not emitting.

Specialized systems include the Czech Ramona (aka KRTP-81 or -81M). The system was first seen in 1979 and deployed in at least 3 countries. It is complicated, with 3 or more stations with 12 hours to emplace, and locates targets by triangulation from the separate stations. The system is difficult to operate, but can track up to 20 targets emitting in a band of 1-8 GHZ. The Tamara (KRTP-84) followed in 1987 and is mounted on a rapid deploying 8x8 truck chassis. With a band of 820 MHz-18 GHz, the Tamara can track 72 targets to a maximum range of 450 km.

Modern systems include the Czech tailored Vera-E and Borap, Chinese DLW002 and YLC-20, and Russian Valeria and Avtobaza (Vol 1, pg 10-12). Below systems are in AD and EW.

UKRAINIAN KOLCHUGA-M

SYSTEM	SPECIFICATIONS		
Alternative Designation	None		
Date of Introduction	2000 for Kolchuga-M		
Proliferation	At least 4 countries. There are reports of sale to Iran		
Components for Complex	2-3 vehicles plus control post		
Crew	2 at the receiving station, 3or 4 at control post		
Platform	6x6 van		
Antenna Type	4 in VHF, UHF, and SHF		
Frequency Range	0.13-18 GHz (to include X and Ku bands		
Azimuth Coverage (°)	360		
Surveillance range (km)	450-620 depending on target altitude and frequencies. The latter figure is for targets at 18.5 km altitude. Manufacturer claims 800 km (may be valid - some frequencies).		
Effectiveness Against Stealth	Reported but not likely		
Maximum Number of Targets Tracked	32		
Range for a Complex (km)	1,000 frontage or 450-600 radius		
Operation Duration Time (hrs)	24		

NOTES

RUSSIAN 85V6 VEGA ORION ELINT SYSTEM

NO PHOTO AVAILABLE

SYSTEM	SPECIFICATIONS		
Alternative Designation	85V6-A or 85V6E		
Date of Introduction	By 2000		
Proliferation	At Least 3 countries		
Components for Complex	3 stations and control post		
Crew	2 per station, 3-5 at the control post		
Platform	URAL 43203 6x6 van, for receiver and for control post		
Power source	Vehicle PTO, or diesel APU on a trailer		
Antennae Type	Spinning omni-directional and dish receiver antenna. The antenna can be manually pointed or set on auto-track		
Frequency Range	0.2 – 18 GHz C-D (up to 40 option)		
Azimuth Coverage (°)	360		
Elevation Coverage (°)	0-20		
Bearing Accuracy (°)	1-2 for .2-2 GHZ, 0.2 for 2 GHz or more		
Maximum Distance Between Stations (km)	30		
Maximum Control Post Separation (km)	20, near ELINT user		
Deployment time (min)	5-10 for station, 40 for system. Receiver stations may make several local moves before the CP moves. Some users will locate the CP near a receiver station		
Report Format	Digital map display plus acoustic alert, RF signal		

	NOTES

RUSSIAN AVTOBAZA GROUND BASED ELINT SYSTEM

SYSTEM	SPECIFICATIONS	ANTENNA	SPECIFICATIONS
Alternative Designations	1L222	Description	Rotating Parabolic Antenna
Date of Introduction	1980-1999	Azimuth	360º
Proliferation	At least 4 countries*	Elevation	18º- 8.5 to 10.2 GHz 30º– 13.4 to 17.5 GHz
Crew	4	Rotation	6 -12 orbits per minute
Power Supply	6V or 15 V DC	Environmental Conditions	
Weight	13.3 t.	Operational Range	Ambient temperature, ° C from -45 to +40 Humidity 98% at temp ≥25 ° C
Frequency Range	GHz to 17.5 GHz	**VARIANTS**	**SPECIFICATIONS**
Power (kW)	12 consumption	Avtobaza-M	Target detection range of up to 400 km (est.) Frequency range: 0.2 to 18 GHz
RECEIVER	**SPECIFICATION**		
Range (km)	150		
Sensitivity of Receiver	-88dB		
Receiver Modes	Side-looking airborne radars (SLAR) used in combat aircraft, targeting radars of air-to-surface weapons, and radars used to guide aircraft flying at extremely low altitudes, early warning and control radars and jammers.		
Operational Range	X and Ku –Band		
Target Data	Target quantity according to frequency, assignment of		

	jamming systems, type of emitting radars and their angular coordinates		
Frequency identification accuracy	± 30MHz		
Accuracy of DF, degrees	Azimuth: 0.5 Elevation: 3		
Target Throughput	Up to 60 targets		
Reaction Time	50 µs		

DESCRIPTION

PASSIVE ELINT SIGNALS INTERCEPT SYSTEM DESIGNED TO INTERCEPT AND LOCATE PULSED AIRBORNE RADARS INCLUDING FIRE CONTROL RADARS, TERRAIN FOLLOWING RADARS AND GROUND MAPPING RADARS AS WELL AS WEAPON (MISSILE) DATA LINKS.

OPERATION

- FREQUENCY RANGE: 8,000 MHZ-17,455 MHZ
- ADJUSTABLE PRIORITIZATION OF TARGET SETS
- UP TO 100 METERS DISTANCE FROM AUTOMATED COMMAND POST (ACP)
- MONITORS 15 TARGETS PER SECOND UP TO 60 TARGETS
- LESS THAN 25 MIN SET UP TIME
- REAL TIME SELF REPORTING STATUS UPDATES
- PROVIDES LOCATION DATA, AND TARGET PROCESSING FOR GROUND-BASED AIRCRAFT RADAR JAMMING SYSTEM

NOTES

IT WAS REPORTED BY AT LEAST ONE SOURCE TO HAVE BEEN MODIFIED TO RECEIVE AND LOCATE EMISSIONS ASSOCIATED WITH SATELLITE TELEPHONES. THE SYSTEM WAS REPORTEDLY PROLIFERATED TO IRAN AND SYRIA IN 2011 -2012.

Airship Support to Air Defense

Airships ("lighter-than-air" craft) have been used in warfare since the 1800s, when balloons offered elevated platforms for military observers. Airships are increasingly used in civilian venues and offer capabilities for military use including air defense. Roles include support to communications, with airship lift for longer range antennae, and airborne mounting of communications retransmission systems. AD electronic warfare and RISTA units can use aerostats to raise recon systems. A simple method would be to attach a jammer round on a cable. A GPS jammer could be mounted on a vehicle-based aerostat or a dirigible moving within protected zones.

Some signal intelligence and communications units have the option of using aerostats to raise antennae for increased operating range. British Allsopp developed the Mobile Adhoc Radio Network (MANET), with three steerable Low Visibility Skyhook Helikites bearing ITT Spearnet radios to 65-m height. They demonstrated that an infantry radio, usually limited to 1 km range, can send video data (with a 15 kg helikite backpack) to a receiver 10 km away. The company claims that antenna altitude could rise up 500 m.

Electronic warfare units can use aerostats to raise antennae on jammers and recon systems. A simple method would be to attach a jammer round on a cable. A GPS jammer could be mounted on a vehicle-based aerostat or on a dirigible moving within protected zones. Artillery units have long used weather balloons in meteorological units to supply data for calculating fire adjustments. Those units also have helium generators for supplying the gas.

The most widely-used role for airships is reconnaissance, including low level aerial surveillance. Airship-mounted camera systems can detect helicopters flying at low altitudes (using forest canopy for cover) earlier than their ground-based counterparts. Some military and civilian forces use large aerostat balloons with cameras for border aerial surveillance. Elevated view offers a long-range unobstructed field of view, and extended viewing duration. Airship-mounted sensor arrays vary from a simple camera or camcorder hung underneath to a day/thermal video-camera or TV transmitting real-time to a palm pilot or laptop, or over a digital net. The Israeli Speed-A stabilized payload system with automated EO/thermal imager and laser rangefinder fits on lightweight airships. Gondolas can have a camera bar, stabilized mount, or even a gimbaled sensor ball (above) with multiple sensors, laser-rangefinder (LRF), auto-track, and 60+ power digital/optical zoom. Navigation can include GPS location, ground-based location with a LRF, or inexpensive in-viewer display.

As airships become better-controlled and more stable, other sensors can be added to the payload.

An airship could be used in reconnaissance units to mount a small light-weight radar antenna such as on the FARA-1E (Vol 1, pg 4-29). The Russian Gepard airship automated platform offers an electric link and 300 kg payload to 2 km. Airships could raise a cordon of light-weight radar antennae over obscured approaches for detection of helicopters and other threats. Because they may be vulnerable to enemy aerial threats, the airships can be motorized with paramotors for remote steering and navigation thus, avoiding a fixed location for easy interdiction. The airships can also be raised and lowered from transport vehicles, which can rapidly relocate.

Another air defense use can be resurrected from the World War II era using modern airships as barrage balloons. They can deny low-level airspace to enemy aircraft by:

- Forcing aircraft to fly at higher altitudes, thereby decreasing surprise and attack accuracy,
- Limiting direction of attack, permitting more economical use of AD assets, and
- Presenting definite mental and material hazards to pilots by cables and airships.

During WWII in 1944, the UK had 3,000 aerostats operating. During the Blitz, 102 aircraft struck cables (66 crashed or forced landings), and 261 V-1 rockets were downed. The blimps were 19 m long. Modern more compact airships offer more flexible options, with fast vehicle-mount winches, powered dirigibles, and lighter and stronger cables. Although modern aircraft have better sensors (such as thermal sights for night use), most airships have no thermal or radar signature and can be camouflaged and concealed for rapid rise with minimal visual signature. Latest recorded catastrophic collision of an aircraft with aerostat cable was 2007 in the Florida Keys. The Iranians have demonstrated ***air mines***, barrage balloons with explosive charges.

The tether cable and loose lines are the main threat to low- flying aircraft. Tether cables are next to impossible to detect in either day or night conditions, and can be steel, Kevlar, PBO or nylon. Type and length of tether material is determined by lift capacity of the balloon. Multiple loose lines and/or tethers may be suspended from the balloon. Short-notice balloon fields can be emplaced in 10-20 minutes, and raised or lowered with fast winches in 1-5. Netting, buildings, and trees can be used to conceal inflated balloons between uses. Smaller (e.g., 1-m) inflated shaped balloons can be used in target shaping, altering appearance of buildings, vehicles, weapons, etc. They can also be raised as AD aerostats.

Although some balloons will use concealment, others will be clearly displayed to divert aircraft, or trigger a response and draw aircraft into air defense ambushes. Captured marker balloons can divert search and rescue aircraft into ambushes. Balloons can be used in deception as decoys to draw aircraft away from high-value targets.

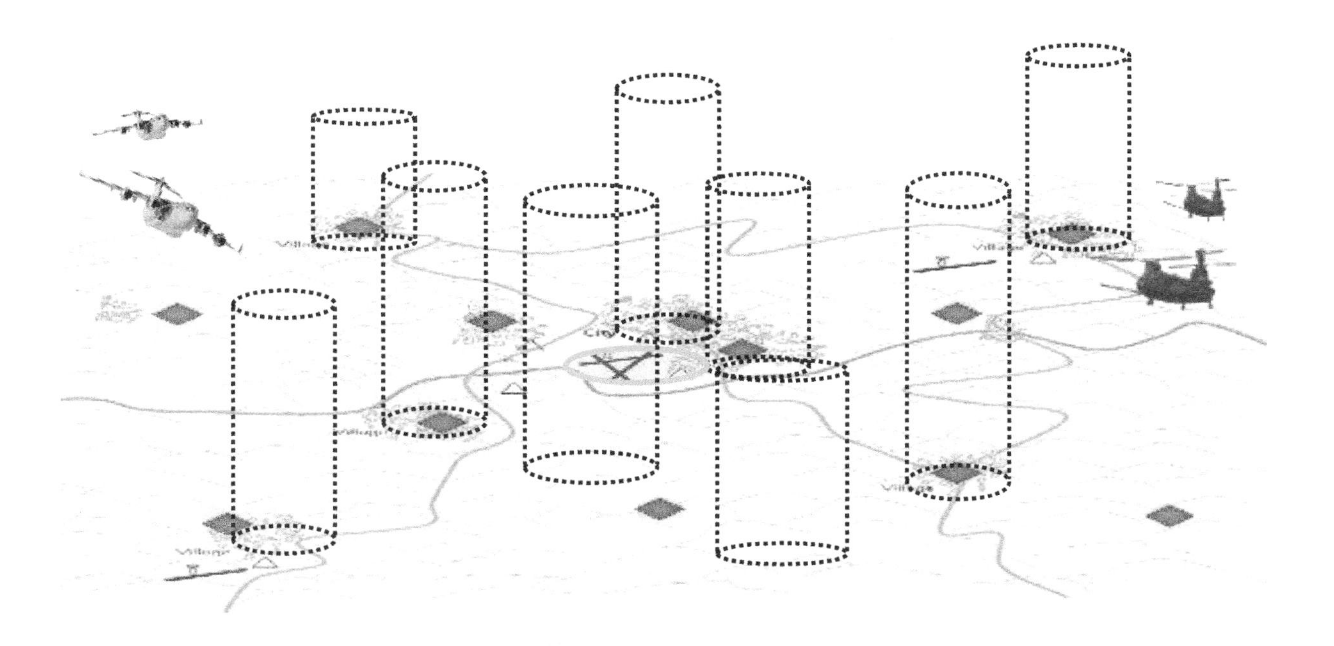

Two areas where airships are most effective in air defense are urban and complex terrain.

Recent Developments in Very Short Range Air Defense (VSHORAD) Systems

VSHORAD systems include a wide variety of technologies defined by mission (AD) and range (to 8 km). These systems are proliferated throughout the battlefield and are used for area defense, site defense, and as multi-role systems for use against a wide variety of targets on the battlefield. They are used by modern regular forces and irregular forces with limited budgets, limited training, and limited mobility assets.

The most widely proliferated VSHORAD threats are weapons throughout the force in the All-Arms Defense. These weapons are primarily used against low-flying aircraft (helicopters, UAVs, etc) which venture into their area and into range of those weapons. These include infantry small arms, vehicle guns, grenade launchers, and missiles. The single most prolific and dangerous category among these weapons is machineguns. Medium (12.7-mm) and heavy (14.5-mm) MGs permit dismounted personnel and any vehicle, boat, or RV to provide protection and/or attack those targets. These can also be used against the growing UAV threat.

All-arms weapons include new multi-role weapons and munitions for use in ground forces, and which can engage aerial targets. Antitank guided missiles (ATGMs) have always been able to engage low-flying aircraft (most of which must fly at slow speeds). However, some ATGMs fly at higher speeds (such as AT-9) for superior intercept. The AT-9 and some others feature an anti-helicopter missile, with proximity fuze and increased lethal radius warhead. The following section also notes other adaptive weapons for the mission. Tactical units can use selected mines, including anti-helicopter mines (pgs 6-29 to 31), to support AD activities.

The most widely fielded VSHORAD weapons for lower-tier forces are AD guns, including MGs, and medium cannons to 57 mm. There are even heavy AD cannons (76-100+ mm, see pg 6-45). With improved fire control (e.g., radars) and improved munitions, some of these remain a viable threat to aircraft flying at 0-6,000 m. Forces are upgrading some ground mounted guns by fitting them on vehicles with modern fire control (6-32, 50). They are also fielding multi-role systems (AD/AT) and infantry fire support vehicles with improved AD guns.

More modern forces have generally chosen a different route. They mounted robust AD capable guns on ground force IFVs and APCs, but equipped AD forces primarily with missiles. The most widely proliferated missiles in any force are man-portable SAMs (MANPADS). These are missiles launched from disposable canisters attached to hand-held gripstocks. They are used not only with dismounted soldiers, but also mechanized units, in missile launcher vehicles, on helicopters, ships, and boats. Some mount MANPADS on support vehicles, e.g., motorcycles, ATVs, light strike vehicles, and even on AD guns (pgs 6-41 and 49). MANPADS have seen upgrades in fire control (EO/thermal and auto-trackers), in warheads (proximity fuzing, larger Frag-HE fills with HMX explosive, KE frangible, etc.), and in missile motor design (high velocity speeds and improved maneuverability). Most MANPADS use IR homing with seekers cooled by an attached battery coolant unit (BCU), with modern upgrades such as two-color IR with improved detectors and needle shockwave dampers for cooler seekers, better clutter rejection for improved lock-on and countermeasure rejection and a probability of hit of up to 85% (90% versus helicopters). Recent guidance modes include SACLOS laser beam rider (LBR on Starstreak) and semi-active laser (SAL) homing to defeat countermeasures with a P-hit of 95% or more. The Lightweight Multi-role Missile variant of Starstreak is due out soon and is offered on a Camcopter UAV combat variant (pg 4-12).

Adaptive Weapons for Air Defense in Close Terrain

Military forces worldwide generally recognize the need to counter aerial threats throughout the battlefield. Fixed-wing threats used to drive the requirements for air defense systems but since the Vietnam War era most countries have increased capabilities throughout the force to counter rotary-wing aircraft. These weapons may not destroy the aircraft but their damage can disrupt the aircraft mission and take them out of action for subsequent missions.

The OPFOR will employ conventional AD weapons against helicopters when available. In some environments however, many AD weapons are less effective, such as in dense terrain or urban areas. In dense terrain helicopters may be spotted at <500m, with concealment or sudden appearance requiring fast reaction, minimum range, altitude, or which limits use of most surface-to-air missiles (SAMs). Helicopter countermeasure systems may degrade SAM performance.

Tactical forces may employ teams and assets in addition to specified air defense assets to counter the helicopter threat, in addition to ground threats, in the area.

--*Tactical security elements* are special-designed units which operate in the OPFOR rear area and use weapons such as machineguns to protect rear area assets from ground and air attack.

--*Air defense observers*. Units will assign AD observers for moving and stationary units. At least one observer team (1-3 people) per platoon is assigned the role of AD observation. Most tactical units are linked into the tactical warning net with an alarm system which can warn of ground and air attacks. The team may be assigned a machinegun or other weapon for the role.

--*Air defense teams.* Infantry forces in close terrain and in dispersed operations may send out teams (2-3 men) against helicopters. These teams can also move with other units for tactical and security missions. A team has to travel fast and light and engage quickly thus the maximum weapon weight recommendation is 20 lbs (9.1 kg). The AD team should employ a weapons mix against air and ground threats. The most common AD weapon is a 7.62 or 12.7-mm MG. An AD team may encounter numerous targets. Systems need ammunition for 2-5 encounters per mission. Equipment needed includes a radio, night vision equipment, and laser rangefinders. These teams can use light vehicles but might be better served with motorcycles or ATVs.

--Combat support and combat service support vehicles with machineguns, medium guns, or automatic grenade launchers will generally not initiate engagements with aircraft, rather have weapons for defense. They may destroy or damage aircraft, force aircraft to break off engagements, and deny aircraft the option for low-altitude flight over wide areas.

--Combat vehicle weapons. Desert Storm demonstrated the capability of helicopters against fighting vehicles. Therefore, AFVs are increasingly addressing that threat with improved weapon systems. Training experience has shown tank main guns with sabot rounds to be a significant threat to rotary-winged aircraft. High-angle-of-fire turrets and air defense sights for LAFV medium guns and machineguns are being fielded and upgraded to address aerial threats. Frangible rounds offer KE-type accuracy and HE-like lethal effects against aircraft. Vehicle guns with programmable-fuze ammunition (such as BMP-3M and T-80UK) can approach the lethality of precision AD systems such as Skyguard. Antitank guided missiles, especially gun-launched (WEG Vol 1 pp 6-41 to 6-45), are a threat to slow-moving or hovering aircraft.

--Anti-helicopter mines or directional mines (such as Claymore or Russian MON series) (See WEG Vol 2, pp. 6-29) can be used. Conventional mines can be adapted with acoustic or multi-sensor units (such as Ajax) to create anti-helicopter mines. RW aircraft obstacle systems can include wire obstacles at LZs and airship nets (armed or unarmed).

Here are a few adaptive weapons for use against aircraft.

System Type	Example
ATGM Launcher	- Short-range systems like Eryx and man-portable ATGMs like Gill, AT-13, AT-7 - Portable systems like European HOT, Russian Kornet, AT-5B
Machineguns	- SQD: Russian 7.62-mm PKM - CO: 12.7-mm w/API, sabot, and frangible
Sniper/Marksman rifle	- 7.62-mm SVD, or .338, with API rounds
Under-barrel grenade launcher	- 40-mm GP-30 HE grenade
Rifle grenades	- BE FN Bullet-thru AV (Anti-vehicle), 3 per rifle
Lightweight grenade launcher	- M79 40-mm grenade launcher
Automatic grenade launcher	- CH 35-mm W-87 w/HEDP, 30-mm AGS-17 (HE) Singapore CIS 40GL, HEDP or airburst munitions
Antitank grenade launcher	- Any ATGLs, esp with longer-range DP or HE grenades - Carl Gustaf M3, w/HEDP grenade, LRF and night sight - German PZF3-T600 or -IT600 with HE and DP grenades
Recoilless rifle	- Yugoslavian M79, US/Swedish M40/M40A1
Antitank disposable launcher	- German Armbrust, Russian RPO-A
Mini-UAVs/Micro-Aerial Vehicles	- With or without warheads, to attack/harass RW aircraft
Air-to-surface rocket launcher	- "C-5K" Iraqi or Chechen launcher with S-5 57-mm rockets

Semi-active laser homing	- Recent ATGLs and ASRs with SAL-H homing munitions

--Improvised rocket launchers. Man-portable air-to-surface rockets of less than 100 mm (Vol 1 pp 14-7) can be launched at low-flying helicopters. Rockets include Russian S-5 series, French 68-mm SNEB, and others. Most improvised launchers lack sights with enough precision. However, some fabricators use fairly standard designs and have employed sights from the Russian RPG-7V ATGL. These sights are adequate for use out to a range of 500 m. To avoid the current problem of high amounts of ash discharge some fabricators added plexiglass shields. With these improvements, launchers for these high velocity rockets with very flat trajectories are a viable threat to helicopters and are claimed to have downed at least one in Iraq.

Air defense teams using man-portable air defense systems (MANPADS) are not adaptive responses but MANPADS can be employed in an adaptive manner. Because of its vulnerability to detection and priority as a target, an AD team needs to be equipped to address multiple targets - air and ground. The Starstreak MANPADS system offers a unique flexibility. It was optimized against helicopters but it can also be employed against FW aircraft, light armored vehicles, and selected other priority targets, such as snipers in bunkers or buildings. Thus a team equipped with Starstreak and other multi-use weapons (e.g., ATGLs, AGLs, machineguns, etc) can be used for a wide array for security, ambush or attack missions. The MANPADS can be linked to MG or cannon fire control, or mounted on reconnaissance vehicles.

Anti-helicopter Mines for Use in Air Defense

The modern attack helicopter, with increasing agility and weapons payload, is able to bring enormous firepower to bear on enemy forces. To counter this threat, some forces employ air defense mines to assist to support air defense ambushes. The intent is less to destroy helicopters, than to: (1) force low-flying helicopters to rise or change course, (2) alert air defenders to trigger the ambush, and (3) distract pilots while engaging them with ground weapons. Some ground-based mines, such as Mon-100 and Mon-200 directional fragmentation mines can be pointed upward for use against helicopters.

Additionally a type of mine—the anti-helicopter mine—was recently developed. By borrowing technologies from side-attack and wide-area landmines, anti-helicopter mines may make use of acoustic fuzing to locate and target potential low-flying targets at significant distances. Their multiple-fragment warheads are more than capable of destroying light-skinned, non-armored targets and damaging any helicopters at closer ranges.

A simple anti-helicopter mine can be assembled from an acoustic sensor, a triggering IR sensor, and a large directional fragmentation mine. More advanced mines use a fairly sophisticated data processing system to track the helicopter, aim the ground launch platform, and fire the kill mechanism toward the target. As the helicopter nears the mines, the acoustic sensor activates or cues an IR or MMW sensor. This second sensor initiates the mine when the helicopter enters the lethal zone of the mine. A typical large fragmentation warhead is sufficient to damage soft targets such as light armored vehicles and aircraft. Alternate warhead designs include high-explosive warheads and single or multiple explosively-formed penetrators.

This data was developed for and incorporated in the Engineer Chapter of Volume 1; see also pg 8-5. OPFOR forces would be expected to deploy mines in Air Defense units to support air ambushes. Therefore, pertinent data was duplicated here to assist the Air Defense planner.

AUSTRIAN ANTI-HELICOPTER MINE HELKIR

SYSTEM	SPECIFICATIONS	FUZE/SENSOR	SPECIFICATIONS
Alternate Designations	None	Types	Dual, Acoustic, and IR
Date of Introduction	In Current Production	Number of Fuze Wells	INA
Proliferation	At Least 1	Resistant to Explosive Neutralization	Yes
Shape	Rectangular	**PERFORMANCE**	**SPECIFICATIONS**
Color	Green	Armor Penetration (mm)	6 @ 50 m or 2 @ 150 m
Case Material	Metal	Effect	Directed Fragmentation
Length (mm)	INA	Effective Range (m)	150
Height (mm)	INA	Target Speed (km/h)	250
Diameter (mm)	INA	Emplacement Method	Manual
Total Weight (kg)	43	Controllable (remotely detonated)	Yes
DETECTABILITY	**SPECIFICATIONS**	Antihandling Device	Yes
Ready	Visual	Self-destruct	INA
EXPLOSIVE COMPOSITION	**SPECIFICATIONS**	**VARIANTS**	**SPECIFICATIONS**
Type	INA	None	
Weight	20		

NOTES

THE HELKIR ANTI-HELICOPTER MINE IS DESIGNED TO ENGAGE NAP-OF-THE-EARTH TARGETS. THE SENSOR IS A DUAL ACOUSTIC-IR. THE ACOUSTIC SENSOR LISTENS FOR A VALID NOISE INPUT AND TURNS ON THE IR SENSOR. THE IR

SENSOR IS LOCATED COAXIALLY TO THE WARHEAD. WHEN A HOT IR SIGNATURE IS DETECTED, THE WARHEAD IS FUNCTIONED.

Anti-helicopter Mines

Country of Manufacture	Number of User Countries	Emplacement Method	Armor Penetration (mm)/ Kill Mechanism	Effective Range (meter) Maximum /Minimum	Detectability/ Composition	Target Velocity (m/s)	Fuze Type/	Warhead Type/Total Weight (kg)	Status
Bulgaria	1	manual	10 @ 100 m	max 200	visual		combined acoustic & Doppler SHF	Total weight: 35 kg	in production
Austria	1	manual	6 @ 50 m 2 @ 150 m		visual		dual acoustic & IR	Total weight: 43 kg	in production
Russia	0	manual		detection 1,000 max 200	visual	100	dual acoustic & IR	Total weight: 12 kg	development
UK	0	manual remote		200/50	visual		dual acoustic & IR	multiple EFP	development

Air Defense/Antitank (ADAT) Vehicles

The battlefield has always held a requirement to fight dispersed and to be able to engage a variety of threats. In the era of large conventional forces, requirements could be met efficiently and inexpensively by task organizing units to meet any fighting requirement. Most weapon systems can be employed against multiple targets. Any machinegun can be employed against aircraft, as well as unarmored and some light armored vehicles. Most forces will include weapons in tactical vehicles to address various threats. But technologies and budgets now permit tactical forces to use systems which can be effective in both air defense and anti-armor missions.

In the Infantry chapter (pages 3-55 to 57), we discussed *infantry ADAT vehicles*. By the 1960s, infantry fire support vehicles were distributed within infantry and dispersed throughout the battlefield. The vehicles had some limited ADAT capability, but their primary role was to carry dismount teams with weapons corresponding to the particular subunit support mission. More capable and responsive vehicles for infantry ADAT, AD, and AT units are available.

Technological changes, force reductions, and increased emphasis on rapid deployment equipment (which may have to fight dispersed) have led to development of more capable *ADAT vehicles*. Improvements in fire control systems and weapons stabilization are crossing over from the antitank arena into air defense. Reverse technologies from air defense systems are also available for antitank and anti-armor roles. The ADAT vehicle has multi-mission capability.

Among the modern specialized systems advertised with this dual capability is the Canadian Air Defense/Antitank System (ADATS). The system features a high-velocity missile launcher on a tracked chassis. It offers responsiveness, high lethality, and lethal SHORAD capability for use in specialized roles or at the division/brigade level.

The German Rheinmetall SkyRanger Advanced Maneuver Support System is advertised as a multi-mission vehicle. With a 35-mm revolver cannon on a Piranha IV wheeled APC chassis; it can defeat aircraft (and vehicles other than tanks) out to a 4,000-m range. Rounds include electronically-fuzed AHEAD (Advanced Hit Efficiency and Destruction, electronically fuzed) rounds against aircraft, some vehicles, and selected ground targets. The highly mobile unit also includes a Bolide SAM launcher vehicle and a radar vehicle on the same chassis.

The Starstreak ADAT application was discussed earlier. Armored Starstreak is the missile launcher vehicle which could be used for multiple roles, including AD and anti-armor use. Now there is another Starstreak application, the Thales Thor remote weapon system. The light-weight (0.5 mt) RWS features a turret with four launchers, modern responsive day/night fire control system, and remote laptop displays and controls. The launchers will accommodate Starstreak and other MANPADS, such as Mistral and Stinger. It also launches ATGMs such as HELLFIRE, TOW, Ingwe (and probably Mokopa), and Spike-LR.

The Multi-purpose Combat Vehicle (MPCV) is a French and German system with an RWS missile launcher mounted atop a VBR combat support vehicle. The launcher in AD configuration holds 4 x IR-homing Advanced Short-Range Air Defense (ASRAD) MANPADS missiles. In the AT configuration it can launch 4 x MILAN-ER ATGMs. The system includes a CCD camera, laser range-finder, and 3rd general thermal sight. The missiles cannot be mixed.

Some ADAT vehicles were designed from the beginning to fulfill the multi-role requirement. Most were modified from existing systems with replacement subsystems or added capabilities. Add-ons, e.g., Strelets remotely operated MANPADS launcher, or the Israeli RWS with the Spike ATGM launcher enable vehicles to perform multiple missions at less cost than special-built designs, but comparable capability. Thus, the BTR-80 APC features a higher angle-of-fire gun to address aircraft and other higher-angle targets. Ukrainian KMDB developed twin 23-mm cannon to replace turrets or fit atop existing turrets and engage fast-moving targets which cannot be engaged by other vehicle guns. The 23-mm round is also affective against light armored vehicles, materiel, and personnel such as snipers firing from high angles.

A Russian developer offers a replacement turret for the PT-76B amphibious tank (and other AFVs). The PT-76E turret uses a 57 mm stabilized auto-cannon from S-60, with modern FCS (Vol 1, pg 6-52). The 57-mm KE round defeats almost all light armored vehicles at 2,000 m, and accurate fires to 3,000+ m. The upgrade converts the tank into an effective AD/anti-armor system with mobility superior to almost all other vehicles and at a fairly low cost.

Most ATGMs can be employed against helicopters. The faster ATGMs, such as gun-launch missiles and those from the Russian 9P149/Shturm-S ATGM launcher vehicle (Vol 1, pg 6-63) are more effective in intercepting a fast-flying helicopter. The 9P149 now features an Ataka missile AD variant with a proximity fuze and frangible rod designed for use against helicopters. Spike-ER, with fiber-optic guidance and IIR-homing option, is advertised as an effective missile for use against tanks and helicopters. Vehicle remote weapon stations include launchers for this missile with range out to 8+ km. Modern RF threat warning systems can warn of attacks from aircraft and ground vehicles, and differentiate the threats. Some of those systems designate direction of threat approach, such as the azimuth warning system 1L15-1.

The ADAT requirement has also driven improvements in ammunition and sensors. Modern Russian tanks can remotely fire their AAMGs using special air defense sights. The Russian FARA-1E radar can be attached to the NSV 12.7-mm MG as a fire control radar against ground and aerial targets. Long range AD sensors such as 3rd gen FLIR on the MPCV offer night range comparable to day sight range. Improvements in AD gun ammunition are discussed on pages 6-35 (MGs) and 6-37 to 38 (medium cannon).

Many air defense systems mount guns and missiles which can easily engage and destroy light armored vehicles. The Russian 2S6M1, Pantsir-S1, and Sosna-R drop-in turret all feature 30-mm twin-tube auto-cannons and high-velocity missiles with kinetic energy effects. The manufacturers claim that these can be effective against aircraft and light armored vehicles. Similarly, the SA-11/SA-11 FO/SA-7 systems are claimed to be effective against ground targets. The 690 or 715 kg missiles (even with only Frag-HE warheads) can destroy any vehicle. But with the cost of SAMs, ADAT systems mostly use guns and ATGMs against ground vehicles.

Current trends indicate that recent technology improvements offer a greater variety of ADAT vehicles. Technologies include gimbaled and gyro-stabilized RWS and OWS, better recoil compensation systems, auto-trackers and stabilized fire control, computer-based integration, radars, EO, acoustics, laser systems, and GPS-based digital C^2. Breakthroughs in ammunition and vehicle drive stabilization offer more responsive precision. In the near term, these capabilities become prevalent, so that forces will increasingly be organized economically to fight dispersed, with the ability to engage air and land force threats with equal deadly effect.

Short-Range Air Defense: Gun and Gun/Missile System Technology Trends

The primary role of air defense continues to be defensive, to deny any adversary the opportunity to use OPFOR air space. A fundamental tenet in that role is to provide area-wide protection. That protection is accomplished with three methods: maintain sufficient inventory, achieve high system mobility, and engage all units to achieve an effective air defense. Methods include use of passive counter-air protective measures and use of lethal counter-air weapons. The focus for many force and weapons designers in recent years has been on missile systems, because of their range and precision against modern aircraft. Gun range limits them to the Very Short-Range Air Defense (VSHORAD) role, but that role is increasingly critical today.

Many countries have significant inventories of air defense guns and are modernizing their inventory of guns. Reasons for this activity are the following:

- Large inventories, offer wide dispersion for area and point protection of assets.
- Guns rarely lose their operability over time. Even older guns can be used.
- Guns are very difficult to put out of action. A vehicle can be killed, and personnel can be killed. But the weapon can usually be brought back into action quickly.
- They are generally less costly to produce, train on, and use than missile systems.
- They can respond to air threats more quickly than missile launchers.
- There is no "dead zone", compared to missile systems. Guns can engage targets down to 0 meters altitude and at a few hundred feet minimum range.
- They are nearly immune to countermeasures.
- They are multi-target systems that can engage a variety of aerial targets (including most likely air threats – helicopters and unmanned aerial vehicles), and a variety of ground threats (including infantry and light armored vehicles).
- They can engage small aerial targets (mini-UAVs, rockets, etc.) which missiles cannot engage.
- The active market in add-on subsystems supports improvements in gun mobility, survivability, fire control, weapon function, ammunition handling, and C^2.
- New types of ammunition increase range, precision, and lethal effects.

New gun systems are being produced but the greatest activity is in the area of upgrading existing gun systems. To examine modernization activities in AD guns, we will look at them from the aspect of three primary factors: mobility, survivability, and lethality.

The most numerous guns used for air defense are not specifically AD guns. These are small arms and general weapons in tactical and supporting units which can engage aerial targets which fly within range. Weapons used in these units to engage aerial targets include grenade and rocket launchers, ATGM launchers, combat shotguns, tiltable mines, and IEDs.

The most numerous gun systems which are effective for air defense are machineguns in 7.62 mm to 14.5mm. These weapons are used for targets of opportunity, especially aerial targets. These can be ground-mounted (shoulder-fired, tripod, or bipod), can be fitted onto a pintle for vehicle mount, or can be integrated into a vehicle fire control system (turret or remote weapon station (RWS) mount, coaxial with a main gun, or fired from a firing port. Most tactical vehicles use machineguns as the vehicle main gun.

Even AD unit missiles and medium guns also use common MGs in supporting units and on combat unit support vehicles

Mobility. The guns, missile systems, and gun/missile systems in AD units are generally towed, porteed, or vehicle mounted. Most towed guns have limitations in mobility. They cannot be towed cross-country and in amphibious crossings as easily as with self-propelled anti-aircraft guns (SPAAGs). There are a few towed guns, like the Russian 37mm M1939 which can be quickly halted, mounted, and fired during a road march. A few developers have marketed towable gun complexes which permit them to be manned and operated during the march (such as the Oerlikon 25mm Diana). These ventures have not found market success because they are still less mobile and responsive than SPAAGs, and are almost as expensive as SPAAGs.

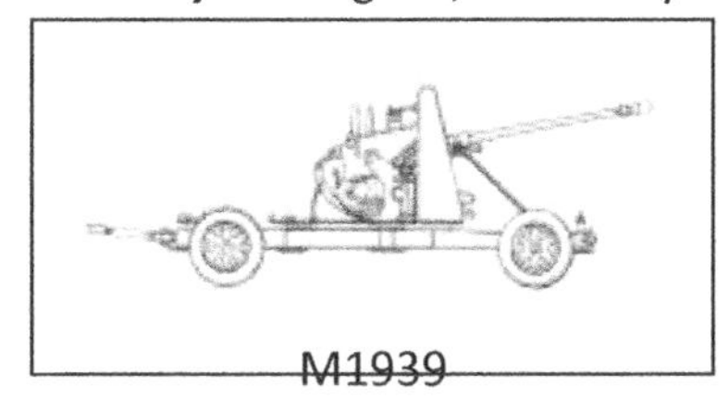
M1939

A new kind of ground mount is the remote-operated modular gun system. An example is the Skyshield 35 35-mm AA gun unit (2 guns, radar, and generator), for use in the Skyguard air defense system. An entire gun unit can be carried on a flatbed truck, hoisted to the ground, and brought into operation in a few minutes. These guns can locate on uneven ground and orient to level with their servo drive, using computer-adjusted fires. Operators can be up to 500 m away.

Some tow systems can be porteed, then dismounted upon arriving at an AD site. Vehicles can operate in locations beyond towed guns. The BTR-ZD in airborne AD units (pg 6-50) transitioned from tow to portee carry. Although portee improves gun mobility, the penalty is that emplacement time may be even greater than normal transition from a towed mount. Thus, after an initial displacement from an airborne LZ, the BTR-ZDs are more likely to mount their ZU-23 AA gun onboard, using a simple method of fitting the gun on top of the hull.

Hull mount is one basic way of converting a vehicle into a SPAAG. Another common mount is in the bed of a "gun truck". An early example was BTR-152 truck-based APC SPAAG variants (Vol 1, pg 3-21), e.g., a 14.5-mm ZPU-2 in the bay. Many insurgent forces and Third-World military forces "technicals" are pick-up and utility trucks with AD guns. Some developers offer trucks with medium guns on flatbed trailers, in highly integrated mobile gun systems. These gun trucks can provide general fire support against all air and ground threats.

Self-propelled anti-aircraft guns (SPAAGs) have been in use well before World War II.

Most early SP systems use AA guns in shielded open turrets, so that crews can easily feed ammunition and slew the guns. Later SPAAGs with auto-cannons, auto-loaders, and integrated wide-aspect FCS, can be responsive and precise without the need for large gun crews and open turrets. To handle the recoil of medium caliber guns (20-75mm), SPAAG chassis are generally heavier than on commercial vehicles. Best-suited chassis for handling gun weight and providing a stable mount for precision fires are tracked, especially modified tank chassis. However, those chassis may be costly, and are less mobile on roads while travelling with wheeled units. A good rule is for the SPAAG to use the same chassis, or more mobile chassis, as the units supported. Thus, SPAAG often use existing chassis (especially APC/IFV or combat support vehicles) also used by tactical units. For instance, the Russian Pantsir gun/missile system (pg 6-64) initially was fitted on a truck chassis; but early sales favored the turret (Pantsir-S1-0) on a BMP-3 IFV chassis. Considerations for some forces include cross-country capability and swim capability, to assure that units can bring their AD systems with them wherever they go. A few new SPAAGs have been offered on the world market; but sales have been slow. Current trends favor using modular AD turrets or RWS which can be fitted to a variety of existing chassis. Other forces are adding gun, FCS, and ammo subsystem upgrades and vehicle conversions to the AA role.

M42 Duster

***Survivability*.** Factors for survivability of AD guns combat are similar to other AD systems and the force in general (see pgs 6-8 to 6-12). Forces are upgrading them to improve survivability. Improved mobility and lethality aid survivability. Use of CCD (including MMW/IR netting) and the low profile inherent in many towed guns still challenge modern air and ground threats.

Two other factors which help counter modern air threats and SEAD are autonomy and integration. Modern guns are increasingly equipped to function effectively as a battery, platoon, or single gun. Thus they can be assigned to tactical units as support. They may have effective links to the AD network, or to direct links with their own forward observers (FOs) or use assigned unit air watches. Attack alerts and azimuth warning receivers like 1L15-1 (pg 6-56) are dispersed to tactical unit CPs and AD guns, to alert them to approaching targets with direction. At the same time that autonomy is improved, AD units have increased integration. Widespread use of comms and improvements such as digital systems, encryption, frequency agility, SATCOM, and redundancy can assure the integrity of C^2 for IADS (pg 6-2), AD units, and links to nearby tactical and supporting units. Vehicles like Sborka (pg 6-15) and Giraffe AMB (pg 6-16) link to IADS and adjacent units to assure that gun crews are aware of air activities in their sector.

FO with LRF

Lethality. The most dramatic upgrades in AD gun capabilities are in the area of lethality. As with other tactical weapons, lethality can be addressed in terms of its components: gun, mount, sensors and fire control, C^2, and ammunition. Modernization continues in all of the components. Conventional wisdom for AD guns is that success means putting more rounds onto the target. Therefore, most gun design improvements focus on longer range, better gun stabilization (and reduced recoil and barrel-whip) for better accuracy, reduced weight for shorter response, and increasing rate-of-fire while decreasing overheating – for more rounds per salvo.

Machineguns. The most proliferated guns used for AD are small-caliber (5.45-14.5 mm), because of the inventory of machineguns in all forces. Because MG size and lower cost separate them from medium-caliber guns, they should be treated separately. The inventory for MGs is so large because they can be ground-mounted and easily added to light vehicles with a pintle mount. All MGs can be used against aerial as well as ground targets.

Machineguns are increasingly available for use on unarmored or lightly armored combat support vehicles, including tactical utility vehicles, motorcycles, and all-terrain vehicles. Vehicle mounts include pintle mounts, remote weapon stations, overhead weapon stations, and turrets. Using economical laptop computer FCS, servo-motors and stabilization, MG add-ons are increasingly being used for vehicle main weapons or as secondary weapons to supplement main weapon fires and provide general and AD security. For more information on MG applications, see the section at Vol 1, *Auxiliary Weapons for Infantry Vehicles*.

A general rule for guns is that AD range can be calculated at 100 times the mm bullet size, in meters. Of course range actually varies by the components noted above, especially ammunition. But under that rule of thumb, a 7.62-mm MG has a 1,000-m AA range, and a 12.7-mm MG ranges about 1,300 meters. Those estimates are pretty close (see Vol 1, pgs 2-16 and 17). Vehicle-mounted with a good FCS, ranges can extend somewhat farther. Better range and penetration usually favors 12.7mm mm over 7.62 mm. The 14.5 mm round is larger than 12.7 mm, with a marginal edge in penetration and range. But superior round capacity, precision, recoil, ease of fit, and rate of fire favor 12.7 mm for use by dismounts and light vehicles.

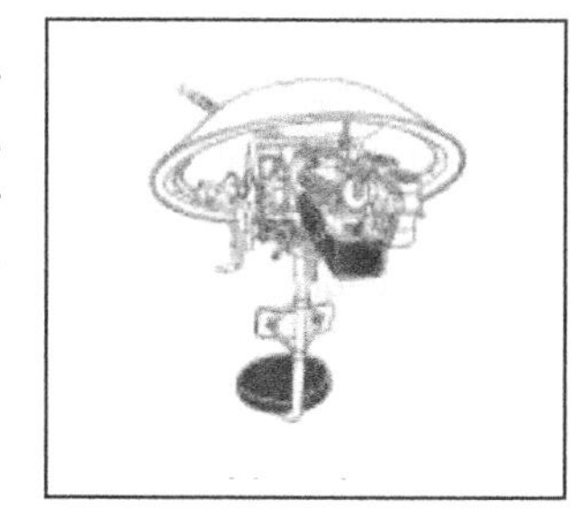

The 14.5 MGs are widely fielded on APCs, such as BTR-80 (Vol 1, pg 6-34). But Russian forces consider it to be obsolescent (Tier 4) for AD guns. Thus they have generally replaced ZPU guns (pg 6-45, on towed mounts of 1, 2, or 4 guns) with 23-mm cannons. Nevertheless, these guns endure and can still be found in more than 45 countries. Improvements available include fire control radars (like SON-9), turret mounts, improved command and radio links, such as an azimuth warning receiver and handheld encrypted radios.

Machineguns in AD units or specifically noted as AD MGs tend to be better equipped to deal with air threats, with features like improved recoil damping, stabilization, twin barrels for higher rate-of-fire, and better fire control. Another modern trend is to chain-drive guns. With chain drive comes more efficient and compact guns, multiple barrels, less recoil, and better precision at range. Air defense MGs often use quick-change barrels and superior air cooling for successive 10-15 round bursts and increased practical rate of fire (100 rounds per minute up to 250-300). Like other MGs, many AA MGs are remote-operated with electronic triggers. Due to their shorter range, MGs will require low technology support, including air watches, forward observers, and links (to nearby units for warning, to AD command nets, and to air warning nets).

Fire control system improvements have caught up with gun and mount technologies. Gun mounted or stabilized remoted day/night ballistic computer sights with EO and LRF are available. The FARA-1E MMW radar (Vol 1 pg 4-29) can be mounted and bore-sighted for immediate fire control. Binocular LRF such as the Sophie-LR or –MF offer thermal day/night use with other functions. For responsive C^2, hand-held radios and the 1L15-1 azimuth warner give alerts and azimuth. In vehicle mounts with good telescopic EO sights, effective gun AA range is extended up to 2,000 m. Russian AD sights offer a high-angle view for the AD role. Lead-angle sights and auto-trackers are available.

FARA-1E on MG

An emerging trend among small-caliber AD guns is the Gatling-type multi-barrel gun. The weapon was modernized in the US 20-mm towed M168 Vulcan cannon and was employed in the 1950s. The M163 AD vehicle was an M113 with the Vulcan cannon. Other countries, including Russia, have fielded Gatling-type guns in 12.7, 20, 25, and 30 mm. The US Dillon Aero M134 fires 7.62-mm ammunition. There are inherent advantages in these guns. The multi-barrel design permits larger salvos against a fleeting enemy before overheating. The flanged barrels reinforce each other to eliminate barrel-whip. They can use chain-drive for maximum recoil dampening with precision fires. Recoil is still significant but it can be damped to a constant amount which permits accurate aiming. The design also reduces halts due to jammed rounds. But Gatling guns have significant limitations. Recoil and system weight can overwhelm light vehicles (and require stopping). The huge ammo requirement can strain logistic assets. Cost per kill is greater. Thus, Gatlings have seen limited use as light vehicle main guns or in vehicle auxiliary AD weapon station upgrades. The greatest limitation for small-caliber Gatlings is insufficient range against aircraft weapons. In the future, if ranges for small-caliber ammunition improve with the guns, higher carry capacity with smaller rounds may make 12.7-mm Gatling-type guns a preferred replacement for MGs (versus medium guns).

GE Miniguns

Ammunition developments are the single greatest factor for improving air defense. Improved ammunition is increasing range, precision, and lethality for all air defense guns. Although small-caliber guns have less variety of rounds than medium/large guns, there are new types. Improved armor-piercing incendiary tracer rounds can extend useable range for MGs. Chinese and Russian 12.7-mm duplex rounds (e.g., Russian 1SLT) have two separate projectiles,

doubling the pattern of projectiles in any salvo fired. These rounds are especially useful against close-in small targets, like UAVs. Several countries make 12.7-mm sabot rounds like the US Olin M903 SLAP round, with greater precision and penetration at maximum range. Frangible rounds are made in calibers 7.62, and 12.7 mm and .50-cal. They fly like KE rounds, and can be ballistically matched to KE rounds (unlike HE), yet are more lethal at the target than KE (like explosive HE rounds). The most lethal mix may be KE and frangible. One problem associated with having more than one type of ammunition on hand is being able to switch between or among them against fleeting targets. With some MGs using box feed, the mounts permit two boxes, left and right, thus an ammunition switch can be very fast.

Medium AD Guns. In order to increase lethality, the best course of action is to go up in gun size to medium guns. As we noted in the range rule, a 12.7-mm MG ranges about 1,300 meters. But a 30-mm gun can range 3,000 m and a 57-mm gun ranges 5,700 m. Emphasis in modern AD guns is on medium calibers (20-75 mm). There are still some larger caliber guns in 76, 85, 100, and 122 mm; but upgrades are limited to adding radars, radios, and azimuth warners (KS-19M2, pg 6-42). Within medium AA guns, calibers are creeping up to better range aerial threats.

Medium-caliber guns (cannons) have seen the greatest variety of upgrades. Medium guns suffer from many of the same problems of MGs like barrel whip, overheating, and recoil. In the 1950s and 60s, most AD cannons were of 20, 23, 37, 40, and 57 mm. Most are still in use today and are recoil/gas-operated. Many are twin guns like the Russian 23-mm ZU-23 and Chinese 37-mm Type 65. Later, 25, 30, and 35 mm auto-cannons have grown in usage. Many use chain-drive. Modern guns like the Russian 30-mm 2A38 and Swiss 35-mm GDF-003 can fire at rates up to 2,400 rds/min (for 2A38), with 25-round bursts. But limited salvo size and practical rate-of-fire still limit fires to avoid over-heating.

GDF-003

The modern gun size that has received the latest technology is 35-mm. With lightweight designs (some less than 100 kg), these guns can be fitted on ground chassis like the GDF-003 (pg 6-43) and vehicle-mounted in modern turrets like the South African LCT35 for IFV or SPAAG. The best AA gun examples are in the 35-mm and 40-mm guns (made by manufactures like Bofors, Oerlikon, and LIW). These modern weapons can range to 4 km accurately, and exploit new round technologies. For instance, the Swedish Skyshield-35 gun uses a compact 35/1000 revolver cannon (with single barrel, rotating cylinders, and linkless rounds in a conveyor feed system). The gun weighs half the weight of the GDF-series 35-mm guns.

A new AD gun technology is the RMK-30 30x173 mm recoilless auto-cannon from Rheinmetall, fielded on the Spanish Pizarro and Austrian Ulan IFVs. The combustible case rounds produce gas blowback, expended out of the cannon rear to <2 feet. Rate of fire for the 100 kg gun varies from 300-800 rds/min. Fitted in a RWS, the gun can mount on nearly all light tactical vehicles. It fires sabot and frangible rounds, and AHEAD-type programmable air-burst rounds to 3,000 m effective range. The gun could replace MGs in light AD and combat vehicles.

Some countries use Gatling-type cannons for AD. The US M163 SPAAG with 20-mm Vulcan gun was followed by the Blazer with a 25-mm Gatling gun on a Bradley chassis, and the LAV AD with Blazer gun on the USMC LAV chassis. A French program fitted the gun on a French chassis with radar FCS. The Chinese M1990 30-mm towed gun features a 4-barrel Gatling system. Nevertheless, because of reasons noted on the page above, Gatling AD guns are not widely fielded. Also, as they increase in caliber, recoil and the ammunition storage burden increase dramatically. Better gun precision and range more than offset the advantages of high-volume fire with Gatlings.

M163

The AD gun mount is a critical consideration in gun system, as noted in the discussion of mobility (pg 6-35). For ground systems, we do not see an auxiliary power unit, like the APUs such as on the GHN-45 artillery cannon, and on the Russian 2A45M AT gun. But some modern guns have lift hooks for rapid mount/dismount. Motor gun drive, such as on the Chinese Type 79, permits faster slew to target, for more precise fires and more salvoes against fleeting targets.

Some SPAAGs have stabilized guns for AA fire on the move. Stabilization kits are available and are fairly inexpensive. Turrets for IFVs and several RWS are easily accurate enough for AA use, and can be fitted on a variety of vehicles. For vehicle mounts, cannon recoil has led some forces to use tank chassis to absorb the load and assure accuracy. The Polish Loara SPAAG features twin 35-mm guns on the PT-91 (T-72 upgrade) chassis. Light turrets such as the Russian Sosna turret (either 30-mm guns or gun/missile system) can fit on IFV/APC chassis of supported units, which means that they offer amphibious or airborne chassis. The German RMK-30/Wiesel can be used with airmobile units. The Bofors TriAD turret fits on IFVs such as the Swedish CV90 and the Piranha APC. With the radar, superior EO, and quick response 40-mm L70 gun, the SPAAG fires programmable 3P HE rounds for lethal fires.

CV90C with TriAD

Slovak BRAM

A few new SPAAGs have been developed. The Rheinmetall SkyRanger is actually a multi-role system, and is discussed on pg 6-29. Recent truck-mounted SPAAGs include the South African Zumlac, with a mine-protected SAMIL (4x4) truck, and a ZU-23 gun on the rear bed. China offers its FAV light strike vehicle with the ZU-23 on the rear and extendable spades. Oerlikon and Skoda proposed a SPAAG with a Tatra T815 8x8 truck, and a Skyshield 35 gun mounted on the rear. A disadvantage with large truck-mounted SPAAGs is that they can be distinguished from other vehicles, making them high-priority targets for destruction. Note that most of the systems mate existing guns and vehicles rather than costly special-design systems.

Improvements in fire control include day/night all-weather EO computer-based sights and monitors with digital transmission capability. Many older AD guns have added target acquisition radars such as AA guns noted at pgs 6-43 to 52. With added onboard computers, radars (and EO TV/thermal sights with auto-tracker for day/night passive operation) and older guns like the ZU-23 (pg 6-49) can be converted into a responsive autonomous weapon like the ZU-23M or ZU-23M1. Vehicles can integrate a FCS from disparate fire control elements (CCD TV day sight, thermal night sight, ballistic computer, voice radio nets, and forward observers, digital C^2 nets in the IADS and other AA and tactical nets, auto-tracker, dual-mode radar, AD net azimuth warning system, laser rangefinder, laser radar, RF detectors, digital displays from remote cameras, robots, UGS, acoustic sensors, UAVs etc). Many of these can also be linked to laptop monitors or FCS displays for ground AA gun systems or transmitted to the unit net or IADS.

The greatest changes for AD guns are in new ammo for longer range and better precision. These rounds for medium guns generally make the previous requirement for higher rates of fire irrelevant to air defense lethality. Air defense guns generally have rounds such as HEI, API-T, and SAPHEI-T. More recent guns use sabot (APFSDS-T) rounds, frangible rounds, and proximity-fuzed HE rounds. These rounds enable many systems, which could not reach beyond 2,000 m without losing velocity and their probability of hit to reach out to 3,000+ m accurately. Most of the older guns can also use these rounds, as well. The Russians offer a 30-mm "CC" round (with 28 sub-projectiles) for use on aircraft guns. It could be a good anti-UAV AD round.

Proximity fuzing permits guns to reach farther and higher and offsets the inaccuracies of HE rounds compared to KE rounds. One proximity-fuzed round is more accurate (because a near miss still detonates the round for a "hit") than ten rounds of HEI. Salvo size and cost per kill are lower with proximity rounds, making existing or older gun systems effective and lethal in the air defense role. However, proximity-fuzed rounds can be counter measured or decoyed when fired in obstructed areas. Environmental clutter such as vehicles and power lines can predetonate the rounds. Swedish Bofors developed the 3P HE round in 40-mm and 57-mm with a 6-way programmable fuze, which can avoid pre-detonation. One of the fuze modes is gated proximity, which desensitizes the round until near-impact time. Even when engaging helicopters flying nap-of-the earth at low altitude, effects of electronic jammers and clutter are negated. The 40-mm round produces a cloud of 2650 fragments. This is a very affordable option as fewer rounds are needed and more costly rounds are selected only for specific targets.

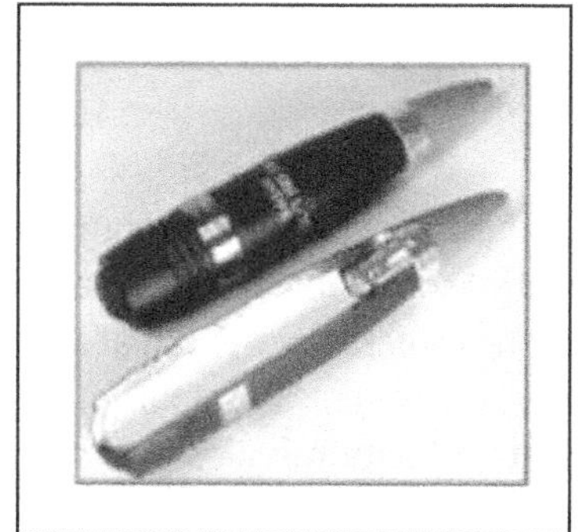

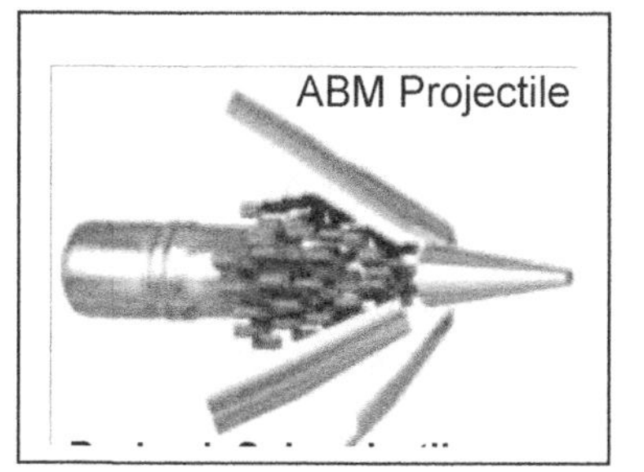

Another round for medium guns is the Swiss Oerlikon AHEAD (Advanced Hit Efficiency And Destruction) round (and similar technology rounds), for use in 30 mm, 35 mm, 40 mm, and 57 mm guns. The rounds, also known as Air Burst Munition (ABM), can range 4000 m, using their electronically-programmed time fuze to dispense a wall of tungsten sub-projectiles at an aerial target 10-40 m away. A 40 mm gun round dispenses 152 sub-projectiles. From a 35-mm gun, 24 AHEAD rounds (1-2 sec) usually assure a kill against a fleeting aircraft. The round can be used against even small targets, like mini- and micro-UAVs, artillery rounds, and rockets, or for top/direct attack against ground vehicles, dismounted troops, and materiel targets. Russian Aynet tank round and BMP-3M HEF round are also programmable, for AD and against ground targets like AT assets.

One of the most lethal AD calibers continues to be 57 mm, in the Russian 57-mm S-60 (pg 6-43 and their variants), and Swedish 57-mm naval guns. The rounds are large enough to deliver substantial bursts out to 6000 m. A variety of upgrades are offered for the guns, and proximity and AHEAD rounds are available for effective fires out to the maximum range.

Improvements in ammunition-handling are keeping pace with the weapons systems. Selected gun systems have multiple ammunition feed systems for the different types of rounds. Cased-telescoped gun systems (and their CT ammunition, round in the photo right) are a recent development - which may supplant existing designs. Cased rounds shorten round length, permitting smaller gun breaches that better fit inside of vehicle turrets and weapon stations. The rounds enable autoloaders to hold more ammo in smaller spaces and more easily manipulate rounds in loading trays. Faster loading and more rounds decrease jams and ammo outages at critical moments.

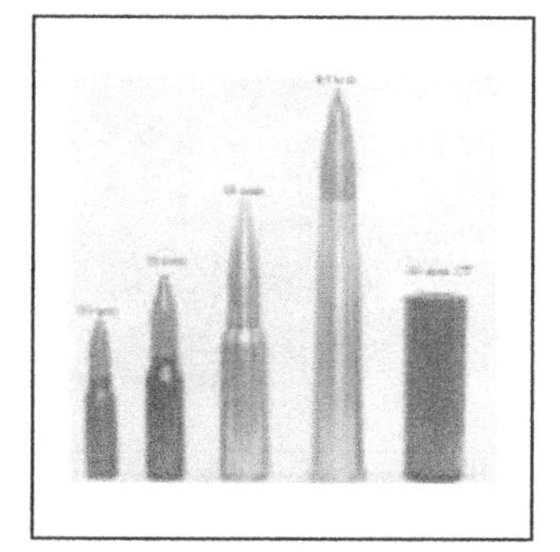

Gun/Missile Systems. Another lethality trend which has reinvigorated SHORAD is widespread fielding of, or conversion to, gun/missile systems. Most SHORAD systems are being converted to having both guns and missiles. Thus the guns, with their links to the AD net and improved FCS, can also serve as platforms for missiles. The guns and missiles can protect each other to provide lethality beyond effective range for most guns, no dead spots for the missiles, and effective lethality despite aircraft countermeasure systems.

A significant amount of SHORAD modernization activity includes gun/missile systems. We have noted some new SPAAGs have been marketed without missile capability. Nevertheless, most new AD gun systems actually fielded are gun/missile systems, such as BRAM, in the photo on pg 6-40. A few systems feature robust SAMs. The Russian 2S6M1 (pg 6-63) was followed by the Pantsir (pg 6-64), with 18-km high-velocity missiles and 30-mm twin auto-cannons. The Ukrainian Donets mounts a ZSU-23-4 turret (with four 23-mm AA guns) on a tank chassis. Also mounted on the turret is an SA-13 missile launcher. China's Type 95 pairs 25-mm guns and QW-2 MANPADS. TY-90 has a 12.7-mm MG and six robust SAMs.

The Russians now offer modular turrets for the robust gun/missile systems. Pantsir-S1-0 turret can be fitted to a wide variety of chassis. They can use IFV/APC chassis, are almost visually indistinguishable from them, and are compatible with the mobility and maintainability of supported units. With existing chassis and indigenous installation, fielding costs are lower.

Another turret, the Sosna-R, uses a twin 30-mm AA gun and Sosna-R 8-km laser beam-rider missile. The turret is lighter and less costly than Pantsir, and fits many combat vehicles. Its range, precision, and responsiveness can challenge aerial systems well beyond gun range.

Several ground-based gun/missile complexes include robust missile systems. Among the best is Skyguard (pgs 6-43 and 6-72), which feeds compatible digital fire control and radar to both guns and missiles. The Chinese PL-11 system is similar. Many countries will co-locate guns and missiles for mutual fires and support. Germany employs a "team" which includes Roland SAMs and Gepard SPAAGs. Similarly, the French army mixes Roland and AMX-13.

Most gun/missile systems use the less costly low-technology approach of pairing guns and man-portable SAMs (aka: MANPADS). Vehicles such as the US Avenger, LAV-AD, and Blazer use Stinger SAMs. China likes this upgrade approach. The recent FAV light strike AD vehicle mounts a ZU-23 gun and twin MANPADS launcher. Russian variants of MT-LB include the MT-LB6MB3 IFSV/APC with 23-mm GSh-23L twin cannon, 30-mm AGL, and 7.62-mm MG. But the MT-LB6MB5 IFSV has a 2A38 twin 30-mm AD gun, MGs, AGL, and SA-18 SAM launchers. The Polish Sopel tracked system mounts a turreted twin 23-mm gun and twin Grom MANPADS launcher. GMW developed a twin Stinger launcher for mounting on the Gepard AA gun. A French-marketed variant of the Blazer turret features a 25-mm Gatling-type gun and four Mistral MANPADS missiles. The turret also has a radar FCS and it can be fitted on LAVs such as M113.

The widely fielded ZSU-23-4 SPAAG (6-7,000) is the subject of various upgrade packages (pg 6-52). Several include adding MANPADS, integrated into the fire control system. The Russian ZSU-23M5 mounts one or two Strelets MANPADS modules (each with two SA-18 missiles). The Polish Biala fits four Grom MANPADS launchers onto the turret. A Ukrainian upgrade includes a swing-up launcher with six SA-18 missiles. Other modernizations include GPS navigation, a new radar for some, ballistic computer and TV FCS with thermal sights, digital communications, NBC protection, side skirts, and smoke grenade launchers. An Iranian version includes an auto-tracker and laser warning system.

Towed gun systems can also be fitted with missile launchers. The widely fielded Russian ZU-23 offers an –M1 upgrade (pg 6-49) with a Strelets two-SAM module, also integrated into gun FCS. The FCS in ZU-23M and ZU-23M1 has TV and thermal sights, LRF, IR auto-tracker, and a ballistic computer. Strelets module (with SA-18 SAMs) can be fitted to many AA systems.

Included in the market for AD guns are turrets, remote weapon stations, and subsystem upgrades for infantry vehicles which enable them to reach similar capabilities as specialized AD guns and gun/missile systems. Developments in this area for infantry vehicles are discussed in Vol 1, pgs 3-12 to 14, and 3-55 to 57. Infantry fire support vehicles in maneuver battalions and below offer mobile and responsive AD and AT support (Vol 1, pages 3-52 to 54). For more discussion of AD guns, see ***Air Defense/Antitank (ADAT) Vehicles,*** in this chapter at pg 6-32.

Swiss 35-mm Towed AA Gun GDF-003/-005, and Skyguard III System

Weapons and Ammunition Types	Typical Combat Load
35-mm automatic cannon	238
FAPDS	119
APFSDS-T	119
(Preferred mix)	
AHEAD	74
FAPDS	74
APFSDS-T	74
(Estimated w/ AHEAD)	

SYSTEM	SPECIFICATIONS	MAIN ARMAMENT AMMUNITION	SPECIFICATIONS
Alternative Designations	Skyguard Gun/Missile Air Defense System (See VARIANTS, Skyguard).	Best Ammunition Mix	See Above
Date of Introduction	Circa 1981-84	Type	HEI-T
Proliferation	At least 3 countries	Tactical AA Range (m)	4,000 (self-destruct)
Crew	3	Tracer Range (m)	3,100+
Carriage	4-wheeled/2-axle towed chassis	Effective Altitude (m)	3,100-4,000
Combat Weight (kg)	6,400	Self-destruct time (sec)	6-12
Travel Position Length Overall (m)	7.8	Type	Semi-armor-piercing HEI-T (SAPHEI-T)
Firing Position Length Overall (m)	8.83	Range (m)	4,000
Length of Barrel (m)	INA	Tactical AA Range (m)	4,000 (self-destruct)
Travel Position Height (m)	2.6	Effective Altitude (m)	4,000 (est)
Firing Position Height (m)	1.72	Self-Destruct Time (sec)	6-12
Travel Position Width Overall (m)	2.26	Penetration (mm, KE)	40 at 1,000m
Firing Position Width Overall (m)	4.49	Type	APDS-T
Max. Towed Speed (km/h)	60	Range (m)	4,000
Emplacement Time (min)	1.5	Tactical AA (m)	4,000
Battery Emplacement Time (m)	15	Tracer Range (m)	2,000
Displacement Time	5	Effective Altitude (m)	4,000 (est)
ARMAMENT	**SPECIFICATIONS**	Penetration (mm, KE)	90 at 1,000 m
Gun Caliber Type	35x228 35-mm autocannon	Type	APFSDS-T
Gun Number of Barrels	2	Range (m)	4,000
Gun Operation	Gas-operated	Tactical AA range	4,000

Gun Rate of Fire (rd/min)	Cyclic: 1,100 (550/barrel) Practical: INA, bursts up to 25 rounds	Tracer Range (m)	3,100-4,000
Gun Loader Type	2x56-rd magazine automatic feed	Effective Altitude (m)	4,000 (est)
Gun Reload time (sec)		Penetration (mm, KE)	90 at 1,000 m
Traverse (°)	360	Type	APFSDS-T
Traverse Rate (°/sec)	120	Range (m)	4,000
Elevation Rate (°/sec)	-5 to +92	Tracer Range (m)	3,100-4,000
Reaction time (sec): INA	INA	Effective Altitude (m)	4,000 (est)
FIRE CONTROL SYSTEM	**SPECIFICATIONS**	Penetration (mm, KE)	115+ at 1,000 m
On-Carriage Sights	Lead-computing optical sight, or GUN KING electro-optical system on GDF-005	Type	Frangible APDS (FAPDS) The round has higher velocity and flat trajectory of a APFSDS-T round (same gun data), and Frag-HE effects. On impact with the target surface, penetrator breaks into 100s of KE fragments.
Off-Carriage		Type	AHEAD (Advanced Hit Efficiency and Destruction), designated AG 35x228. The AHEAD round uses a programmable time fuze and HE charge to dispense a cloud of 152 pellets (3,800 from a 25-round burst) at or in the path of a target helicopter, LAV, or soft target. Other fuze modes include proximity and PD.
Name	Skyguard radar and CP system Platform: Towed compartment		
Platform	Towed compartment		
Sights	SEC-Vidicon TV Tracking System		
Range (km)	25 day only		
Laser Rangefinder	Yes		
Search and Track Radars			
Name	Skyguard Mk II (SW)		
Function	Dual mode doppler MTI		
Detection Range (km)	25-45		
Tracking Range (km)	25		
Frequency	8-20 GHz, I/J Band		
Rotation Rate/min	60		
Mean Power (W)	200		
Link	System uses a wire link among major components. Digital data is invulnerable to ECM, frequency hops.		

OTHER FIRE CONTROL

GUNS ARE LINKED TO BATTERY/BATTALION NETS AND THE IADS, AND RECEIVE DIGITAL ALERTS OF APPROACHING AIRCRAFT. GUNS, BATTERY, AND BATTALION USE AIR WATCHES AND FORWARD OBSERVERS FOR FAST RESPONSE.

VARIANTS

SKYGUARD: SYSTEM/COMPLEX DESCRIBED FOR THE OPFOR HAS A RADAR, 2 ASPIDE MISSILE LAUNCHERS AND GENERATORS. AD COMPLEXES CAN VARY WIDELY. SINCE THEY ARE ORGANIZED AROUND THE SKYGUARD RADAR/CP UNIT, THEY MAY BE GUNS ONLY OR MISSILE LAUNCHERS ONLY. THE MOST EFFECTIVE AD ARRANGEMENT IS THE ONE NOTED ABOVE, AS A GUN/MISSILE SYSTEM.

GDF-001: SYSTEM HAS A SIMPLE SIGHT.

GDF-002: SYSTEM LINKS TO SKYGUARD.

GDF-003: ADDS GUN SYSTEM UPGRADES.

GDF-005: UPGRADE (FOR -003 WITH NDF-C KIT) HAS GUN KING 3-D AUTONOMOUS SIGHT SYSTEM, ONBOARD POWER SUPPLY AND AUTO-LOADER. CAN FIRE AHEAD ROUNDS.

NOTES

ORIGINAL MK I RADAR RANGE WAS 20 KM. SYSTEM CAN ALSO BE USED AGAINST GROUND TARGETS.

Russian 37-mm Towed AA Gun M-1939

	Weapons and Ammunition Types	Typical Combat Load
	1 x 37-mm AA gun HE HE-FRAG-T AP AP-T HVAP HVAP-T HEI-T	500

SYSTEM	SPECIFICATIONS	AMMUNITION	SPECIFICATIONS
Alternative Designation	None	Type	HE, HE-FRAG-T, AP, AP-T, HVAP, HVAP-T, HEI-T
Date of Introduction	1939 (61-K)	Max Range (m)	8,500
Description		Max Effective Range (slant) (m)	3,500
Crew	8, 4 (est) while traveling	Max Effective (ground targets) (m)	3,500
Carriage	Four Wheels	Max Altitude (m)	6,000
Combat Weight (kg)	2,050	Max Effective Altitude (m)	3,000
Length Overall (m)	6.04	Min	0
Length of Barrel (m)	2.73	Armor Penetration (mm)	55 @ 500m
Height Overall (m)	2.11	HE Projectile Weight (kg)	0.74
Width Overall (m)	1.95	AP Projectile Weight (kg)	,77
Prime Movers	Utility Vehicles, Small, and Medium Trucks	HE-FRAG-T Projectile Weight (kg)	0.73
Automotive Performance		HVAP Projectile Weight (kg)	0.62
Max Towed Speed (km/h)	60	HEI-T Projectile Weight (kg)	INA
Cross Country (km/h)	25	HE Muzzle Velocity (m/s)	880
Fording Depth (m)	0.7	AP Muzzle Velocity (m/s)	880
Emplacement Time (sec)	8.5 while traveling. Gun can be fired from a halt without dropping trails. 30 full emplace, to drop trails.	HE-FRAG-T Muzzle Velocity (m/s)	880
Displacement Time (sec)	8 sec while traveling. 30 from full emplacement	HVAP Muzzle Velocity (m/s)	960
ARMAMENT	SPECIFICAITONS	HEI-T Projectile Weight (kg)	INA

Caliber, Type	37-mm Rifled	Self-Destruct (sec)	8 to 12
Number of Barrels	1	Self-Destruct Range (m)	3,700 to 4,700
Breech Mechanism	Rising Block		
Cyclic Rate of Fire (rd/min)	180		
Practical Rate of Fire (rd/min)	80		
Clip Capacity (rds)	5, gun magazine holds 2 clips for 10 rounds		
Loader Type	Manual Gravity Feed		
Reaction Time (sec)	4.5, 4 to stop and fire during a move (without radar)		
Reload Time (sec)	2 per clip		
Traverse (°)	360		
Traverse Rate (°/sec)	61		
Elevation (°) (-/+)	-5/+85		
Elevation Rate: (°/sec)	22		

FIRE CONTROL

SIGHTS WITH MAGNIFICATION: AZP-37 OPTICAL SIGHT, ALSO STEREOSCOPIC RANGEFINDER, COMMANDER'S TELESCOPE.

OTHER FIRE CONTROL: THE GUN IS LINKED TO THE BATTERY NET, AND RECEIVES ANALOG VOICE RADIO ALERTS OF APPROACHING AIRCRAFT, INCLUDING DIRECTION, ALTITUDE, AND AIRCRAFT TYPE. GUNS, BATTERIES, AND BATTALIONS USE AIR WATCHES AND FORWARD OBSERVERS. ALSO AVAILABLE ARE RF 1L15-1 OR SIMILAR AZIMUTH WARNERS TO PROVIDE ALERTS WITH APPROACH DIRECTION, TO READY THE GUNS FOR FAST RESPONSE.

OFF-CARRIAGE FIRE CONTROL SYSTEMS: SEVERAL DIRECTORS CAN BE USED WITH TELESCOPIC SIGHT, AND WITH AN ADDED LASER RANGE-FINDER.

RADAR: CHINESE TYPE 311 OPTIONAL. THIS IS A CONTINUOUS WAVE FIRE CONTROL RADAR WAS DESIGNED AND PRODUCED TO SUPPORT 37-MM AND 57-MM GUNS. THE I/J-BAND TRAILER-MOUNT RADAR WITH COMPUTER AUTOMATION CAN CONDUCT SURVEILLANCE AND TARGET ACQUISITION. IT HAS AT LEAST THREE VARIANTS, WITH RANGES OF 30 KM (311-A), 35 KM (311-B), AND 40 KM (311-C). TARGET TRACKING RANGE IS 25 KM FOR THE -A VARIANT. EMPLACEMENT TIME IS 15 MINUTES. RADAR GIVES USER WEAPONS A NIGHTTIME AND ADVERSE WEATHER CAPABILITY. THE RADAR HAS BEEN EXPORTED.

VARIANTS

M-1939 IS A DERIVATIVE OF THE BOFORS L60.

TYPE 55: CHINESE COPY OF ORIGINAL GUN

TYPE 65: CHINESE TWIN BARREL VERSION, AND OTHER VARIANTS, SEE PAGE 6-44.

TYPE 74: A CHINESE TWIN GUN WITH A HIGHER RATE OF FIRE (360-380). THE TYPE 311 RADAR IS OFTEN USED WITH THIS GUN SYSTEM. MAX EFFECTIVE RANGE AND ALTITUDE WITH THESE ARE 4,700 M

NOTES

THE M-1939 IS A TOWED 37-MM ANTIAIRCRAFT GUN MOUNTED ON A FOUR-WHEELED CARRIAGE. NORMAL EMPLACEMENT REQUIRES THE WHEELS TO BE REMOVED OR RAISED AND A JACK PLACED UNDER EACH AXLE FOR SUPPORT PRIOR TO FIRING. THE ROUNDS ARE GRAVITY FED INTO THE VERTICALLY OPENING SLIDING BREECH WITH THE EMPTY CARTRIDGES AUTOMATICALLY EXTRACTED.

WHEN USED WITHOUT RADAR, THE M-1939 IS CONSIDERED TO BE EFFECTIVE ONLY DURING DAYLIGHT AND IN FAIR WEATHER.

Russian 100-mm Towed AA Gun KS-19M2

	Weapons and Ammunition Types	Typical Combat Load
	100-mm Gun Frag-HE AP-T APC-T	100

SYSTEM	SPECIFICATIONS	FIRE CONTROL	SPECIFICATIONS
Alternative Designation	None	On-Carriages	
Date of Introduction	1949	PO-IM Telescope Field of View (o)	14
Proliferation	At Least 20 Countries	PO-IM Telescope Power	5x
Crew	15	PO-IM Telescope Range (m)	3,500
Carriage	Towed 2-axle, 4-wheel carriage	PG Panoramic Telescope Field of View (o)	10
Combat Weight (kg)	11,000	PG Panoramic Telescope Power	4x
Length Overall (m)	9.3	Off-Carriage	
Travel Position (m)	INA	Rangefinder	D-49 (off carriage)
Firing Position (m)	7.62	Radar Name	SON-9/SON-9A (FIRE CAN)
Width Overall (m)	2.32	Radar Function	Fire Control
Prime Mover	Towing vehicle AT-S or AT-T	Radar Detection Range (km)	80
Automotive Performance		Radar Tracking Range (km)	35
Max. Towed Speed (km/h)	35	Radar Frequency	2.7-2.9 GHz (E/F-band)
Emplacement Time (min)	7	Radar Peak Power (kW)	300
Displacement Time (min)	6	PUAZO 6-19 or 6-19M	fire control director
ARMAMENT	**SPECIFICATIONS**	Other Fire Control	
GUN		**VARIANTS**	**SPECIFICATIONS**
Caliber, Type	100-mm Gun	Type 59	Chinese Variant

Number of Barrels	1	MAIN ARMAMENT AMMUNITION	SPECIFICATIONS
Service Life of Barrel (rds)	2,800	Types	Frag-HE, AP-T, APC-T
Rate of Fire (rd/min)	Maximum: INA Practical: 10-15	Range With On-Carriage Sight (m)	3,500
Loader Type	Manual	Range With Off-Carriage Radar (m)	12,600
Reload Time (min)	INA	Max Altitude (m)	14,500
Traverse (o)	360	Max Effective Altitude (m)	13,700
Traverse Rate (o/sec)	20	With On-Carriage Sight Altitude (m)	3,500
Elevation (o) (-/+)	-3 to 89	Min	0
Elevation Rate (o/sec)	12	Frag-HE Projectile Weight (kg)	15.61
Reaction Time (sec)	30	AP-T Projectile Weight (kg)	15.89
		APC-T Projectile Weight (kg)	16
		Muzzle Velocity (m/s)	900-1,000
		Fuze Type	Proximity and Time
		Self-Destruct (sec)	30

OTHER FIRE CONTROL

THE GUN IS LINKED TO THE BATTERY NET WHICH RECEIVES ANALOG VOICE RADIO ALERTS FOR APPROACHING AIRCRAFT, INCLUDING DIRECTION, ALTITUDE, AND DIRECTION. GUNS, BATTERIES, AND BATTALIONS USE AIR WATCHES AND FORWARD OBSERVERS. ALSO AVAILABLE IS 1L15-1 OR SIMILAR RF AZIMUTH WARNERS TO PROVIDE ALERTS WITH APPROACH DIRECTION, TO READY THE GUNS FOR FAST RESPONSE.

NOTES

THE KS-19M2 MAY ALSO BE EMPLOYED IN A GROUND SUPPORT ROLE.

Russian 57-mm Towed AA Gun S-60

	Weapons and Ammunition Types	Typical Combat Load
	57-mm Gun	200
	FRAG-T	
	APC-T	

SYSTEM	SPECIFICATIONS	FIRE CONTROL	SPECIFICATIONS
Alternative Designation	None	On-Carriages	
Date of Introduction	1950	Optical mechanical computing sight AZP-57 (m)	
Proliferation	At Least 46 Countries	Target Range (m)	
Primary Components		Direct Fire Telescope	
Description		Off-Carriage	
Crew	7	Rangefinder	D-49
Carriage	Four Wheel	Radar Name	Son-9/Son-9A (NATO FIRE CAN)
Weight (kg)	4,500	Radar Function	Fire Control
Length Overall (m)		Radar Detection Range (km)	80
Travel Position (m)	8.50	Radar Tracking Range (km)	35
Firing Position (m)	8.84	Radar Frequency	2.7-2.9 GHz
Length of Barrel (m)	4.39	Radar Frequency Band	E
Height (m)		Radar Peak Power (kW)	300
Overall		Alternative Radar RPK-1/FLAP WHEEL Range (km)	34
Travel Position	2.37	Type 311	See below
Firing Position	6.02	**VARIANTS**	**SPECIFICATIONS**
Width Overall		Type 59	Chinese Variant
Travel Position	2.08	SZ-60	Hungarian Licensed-Built Variant
Firing Position	6.9	**MAIN ARMAMENT AMMUNITION**	**SPECIFICATIONS**
Prime Mover	Ural-375D	Type	57x348 SR, FRAG-HE, APC-T
Max. Towed Speed (km/h)	60	Preferred Round	UBR-281U APHE
Emplacement Time (min)	1	Max Effective Range (m)	4,000 on-carriage sight 6,000 w/off-carriage radar

Displacement Time (min)	3	Max Effective Altitude (m)	4,300 on-carriage sight 6,000 w/off-carriage radar
ARMAMENTS	**SPECIFICATION**	Min	0
Caliber, Type	57-mm automatic cannon	Projectile Weight (kg)	FRAG-T: 2.81 APC-T: 2.82
Number of Barrels	1 each	Muzzle Velocity (m/s)	1,000
Service Life of Barrel (rds)	INA	Fuze Type	FRAG-T: Point detonating APHE: Base detonating
Cyclic Rate of Fire (rd/min)	105-120	Self-Destruct (sec)	13-17
Practical Rate of Fire (rd/min)	70	Penetration (mm CE)	130 mm at 1000m, APHE
Loader Type	4 Round Clip, Manual		
Reload Type (sec)	4-8		
Traverse (o)	360		
Traverse Rate (o/sec)	40		
Elevation (o) (-/+)	-4 to +87		
Elevation Rate (o/sec)	34		
Reaction time (sec)	4.5		

PRIMARY COMPONENTS

BATTERY USUALLY HAS 6 GUNS, A FIRE-CONTROL RADAR AND A FIRE-CONTROL DIRECTOR. MOBILITY NEEDS AND ORGANIZATIONAL SUBORDINATION DETERMINES VEHICLES AND OTHER EQUIPMENT AVAILABLE.

OTHER FIRE CONTROL

THE GUN IS LINKED TO THE BATTERY NET, AND RECEIVES ANALOG VOICE RADIO ALERTS OF APPROACHING AIRCRAFT, INCLUDING DIRECTION, ALTITUDE, AND TYPE. GUNS AND BATTERY/ BATTALION USE AIR WATCHES AND FORWARD OBSERVERS. ALSO USED BY TIER 1-3 UNITS ARE RF 1L15-1 OR SIMILAR AZIMUTH WARNERS TO PROVIDE ALERTS WITH APPROACH DIRECTION, FOR FAST AA RESPONSE.

NOTES

THE S-60 ALSO HAS AN AMMUNITION READY RACK THAT CAN HOLD 4 FOUR-ROUND CLIPS NEAR AMMUNITION FEED MECHANISM ON LEFT SIDE OF THE BREECH. THE S-60 CAN ALSO BE USED IN A GROUND SUPPORT ROLE. THE S-60 CAN BE FIRED WITH WHEELS UP, OR WITH WHEELS ON THE GROUND.

FIRE CONTROL RADARS SUCH AS THE CHINESE TYPE 311 CAN BE USED WITH THIS WEAPON. THE CHINESE TYPE 311 CONTINUOUS WAVE FIRE CONTROL RADAR WAS DESIGNED AND PRODUCED TO SUPPORT 37-MM AND 57-MM GUNS. THE I/J-BAND TRAILER-MOUNTED RADAR WITH COMPUTER AUTOMATION CAN CONDUCT SURVEILLANCE AND TARGET ACQUISITION. IT HAS AT LEAST THREE VARIANTS, WITH RANGES OF 30 KM (311-A), 35 KM (311-B), AND 40 KM (311-C). TARGET TRACKING RANGE IS 25 KM FOR THE -A VARIANT. EMPLACEMENT TIME IS 15 MINUTES. THE RADAR GIVES USER WEAPONS A NIGHTTIME AND ADVERSE WEATHER CAPABILITY. THIS RADAR HAS BEEN EXPORTED.

Chinese 37-mm Towed AA Gun Type 65

	Weapons and Ammunition Types	Typical Combat Load
	37-mm automatic cannons	400

SYSTEM	SPECIFICATIONS	ARMAMENT	SPECIFICATIONS
Alternative Designations	INA	Gun	
Date of Introduction	Circa 1965	Caliber, Type	37-mm automatic gun
Proliferation	At Least 7 Countries	Number of Barrels	2
Description		Operation	Recoil
Crew	5 to 8	Service Life of Barrel (rds)	2,500+
Carriage	4 Wheeled/ 2 Axle Towed	Barrel Change Time (min)	2-3
Combat Weight (kg)	2,700	Cyclic Rate of Fire (rd/min)	Cyclic: 320-360 (160-180/barrel)
Length Overall (m)	5,490	Practical Rate of Fire (rd/min)	80
Travel Position	6,036	Loader Type	Two 5-Round Clips
Firing Position	INA	Reload Time (sec)	4-8
Length of Barrel (m)	2,729	Traverse (°)	360
Height (m)	2,080	Traverse Rate (°/sec)	INA
Overall	INA	Elevation (°)	-5 to 85
Travel Position	2,105	Elevation Rate (°/sec)	INA
Firing Position	INA	Reaction time (sec)	INA
Width Overall (m)	1,901	**MAIN ARMAMENT AMMUNITION**	**SPECIFICATION**
Primer Mover	INA	Types	AP-T, HE-T, HEI-T
Automotive Performance		Max Effective (Slant) Range (m)	3,500
Max Towed Speed (km/h)	60; 25 Cross Country	Max Effective (grnd targets) Range (m)	3,500
Emplacement Time (min)	1 (est)	Max Effective Altitude	3,000

Displacement Time (min)	3 (est)	Min Altitude	0
Fording Depth (m)	0.7	Self-destruct time (sec)	8-12
Tuning Radius (m)	8	Self-destruct range (m)	3,700-4,700

FIRE CONTROL

SIGHTS W/MAGNIFICATION: OPTICAL MECHANICAL COMPUTING SIGHT

AZIMUTH WARNING RECEIVER: 1L15-1

OFF-CARRIAGE RADAR: OPTIONAL. THE CHINESE TYPE 311 CONTINUOUS WAVE I/J-BAND FIRE CONTROL RADAR WAS DESIGNED AND PRODUCED TO SUPPORT 37-MM AND 57-MM GUNS. THE TRAILER-MOUNT RADAR WITH COMPUTER AUTOMATION CAN CONDUCT SURVEILLANCE AND TARGET ACQUISITION. IT HAS AT LEAST THREE VARIANTS, WITH RANGES OF 30 KM (311-A), 35 KM (311-B), AND 40 KM (311-C). TARGET TRACKING RANGE IS 25 KM FOR THE -A VARIANT. EMPLACEMENT TIME IS 15 MINUTES. RADAR GIVES USER WEAPONS NIGHT-TIME AND ADVERSE WEATHER CAPABILITY. THIS RADAR HAS BEEN EXPORTED.

OTHER FIRE CONTROL:

THE GUN IS LINKED TO THE BATTERY NET WHICH RECEIVES ANALOG VOICE RADIO ALERTS FOR APPROACHING AIRCRAFT, INCLUDING DIRECTION AND ALTITUDE. GUNS AND BATTERY/ BATTALION HAVE AIR WATCHES AND FORWARD OBSERVERS.

VARIANTS

CHINESE DIRECT COPY OF THE SOVIET TWIN BARREL EXPORT VERSION OF THE M-1939.

TYPE 65: A CHINESE TWIN-BARRELED VARIANT OF RUSSIAN M-1939 AD GUN.

TYPE 74 IS A SIMILAR CHINESE TWIN GUN WITH A HIGHER RATE OF FIRE (360-380). THE TYPE 311 RADAR IS OFTEN USED WITH THIS GUN SYSTEM. MAX EFFECTIVE RANGE AND ALTITUDE WITH THESE ARE 4,700 M.

TYPE P793 IS A TYPE 74 ON AN IMPROVED CARRIAGE WITH A GALILEO ELECTRO-OPTICAL FCS, AND AN ELECTRIC MOTOR FOR VERTICAL AND HORIZONTAL SLEWING. THE GUN CAN BE EMPLOYED ON AN SP TRACKED VEHICLE MOUNT.

M1985: NKPA HAS MOUNTED THE DUAL 37-MM TYPE 65 GUN ON AN OPEN TURRET VTT APC CHASSIS. SLANT RANGE AND EFFECTIVE ALTITUDE ARE 2,500 M WITH AN OPTICAL SIGHT. GROUND TARGET RANGE IS 3,500 M. THIS SYSTEM APPEARS TO SOMETIMES BE CONFUSED WITH THE M1992 SPAAG, WHICH HAS 30-MM GUNS. THERE IS NO 37-MM SPAAG CALLED M1992.

TYPE 88 IS A CHINESE SPAAG WITH THE TYPE P793 GUN ON THE TYPE 69-III TANK CHASSIS. THE VEHICLE HAS AN ELECTRO-OPTICAL FIRE CONTROL SYSTEM, IFF, AND FIRE CONTROL RADAR WITH A RANGE OF 15 KM.

NOTES

STRENGTHS: HIGHLY RELIABLE, RUGGED AND SIMPLE TO OPERATE. AMMUNITION IS INTERCHANGEABLE AMONG TYPES 55, 65, AND 74 AA GUNS.

WEAKNESSES: SHORT RANGE, SMALL PROJECTILE. TYPE 65 HAS NO ORGANIC RADAR. BECAUSE IT LACKS A RADAR AND POWERED GUN LAYING MOTORS, THE TYPE 65 AND MOST OTHER TOWED 37-MM GUNS, WHEN USED WITHOUT A RADAR, ARE CONSIDERED TO BE EFFECTIVE ONLY DURING DAYLIGHT AND IN FAIR WEATHER. THE TYPE 74 AND OTHER LATER SYSTEMS ADD RADARS TO CORRECT THAT WEAKNESS.

ALSO AVAILABLE ARE RF 1L15-1 OR SIMILAR AZIMUTH WARNERS TO PROVIDE ALERTS WITH APPROACH DIRECTION, TO READY THE GUNS FOR FAST RESPONSE.

Russian 14.5-mm Heavy Machinegun ZPU-4

	Weapons and Ammunition Types	Typical Combat Load
	4 barreled KPV 14.5-mm heavy machinegun AP-T API API-T HEI HEI-T	4,800 rds (1,200 rds/barrel)

SYSTEM	SPECIFICATIONS	FIRE CONTROL	SPECIFICATIONS
Alternative Designation	None	On-Carriage	Optical mechanical computing sight Telescope, ground targets
Date of Introduction	1949	Off-Carriage	Generally, there is no organic radar except with variants NK Type 56 and M1983. Many radars are available Optional Radar: SON-9/SON-9A, aka FIRE CAN (NATO)
Proliferation	At Least 45 Countries	Function	Fire Control
Description		Detection Range (km)	80
Crew	5	Tracking Range (km)	35
Carriage	4 wheeled/2 axle towed chassis	Frequency	2.7-2.9 GHz
Combat Weight (kg)	1,810	Frequency Band	E
Length Overall (m)		Peak Power (kW)	300
Travel Position	4.53	**MAIN ARMAMENT AMMUNITION**	**SPECIFICATIONS**
Firing Position	4.53	Types	API, API-T, HEI, AP-T, HEI-T
Length of Barrel (m)	1,348	Max Range (m)	8,000
Height (m)		Max. Effective (slant) Range (m)	1,400
Overall	INA	Max Altitude (m)	5,000
Travel Position	2.13	Max Effective Altitude	0-1,400
Firing Position	INA	Name	BZT-44M API-T
Width Overall (m)	1.72	Max Range (m)	8,000
Prime Mover	INA	Max. Effective (slant) Range (m)	2,200
		Max Altitude (m)	5,000
Automotive Performance		Effective Altitude (m)	0-2,200
Max. Towed Speed (km/h)	35		

Emplacement Time (min)	2		
Displacement Time (min)	2		
ARMAMENT	**SPECIFICATIONS**		
Gun			
Caliber, Type	14.5 mm machinegun		
Number of Barrels	4		
Service Life of Barrels (rds)	INA		
Max Rate of Fire (rd/min)	2,200-2,400 (600/barrel)		
Practical Rate of Fire (rd/min)	600 (150/barrel)		
Loader Type	Belt of 150 rds		
Reload Time	15		
Traverse (°)	360		
Traverse Rate (°/sec)	48		
Elevation (°)	-8 to +90		
Elevation Rate (°/sec)	29		
Reaction time (sec)	8		
The ZPU-4 can be fired from a brief stop (<10 sec) with wheels in travel position.			

OTHER FIRE CONTROL

THE GUN IS LINKED TO AD NETS, AND RECEIVES ANALOG VOICE RADIO ALERTS OF APPROACHING AIRCRAFT, E.G. TYPE, ALTITUDE, AND DIRECTION.

GUNS AND AD BATTERY/BATTALION HAVE AIR WATCHES AND FORWARD OBSERVERS.

UNITS CAN ADD RF 1L15-1 OR SIMILAR AZIMUTH WARNERS TO PROVIDE ALERTS WITH APPROACH DIRECTION, FOR FAST AA RESPONSE.

VARIANTS

ZPU-4: THE MEMBER OF THE ADA GUN FAMILY (ZPU-1, ZPU-2) WITH THE HIGHEST RATE OF FIRE.

TYPE 56: CHINESE AND NK VARIANT. IT IS USUALLY USED WITH DRUM TILT FIRE CONTROL RADAR.

M1983: NK SP VERSION WITH A ZPU-4 TYPE GUN ON A VTT-323 APC CHASSIS, WITH AN OPEN TURRET, AND A MANPADS LAUNCHER. IT ALSO TOWS DRUM TILT FIRE CONTROL RADAR.

MR-4: ROMANIAN SINGLE AXLE VARIANT

VTT-323: NORTH KOREAN APC WITH A TWIN ZPU GUN.

NOTES

IT MAY ALSO BE EMPLOYED IN A GROUND SUPPORT ROLE.

STRENGTHS: HIGHLY RELIABLE, RUGGED AND SIMPLE TO OPERATE. IT HAS QUICK-REACTION TIME, IS WIDELY DEPLOYED, AND HAS AN EXPLOSIVE ROUND.

WEAKNESSES: THE SHORT-RANGE SMALL PROJECTILE REQUIRES A DIRECT HIT.

Russian 23-mm Towed AA Gun ZU-23

	Weapons and Ammunition Types	Typical Combat Load
	2 x 23-mm AA guns	2,400
	HE-I HEI-T API-T APDS-T FAPDS TP	
	See best mix below.	

SYSTEM	SPECIFICATIONS	MAIN ARMAMENT AMMUNITION	SPECIFICATIONS
Alternative Designation	ZU-23-2	Can fire the same ammunition as ZSU-23-4. Best mix for modern versions (ZU-23M and ZU-23M1) is 1,200 APDS-T and 1,200 FAPDS. Rounds are ballistically matched and no HEI is required.	
Date of Introduction	1962	Type	APDS-T and Oerlikon FAPDS-T (Frangible APDS-T). NOTE: FAPDS-T is ballistically matched to the APDS-T round.
Proliferation	At Least 50 Countries	Max Effective Range (m)	2,500+
Crew	5	Max Effective Altitude (m)	1,500+
Carriage	Two-Wheeled	Projectile Weight (kg)	INA
Combat Weight (kg)	950	Muzzle Velocity (m/s)	1,180
Travel Position Length Overall (m)	4.57	FUZE TYPE	API-T: BASE IGNITING
Firing Position Length Overall (m)	4.60	Self-Destruct (sec)	11
Length of Barrel (m)	2.01	Penetration (mm KE)	19 @ 1000 m API-T INA for APDS-T 16+ @ 1500 m, FAPDS-T (helicopter simulant laminate array)
Travel Position Height Overall (m)	1.87	Type	23x152 HE-I, HEI-T, API-T, TP

Firing Position Height Overall (m)	1.28	Max Effective Range	2,500, 2,000 against light armored ground targets such as LAVs
		Max Effective Altitude (m)	1,500
Travel Position Width Overall (m)	1.83	Min Altitude (m)	Min: 0
Firing Position Width Overall (m)	2.41	HE-I Projectile Weight (kg)	0.18
Prime Movers	MTLB-T, GAZ-69 4 x 4 truck, BMD-2, BMD-3, BTR-3	HEI-T Projectile Weight (kg)	0.19
Automotive Performance		Muzzle Velocity (m/s)	970
Max. Towed Speed (km/h)	70	HE-I Fuze Type	Point detonating
Emplacement Time (min)	15-20 Can fire from travel position in emergencies.	HEI-T Fuze Type	Point detonating
Displacement Time (min)	35-40	Self-Destruct (sec)	11
ARMAMENT	**SPECIFICATIONS**		
Caliber, Type	23-mm, gas-operated gun, 2A14 or 2A14M		
Number of Barrels	2		
Breech Mechanism	Vertical Sliding Wedge		
Cyclic Rate of Fire (rd/min)	1,600-2,000		
Practical Rate of Fire (rd/min)	400 in 10-30 rd bursts		
Feed	50-rd ammunition canisters fitted on either side of the upper mount assembly		
Loader Type	Magazine		
Reload Time	15		
Traverse (°)	360		
Traverse Rate (°/sec)	INA		
Elevation (°)	-10°to +90°		
Elevation Rate (°/sec)	54		
Reaction time (sec)	8 (est)		
FIRE CONTROL	**SPECIFICATIONS**		
Sights with Magnification	Optical mechanical sight for AA fire. Straight tube telescope for ground targets.		
Range (m)	2,000		
Azimuth warning receiver	1L15-1		

OTHER FIRE CONTROL

GUN LINKED TO BATTERY NET WHICH RECEIVES ANALOG VOICE RADIO ALERTS FOR APPROACHING AIRCRAFT, INCLUDING DIRECTION, ALTITUDE, AND DIRECTION. FIRE CONTROL RADARS CAN BE USED OFF-CHASSIS WITH THE SYSTEM. A SIMPLE OPTIONAL ADDITION IS THE FARA-1 BSR. IT CAN BE ATTACHED AND BORE-LINED TO THE GUN.

GUNS AND AD UNITS USE AIR WATCHES AND FORWARD OBSERVERS.

VARIANTS

ZU-23-2M: RUSSIAN UPGRADE VARIANT REPLACES OPTICAL SIGHT WITH AN EO FIRE CONTROL SYSTEM EMPLOYING A BALLISTIC COMPUTER WITH DAY TV, THERMAL NIGHT CHANNEL, A LASER RANGEFINDER, AND AN AUTO-TRACKER. HIT PROBABILITY INCREASES 10-FOLD OVER THE ZU-23.

ZU-23-2M1: UPGRADE ADDS A TWIN MANPADS LAUNCHER (SA-16 OR SA-18), WHICH CAN AIM, TRACK, AND LAUNCH USING ABOVE FCS. THE FCS ALSO ADDS A DIGITAL MONITOR. OPERATOR CAN USE MANPADS AT RANGE OUT TO 6,000 M, AND THEN SHIFT TO GUN WHEN THE TARGET IS IN GUN RANGE.

ZUR-23-2KG JODEK-G: POLISH UPGRADE AND EXPORT VERSION OF ZU-23-2M1 WITH FAPDS-T ROUNDS AND GROM MISSILES.

BTR-ZD: IS BTR-D WITH TOWED OR PORTED ZU-23 AND MANPADS. THE BTR-ZD IMPROVED IS A BTR-D WITH PORTEED ZU-23M1 GUN AND SA-18S MANPADS.

NOTES

THIS IS A HIGHLY MOBILE AIR-DROPPABLE SYSTEM. THE ZU-23 CAN ALSO BE USED IN A GROUND SUPPORT ROLE AGAINST PERSONNEL AND LIGHT ARMORED VEHICLES.

Russian 23-mm SP AA Gun System BTR-ZD/BTR-ZD Improved

	Weapons & Ammunition Types	Typical Combat Load
	2 x 23-mm AA guns HE-I or HEI-T API-T, or FAPDS	2,400
	1 x SAM Launcher SA-18	Missiles 5
	BTR-ZD Improved 2 x SAM Launcher SA-18S	Missiles 10

SYSTEM	SPECIFICATIONS	MAIN ARMAMENT	SPECIFICATIONS
Alternative Designation	BTR-3D, incorrect name from translation error	Can fire the same ammunition as ZSU-23-4. Best mix for modern versions (ZU-23M and ZU-23M1) is 1,200 APDS-T and 1,200 FAPDS. Rounds ballistically matched. No HEI required.	
Date of Introduction	1979-1980	Type	APDS-T and Oerlikon FAPDS-T (Frangible APDS-T). NOTE: FAPDS-T is ballistically matched to the APDS-T round.
Proliferation	At Least 1 Country	Range (m)	0-2,500+ Effective
Crew	7, 2 for vehicle and 5 for gun	Altitude (m)	0-1,500+ Effective
Combat Weight (mt)	8 (est)	Projectile Weight (kg)	0.189 API-T
Chassis	BTR-D APC chassis	Muzzle Velocity (m/s)	1,180
Chassis Length Overall (m)	5.88	Fuze Type	API-T: Base igniting
Height Overall (m)	6.3	Self-Destruct (sec)	11
Width Overall (m)	2.63	Penetration (mm)	19 @ 1000 m API-T 16+ @1500m FAPDS-T (helicopter simulant laminate array)
Automotive Performance	See BTR-D. The BTR-ZD is one of only a few SP air defense systems which can swim.	Type	23x152 HE-I, HEI-T, API-T, TP
Radio	R-123	Max Effective Range (m)	2,500, 2,000 against light armored ground targets such as LAVs
Protection	See BTR-D	Altitude (m)	0-1,500
ARMAMENT	SPECIFICATION	HE-I Projectile Weight (kg)	0.18

Caliber, Type	23-mm, gas-operated	HEI-T Projectile Weight (kg)	0.19
Name	ZU-23	Muzzle Velocity (m/s)	970
Number of Barrels	2	HE-I Fuze Type	Point detonating
Breech Mechanism	Vertical Sliding Wedge	HEI-T Fuze Type	Point detonating
Rate of Fire (rd/min)	Cyclic: 1,600-2,000 Practical: 400 in 10-30 rd bursts	Self-Destruct (sec)	11
Loader Type	Magazine	Missiles	
Reload Time (sec)	15	Name	SA-18 Tier 2, SA-18S Tier 1
Traverse (°)	360	Range (m)	500-6,000+
Traverse Rate (°/sec)	INA	Altitude (m)	10 (0 degraded Ph) - 3,500
Elevation (°)	-10°to +90°	Other Missiles	Tier 3 is SA-16, 4 is SA-14
Elevation Rate (°/sec)	54		
Reaction time (sec)	8 (est)		
Fire on the Move	No, in 8 sec stop		
Missile Launcher	Use SAM noted for each tier. For Tier 2 use SA-18. For Tier 1 use SA-18S.		

FIRE CONTROL

SIGHTS WITH MAGNIFICATION: OPTICAL MECHANICAL SIGHT FOR AA FIRE STRAIGHT TUBE TELESCOPE FOR GROUND TARGETS OPTIONAL SIGHTS: SEE ZU-23M/ZU-23M1 BELOW MISSILE SUPPORT EQUIPMENT: GUN/LAUNCHER HAS A NIGHT SIGHT (THERMAL, MOWGLI-2 2 GEN II, OR II NIGHT VISION GOGGLES). ONE MAN OPERATES A 1L15-1AZIMUTH PLOTTING BOARD AND PELENGATOR RF DIRECTION-FINDER.

OTHER FIRE CONTROL

FIRE CONTROL RADARS CAN BE USED OFF-CHASSIS. A SIMPLE OPTIONAL ADDITION IS THE FARA-1 OR MT-12R MMW BSR. IT CAN BE ATTACHED AND BORE-LINED TO THE GUN. GUNS USE AIR WATCHES AND FORWARD OBSERVERS, AND ARE LINKED TO AD NETS.

VARIANTS

BTR-ZD: CAN TOW OR PORTEE-MOUNT THE SYSTEM. USUALLY, THE VEHICLE AND GUN ARE LANDED APART. THE GUN IS TOWED OUT OF THE LANDING ZONE, THEN MOUNTED ON THE VEHICLE. VEHICLE HOLDS 2 SAM LAUNCHERS. IN THE EARLIEST UNITS, THE VEHICLE HAD NO AA GUN, RATHER HAD 6 MANPADS LAUNCHERS, RELOAD RACKS, AND LAUNCH CREWS (1-2).

TIER CONFIGURATIONS INCLUDE EMPLOYING UPDATED VERSIONS OF THE GUN SYSTEM AND SAMS. IN EARLY VERSIONS (TIERS 2 - 4), THE SAM LAUNCHERS ARE SHOULDER-MOUNTED. IN THE LATEST VERSION (TIER 1), THEY ARE MOUNTED ON THE GUN. THE SAMS USUALLY LAUNCH FIRST AT APPROACHING TARGETS.

ZU-23M: REPLACES OPTICAL SIGHT WITH AN EO FIRE CONTROL SYSTEM EMPLOYING A BALLISTIC COMPUTER WITH DAY TV, THERMAL NIGHT CHANNEL, LASER RANGEFINDER, AND AUTO-TRACKER. HIT PROBABILITY INCREASES 10-FOLD OVER THE ZU-23.

FOR OPFOR SIMULATIONS, THIS IS THE TIER 2 AIRBORNE (ABN) SPAAG CAPABILITY.

ZU-23M1: UPGRADE MOUNTS A STRELETS SA-18 /18S/24 MANPADS LAUNCHER, WHICH CAN AIM, TRACK, AND LAUNCH WITH THE ZU-23M FCS. THE FCS ADDS A DIGITAL MONITOR. A SINGLE OPERATOR CAN USE THE MISSILE AT RANGES OUT TO 6,000+ M, THEN SHIFT TO GUN WHEN THE TARGET IS IN RANGE.
NOTES
VEHICLE MOUNT ARRANGEMENTS CAN BE EXECUTED IN THE FIELD. SIMILAR AD HOC MOUNTING OF AD GUN, MACHINEGUN, ROCKET, OR GRENADE LAUNCHERS IS USED BY PARAMILITARY FORCES WITH COMMERCIAL OR MILITARY TRUCKS, PICK-UP TRUCKS, CARS OR UTILITY VEHICLES TO CREATE "TECHNICALS". WHEN THE GUN IS MOUNTED ON THE VEHICLE, IT CAN TOW A TRAILER WITH ADDITIONAL AMMO AND SUPPLIES. THE GUN CAN ALSO BE USED IN A GROUND SUPPORT ROLE, INCLUDING USE FOR HIGH-ANGLE FIRE IN URBAN AND DEFILADE ENVIRONMENTS.

German/Swiss 35-mm SP AA Gun System Gepard

	Weapons and Ammunition Types	Typical Combat Load
	2 x 35-mm cannons HEI-T SAPHEI-T FAPDS APDS-T/APFSDS-T	680

SYSTEM	SPECIFICATIONS	MAIN ARMAMENT AMMUNITION	SPECIFICATIONS
Alternative Designation	5PFZ-B2L Upgrade variant known as FlakPz 1A2	Type	HEI-T
Date of Introduction	1976 Original	Tactical AA range (m)	3,500 (self-destruct)
Proliferation	At Least 5 Countries	Tracer Range (m)	3,500
Crew	3	Effective Altitude (m)	3,100
Combat Weight (mt)	46	Min Altitude (m)	0
Chassis	Leopard 1 tank chassis	Self-Destruct (sec)	6-12
Chassis Length Overall (m)	7.16	Type	Semi-armor-piercing HEI-T (SAPHEI-T)
Radar Up Height (m)	4.23	Range (m)	4,000
Radar Down Height (m)	3.01	Tactical AA Range (m)	3,500 (self-destruct)
Width Overall (m)	3.25	Effective (m)	3,500 (est)
Engine Type	830-hp Diesel	Self-Destruct Time (sec)	6-12
Max Road Speed Cruising Range (km)	65	Penetration (mm KE)	40 to 1,000 m
Fording Depths (m)	2.25	Typ3	APDS-T
Auxiliary power unit has 90-hp engine	Type	Range (m)	4,000
ARMAMENT	**SPECIFICATIONS**	Tactical AA Range (m)	3,500
Caliber, Type	35x228 gun, KDA	Tracer Range (m)	2,000
Number of Barrels	2	Effective Altitude (m)	3,100
Rate of Fire (rd/min)	1,100 (550/barrel	Penetration (mm KE)	90 at 1,000
Reaction Time (sec)	6-10	Type	APFSDS-T
Ammunition Loader	Twin Belt	Range (m)	4,000
Reload Time (sec)	INA	Tactical AA Range (m)	3,500
Elevation (°)	-10 to +85°	Tracer Range (m)	INA
Fire on the Move	Yes (est)	Effective Altitude (m)	3,100

FIRE CONTROL	SPECIFICATIONS	Penetration (mm KE)	115+ at 1,000 m
FC System	EADS digital computer-based FCS	Type	Frangible APDS (FAPDS) for upgrades. On impact with the target surface, the penetrator breaks into several KE fragments. The round has Frag-HE effects with the higher velocity and flat trajectory of a sabot round.
Sights With Magnification	Stabilized video sights for -1A2 upgrade	Other Ammunition Type	HEI
Magnification	INA		
Field of View (°	INA		
Night Sights	Thermal for -1A2 upgrade		
IFF	Yes, MSR-400		
Navigation System	Computerized		
Laser Rangefinder	ND Yag (1.06μ)		
Linked to Air Defense Net	Yes		
Radars			
Name	INA, Siemens Manufacture		
Function	Target Acquisition		
Detection Range (km)	15		
Tracking Range (km)	INA		
Frequency Band	S		
Search on the Move	Yes		
Name	INA		
Function	Fire Control		
Detection Range (km)	15		
Tracking Range (km)	15		
Frequency Band	Ku		

ARMORED COMMAND VEHICLE

SYSTEM WILL LINK TO AN ACV WHICH MAY HAVE A RADAR FOR EW AND TARGET ACQUISITION. FOR EXAMPLE, SEE SBORKA ACV AND RADAR (PG 6-15).

OTHER RADARS

LINKS TO INTEGRATED AIR DEFENSE SYSTEM (IADS) FOR EARLY WARNING AND TARGET ACQUISITION DATA FROM RADARS: GIRAFFE AMB AT SEPARATE BRIGADE AND DIVISION, LONG TRACK OR SIMILAR EW/TA RADAR ECHELONS ABOVE DIVISION, AND RADARS IN SAM UNITS, E.G., SA-10.

OTHER FIRE CONTROL

GUNS USE AIR WATCHES AND FORWARD OBSERVERS AND ARE LINKED TO AD NETS

<u>VARIANTS</u>

GEPARD 1A2: UPGRADE VARIANT WITH NEW FCS, INCLUDING STABILIZED THERMAL SIGHT AND VIDEO AUTO-TRACKER, INTEGRATED C2, INCREASED RANGE, REDUCED REACTION TIME, AND FAPDS.

GEPARD CA1: DUTCH VARIANT (ALSO CALLED 95 CHEETAH) USES SIGNAAL I-BAND MTI RADAR AND DUAL I-BAND K-BAND TRACKING RADARS.

PRTL-35MM GWI: UPGRADE DUTCH VARIANT, WITH UPGRADES SIMILAR TO 1A2 AND NEW RADIOS, BUT WITH DIFFERENT RADARS. RANGE WITH FAPDS IS CLAIMED TO BE 3,500-4,500.

NOTES

KMW IS DEVELOPING AN UPGRADE WITH 2X STINGER MANPADS MISSILE LAUNCHERS ADDED TO A GUN, AND INTEGRATED WITH THE FCS.

Russian 23-mm SP AA Gun ZSU-23-4

		Weapons and Ammunition Types	Typical Combat Load
		4x 23-mm AA guns	2,000
		HE-I HEI-T API-T APDS-T FAPDS TP See best mix below.	
SYSTEM	**SPECIFICATIONS**	**MAIN ARMAMENT AMMUNITION**	**SPECIFICS**
Alternative Designation	Shilka	Can fire the same ammunition as ZU-23. Best mix for modern versions (ZU-23M and ZU-23M1) is 1,200 APDS-T and 1,200 FAPDS. Rounds ballistically matched. No HEI required.	
Date of Introduction	1965	Type	APDS-T and Oerlikon FAPDS-T (Frangible APDS-T). NOTE: FAPDS-T is ballistically matched to the APDS-T round.
Proliferation	At Least 28 Countries	Max Effective Range (m)	2,500+
Crew	4	Max Effective Altitude (m)	1,500+
Combat Weight (mt)	20.5	Projectile Weight (kg)	INA
Chassis	GM-575 Tracked, six road wheels, no track support rollers	Muzzle Velocity (m/s)	1,180
Length (m)	6.5	Fuze Type	None
Height (m)	Radar up: 3.75 Radar down: 2.60	Self-Destruct (sec)	11
Width (m)	3.1	Penetration (mm KE)	INA APDS-T, 16+ @ 1500m FAPDS-T (helicopter simulant laminate array)
Automotive Performance		Type	23x152 HE-I, HEI-T, API-T, TP
Engine Type	V6R-1 diesel	Max Effective Range (m)	2,500, 2,000 against light armored ground targets such as LAVs
Cruising Range (km)	450	Max Effective Altitude (m)	1,500
Max Road Speed (km/h)	50	Min Altitude (m)	0
Radio	R-123	HE-I Projectile Weight (kg)	0.18

Protection NBC Protection System	Yes	HEI-T Projectile Weight (kg)	0.19
ARMAMENT	**SPECIFICATION**	API-T Projectile Weight (kg)	0.189
Caliber, Type, Name	23-mm liquid-cooled AA 2A7/2A7M	TP Projectile Weight (kg)	0.18
Cyclic Rate of Fire (rd/min)	850-1,000	Muzzle Velocity (m/s)	970
Practical Rate of Fire (rd/min)	400, in 10-30 rd bursts	HE-I Fuze Type	Point detonating
Reload Time (min)	20	HEI-T Fuze Type	Point detonating
Elevation (°)	-4° to +85°	API-T Fuze Type	Base igniting
Fire on the Move	Yes	Self-Destruct (sec)	11
Reaction Time (sec)	12-18	Penetration (mm KE)	19 @ 1000 m API-T
FIRE CONTROL	**SPECIFICATIONS**		
Sights with Magnification			
Day and Night Vision Devices			
Driver Periscope	BMO-190		
Driver IR Periscope	INA		
Commander Periscope	TPKU-2		
Commander IR Periscope	TKH-ITC		
IFF	INA		
Radar	1RL33M1		
Name	GUN DISH		
Function	Acquisition and Fire Control		
Detection Range (km)	20		
Tracking Range (km)	13		
Frequency	14.8 to 15.6 GHz		
Frequency Band	J		
RPK-2	Optical-mechanical computing sight and part of FC subsystem		
Armored Command Vehicle			
Name	Sborka (9S80-1 or PPRU-M1)		
Chassis	MTLB-U		
Radar	DOG EAR (use in OPFOR units)		
Function	Target Acquisition		
Frequency	F/G band		
Range (km)	80 detection, 35 tracking ACV links to supported tactical unit nets.		

Other Radars	Using the above ACV, if an Integrated Air Defense System (IADS) is available, ZSU-23-4 links indirectly for early warning and target acquisition data from radars.		
Other Fire Control	Guns use air watches and forward observers, and are linked to AD nets		

VARIANTS

ZSU-23-4M4: RUSSIAN MODERNIZED GUN/MISSILE VEHICLE WITH 2 STRELETS LAUNCH MODULES (4 MISSILES) WITH AN UPGRADE RADAR, AND COMPUTER-BASED FCS WITH CCD TV SIGHT AND NIGHT CHANNEL.

DONETS: UKRAINIAN ZSU-23-4 UPGRADE, WITH A NEW RADAR SYSTEM REPLACING GUN DISH, PLUS A SENSOR POD BELIEVED TO INCLUDE DAY/NIGHT CAMERA, AND A LASER RANGEFINDER. MOUNTED ABOVE THE RADAR/SENSOR POD IS A LAYER OF 6 RUSSIAN SA-18 MANPADS LAUNCHERS.

BIALA: POLISH UPGRADE WITH THERMAL SIGHT, GROM MANPADS, FAPDS-T.

NOTES

AMMUNITION IS NORMALLY LOADED WITH A RATIO OF THREE HE ROUNDS TO ONE AP ROUND. ZSU 23-4 IS CAPABLE OF ACQUIRING, TRACKING AND ENGAGING LOW-FLYING AIRCRAFT (AS WELL AS MOBILE GROUND TARGETS WHILE EITHER IN PLACE OR ON THE MOVE). RESUPPLY VEHICLES CARRY AN ESTIMATED ADDITIONAL 3,000 ROUNDS FOR EACH OF THE FOUR ZSUS IN A TYPICAL BATTERY.

Russian 57-mm Self Propelled SP AA Gun ZSU-57-2

		Weapons & Ammunition Types	Typical Combat Load
		Twin 57-mm automatic cannons Frag-HE AP-T APC-T	300
SYSTEM	**SPECIFICATIONS**	**Armament**	**Specification**
Alternative Designation	None	Gun, Caliber, Type	57-mm recoil-operated air-cooled cannons, S-68
Date of Introduction	1955	Number of Barrels	2
Proliferation	At Least 16 Countries	Rate of Fire (rd/min)	Cyclic: 210-240 (105-120/gun) Practical: 140 (70/gun)
Description		Loader Type	Two 5-round clips, manual, 10 rds
Crew	6	Reload Time (sec)	4-8
Carriage	4 road wheels/T-54 modified chassis	Traverse (o)	360
Combat Weight (mt)	28.0	Traverse Rate (o/sec)	30
Length Overall (m)	8.4	Elevation (o)	-5 to +85
Length of Barrel (m)	INA	Elevation Rate (o/sec)	20
Height Overall (m)	2.75	**VARIANTS**	**SPECIFICATIONS**
Width Overall (m)	3.270	Type 80 Chinese variant on Type 69-II main battle tank chassis	
Prime Mover	A shortened T-54 chassis with thinner armor and only four road wheels.	**Main Armament Ammunition**	**Specifications**
Emplacement Time (min)	N/A	Types	57 x 348 SR APHE, Frag-T, APC-T, HVAP-T, HE-T. Uses same ammo as the towed single S-60
Displacement Time (min)	N/A	Max Effective Range (m)	4,000
Engine Power (hp)	520	Max Effective Altitude (m)	4,237 at 65°
Max Road Speed (km/h)	50	Min Altitude (m)	0
Cruising Range (km)	400	Frag-T Projectile Weight (kg)	2.81
Fording Depth (m)	1.4	APC-T Projectile Weight (kg)	2.82
Armor Protection	13 mm front hull and turret	HE-T Projectile Weight (kg)	2.85
		Muzzle Velocity (m/s)	1,000
		Frag-T Fuze Type	Point Detonating
		APC-T Fuze Type	Base Detonating Fuze
		HE-T Fuze Type	(Yugoslavian, impact [super quick] action with pyrotechnical self-destruct)
		Self-Destruct Time (sec)	13-17
		Armor Penetration (mm CE)	130 at 1,000m, APHE 96 APC-T at 1,000 m

FIRE CONTROL

SIGHTS W/MAGNIFICATION: OPTICAL MECHANICAL COMPUTING REFLEX SIGHT (NOT RADAR CONTROLLED) LATER VARIANTS WERE FITTED WITH A MORE SOPHISTICATED SIGHTING SYSTEM, IDENTIFIED BY TWO SMALL PORTS IN FORWARD UPPER PORTION OF THE TURRET.

OTHER FIRE CONTROL:

ABSENCE OF A TRACKING RADAR, A NIGHT VISION DEVICE, AND AN ENCLOSED TURRET MAKES THIS A DAYLIGHT, FAIR WEATHER WEAPON SYSTEM ONLY. OFF-CARRIAGE RADARS, SUCH AS THE SON-9/SON-9A (NATO FIRE CAN), RPK-1/FLAP WHEEL, OR TYPE 311 CAN BE USED THE GUN IS LINKED TO THE BATTERY NET WHICH RECEIVES ANALOG VOICE RADIO ALERTS FOR APPROACHING AIRCRAFT, INCLUDING DIRECTION, ALTITUDE, AND DIRECTION. GUNS AND BATTERY/ BATTALION HAVE AIR WATCHES AND FORWARD OBSERVERS.

NOTES

THE ZSU-57-2 CAN BE EMPLOYED IN A GROUND SUPPORT ROLE. NO NBC SYSTEM AND NO AMPHIBIOUS CAPABILITY. FUEL DRUMS CAN BE FITTED ON REAR OF HULL. THE GUN HAS AUTO-TRAVERSE WITH MANUAL BACKUP.

Manportable Air Defense Systems (MANPADS) and Trends

In units with dismounted teams or squads, the most effective air defense asset is the MANPADS launcher. This system requires moderate training to reach proficiency and offers high probabilities of hit and kill. Although its cost is greater than most small arms, it offers an asymmetric possibility of causing enemy damage compared to cost and it retains its effectiveness over time. With infrared (IR) homing guidance (pg 6-55), a MANPADS permits the user to engage its target and quickly displace to avoid a lethal response. These systems also offer insurgents and others an asset to down costly military and civilian aircraft targets.

Basic components of legacy MANPADS include the following: gripstock with attached sight, inserted battery coolant unit (BCU), launch tube (canistered) with sling and missile, and possible protective pads. Missile components are the warhead with seeker, missile body with propulsion motor and guidance fins, and an eject motor for soft launch. A 2-3 man team usually includes operator, assistant with support equipment and spare missiles, and perhaps a loader/transporter. Support assets can provide warning of aircraft approach. Once a target is in sight, the launcher signals lock-on with visual/audio cues. A trigger squeeze launches the missile. MANPADS seekers detect engine heat, so they offer longer range at target side and rear aspects.

Aircraft have added many changes and countermeasures to defeat MANPADS. They include IR flare dispensers, low detection designs, and evasive tactics (e.g., terrain flying modes, use of terrain defilade, and night missions). MANPADS have also added improvements since their inception (1960s). Gripstocks added optical or electro-optical sights to augment post-and-blade sights. Support assets include radio links to forward observers, AD nets, azimuth plotting boards which provide alerts and flight path warning, even helmet-mounted RF receivers to give operators warning and direction (pg 6-56). Manportable surveillance radars (e.g., Fara-1E, Vol 1, pg 4-29) can mount nearby, or link with RF display units (e.g., Chinese QW-1A).

The greatest improvements are in the missile designs. Missile motors and eject motors now use faster low-smoke propellants. These and improved steering systems mean reduced warning time, higher speeds, and higher G-force turns to challenge aircraft evasion capability. Seekers shifted to improved detector arrays for 2-color vs earlier 1-color IR. Microcircuits and superior filter algorithms offer higher contrast and less background clutter. Some added spikes to reduce lens heating, and mirrors to widen field of view. These mean higher hit probability (Ph) and greater IR flare rejection. Ground target discrimination allows MANPADS to fly lower (0-10 m) against helicopters flying in nap-of-the-earth mode. Warheads have greatly improved in lethality, with HMX replacing RDX, fuel detonation fuzing, and proximity fuzes to increase Ph. Warheads are growing, from earlier HE ones at <1 kg, to Frag-HE/frangible rod designs to 3 kg.

Given the number of older MANPADS around the world, especially those based on the Russian SA-7a, industries offer refurbishing and upgrades for older systems. Although systems, when stored well, can last for decades, the BCUs may need replacement. New ones are available. In addition, the Russian firm LOMO offers a replacement 9E46M 2-color IR seeker to replace the older 1-color seeker (pg 6-56). With that seeker, SA-7 missile and foreign copies can approach the higher Ph (including flare rejection) of more modern systems. Other parts, e.g., eject motors and sights can be replaced or upgraded. Variants have added changes. The Chinese incorporated many changes to fielded systems. Others then modified those designs.

Some producers countered aircraft CM and improved effectiveness of their MANPADS with new guidance modes. British designers developed Blowpipe system in the early 1970s. But its RF command line-of-sight (CLOS) guidance proved difficult to keep on target. The Javelin RF semi-automatic CLOS (SACLOS) system appeared in the mid-80s. It requires only that the crosshairs stay on target. These systems are unaffected by IR flares and have good head-on range, but are subject to RF jamming. In 1976 Sweden fielded the laser beam-rider (LBR) RBS-70 SACLOS system. RBS-90 offered improved fire control and support. British Starstreak (pg 6-60) also adopted LBR guidance. Now British Shorts offers a Javelin upgrade to LBR with the Starburst system. LBR SACLOS systems cannot be jammed or decoyed. But, all SACLOS systems are challenged to stay fixed on evading aircraft. Starstreak II added an auto-tracker to maintain lock until impact. SACLOS launchers also cannot move until impact. Starstreak's answer is high velocity and long range, to kill aircraft beyond their range. Later seekers were fielded with IR/ultraviolet (UV) homing, and with semi-active laser-homing (SAL-H) guidance.

As new MANPADS have emerged with new features, weight for some has crept up considerably. Several have grown beyond weight limit for truly manportable systems. A more accurate term for these is "portable". It means that several are not generally shoulder-launched, rather are launched from vehicles or pedestal ground launchers. RBS-70 and 90, and Mistral fit on these launchers. Other makers also offer pedestal launchers for true MANPADS, for convenient use by MANPADS teams. The Russian Djigit (pg 6-58) is a pivoting twin launcher with separate triggers and a convenient EO sight system. Pedestal launchers can also mount on vehicles, such as on truck beds, for easy mobile conversions. Also modular launch pods, e.g., Mistral Albi system and Russian Strelets offer multiple launchers with superior FCS. MANPADS can be mounted on helicopters, guns and vehicles, and exploiting their superior FCS and links.

Selected MANPADS Systems

Name	Export Country	Weight (kg)	Range (km)	Altitude (km)	Whd (kg) Prx Fz (P)	Guidance	Remarks
Red Eye	US	13.1	.2-5.5	.05-2.7	Frag 2	1-C IR-H	Obsolete
Blowpipe	UK	20.7	.6-3	.01-2	Frag 2.2 P	RF CLOS	Low hit probability
Javelin	UK	24.3	.3-5.5	.01-3	Frag 2.7 P	RF SACLOS	Blowpipe upgrade
Starburst	UK	20.9	.3-5.5	0-3	Frag 2.7 P	LBR SACLOS	Javelin upgrade
RBS-70	SW	86.5	.2-7	0-4	Frag .9 P	LBR SACLOS	Mk 2 and RBS-90
HN-5/5B	CH	15	.8-4.4	.05-2.5	HE .6	1-C IR-H	SA-7 variant
FN-6	CH	17	.5-5.5	.015-3.8	INA	2-C IR/UV-H	Mistral variant
QW-1	CH	16.5	.5-5	.03-4	Frag .57	2-C IR-H	SA-7b variant
QW-2	CH	18.4	.5-6	.01-4	Frag 1.4	2-C IR-H	SA-14/16 variant
QW-3	CH	29.7	.8-8	.004-5	HE rod	IR SAL-H	Fielding INA
Sakr Eye	EG	18	.8-4.4	.05-2.4	Frag	1-C IR-H	SA-7 variant
Anza Mk I	PK	15	1.2-4.2	.05-2.5	Frag .37	2-C IR-H	HN-5 variant

Anza Mk II	PK	16.5	.5-5	.03-4	Frag .55	2-C IR-H	QW-1 variant
Anza Mk III	PK	18	.5-7	.01-3.5	Frag 1.4	2-C IR-H	QW-2 variant
Misagh-1	IR	16.9	.5-5	.3-4	Frag 1.4	2-C IR-H	QW-1 variant

Developments continue. New thermal night sights are offered. One particularly active area for MANPADS is in support assets. Passive support systems include acoustic detection systems, unattended ground sensors, and remote passive IR camera systems. Radars and links to IADS can link directly with MANPADS or indirectly through the transport/fire support vehicles.

Russian Man-portable SAM System SA-7b/GRAIL

		Weapons & Ammunition Types	Typical Combat Load
		ready missile	1 Normal Dismount 2 From AD Vehicle 5
SYSTEM	SPECIFICATIONS	Length (m)	1,40
Alternative Designation	9K32M Strela-2M	Diameter (mm)	70
Date of Introduction	1972	Weight (kg)	9.97
Proliferation	Worldwide	Missile Speed (m/s)	580
Target	FW, heli	Propulsion	Solid fuel booster and solid fuel sustainer rocket motor. Guidance: Passive 1-color IR homing (operating in the edium IR range)
Crew	1, Normally 2 with a loader	Guidance	Passive 1-color IR homing (operating in the medium IR range)
ARMAMENT	SPECIFICATIONS	Seeker Field of View (°)	1.9°
Launcher		Tracking Rate (°/sec)	6°
Name	9P54M	Warhead Type	HE
Length (m)	1.47	Warhead Weight (kg)	1.15
Diameter (mm)	70	Fuze Type	Contact (flush or grazing)
Weight (kg)	4.71	Probability of Hit (Ph%)	30 FW/40 heli
Reaction Time (acquisition to fire) (sec)	5-10	Self-Destruct (sec)	15
Time Between Launches (sec)	INA	Countermeasure Resistance	The seeker is fitted with a filter to reduce effectiveness of decoy flares and to block IR emissions.
Reload Time (sec)	6-10	FIRE CONTROL	SPECIFICATIONS
Fire on the Move	Yes, in short halt	Sights with Magnification	Launcher has a sighting device and a target acquisition indicator. The gunner visually identifies and acquires the target.
Missile		Field of View (o)	INA
Name	9M32M	Night Sight	None Standard
Range (m)	500-5,000	Acquisition Range (m)	INA

Max Altitude (m)	4,500	IFF	Yes (see NOTES)
Min Altitude (m)	18, 0 with degraded Ph		

VARIANTS

THE MAIN DIFFERENCE BETWEEN THE SA-7 AND SA-7B IS THE IMPROVED PROPULSION OF THE SA-7B. THIS IMPROVEMENT INCREASES THE SPEED AND RANGE OF THE NEWER VERSION.

SA-N-5: NAVAL VERSION

HN-5A: CHINESE VERSION

NATIONAL WAR COLLEGE PHOTO

STRELA 2M/A: YUGOSLAVIAN UPGRADE

SAKR EYE: EGYPTIAN UPGRADE

STRELA-2M2: SA-7/7B AND STRELA-3 /SA-14 MISSILES CONVERTED WITH A LOMO UPGRADE 2-COLOR IR SEEKER FOR DETECTION/IRCM RESISTANCE SIMILAR TO SA-18.

SA-7B CAN BE MOUNTED IN VARIOUS VEHICLES, BOATS, AND VESSELS IN FOUR, SIX, AND EIGHT-TUBE LAUNCHERS. IT CAN ALSO MOUNT ON HELICOPTERS (INCLUDING MI-8/17, MI-24/35 AND S-342 GAZELLE).

NOTES

THIS MISSILE IS A TAIL-CHASING HEAT (IR) SEEKER THAT DEPENDS ON ITS ABILITY TO LOCK ON TO HEAT SOURCES OF USUALLY LOW-FLYING FIXED- AND ROTARY-WING AIRCRAFT. WHEN LAUNCHED TOWARD A RECEDING AIRCRAFT, THE MANPADS CAN BE USED TO SCAN THE DIRECTION AND LOCK ON WITHOUT THE TARGET BEING VISUALLY ACQUIRED IN THE SIGHTS.

AN IDENTIFICATION FRIEND OR FOE (IFF) SYSTEM CAN BE FITTED TO THE GUNNER/OPERATOR'S HELMET. FURTHER, A SUPPLEMENTARY EARLY WARNING SYSTEM CONSISTING OF A PASSIVE RF ANTENNA AND HEADPHONES CAN BE USED TO PROVIDE EARLY CUE ABOUT THE APPROACH AND ROUGH DIRECTION OF AN ENEMY AIRCRAFT.

THE GUNNER MAY HAVE AN OPTIONAL 1L15-1 PORTABLE ELECTRONIC PLOTTING BOARD, WHICH WARNS OF LOCATION AND DIRECTION OF APPROACHING TARGET(S) WITH A DISPLAY RANGE OF UP TO 12.5 KM.

A VARIETY OF NIGHT SIGHTS ARE AVAILABLE, INCLUDING 1 GEN II (2,000-3,500), 2 GEN II (4,500), AND THERMAL SIGHT (5,000-6,000). BRITISH RING SIGHTS PERMIT II NIGHT SIGHT TO BE MOUNTED TO ANY MANPADS.

Russian Man-portable SAM System SA-14/GREMLIN

		Weapons & Ammunition Types	Typical Combat Load
		ready missiles	One-man 1 Normal Dismount 2 From AD Vehicle 5
SYSTEM	**SPECIFICATIONS**	**FIRE CONTROL**	**SPECIFICATIONS**
Alternative Designation	9K34 Strela-3	Sights with magnification	Launch tube has simple sights
Date of Introduction	1978	Gunner	
Proliferation	Worldwide	Gunner Field of View (°)	INA
Target	FW, heli	Gunner Acquisition Range (m)	INA
Description		Gunner Night Sight	None standard, but available
Crew	1, Normally 2 with a loader	Acquisition Range (m)	6,000
ARMAMENT	**SPECIFICATIONS**	IFF	Yes
Launcher			
Name	9P59		
Length (m)	1.40		
Diameter (mm)	75		
Weight (kg)	2.95		
Reaction Time (sec)	25		
Fire on the Move	Yes, in short halt		
Missile			
Name	9M36 or 9M36-1		
Max Range (m)	6,000		
Min Range (m)	600		
Max Altitude (m)	6,000		
Min Altitude (m)	10; 0 with degraded Ph		
Length (m)	1.4		
Diameter (mm)	75		
Fin Span (mm)	INA		
Weight (kg)	10.3		
Missile Speed (m/s)	600		
Propulsion	2-stage solid-propellant rocket		
Guidance	1-color passive IR homing		

Seeker Field of View	INA		
Tracking Rate	INA		
Warhead Type	Frag-HE		
Warhead Weight (kg)	1.0		
Fuze Type	Contact/grazing		
Probability of Hit (Ph%):	50 FW/50 heli		

VARIANTS

IGLA-M/ 9M39 (SA-N-8): NAVAL VERSION

A LOMO SEEKER CAN UPGRADE SA-7/STRELA-2 AND STRELA-3/SA-14 MISSILES TO STRELA-2M2, WITH NEAR SA-18 CAPABILITY.

NOTES

THE GUNNER MAY HAVE AN OPTIONAL PORTABLE ELECTRONIC PLOTTING BOARD, WHICH WARNS OF LOCATION AND DIRECTION OF APPROACHING TARGET(S) WITH A DISPLAY RANGE OF UP TO 12.5 KM. A VARIETY OF NIGHT SIGHTS ARE AVAILABLE, INCLUDING 1 GEN II (2,000-3,500), 2 GEN II (4,500), AND THERMAL SIGHT (5,000-6,000). BRITISH RING SIGHTS PERMIT II NIGHT SIGHT TO BE MOUNTED TO ANY MANPADS. GIVEN WARNING ON APPROACH AZIMUTH AT NIGHT, OR LAUNCHED TOWARD A RECEDING AIRCRAFT, THE MANPADS CAN BE USED TO SCAN THE DIRECTION AND LOCK ON WITHOUT THE TARGET BEING VISUALLY ACQUIRED IN THE SIGHTS.

Russian Man-portable SAM System SA-16/GIMLET

SA-16 missile, and launcher with protective pad and missile cap for transport

Weapons & Ammunition Types	Typical Combat Load
One-man	1
Normal Dismount	2
From AD Vehicle	5

SYSTEM	SPECIFICATIONS	FIRE CONTROL	SPECIFICATIONS
Alternative Designation	9K310 Igla-1	Sights with Magnification	Front hooded ring, rear optical
Date of Introduction	1981	Gunner	
Proliferation	At Least 34 Countries	Day Sight	Field of View (o): INA Acquisition Range (m): 5,200+
Target	FW, heli, cruise missile, UAV	Night Sight	Ring mount with II NVG Field of View (o): INA Acquisition Range (m): 3,500
Description		Note	To portray the system as a 2nd Tier MANPADS, include gen2 II night sight. For a 3rd Tier system, gen1 II sight may be used.
Crew	1, Normally 2 with a loader	Other Acquisition Aids	Aircraft approach warn system: Vehicle alarm Azimuth warn system: 1L15-1 plotting board Other: Pelengator RF direction-finder system These will be found in Tier 2 mech infantry units, and in Tier 3 at brigade level.
ARMAMENT	**SPECIFICATIONS**	IFF	Yes
Launcher			
Name	9P322 launch tube 9P519 launcher gripstock		
Length (m)	1.708		
Diameter (m)	0.08 tube, 0.33 overall		

Weight (kg)	7.1		
Reaction Time (sec)	5-7 Seconds		
Time Between Launches	INA		
Reload Time (sec)	<60		
Fire on the Move	Yes, in short halt		
Missile			
Name	9M313		
Max Range (m)	5,200 other aspects; 4,500 approaching		
Min Range (m)	600		
Max Altitude	3,500 receding slow 3,000 slow approach 2,500 receding fast 2,000 fast approach		
Min Altitude (m)	10; 0 w/ degraded Ph		
Length (mm)	1,593		
Diameter (mm)	72		
Weight (kg)	10.8		
Missile Speed (m/s)	570		
Propulsion	Solid fuel booster and dual-thrust solid fuel sustainer rocket motor.		
Guidance	Passive 2-color IR homing		
Seeker Field of View	80° Unusually wide FOV permits the missile to respond more quickly to maneuvering targets, such as helicopters.		
Tracking Rate	INA		
Warhead Type	Frag-HE. Also, fuel residue is ignited to enhance warhead blast		
Warhead Weight (kg)	1.27		
Fuze Type	Contact		
Probability of Hit (Ph%)	60 FW/70 heli		
Self-Destruct (sec)	14-17		
Countermeasure Resistance	See Notes		

VARIANTS

THE SA-16 IS A VARIANT OF THE IGLA (SA-18) DESIGN. BECAUSE OF DELAYS IN THE IGLA PROGRAM, THE IGLA-1WITH A SIMPLER AND SLIGHTLY LESS CAPABLE SEEKER WAS RUSHED INTO PRODUCTION AND FIELDED 2 YEARS PRIOR TO ITS PROGENITOR. THE SA-16 IS DESIGNED ESPECIALLY TO BE ABLE TO ENGAGE HELICOPTERS.

IGLA-1E: RUSSIAN EXPORT VARIANT. UNLIKE THE BASE SYSTEM, FUEL REMNANTS ARE NOT FUZED ALONG WITH THE WARHEAD. IFF INTERROGATOR CAN BE TAILORED TO CUSTOMER SPECIFICATIONS.

IGLA-1M: EXPORT VARIANT SIMILAR TO -1E, BUT LACKING AN IFF INTERROGATOR.

SPECIALIZED APPLICATIONS INCLUDE AN LUAZ UTILITY CARRIER DESIGNED FOR A MANPADS FIRING UNIT. THE VEHICLE HAS A RACK FOR MOUNTING FIVE 9P322 SA-16 LAUNCHER TUBES. THIS RACK COULD BE USED IN OTHER MAN-PORTABLE AD UNIT VEHICLE APPLICATIONS.

DJIGIT: RUSSIAN TWIN LAUNCHER COMPLEX MOUNTED ON A RAIL FRAME WITH OPERATOR'S SEAT AND TRIPOD. MISSILES CAN BE SIMULTANEOUSLY LAUNCHED USING CENTRALLY MOUNTED SIGHT. A HUNGARIAN MOUNT WITH THIS SYSTEM ON A GAZ-630 4X4 TRUCK IS CALLED IGLA-1E.

STRELETS IS A TWIN MISSILE LAUNCH MODULE TO MOUNT SA-16/SA-18/SA-24 SAMS ON GUNS, PLATFORMS AND VEHICLES.

NOTES

LAUNCHER DEPLOYMENT TIME IS 5-13 SECONDS. MISSILES ARE PRELOADED IN THE LAUNCH TUBE FOR QUICK LOADING TO THE GRIPSTOCK. A TUBE CAN BE USED UP TO FIVE TIMES. THE MISSILE IS COOLED BY A DISPOSABLE BOTTLE OF REFRIGERANT. THE BOTTLE AND LAUNCHER BATTERY ARE USEABLE FOR 30 SECONDS AFTER ACTIVATION. BECAUSE THE NOSE EXTENDS PAST THE LAUNCHER TUBE, THE NOSE IS PROTECTED WITH AN EXTENDED CAP WHICH IS REMOVED BEFORE LAUNCHING.

ONCE THE OPERATOR REACHES THE LAUNCH AREA, HE WILL OFTEN REMOVE THE PROTECTIVE PAD, AND WILL REMOVE THE MISSILE CAP PRIOR TO USE.

MAXIMUM SPEED FOR TARGETS ENGAGED VARIES FROM 320 M/S REAR ASPECT, RECEDING TARGETS, TO 360-400 M/S HEAD-ON, APPROACHING TARGETS.

THE GUNNER MAY HAVE AN OPTIONAL PORTABLE ELECTRONIC PLOTTING BOARD, WHICH WARNS OF LOCATION AND DIRECTION OF APPROACHING TARGET(S) WITH A DISPLAY RANGE OF UP TO 12.5 KM. FOR TIER 1 AND TIER 2 OPFOR SIMULATIONS AND UNITS OPERATING FROM VEHICLES, THIS SYSTEM AND PELENGATOR ARE LIKELY.

MISSILE SEEKER FEATURES A TWO-COLOR SEEKER WITH IMPROVED PROPORTIONAL CONVERGENCE LOGIC, AND AN *IGLA* (NEEDLE) DEVICE ON THE SEEKER, WITH MIRROR AND TRIPOD TO COOL THE SEEKER AND FACILITATE MORE RIGOROUS G-LOAD TURNS WITH REDUCED SEEKER WARMING. WITH THESE FEATURES, THE SA-16 OFFERS SUPERIOR MANEUVER AND COUNTERMEASURE RESISTANCE OVER THE PREVIOUS MANPADS, AND A BASE LEVEL OF PRECISION AGAINST MANEUVERING AIRCRAFT THAT IS SIMILAR TO THE SA-18. NEVERTHELESS, THIS MISSILE IS MORE VULNERABLE TO EO/IR DECOY COUNTERMEASURES THAN THE LATER SA-18.

Russian Man-portable SAM System SA-18/GROUSE, and SA-24/Igla-Super

 SA-18/Igla  Vehicle with SA-18 for AD fire support		**Weapons & Ammunition Types** **ready missiles**	**Typical Combat Load** **One-man 1** **Normal Dismount 2** **From AD Vehicle 5**
SYSTEM	**SPECIFICATIONS**	**FIRE CONTROL**	**SPECIFICATIONS**
Alternative Designation	**9K38 Igla**	**Sights with Magnification**	
Date of Introduction	**1983**	**Gunner Day sight Acquisition Range (m):**	**6,000+**
Proliferation	**At least 6 countries** **At least 4 countries for SA-24**	**Gunner Night Sight Mowgli-2 2 gen II Acquisition Range (m)**	**4,500**
Target	**FW, heli, CM, UAV**	**Other Acquisition Aids**	
Description		**Pelengator RF DF system**	**See Notes**
Crew	**Crew: 1, Normally 2 with a loader**	**IFF**	**Yes**
ARMAMENT	**SPECIFICATIONS**		
Launcher	**The launcher can launch either SA-18 or SA-16 missiles.**	**9S520**	**Package with night sight, aircraft approach warning system, vehicle alarm, and1L15-1 azimuth plotting board.** **An SA-18 battery at brigade/division usually has a Sborka ACV.**
Name	**9P39**		
Length (m)	**1.708**		
Diameter	**INA**		

Weight (kg)	1.63		
Reaction Time (sec)	6-7		
Time Between Launches	16		
Reload Time (sec)	10		
Fire on the Move	Yes, in short halt		
Missile			
Name	9M39		
Range (m)	500-6,000+		
Max Altitude (m)	3,500		
Min Altitude (m)	10; 0 with degraded Ph		
Length (mm)	1,708		
Diameter (mm)	70		
Weight (kg)	10.6		
Missile Speed (m/s)	Mach 2 (570 m/s) mean velocity		
Propulsion	Solid fuel booster and dual-thrust solid fuel sustainer rocket motor.		
Guidance	2-color IR/UV homing		
Seeker Field of View	INA		
Tracking Rate	INA		
Warhead Type	Frag HE		
Warhead Weight (kg)	1.27		
Fuze Type	Impact and Proximity		
Probability of Hit (Ph%)	70 FW, 80 heli		
Self-Destruct (sec)	15		
Countermeasure Resistance	Seeker resists and degrades all pyrotechnic and electronically operated IR CM		

VARIANTS

IGLA-D: LAUNCHER USED IN AIRBORNE FORCES. IT CAN BE SEPARATED IN TWO PARTS FOR EASIER PORTABILITY, BUT THIS ADDS 60 SECONDS TO THE REACTION TIME.

IGLA-N: INCREASED LETHALITY DUE PRIMARILY TO THE WARHEAD MASS INCREASED TO 3.5 KG, AND CAN BE SEPARATED IN TWO PARTS.

IGLA-V: AIR-TO-AIR VERSION

IGLA-1 (SA-16): ECONOMICAL VARIANT OF THE IGLA MANPADS ESPECIALLY SUITED FOR OUT-MANEUVERING HELICOPTERS.

STRELA-2M2: UPGRADE VERSION SA-7/STRELA-2 MISSILE WITH IMPROVED LOMO SEEKER GIVES IT NEAR SA-18 CAPABILITY.

SA-24/IGLA-SUPER (IGLA-S): IMPROVED MISSILE WITH LASER PROXIMITY/PD FUZE, A HEAVIER EXPLOSIVE CHARGE AND SEGMENTING ROD (2.5 KG) WARHEAD WITH INCREASED FRAGMENTATION EFFECTS. PROPORTIONAL NAVIGATION

FURTHER RESISTS FLARES AND OTHER IRCM. THUS, THE MISSILE GREATLY INCREASES P-HIT AND P-KILL EVEN AT LOW ALTITUDES AND AGAINST CM. LAUNCHER NOSE IS MODIFIED TO FIT. IT HAS BEEN EXPORTED TO SEVERAL COUNTRIES. THE SAM FITS A SA-16/18 GRIPSTOCK WITH TRIGGER CHANGE.

GROM-1: POLISH COPY OF SA-18

DJIGIT: RUSSIAN TWIN LAUNCHER PEDESTAL MOUNTED ON A RAIL FRAME WITH OPERATOR'S SEAT AND TRIPOD.

STRELETS IS A TWIN MISSILE LAUNCH MODULE AND COOLANT UNIT, WITH TWO LAUNCHERS MOUNTED AND REMOTELY LINKED TO A SIGHTING AND LAUNCH CONTROL SYSTEM. THE STRELETS MOUNTS DUAL SA-16/SA-18/SA-24 SAMS ON GUNS, PLATFORMS AND VEHICLES, AND INTEGRATE THEM INTO ROBUST FCS AND COMPLEXES. IT CAN LAUNCH TWO MISSILES SIMULTANEOUSLY AT A SINGLE TARGET. STRELETS IS USED AS A PAIR, OR CAN BE LINKED FOR 4-LAUNCHER, 8-LAUNCHER OR OTHER ARRANGEMENTS. AN EARLY APPLICATION IS THE ZU-23M1 AIR DEFENSE GUN/MISSILE SYSTEM WITH A LAUNCH MODULE MOUNTED ON THE TOWED GUN CHASSIS AND LINKED TO A GUN-MOUNT FCS ON A NOTEBOOK COMPUTER WITH FLIR NIGHT SIGHT.

SA-18 LAUNCHER VEHICLES: RUSSIA, FOLLOWING A TREND IN AD SYSTEMS, DEVELOPED A VARIETY OF MOUNTS FOR LAUNCHERS ON AD GUNS AND VEHICLES.

THE DJIGIT TWIN-LAUNCHER CAN BE MOUNTED ON A TUV TO FORM A LOW-COST AD LAUNCHER VEHICLE WITH REMOTE SIGHTING AND DUAL MISSILE LAUNCH CAPABILITY.

THE FENIX AIR DEFENSE SYSTEM CONSISTS OF THE VODNIK TUV WITH AN IR AUTO-TRACKER PASSIVE FCS AND FOUR STRELETS LAUNCHER MODULES (8 MISSILES).

IGLA SAM SYSTEM TURRET FOR MOUNT ON APC, IFV, OR OTHER CHASSIS FEATURES AN SA-13 TYPE 1-MAN TURRET WITH EO FCS AND 4 STRELETS (8 LAUNCHERS). THE TURRET HAS BEEN DISPLAYED ON MT-LB AND BRDM-2.

LUAZ/IGLA FEATURES STRELETS LAUNCHERS ON THE AMPHIBIOUS TUV, AS AN ALL-TERRAIN AD VEHICLE.

A MODERNIZED ZSU-23-4 SP GUN IS NOW A GUN/MISSILE VEHICLE WITH 2 LAUNCH MODULES (4 MISSILES) LINKED TO A COMPUTER-BASED FCS WITH LLLTV SIGHT.

NOTES

IN TIER 1 AND 2 UNITS, PELENGATOR RF HELMET-MOUNT DIRECTION-FINDER SYSTEM PERMITS THE MISSILE OPERATOR TO SLEW TO TARGET, AND RANGES 20+ KM.

AVAILABLE NIGHT SIGHTS INCLUDE 1-3 GEN II AND THERMAL SIGHTS. BRITISH RING SIGHTS PERMIT AN II NIGHT SIGHT TO BE MOUNTED TO ANY MANPADS.

British Air Defense/Anti-Armor (High Velocity) Missile System Starstreak

Starstreak Lightweight Multiple Launcher

		Weapons & Ammunition Types	Typical Combat Load
		Ready missiles	Dismount 3 Team in Vehicle 5
SYSTEM	**SPECIFICATIONS**	Missile	
Alternative Designation	Manportable is Shoulder-Launched (SL) Starstreak	Name	Starstreak
Date of Introduction	1997 vehicle (SP HVM), 2000 man-portable (-SL)	Range (m)	300-7,000 max (guided)
Proliferation	2-6 Countries	Altitude (m)	0-5,000
Target	FW, heli, ground vehicles	Length (mm)	1,400
Description	(SL configuration)	Diameter (mm)	127
Crew	2 with a loader (one possible)	Weight (kg)	14.0
ARMAMENT	**SPECIFICATIONS**	Max Missile Speed (m/s)	1,364 m/s, Mach 4
Launcher		Propulsion	Canister launch booster, bus missile, and 3 darts (sub-missiles)
Name	Aiming Unit	Flight Time to Max Range (sec)	5-7
Dimensions (m)	See Missile	Guidance	Laser beam rider SACLOS
System Weight (kg)	24.3 with missile	Warhead Type	Three 25-mm darts- tungsten KE tip and case & HE fill
Reaction Time (sec)	<6	Penetration (mm KE)	120+ all LAVs (Equal to 3 x 40-mm APFSDS-T rds) HE detonates after for frangible effects
Time Between Launches	<30 sec	Fuze Type	Contact with time delay
Reload Time (sec)	<25 sec est	Probability of Hit (%)	60 FW, >95 heli (each dart 67 vs heli)
Fire on the Move	Yes, in short halt	Self-Destruct (sec)	Yes, INA
		Fire control	**SPECIFICATIONS**
		Sights with Magnification	
		Day sight	Avimo stabilized optical sight with lead bias system

		Field of View (o)	INA
		Acquisition Range (m)	7000+
		Night sight Thales clip-on thermal sight Acquisition Range (km)	4-5 est

OTHER MISSILES

STARSTREAK II: IMPROVED MISSILE HAS 8-KM RANGE AND BETTER PRECISION. FIELDED 2010.

LIGHTWEIGHT MULTI-ROLE MISSILE/LMM: A MULTI-ROLE MISSILE OPTION WITH A SINGLE 3-KG TANDEM (HEAT/HE) WARHEAD AND PROXIMITY FUZE. AT 13 KG, THE LOWER-COST MISSILE FLIES 8-KM AT 1.5 MACH. IT IS DUE IN 2013, AND WAS SUCCESSFULLY LAUNCHED BY A CAMCOPTER S-100 UCAV VARIANT. OTHER PROJECTED UPGRADES ARE SEMI-ACTIVE LASER-HOMING AND/OR DUAL-MODE (LBR/SAL-H).

OTHER ACQUISITION AIDS

ADAD: BRITISH PASSIVE THERMAL IR SCANNERS ON REMOTE TRIPOD OR VEHICLE MOUNT WITH 240 O FOV AUTOMATIC CUEING.

MISSILE TEAM EMPLOYS AN AZIMUTH PLOTTING BOARD (E.G., RUSSIAN 1L15-1), FOR DIRECTION OF APPROACH ON AERIAL TARGETS.

VARIANTS

THE MOST COMMON LAUNCHER USED IS–SL.

STARBURST: JAVELIN SAM LAUNCHER ADAPTED FOR STARSTREAK LBR GUIDANCE- IN PRODUCTION

LIGHTWEIGHT MULTIPLE LAUNCHER (LML): PEDESTAL LAUNCHER FOR THREE MISSILES (ABOVE). THE LAUNCHER CAN ALSO MOUNT ON A LIGHT VEHICLE, E.G., TUV. A DEMONSTRATOR IS LML ON A PANHARD TACTICAL TRUCK.

STARSTREAK II: IMPROVED LAUNCHER USES STARSTREAK OR STARSTREAK II MISSILE. IT HAS AN AUTO-TRACKER FOR HANDS-FREE GUIDANCE. IT WAS FIELDED IN 2010.

STARSTREAK LIGHTWEIGHT VEHICLE (LWV): LAND ROVER TRUCK CONVERTED INTO AN SP SAM SYSTEM WITH A 6-CANISTER LAUNCHER, ADAD AUTO-TRACKER, AND TV/ THERMAL FCS. THIS LAUNCHER CAN BE MOUNTED ON OTHER VEHICLES.

ARMORED STARSTREAK OR (SP HVM): VEHICLE IS A STORMER TRACKED APC CHASSIS, WITH AN 8-MISSILE LAUNCHER. THE PASSIVE IR FIRE CONTROL SYSTEM USES ADAD, AN AUTO-TRACKER AND THERMAL SIGHT. THE LAUNCHER CAN BE MOUNTED ON OTHER VEHICLES.

SEASTREAK: SINGLE-STAGE MISSILE NAVAL VARIANT IN A 12-MISSILE LAUNCHER, WITH MM-WAVE RADAR FCS.

OPTIONAL USE: AS A LOW-COST AIR DEFENSE/ANTI-ARMOR (MULTI-ROLE) SYSTEM, STARSTREAK CAN BE EMPLOYED AGAINST GROUND TARGETS, SUCH AS LIGHT ARMORED VEHICLES, AND SNIPERS IN BUNKERS OR BUILDINGS. THE MISSILE AND ITS DARTS, WITH A UNIQUE COMBINATION OF PENETRATOR AND FOLLOWING FRAG-HE, HAVE BEEN SUCCESSFULLY TESTED AGAINST VEHICLE TARGETS. WITH A MISSILE COST OF 1/2 TO 1/3 OF COMPETING MANPADS, THE SYSTEM COULD BE USED AS A FIRE SUPPORT ASSET TO COMPLEMENT ATGM LAUNCHERS AND VEHICLE WEAPONS.

THOR: BRITISH MULTI-MISSION AIR DEFENSE SYSTEM IS A RWS, WITH 4 MISSILE LAUNCHERS, TV, FLIR, AND AN AUTO-TRACKER. WEIGHING .5 MT, IT MOUNTS ON TRUCKS, VANS, TUVS, APCS, ETC., WITH A REMOTE OPERATOR. DESIGNED FOR STARSTREAK, LAUNCHERS, IT CAN ALSO MOUNT OTHER MANPADS, AND ATGMS, SUCH AS INGWE, TOW, HELLFIRE, MOKOPA, SPIKE, ETC.

NOTES

GROUND-BASED AD SYSTEM OPTIMIZED FOR USE AGAINST ARMORED HELICOPTERS AND LOW FLYING FIXED-WING AIRCRAFT. MISSILE EMPLOYS SMOKELESS PROPELLANT FOR MINIMAL SIGNATURE. FLIGHT TIME (5-8 SEC) AND LBR GUIDANCE MAKE IT ESSENTIALLY IMMUNE TO COUNTERMEASURES. BECAUSE OF THE HIGH VELOCITY, THE SYSTEM EXCEEDS THE HIT PROBABILITY OF COMPETING SYSTEMS AGAINST HIGH -SPEED AIRCRAFT ON RECEDING FLIGHT PATHS.

THE STARSTREAK'S LOWER COST AND CAPABILITIES AS A MULTI-ROLE MISSILE SYSTEM OFFERS VARIED USES. TWO CONSIDERATIONS ARE THE SEMI-AUTOMATIC COMMAND LINE-OF-SIGHT (SACLOS) GUIDANCE AND CONTACT FUZES WHICH MAKE IT LESS EFFECTIVE AGAINST AGILE FIXED-WING AIRCRAFT FROM SOME ASPECTS. THUS A MORE PRACTICAL COURSE WOULD BE TO REPLACE 33-50% OF THE MANPADS. WITH THE LOWER COST OF STARSTREAK AND ITS MULTI-ROLE CAPABILITY, IT COULD REPLACE A PORTION OF THE EXPENSIVE SINGLE-ROLE MANPADS WITH STARSTREAKS. FOR INSTANCE, AN 18-MANPADS BATTERY COULD BE REDUCED 33% TO 12 MANPADS WHILE ADDING 12 STARSTREAKS, WITH THE LATTER USED AS A MULTI-ROLE SYSTEM. WITH 50% OF THE MANPADS REPLACED, THE MIX WOULD BE 9 MANPADS AND 18 STARSTREAKS. ADDED ANTI-ARMOR CAPABILITY IS A BONUS. SUBSTITUTION COULD VARY WITH THE EXPECTED ADVERSARY TARGET MIX.

U.S. Man-portable SAM System Stinger

	Weapons & Ammunition Types	Typical Combat Load
	Ready missiles	
		One-man 1
		Dismount 2
		From AD Vehicle 5

SYSTEM	SPECIFICATIONS	FIRE CONTROL	SPECIFICATIONS
Alternative Designation	FIM-92A Basic Stinger	Sights with Magnification	
Date of Introduction	1981	Day Sight	Ring and bead, most launchers Optical sight with lead bias available.
Proliferation	At least 22 countries, base and all variants	Day Sight Field of View (°)	INA
Crew	1, Normally 2 with a loader		Acquisition Range (m):
System	Grip-stock (with battery coolant unit, IFF, impulse generator, and seeker redesign), missile, night sight, radio and other acquisition aides	Day Sight Field of View (°)	INA
ARMAMENT	**SPECIFICATIONS**	Day Sight Acquisition Range (m)	4000+
Launcher		Night Sight	Optional AN/PAS-18, Wide-Angle Stinger Pointer System (WASP) thermal sight.
Name	Stinger grip-stock	Night Sight Field of View (°)	20° x 12°
Length	1.52+ launch tube	Night Sight Acquisition Range (km)	20-30 side or tail aspect, 10 head-on aspect
Diameter	INA	IFF	AN/PPX-1 trigger-activated on grip-stock, with battery belt-pack
System Weight	15.2 launch-ready 2.6 belt-pack IFF	Target Alert Display Set (TADDS)	US portable graphic display set w/audio alert, VHF radio, and IFF.
Reaction Time (sec)	6 tracking and missile activation (3-5 cooling)	ADAD	British passive thermal IR scanners on remote tripod or vehicle mount with 240° FOV automatic cueing system.

Time Between Launches (sec)	INA	Radar Equipment Providing Omni-directional Reporting of Targets at Extended Ranges (REPORTER)	German/Dutch EW system with I/J band radar and IFF. Range: 40 km. Altitude: 15-4000 m.
Reload Time (sec)	<10	Several U.S. and foreign radars are available for use with Stinger.	
Fire on the Move	Yes, in short halt		
Missile			
Name	FIM-92A		
Max Range (m)	4,000+		
Min Range (m)	200		
Max Altitude (m)	3,500		
Length (mm)	1.52		
Diameter (mm)	70		
Weight (kg)	10.0		
Target Maneuver Limit	Up to 8 g		
Missile Speed (m/s)	745 m/s, Mach 2.2		
Propulsion	Solid fuel, dual-thrust (ejector motor and sustainer motor)		
Guidance	Cooled 2nd gen passive IR homing (4.1-4.4 μm)		
Seeker Field of View	INA		
Tracking Rate	INA		
Warhead Type	Frag-HE		
Warhead Weight (kg)	1.0		
Fuze Type	Contact with time delay		
Probability of Hit (Ph%)	INA		
Self-Destruct (sec)	20		

VARIANTS

STINGER-PASSIVE OPTICAL SEEKER TECHNIQUE (POST) / FIM-92B: LIMITED PRODUCTION UPGRADE IN 1983 ADDED AN IR/UV SEEKER WITH IMPROVED SCAN TECHNIQUE IMPROVED FLARE CM RESISTANCE. SEEKER ADDS TARGET ADAPTIVE GUIDANCE (TAG), WHICH SHIFTS IMPACT POINT FROM THE EXHAUST PLUME TO A MORE CRITICAL AREA OF THE TARGET. MAX RANGE INCREASES TO 4,800 M, AND MAX ALTITUDE INCREASES TO 3,800 M.

STINGER-REPROGRAMMABLE MICRO-PROCESSOR: (RMP) / FIM-92C: PRODUCTION BEGAN IN 1989. THE UPGRADE PERMITS UPLOADING NEW CCM SOFTWARE. EXPORT VERSION LACKS REPROGRAM CAPABILITY BUT USES AN EMBEDDED IRCM PROGRAM.

THE MANPADS HAS BEEN ADAPTED FOR LAUNCH FROM APC OR IFV CHASSIS. IT HAS ALSO BEEN ADAPTED FOR LIGHT UTILITY VEHICLES AND COMBAT SUPPORT VEHICLES, SUCH AS THE GERMAN WIESEL-BASED FLIEGERFAUST-2 (FLF-2). A VARIETY OF AIR DEFENSE LAUNCHER SYSTEMS CAN USE STINGER, MISTRAL, OR OTHER MANPADS.

PEDESTAL MOUNTED STINGER (MULTIPLE LAUNCHER WITH STINGER MANPADS AND INTEGRATED FCS).

DUEL MOUNTED STINGER IS A DANISH EASILY MOUNTED TRIPOD LAUNCHER WITH OPERATOR SEAT AND CONSOLE, WHICH CAN BE MOUNTED ON BOAT OR TRUCK BED.

AN AIRCRAFT MOUNT IS AIR-TO-AIR STINGER - ATAS.

NOTES

A NUMBER OF U.S. UPGRADES AND STINGER APPLICATIONS ARE IN DEVELOPMENT.

French MANPADS Launcher Vehicle Albi/Man-portable SAM System Mistral 2

		Weapons & Ammunition Types	Typical Combat Load
 Albi with Mistral 2	 Mistral on Tripod Launcher		
		Mistral 2 missiles	
		On launcher	8
		Normal reload	2
		Added reload (est)	4
			2
		7.62-mm Machinegun	
		API-T	1200

SYSTEM	SPECIFICATIONS	FIRE CONTROL	SPECIFICATIONS
Alternative Designation	VBR Mistral	Sights with Magnification	
Date of Introduction	2000-2001 Albi and Mistral 2, 1988 original Mistral	Day Sight	EO/IR sight: Range (m): 6,000 or more
Proliferation	25+ countries for missile, at least 2 for launcher vehicle	Night Sight	Alis or MATIS thermal sight Range (m): 5,000-6,000
Target	FW, heli, CM, UAV	Other Acquisition Aides	
Description	System includes Mistral Coordination Post and up to 12 fire units	Weapon Terminal links to alert system and provides azimuth of approaching aircraft.	French Army Samantha digital alert system with GPS or export Aida terminal linking to MCP
Launcher Vehicle		IFF	Thompson SB14 on MCP or other
Description	Tactical utility vehicle with foldable MANPADS launcher turret	**ASSOCIATED VEHICLES/RADARS**	**SPECIFICATIONS**
Name	Albi for turret, and vehicle system	Name	Samantha aircraft warning station
Crew	2-3; driver, gunner, assistant gunner	Chassis	VBL
Chassis	VBL tactical utility vehicle	Radar	Griffon TRS 2630
Vehicle Description	See VBL	Function	Target acquisition radar
Automotive Performance	See VBL	Band	S
Radio	INA	Range (km)	15-20
Protection	See VBL	Name	Mistral Coordination Post (export)
ARMAMENT	**SPECIFICATIONS**	Chassis	VBL or other, such as Unimog truck
Launcher		Radar	SHORAR
Name	Albi twin launcher on turret	Function	Alerting radar, target acquisition
Reaction Time (sec)	5 stopped, 3 with	Range (km)	25

	warning and azimuth from terminal		
Time Between Launches (sec)	<5		
Reload Time (min)	<1.5		
Fire on the Move	No, stop or short halt		
Launcher Elevation (°)	0/+80		
Emplace/Displace Time (min)	0.08		
Missile			
Name	Mistral 2		
Max Range (m)	6,000		
Min Range (m)	600		
Max Altitude (m)	3,000		
Min Altitude (m)	5, 0 with degraded Ph		
Length (mm)	1.86		
Diameter (mm)	90		
Weight (kg)	18.7		
Missile Speed (m/s)	870 (Mach 2.7)		
Maximum Target Speed (m/s)	INA		
Propulsion	Solid motor plus booster motor		
Guidance	Passive IR/UV homing with digital multi-cell pyramidal seeker		
Warhead Type	HE with Tungsten Balls		
Warhead Weight (kg)	3		
Fuze Type	Laser proximity/contact		
Probability of Hit (Ph%)	70 FW, 80 heli		
Self-Destruct (sec)	INA		
Countermeasure resistance	Mistral 2 resists nearly all IR countermeasures.		
Auxiliary Weapon			
Caliber, Type, Name	7.62-mm MG, AAT 52		
Cyclic Rate of Fire (rd/min)	900 cyclic, in bursts		
Practical Rate of Fire (rd/min)	250 (est)		
Loader Type	200-rd Magazine		
Ready/Stowed Rounds	200/1,000		
Fire on the Move	Yes		

VARIANTS

THE MISTRAL PORTABLE LAUNCHER EMPLOYS TRIPOD, SEAT, AND SINGLE LAUNCHER STAND. ORIGINAL MISTRAL 1 MISSILE WAS MORE VULNERABLE TO IR COUNTERMEASURES.

ALAMO: CYPRIOT MOUNT OF SINGLE MISTRAL LAUNCHER ON 4X4 TUV.

ALBI CAN BE MOUNTED ON A VARIETY OF VEHICLES.

ASPIC: 4-MISSILE LAUNCHER FOR VEHICLE MOUNT

ATLAS: A TWIN LAUNCHER ON A PORTABLE STAND. HUNGARY PURCHASED UNIMOG 4X4 LIGHT TRUCKS WITH ATLAS PLATFORM-MOUNTED LAUNCHERS. THE LAUNCHERS CAN BE QUICKLY REMOVED FROM A VEHICLE AND GROUND MOUNTED.

ONE BLAZER AD VEHICLE VARIANT USES MISTRAL AND 25-MM AUTO-CANNON.

GUARDIAN IS HMMWV W/MISTRAL LAUNCHERS.

SANTAL: TURRET 6-MISSILE LAUNCHER, FOR USE ON ARMORED VEHICLES.

AIR-TO-AIR MISTRAL (ATAM): TWIN MISSILE POD FOR USE ON HELICOPTERS.

THE FRENCH NAVY USES A VARIETY OF LAUNCHER CONFIGURATIONS, E.G., SADRAL, SIMBAD, SIGMA, TETRAL, AND LAMA.

FN-6: RECENT CHINESE MANPADS-A LIKELY COPY OR VARIANT OF MISTRAL ON A LIGHTWEIGHT MAN-PORTABLE LAUNCHER. IT WILL BE EXPORTED TO MALAYSIA AND OTHER COUNTRIES. YTIAN/TY-90 IS AAM/VEHICLE LAUNCH VERSION OF MISTRAL WITH 8- LAUNCHER TURRET, 3-D RADAR, AND EO.

TURRET FITS ON LAV, TUV, OR TOW CARRIAGE.

THE MISTRAL HAS BEEN EVALUATED AND TESTED AS AN UPGRADE MANPADS OPTION FOR A VARIETY OF LAUNCHERS ON VERY SHORT RANGE AIR DEFENSE (VSHORAD) VEHICLES, AND AS AN AIR-TO-AIR MISSILE FOR USE ON HELICOPTERS.

NOTES

THIS SYSTEM IS AN IDEAL VSHORAD VEHICLE TO PROVIDE MOBILE AND RESPONSIVE AD FOR AIRBORNE, AMPHIBIOUS, MOTORIZED, AND RAPID RESPONSE FORCES. VEHICLES ARE FAIRLY VULNERABLE NEAR FRONT LINES, BUT OFFER FLEXIBLE PROTECTION FOR DEEPER BRIGADE HIGH-VALUE ASSETS. THEY OFFER A LOWER-COST BUT LESS EFFECTIVE SUBSTITUTE FOR SYSTEMS SUCH AS 2S6M. AN ALBI COULD REPLACE A MANPADS SQUAD (APC/IFV, TUV, ETC, AND TWO MANPADS LAUNCHERS).

ALBI RESPONSE TIME MOVING IS 15 SEC AFTER STOP. HOWEVER, MOST OF THE TIME, THE VEHICLE IS STOPPED AND CONDUCTING OVERWATCH RATHER THAN MOVING. ALSO, THANKS TO THE MISSILE WARNING SYSTEM, THE VEHICLE HAS AMPLE TIME TO BE STOPPED AND READY TO LAUNCH PRIOR TO AIRCRAFT APPROACH. WITH A TWO-MAN CREW,

THE MISSILE RELOAD CAPACITY IN THE REAR CAN BE INCREASED TO 10 OR MORE. A 3-MAN CREW WITH 8 MISSILES IS A RATIONAL COMPROMISE, PERMITTING THE THIRD CREWMAN TO MONITOR THE WEAPON TERMINAL TO RAPIDLY RESPOND TO ALERTS, AND TO ASSIST IN RELOADING THE LAUNCHERS.

Russian 30-mm SP AA Gun/Missile System 2S6M1

	Weapons & Ammunition Types	Typical Combat Load
	2 x 30-mm twin-barrel cannons	1,904
	Frangible APDS	
	AP-T, APDS	
	Frag-T	
	HE-I	
	API	
	SA-19/GRISON	10 On Launchers 8 Stowed Inside 2

SYSTEM	SPECIFICATIONS	FIRE CONTROL	SPECIFICATIONS
Alternative Designation	2K22M, Tunguska-M, Tunguska-M1	Sights with Magnification	
Date of Introduction	1990	Gunner Sights	Day: Stabilized EO sight 1A29M Magnification: 8x Field of View (°): 8o Night: 1TPP1 thermal sight Range: 18 km, 6 ground targets
Proliferation	At Least 2 Countries	Commander's Day/Night Sight	IR
Target	FW, heli, cruise missile (CM), and UAV, as well as ground targets	IFF	Yes
Crew	4 (cdr, radar op, gunner, driver)	Radars	HOT SHOT
Combat Weight (mt)	34	Name	1RL144 (TAR)
Chassis	GM-352M tracked vehicle	Function	Target Acquisition
Chassis Length Overall (m)	7.93	Detection Range (km)	18-20
Height (m)	TAR up: 4.02 TAR down: 3.36	Tracking Range (km)	INA
Width Overall (m)	3.24	Frequency	2-3 GHz (E Band)

Engine Type	V-12 turbo diesel	Name	1RL144M (TTR)
Cruising Range (km)	500	Function	Fire Control
Speed (km/h)	Max. Road: 65 Max. Swim: INA	Detection Range (km)	16
Fording Depths (m)	INA	Tracking Range (km)	INA
Radio	R-173	Frequency	10-20 GHz (J band)
Protection NBC Protection System	Yes	Armored Command Vehicle	
Altitude	Max. Altitude: 6000 for 2S6M1 Min. Altitude: 0 for 2S6M1 0 w/ degraded Ph 2S6M	Name	Sborka AD ACV
Dimensions	Length (m): 2.83	Chassis	MTLB-U
Weight (kg)	57 (in container)	Radar	DOG EAR (use in OPFOR units)
Missile Speed (m/s)	600-900	Function	Target Acquisition (EW to 80 km)
Guidance	Radar SACLOS	Frequency	F/G band
Seeker Field of View (°)	INA	Range (km)	80 detection, 35 tracking ACV also links to supported tactical unit nets.
Tracking Rate	INA	Other Radars	Links to Integrated Air Defense System (IADS) for early warning and target acquisition data from radars: Giraffe AMB at Separate Brigade and Division, LONG TRACK or similar EW/TA radar echelons above division, and radars in SAM units, e.g., SA-10.
Warhead Type	Frag-HE	**MAIN ARMAMENT AMMUNITION**	**SPECIFICATIONS**
Warhead Weight (kg)	9	Types	Frangible APDS-T is the preferred round. Other Rounds: AP-T, APDS, Frag-T, HE-I, API
Fuze Type:	Proximity, 5 m radius	Type	Frangible APDS-T
Probability of Hit (Ph%)	65 FW, 80 heli	Range (m)	Max: 4,000 Min: 0
Simultaneous Missiles per target	2	Altitude (m)	Max: 3,000 Min: 0
Self-Destruct (sec)	INA	Penetration (mm KE)	25 at 60° 1,500 m, APDS
System Reaction Time (sec)	6-12		

Fire on Move	Yes, short halt or slow move		

VARIANTS

2S6: PRE-PRODUCTION DESIGN MOUNTING 4 MISSILES

2S6M: FIELDED SYSTEM BEFORE UPGRADES.

2S6M1: UPGRADE VERSION WITH IMPROVED FCS AND DIGITAL C2 INTEGRATION, 9M11-1M MISSILE, IMPROVED ECM RESISTANCE, AND 0 M MIN ALTITUDE.

UPGRADE 9M311-1M MISSILE HAS A PULSE CODED XENON BEACON FOR RESISTANCE TO IRCM, A NEW RF PROXIMITY FUZE, IMPROVED KINETICS FOR A 10 KM RANGE TO ALL TARGETS, AND OPERATING ALTITUDES OF 0 - 6000 M WITH HIGH PRECISION AND HIGH PH.

THE MISSILE MAY BE SUITABLE AS AN UPGRADE ON EXISTING 2S6M LAUNCHERS.

NOTES

MAIN OPERATING MODE IS RADAR MODE, WITH DAY/NIGHT CAPABILITY. OTHER MODES OFFER REDUCED RADAR SIGNATURE. THERMAL SIGHT LISTED IS OPTIONAL, REPRESENTING A RATIONAL UPGRADE TO EXISTING 2S6M AND IS STANDARD ON 2S6M1SYSTEM.

Russian Gun/Missile System Pantsir-S1 and Pantsir-S1-0

Pantsir-S1-0 System with Unified Turret on BMP-3 Chassis

Weapons & Ammunition Types	Typical Combat Load
2 x 30-mm twin-barrel auto-cannons Mix of FAPDS-T and APFSDS-T	1,400
57E6-E Missiles	Total Pantsir -S1 12 -S1-0 12 On Launchers 8 Spares inside 4

SYSTEM	SPECIFICATIONS	FIRE CONTROL	SPECIFICATIONS
Alternative Designation	SA-22E. Other spellings: Pantsyr, Pantzyr, Pantzir.	Sights with Magnification	
Date of Introduction	By 2004	Gunner	1TPP1 stabilized day/night, dual channel thermal sight
Proliferation	At least 3 countries, with tracked version under export contract	Field of View (°)	1.8 x 2.6
Target	FW, heli, CM, ASM, UAV, guided bomb	Acq Range (km)	18 air targets, 4-6 grd Commander's position IR day/night sight Auto-tracker: Dual Infrared/video tracker
Primary Components	System (battery) has a command post, up to 6 combat vehicles (gun/missile launch vehicles), and 73V6-E transloaders (1 per 2 CVs).	Commander's position IR day/night sight Auto-tracker	Dual Infrared/video tracker
Combat Vehicle Description		IFF	Yes
Crew	3 (cdr, gunner, driver)	Countermeasure Resistance	Passive acquisition modes. Resists IR and most RF SAM CM and suppression systems.
Combat Weight (mt)	20 est	Radars	
Chassis	BMP-3 (and see VARIANTS)	Name	INA, 3D Phased Array
Chassis Length Overall (m)	6.73	Function	Target Acquisition
Height (m)	INA	Detection Range (km)	36-38
Width Overall (m)	3.15	Frequency Band	INA
Automotive Performance	Performance data based on BMP-3	Simultaneous Target Detection	20 Targets

Engine Type	500-hp diesel	Name	1RS2-1E for export version
Cruising Range (km)	600	Function	Fire control and guidance
Speed (km/h)	Max. Road: 65-70 est Max. Swim: 10 est	Tracking Range (km)	24-30
Fording Depths (m)	Amphibious	Scan Sector	90° x 90°
Radio	R-173, R-173P	Frequency Band	Ku and Ka
Protection		Signal Processing	Digital
NBC Protection System	Yes	Guidance Channels	Two simultaneous
ARMAMENT	**SPECIFICATIONS**	C3 Modes	Netted, battery, autonomous
Gun		Target Handling Rate	Up to 2 targets/min Up to 12/min btry
Caliber, Type, Name	30-mm, 30x165 2A38M auto-cannon	Name	Ranzhir ACV or Sborka ACV
Rate of Fire (rd/min)	4,800 (2 twin guns)	Chassis	MTLB-U
Reload Time (min)	15-16 min, gun ammunition and missiles	ACV also links to supported tactical unit nets.	
Elevation (°)	-5 to + 87	Other Radars	Links to Integrated Air Defense System (IADS) for early warning, and data from target acquisition radars, esp. Giraffe AMB or LONG TRACK at Separate Brigade and Division, EW/TA radar echelons above division, and radars in SAM units, e.g., SA-10.
Fire on Move	Yes	**MAIN ARMAMENT AMMUNITION**	**SPECIFICATIONS**
Missile		An optimized mix uses 2 rounds, with each having similar ballistics. The below rounds offer flat trajectory, long range, armor penetration, high P-hit, and frangible round (KE/'CE) effects.	
Name	57E6-E/9M335/SA-22E	Type	Frangible APDS-T
Range (m)	Max. Range: 12,000 below 1,500 m 18,000 above 1,500m Min. Range: 1,500	Range (m)	200-4,000
Altitude (m)	Max. Altitude: 10,000 Min. Altitude: 5, 0 with degraded Ph	Altitude (m)	0-3,000
Dimensions	Length (m): 3.2 in canister Diameter (mm): 170/90 second stage	Type	APFSDS-T, M929
Weight (kg)	65, 85 in container	Range (m)	200-2,500+
Missile Speed (m/s)	1,300	Altitude (m)	0-3,000

Guidance	Radar SACLOS, ACLOS, Home-on-Jam	Penetration (mm CE)	45 (RHA) 2,000 m
Seeker Field of View (°)	INA	Other Ammuntion Types	Earlier 30 x 165 rounds: Frag-HE and HEI-T, API, API-T, APDS
Warhead Type	Fragmenting rod and HE		
Warhead Weight	16		
Fuze Type	Proximity, PD, and KE impact		
Probability of Hit (Ph%)	80 undegraded		
Simultaneous Missiles	3 (1-3 per target)		
Self-Destruct (sec)	INA		
System Reaction time (sec)	5-6		
Fire on Move	Yes, short halt or slow move		
Simultaneous Targets	2 per vehicle		

VARIANTS

PANTSYR-S1: THE GUN/MISSILE SYSTEM MODULE CAN BE MOUNTED ON VARIOUS CHASSIS. THE EARLY VERSION IS MOUNTED ON A URAL-5323 TRUCK, USED FOR SITE DEFENSE OF STATIONARY TARGETS. IT HAD THREE RADARS AND 2A72 GUN. A PRODUCTION VERSION HAS NEWER RADARS, GUNS, AND 12 MISSILE LAUNCHERS.

PANTSIR-S1-0: "UNIFIED ARMAMENT TURRET" WITH 8 LAUNCHERS (12 SAMS) AND 2 GUNS MOUNTS ON VARIOUS CHASSIS (E.G., TRUCKS, BTR-80, BMP-3, BMD-3, TRAILERS, AND STANDS). RUSSIANS NOW OFFER THIS VARIANT ON THE 2S6 CHASSIS. A LOW COST VERSION HAS MISSILES AND ONLY EO GUIDANCE

NOTES

THE GUNS CAN BE USED TO ENGAGE GROUND TARGETS, PRIMARILY FOR SELF-DEFENSE.

French SAM System Crotale 5000 and Chinese FM-90

		Weapons & Ammunition Types	Typical Combat Load
		R440 missile canisters	
		On launchers	8
		Onsite resupply	4
			4+

SYSTEM	SPECIFICATIONS	FIRE CONTROL	SPECIFICATIONS
Alternative Designation	TSE 5000	Sights with Magnification	
Date of Introduction	4000 in 1988	Day Camera	TV tracker, low elevation Range (km): 14.0
Proliferation	At least 9 countries	Optical Sight	back-up binocular tracker
Target	FW, heli, CM, ASM also ARM for FM-90	Day/Night Camera	Thermal sight is on most Crotale 4000, all HQ-7 and FM-90 Field of view (°): 8.1/2.7 Elevation (°): 5.4/1.8 Range (km): 19.0
Description	Battery has 2 platoons (4 TELARs), tech, and resupply vehicles	Missile Tracker	IR, for remote control
TELAR	P4R 4x4	Countermeasures	Digital C2 and ECM
Crew	3 launcher vehicle	IFF	Yes, dipole on ACU (See Notes)
Combat Weight (mt)	15.0	Radar	
Length (m)	6.22	Name	Mirador IV pulse doppler
Height (m)	3.41	Function	Target acquisition, surveillance
Width (m)	2.72	Antenna Rotation Rate (rpm)	60
Engine Type	INA	Detection Range (km)	18.5
Cruising Range (km)	600	Altitude Coverage (m)	0 - 4,500
Max Road Speed (km/h)	70	Target Detection	30 targets per rotation
Fording Depths (m)	0.68	Multiple Target Tracking	12 targets
Radio	INA	Frequency Band	E
Protection		Radar	
Armor Protection (mm)	3-5	Name	INA, on launcher vehicle
NBC Protection System	No	Function	Fire Control
Armament	Specifications	Targets Tracked	1

Launcher		Missile Guidance, Simultaneous	2
Name	Crotale	Detection Range (km)	17
Weight (mt)	INA	Altitude Coverage (m)	0 – 5,000
Set-up Time (min)	5	Frequency (GHz)	12-18
Reaction Time (sec)	6.5	Frequency Band	J, Monopulse
Time Between Launches (sec)	2.5	Associated Radar	I-band (8-10 GHz) cmd
Reload Time (min)	2	Other Assets	The SAM system links to the IADS to get digital AD data and warnings. Associated radar for EW and TA data is radar at Brigade and Division Tier 1 and 2. System can also pass data to the net.
Fire on Move	No		
Missile			
Name	R440		
Range (m)	Max: 10,000, 14,600 heli 15,000 FM-90 17,000 ARM mode FM-90 Min. Range: 500		
Altitude (m)	Max. Altitude: 5,000 Min. Altitude: 15, 7 w/blast radius		
Dimensions (mm)	Length: 2890 Diameter: 150		
Weight (kg)	84, 100 with canister		
Missile Speed (m/s)	750		
Maneuver Capability (Gs)	27		
Propulsion	Solid propellant motor		
Guidance	RF CLOS		
Warhead Type	Focused frag-HE, 15 kg		
Lethal Radius (m)	8, proximity fuze		
Probability of Hit (Ph%)	80 FW, heli		
Simultaneous Missiles	2 per target		

<u>VARIANTS</u>

SYSTEM IS MOUNTED ON VEHICLES, SHELTER, SHIPS CROTALE 1000: INITIAL VERSION 1971 W/CABLE LINK CROTALE 2000: VARIANT WITH TV AND IFF. CROTALE 3000: VARIANT HAS TV AUTO-TRACKER.

CROTALE 4000: HAS RADIO DATA LINK AND THERMAL

CROTALE 5000: ADDS IR AUTO-TRACKER, AND NEW SURVEILLANCE ANTENNA. THE LAUNCHER CAN ADD 2 MISTRAL MISSILES.

CROTALE IMPROVED: AN AIR FORCE UPGRADE HAS PLANAR RADAR, IMPROVED ECCM.

CROTALE NAVAL: FEATURES A DOPPLER-FUZED R440N MISSILE. CROTALE-S SYSTEM FOR SAUDI ARABIA IS A PASSIVE ALL-WEATHER SYSTEM, WHICH CAN BE FITTED TO PREVIOUS NAVAL SYSTEMS.

CACTUS: SAUDI VARIANT FOR SAHV-3 MISSILE.

FM-80/HQ-7: CHINESE IMPROVED VERSION WITH E/F-BAND TA RADAR, EO RANGE OF 15 KM, IR LOCALIZER AND HQ-7 MISSILE RANGE OF 12 KM.

SHAHAB THAQUEB: IRANIAN FM-80 VARIANT WITH THE 45KM SKYGUARD RADAR (25 TRACKING) /CP UNIT. RANGE IS 12 KM. ECCM DEFEATS ALL CM.

FM-90: CHINESE 1998 FIELDED AND EXPORTED UPGRADE WITH: NEW DIGITAL C2, THERMAL SIGHT, DUAL BAND TA TRACKING RADAR (RANGE 25 KM). A NEW FASTER MISSILE HAS A RANGE OF 15 KM IN EO/ RADAR MODES, A NEW FUZE SYSTEM, AND 17 KM RANGE IN ANTI-RADIATION MISSILE MODE. MAX ALTITUDE IS 6 KM. DIGITAL ECCM HAS NEAR JAM-PROOF FCS. LAUNCHER CAN ENGAGE THREE SIMULTANEOUS TARGETS IADS LINK CAN FEED REMOTE FC RADAR GUIDANCE.

SHAHINE: UPGRADE HAS R460 15-KM MISSILE ON AMX-30 TANK CHASSIS. SHAHINE 2 FEATURES RADAR RANGE TO 19.5 , M3.5 VELOCITY, AND 5-M MINIMUM ALTITUDE (SLOW MOVERS). THE RADAR CAN RACK 40 TARGETS AND ASSIGN 12 PER BATTERY.

NOTES

THE ALL-WEATHER SYSTEM IS DEPLOYED IN PLATOONS. A PLATOON INCLUDES AN ACQUISITION AND COORDINATION UNIT (ACU) VEHICLE AND 2-3 "FIRING UNITS" (LAUNCHER VEHICLES). A BATTERY INCLUDES TWO PLATOONS. BATTERY RELOADS ARE DELIVERED ON TRUCKS. AN ACU USES THE SAME P4R CHASSIS AND A SURVEILLANCE RADAR, IFF INTERROGATOR, BATTLE MANAGEMENT COMPUTER, DIGITAL RF DATA LINK, AND VHF RADIOS. WITH RF DATA LINK, INTERVAL CAN BE UP TO 10 KM BETWEEN ACUS, AND UP TO 3 KM BETWEEN ACU AND LAUNCHER VEHICLES. OFF-CHASSIS REMOTE CONTROL SYSTEM CAN BE USED TO GUIDE THE MISSILE.

European SAM System Crotale-New Generation

XA-181 SAM Launcher Vehicle

Weapons & Ammunition Types	Typical Combat Load
VT-1 missile canisters	8

SYSTEM	SPECIFICATIONS	FIRE CONTROL	SPECIFICATIONS
Alternative Designations	Crotale-NG, XA-181 (Finnish Launcher vehicle) This is not a modification to Crotale. It is a completely new modular system.	Sights with Magnification	
Date of Introduction	1991-1992	Day Camera	Mascot, CCD TV Field of view (°): 2.4 Elevation (°): 1.8 Range (km): 15
Proliferation	At least 5 countries, all variants	Night Camera	Night Camera: Castor, thermal Field of view (°): 8.1/2.7 Elevation (°): 5.4/1.8 Range (km): 19
Target	FW, heli, CM, ASM, UAV	Missile Tracker	IR missile localizer on CCD camera for passive TV tracking
Description		IFF	Yes
TELAR	XA-181 is XA-180 (PASI) 6x6 APC with Crotale NG launcher system	Radar	
Crew	4	Name	TRS 2630 Griffon
Combat Weight (mt)	23.0 launch-ready	Function	Target acquisition
Length (m)	7.35	Antenna	Planar array

			Detection Range (km): Aircraft: 20
Height (m)	2.3 for vehicle hull +2-3 m	Hovering Rotary Wing Aircraft	11
Width (m)	2.9	Altitude Coverage (m)	0-5000
Automotive Performance		Multiple Target Tracking	Automatic track-while-scan for up to 8 targets.
Engine Type	240-hp diesel	Frequency Band	S
Cruising Range (km)	800	ECCM	Low sidelobes, wide-band frequency agility, search on the move capability
Max Road Speed (km/h)	80	Radar	
Swim Capability	No	Name	
Radio	INA	Function	Fire Control, tracking
Protection		Detection Range (km)	30
Armor Protection (mm)	6-12mm	Frequency (GHz)	35 doppler TWT (travelling wave tube)
NBC Protection System	No	Frequency Band	Ku
ARMAMENT	**SPECIFICATIONS**	ECCM	Wideband frequency agile
Launcher	TELAR	Other Assets	The SAM system links to the IADS to get digital AD data and warnings. Associated radar for EW and TA data is radar at Brigade and Division Tier 1 and 2. System can also pass data to the net.
Name	VL-VT-1		
Weight (mt)	4.8		
Reaction Time (sec)	<6		
Time Between Launches (sec)	1-2		
Reload Time (min)	10		
Fire on Move	No		
Missile			
Name	VT-1		
Range (m)	Max. Range: 11,000 Min. Range: 500		
Altitude (m)	Max. Altitude: 6,000 Min. Altitude: 5 0 with degraded Ph		
Dimensions (mm)	Length: 2300 Diameter: 170		

Weight (kg)	75		
Missile Speed (m/s)	1.250		
Maneuver Capability (Gs)	35		
Propulsion	Solid propellant motor		
Guidance	RF CLOS		
Warhead Type	Focused frag-HE, 14 kg		
Lethal Radius (m)	8		
Fuze Type	Proximity		
Probability of Hit (Ph%):	80 FW, heli		
Simultaneous missiles	2 per target		

VARIANTS

SYSTEM IS IN A MODULAR POD, DESIGNED TO FIT ON SHIPS, VEHICLES, AND ON STATIONARY PLATFORMS. THE MODULAR ALL-WEATHER SYSTEM INCLUDES ACQUISITION, TRACKING, LAUNCH, AND SUPPORTING COMPUTER UNITS INTEGRATED ON ONE VEHICLE, FOR MANAGEMENT BY A SINGLE SYSTEM OPERATOR.

VEHICLE PLATFORMS INCLUDE APCS, E.G., M113, KOREAN IFV, PIRANHA 10X10, AND THE XA-180 AS NOTED.

THE SYSTEM CAN BE RETROFITTED ONTO EXISTING CROTALE LAUNCHER VEHICLES.

PEGASUS: SOUTH KOREAN SYSTEM WITH A

DIFFERENT MISSILE

NOTES

RUSSIAN FAKEL VL-VT-1 LAUNCHER GIVES THE VT-1 HYPERVELOCITY MISSILE (HVM) VERTICAL 40-M RISE BEFORE PITCH-OVER TO TARGET. IT PERMITS 360° LAUNCH WITHOUT NEED TO RE-ORIENT THE VEHICLE, AND A SHORTER REACTION TIME.

Russian SAM System SA-8b/GECKO Mod 1 and SA-8P/Sting

		Weapons & Ammunition Types	Typical Combat Load
		SA-8b in canisters	6
SYSTEM	**SPECIFICATIONS**	Guidance	RF CLOS
Alternative Designation	Osa-AKM Osa-AKM-P1for Polish upgrade	Warhead Type	Frag-HE
Date of Introduction	1973, 1980 for AKM	Warhead Weight (kg)	16
Proliferation	At least 25 countries	Fuze Type	Contact and Proximity
Target	FW, heli, CM, ASM, UAV, bomb	Probability of Hit (Ph%):	80 FW, 65 heli 65 against heli w/EO
Description	Battery includes 4 TELARS, 2 TZM transporter -loaders, PU-12M battery CP, 9V914 survey vehicle, maintenance vehicle, 9V242-1 test station, and ground set	Simultaneous missiles	2 per Target
Launcher Vehicle		Self-Destruct (sec)	25-28
Name	9A33BM3 for updated version	Performance	
Description	TELAR	With Radar	Note: Primary mode with higher probabilities of hit and kill for targets above 25 m. Aircraft can be sighted to max altitude Range (m): 1,500-10,000 Altitude (m): 25-5,000 Preferred (passive) mode for use vs low flyers and ECM. Range (m): 2,000-6,500 Altitude (m): 10-5,000 FW 0-5,000 helicopters
Chassis	BAZ-5937 6x6 vehicle	With EO Sight	Preferred (passive) mode for use vs low flyers and ECM. Range (m): 2,000-6,500 Altitude (m): 10-5,000 FW

			0-5,000 helicopters
Crew	3		
Combat Weight (mt)	9		
Length (m)	9.14	FIRE CONTROL	SPECIFICATIONS
Height (m)	4.2 TA radar folded down	Sights with Magnification	Secondary mode. Electro-optical LLLTV with EO IR assist, for low flyers and target tracking in low visibility, heavy ECM environment EO system day/night range (km): 6
Width (m)	2.75	Onboard Radar System	
Automotive Performance		Name	LAND ROLL
Engine Type	D20K300 diesel	Function	Dual (TA and FC)
Cruising Range (km)	250	Can System Operate Autonomously	Yes
Speed (km/h)	60 max road 30 off-road 10 cross-country	Radar Antenna	
Max Swim	8	Function	Search (target acquisition)
Radio	R-123M	Detection Range (km)	45 in -AKM
Protection		Tracking Range (km)	20-25
Armor (mm)	None	Frequency	6-8 GHz
NBC Protection System	Yes	Frequency Band	H
ARMAMENT	SPECIFICATIONS	Radar Antenna	
Launcher		Function	Fire control (monopulse TTR)
Name	9P35M2	Detection Range (km)	20-25
Dimensions	Length (m): 3.2 Diameter (mm): INA	Tracking Range (km)	20-25
Weight (mt)	35	Frequency	14.2-14.8 GHz
Reaction Time (sec)	18-36	Frequency Band	J
Time Between Launches (sec)	4	Radar Antenna	
Reload Time (min)	No	Function	Fire control (missile guidance)
Fire on Move	No	Frequency Band	I
Emplacement Time (min)	4 or less	Counter-countermeasures	2-channel FH agile
Displacement Time (min)	<4 (est.)	Other Radars	Associated radar for EW and TA data is Giraffe AMB at Separate Brigade and Division Tier 1 and 2, or LONG TRACK at Tier 3 and 4. The SA-8b can also link to the IADS to get analog AD data

			from: Sborka AD battery ACV, radars in echelon above division SAM units (e.g., SA-10).
Missile			
Name	9M33M3 latest fielded		
Dimensions (mm)	Length: 3158 Diameter: 209.6		
Weight (kg)	170		
Missile Speed (m/s)	1,020		
Propulsion	Solid propellant motor		

VARIANTS

SA-8A: INITIAL PRODUCTION MODEL THAT CARRIED FOUR MISSILES ON EXPOSED RAILS.

OSA-1T, SA-8B MOD 1: BELORUSSIAN SYSTEM ON MZKT-69222 CHASSIS, WITH A VARIETY OF UPGRADES (E.G., NIGHT SIGHTS, INTEGRATED DIGITAL

C3 AND IMPROVED MISSILES) ARE AVAILABLE. RANGE IS 1.5-14 KM. WARHEAD LETHALITY IS INCREASED 25%. ALTITUDE IS 100-5,000 M.

T-38/STILET, WITH OSA-1T MISSILE, RANGE OF 12 KM, ALTITUDE 8,000 M. P-HIT /KILL IS 85%.

SA-8P/OSA-AKM-P1/STING: POLISH UPGRADE WITH SIC 12/TA FCS (TV DAY SIGHT, 3RD GEN FLIR SIGHT, IR AUTO-TRACKER, AND LRF. PASSIVE EO RANGE IS 40 KM. REGA-2 AUTOMATED C2 HAS INERTIAL AND GPS NAV. DIGITAL SYSTEM LINKS TO MODERN IADS NETS. DAY/NIGHT RANGE WITH THE OSA-1T MISSILE IS 12,000 M, ALTITUDE 0-8,000 M. THE FIRST SCHEDULED EXPORT CUSTOMER IS INDIA.FUTURE GOAL IS TO ADD FIRE-AND-FORGET MISSILES.

NOTES

THIS IS ONE OF THE LONGEST-RANGE FIELDED AMPHIBIOUS SYSTEMS IN THE WORLD. THIS SYSTEM IS ALSO AIR-TRANSPORTABLE AND CROSS-COUNTRY CAPABLE. ONE TRANSLOADER VEHICLE (CARRYING 18 MISSILES BOXED IN SETS OF THREE) SUPPORTS TWO TELARS.

Russian SAM System SA-9/GASKIN

Weapons & Ammunition Types	Typical Combat Load
9M31M missiles	6
Ready	4
With Add-on racks	+2

SYSTEM	SPECIFICATIONS	Missile	
Alternative Designation	Strela-1M	Name	9M31
Date of Introduction	1968	Range (m)	Max. Range: 4,200 (6,100 tail aspect) Min. Range: 800
Proliferation	At Least 30 Countries	Altitude (m)	Max. Altitude: 3,500 Min. Altitude: 30 0 with degraded Ph
Target	FW, heli	Dimensions	Length: 1.80 Diameter: 120
Description	An SA-9 platoon complex (9K31) includes four 9A31M TELs. One SA-9a TEL (aka BRDM-2A1) mounts a passive RF direction-finder system (see FIRE CONTROL). Three SA-9b TELs (BRDM-2A2) do not. Platoon ACV is the PU-12M or PPRU CP vehicle. The complex includes resupply vehicles.	Weight (kg)	32
Launcher Vehicle		Missile Speed (m/s)	580
Name	9A31M	Propulsion	Single-stage solid propellant
Description	Transporter-Erector-Launcher	Guidance	Photo contrast IR-homing, 1-3µm
Crew	3	Warhead Type	Frag-HE
Chassis	BRDM-2	Warhead Weight (kg)	2.6
Combat Weight (mt)	7.0	Fuze Type	Proximity and Contact
Length (m)	Launch position: 5.8 Travel position: 5.8	Probability of Hit (Ph%):	60 FW, 70 heli
Height (m)	TEL up: 3.8 TEL down: 2.3	Simultaneous missiles	2 per Target
Width (m)	2.4	Self-Destruct (sec)	Yes

Automotive Performance		Auxiliary Weapon	None
Engine Type	V-8 gasoline	**FIRE CONTROL**	**SPECIFICATIONS**
Cruising Range (km)	750	Sights with Magnification	
Speed (km/h)	Max. Road: 100.0 Max Swim: 10	Elecro-optical/ Infrared System	Day Range (m): 6,500 Night Range (m): 2,000 tail chase only
Radio	INA	Navigation	Inertial
Protection		IFF	INA
Armor (mm)	14 Front	RF Direction-Finder	The FLAT BOX-A passive system uses several Pelengator sensors mounted on the vehicle to detect aircraft navigation signals for early warning and DF of approach azimuth. Detection range is up to 30 km. Many forces with this older air defense system are not proficient in using the RF DF system.
NBC Protection System	Collective	**ASSOCIATED VEHICLES/ RADARS**	**SPECIFICATIONS**
ARMAMENTS	**SPECIFICATIONS**	Name	PPRU-1/Ovod AD ACV
Launcher		Chassis	MTLB-U
Name	9P31	Radar	DOG EAR (use in OPFOR units)
Reaction Time (sec)	6	Function	Target Acquisition
Time Between Launches (sec)	5	Frequency	F/G band
Reload Time (min)	5	Range	80 detection, 35 tracking
Fire on Move	No, stop or short halts	Other Radars	The SA-9 can also link to the IADS to get analog AD data and warnings.
Emplacement Time (min)	<2.0	Radar: Gundish	In the earlier unit configuration, an SA-9 platoon is employed in an AD battery/ battalion with ZSU-23-4 SPAA guns. The radar on those systems supports the SA-9 platoon by providing detection and warning. Some of the users employ truck-mounted J-band GUN DISH acquisition radar in the platoons, instead of the Pelengator system.
Displacement Time (min)	<2.0		

VARIANTS

UPGRADE 9M31M MISSILE HAS A 1-5 μM SEEKER WITH IMPROVED RANGE (8 KM ALL ASPECT, 11 KM AGAINST SLOW MOVERS AND TAIL CHASE). ALTITUDE INCREASES TO 6,100M. NIGHT RANGE IS 4,000+ M. THE IMPROVED AND COOLED

SEEKER MAKES THIS MISSILE FAIRLY RESISTANT TO IR COUNTERMEASURES. SYSTEM WITH THIS MISSILE IS CALLED GASKIN MOD 1.

TARGET: FW, HELI, CM, UAV

NOTES

GENERALLY, THE SYSTEM WOULD BE EXPECTED TO HAVE THE FLAT BOX-A BUT NOT THE GUN DISH RADAR IN THE PLATOON. THE INSENSITIVE MISSILE SEEKER WAS DIFFICULT TO LOCK ON TARGET AND WAS FAIRLY EASILY COUNTERMEASURED FROM ANY ASPECT EXCEPT THE TAIL ASPECT.

SYSTEM CAN USE THE SBORKA PPRU-M1 UPGRADE ACV. HOWEVER, THE ABOVE SYSTEM MATCHES THE LOWER TIER TECHNOLOGY AND EARLIER FIELDING OF SA-9.

Russian SAM System SA-13b/GOPHER

Weapons & Ammunition Types	Typical Combat Load
9M333 missiles	8
Ready	4
Reload	4
7.62-mm MG RPK	2,000

SYSTEM	SPECIFICATIONS	Propulsion	Single-stage solid propellant
Alternative Designation	Strela-10M3, 9K35M3	Guidance	Photo-contrast or dual-band IR-H
Date of Introduction	1981	Warhead Type	HE with fragmenting rod
Proliferation	At Least 22 Countries	Warhead Weight (kg)	5 (4 m lethal radius)
Target	FW, heli, CM, selected UAV	Fuze Type	Laser proximity (3 m), contact
Description	Battery has 6 TELARs, Sborka ACV (CP/radar vehicle), and truck.	Probability of Hit (Ph%):	60 FW, 70 heli
Launcher Vehicle		Simultaneous missiles	2 per target
Name	9A34M3/ 9A35M3 (see NOTES)	Self-Destruct (sec)	29
Description	TELAR/Platoon Cmd TELAR	Countermeasure Resistance	System resists nearly all IR countermeasures.
Crew	3	Auxiliary Weapons	
Chassis	MT-LB	Caliber, Type, Name	7.62-mm MG, RPK
Combat Weight (mt)	12.3	Rate of Fire (rds/min)	600/150 practical, bursts
Length (m)	Launch position: 6.45 Travel position: >6.45	Loader Type	40/75-rd magazine
Height (m)	TAR up: 3.8 TAR down: 2.22	Ready/Stowed Rounds	1000/1000
Width (m)	2.85	Fire on Move	Yes
Automotive Performance		**FIRE CONTROL**	**SPECIFICATIONS**
Engine Type	290-hp diesel	Sights with Magnification	
Cruising Range (km)	500	Electro-optical/IR system with auto-slew,	Range (km): 10 helicopter, 5 FW

		electro-mechanical aiming, and auto-tracker	Night Sight: passive IR, Strizh TV/thermal, video display Range (m): 6,000 IR, 12,000 thermal
Speed (km/h)	Max. Road: 61.5 Max Swim: 6	IFF	1RL246-10-2/PIE RACK (RF)
Radio	INA	Onboard Radar	
Protection		Name	9S86/SNAP SHOT on 9A34M3
Armor (mm)	7.62-mm anti-bullet	Function	Range Only
NBC Protection System	Yes	Detection Range (km)	10
ARMAMENTS	**SPECIFICATIONS**	Frequency Band	K-Band
Launcher		Other Onboard Sensors	9S16/FLAT BOX -B passive radio DF system. Range is 30 km.
Name		**Associated vehicles/ radars**	**Specifications**
Reaction Time (sec)	7-10	Name	Sborka AD ACV
Time Between Launches (sec)	<5	Chassis	MTLB-U
Reload Time (min)	3	Radar	DOG EAR
Fire on Move	No, stop or short halts	Function	Target Acquisition
Launcher Elevation (°)	-5/+80	Frequency	F/G band
Emplacement Time (min)	0.67	Range	80 detection, 35 tracking
Displacement Time (min)	<1.0	Previous Battery	PU-12M
Auxiliary Power Unit	Yes, gasoline power		
Note	The SA-13 can launch SA-9 SAMs, and can mix the SAMs.		
Missile			
Name	9M333/Strela-10M3		
Range (m)	Max. Range: 5,000, fly-out to 7,000+ m Min. Range: 800		
Altitude (m)	Max. Altitude: 3,500 Min. Altitude: 10, 0 with degraded Ph		
Dimensions	Length: 2,223 Diameter: 120		
Weight (kg)	42		
Missile Speed (m/s)	Up to 800/517 average		
Max Target Speed (m/s)	420		

VARIANTS

SA-13A: EARLIER SYSTEM WITH SA-9 MISSILE - 7 KM RANGE, BUT LOWER OVERALL LETHALITY.

MISSILE VARIANTS: STRELA-10M HAS UNCOOLED LEAD SULPHIDE (PBS) IR SEEKER. STRELA-10M2 HAS UNCOOLED PBS SEEKER OR COOLED INDIUM ANTIMONIDE MID-IR SINGLE-MODE SEEKER. STRELA-10M3 DETECTION RANGE 10 KM DAY/NIGHT, ENGAGE UAVS TO 4,000M.

CZECH SNAP SHOT RADAR: VERSION WITH HEIGHT ADJUSTMENT CAPABILITY, AND IMPROVED AUTOMATION AND COMMUNICATIONS

SAVA: YUGOSLAV VARIANT OF STRELA-10M/ SA-13A ON A BVP M80A IFV CHASSIS.

STRIJELA-10CROAL: CROATIAN VARIANT WITH A TAM 150.B 6X6 VEHICLE CHASSIS, TV-BASED FIRE CONTROL AND THERMAL NIGHT SIGHT.

9A34A: UPGRADE TELAR WITH THERMAL SIGHT, BETTER INTEGRATED C^2, IMPROVED FCS, AND A PKM MACHINEGUN. DETECTION RANGE WITH THE FCS IS 10-12 KM.

MUROMTEPLOVOZ OFFERS A LAUNCHER VEHICLE WITH THE LAUNCHER ON A BTR-60 CHASSIS.

NOTES

THE SA-13A REPLACED SA-9 WITH AN UPDATED LAUNCHER MOUNTED ON A DIFFERENT CHASSIS. THE MT-LB HULL OFFERS HALF THE PROTECTION OF THE SA-9 BRDM-2 CHASSIS, BUT WITH MORE MOBILITY. THE BATTERY SET USES CENTRALIZED DIGITAL TARGET WARNING NET; BUT EACH LAUNCHER MUST INDIVIDUALLY ACQUIRE AND LAUNCH AGAINST TARGETS. ASSOCIATED EQUIPMENT INCLUDES A 9V915M MAINTENANCE VEHICLE, 9I11 EXTERNAL POWER SUPPLY SYSTEM, AND A 9V839M TEST VEHICLE. THE PLATOON CMD LAUNCHER (9A35M/TELAR-1) HAS A FLAT BOX -B, AND CAN PASS DATA TO THE OTHER LAUNCHERS (9A34M/TELAR-2.

Russian SAM System SA-15b/GAUNTLET

	Weapons & Ammunition Types	Typical Combat Load
	Ready missiles	8

SYSTEM	SPECIFICATIONS	FIRE CONTROL	SPECIFICATIONS
Alternative Designation	9K331 Tor-M1	Sights with Magnifications	
Date of Introduction	1990	Electro-optical (EO) television system with IR auto-tracker Range (km)	20
Proliferation	At Least 5 Countries	IFF	Yes
Target	FW, heli, CM, ASM, UAV, bomb	Radar	
Description	Battery system includes 4 TELARs a CP vehicle, transloaders, and maintenance vehicles	Name	SCRUM HALF
Launcher Vehicle		Function	Target acquisition (TAR)
Name	9A331	Detection Range (km)	25+
Description	TELAR	Tracking Range (km)	25
Crew	3	Targets Tracked	10
Chassis	GM-355 tracked vehicle	Frequency Band	G/H-band 3D doppler, Stabilized for use on move
Combat Weight (mt)	34	Target Detection Time (sec)	1.5-3.0
Length (m)	7.5	Radar	
Height (m)	5.1 (TAR up)	Name	INA, sometimes called "Tor" Also SCRUM HALF, some sources
Width (m)	3.3	Function	Dual - acquisition and fire control (includes tracking and guidance)
Automotive Performance		Detection Range (km)	25+

Engine Type	V-12 Diesel	Tracking Range (km)	25, farther with slower reaction time
Cruising Range (km)	500	Targets Engaged Simultaneously	2
Speed (km/h)	Max. Road: 65	Frequency Band	J/K-band Doppler phased array
Radio	INA	**ASSOCIATED VEHICLES/ RADARS**	**SPECIFICATIONS**
Protection		Name	Sborka AD ACV
Armor (mm)	Small Arms (est)	Chassis	MTLB-U (same as Ranzhir)
NBC Protection System	Yes	Radar	DOG EAR
Armaments		Function	Target Acquisition
Launcher		Frequency	F/G band
Name	INA, Vertical Launch	Range	80 detection, 35 tracking
Dimensions	Length (m): INA Diameter (mm): INA		
Weight (kg)	INA	Name	Ranzhir/Rangir/9S737 AD ACV
Reaction Time (sec)	3-8, +2 halt from move	Chassis	MTLB-U
Time Between Launches (sec)	see NOTES	Radar	None, via radar reports from SA-15b
Reload Time (min)	10	Other Assets	Associated radar for EW and TA data is Giraffe AMB at Separate Brigade and Division Tier 1 and 2. It links to the IADS to get digital AD data from: Sborka AD battery ACV, radars in echelon above division SAM units (e.g., SA-10). The SA-15b can also pass data to the net.
Fire on Move	Yes		
Emplacement Time (min)	5		
Displacement Time (min)	Less than 5		
Missile			
Name	9M331		
Range (m)	Max. Range: 12,000 Min. Range: 1,000		
Altitude (m)	Max. Altitude: 6,000 Min. Altitude: 10 0 with degraded Ph		
Dimensions	Length: 2,900 Diameter: 235		
Weight (kg)	167		
Missile Speed (m/s)	850		
Propulsion	INA		

Guidance	Command		
Warhead Type	Frag-HE		
Warhead Weight (kg)	15		
Fuze Type	RF Proximity		
Probability of Hit (Ph%):	90 FW, 80 heli		
Simultaneous missiles	2 per target		

VARIANTS

SA-N-9: NAVAL VERSION

TOR-M1T: VERSIONS ON THE GROUND OR TOWED TRAILERS. THE CREW SITS 50 M AWAY FROM THE ANTENNA/LAUNCHER TRAILER. THE -M1TA HAS A BOX-BODY (BB) CREW TRUCK. THE –M1TB HAS A BB TRAILER. A GROUND-MOUNT VERSION IS TOR-M1TS. ONLY DIFFERENCES ARE EMPLACE/ DISPLACE TIMES, AND 0 VERSUS 1, OR 2 TRUCKS.

TOR-M2: VERSION WITH LAUNCHER ON ARMORED KAMAZ 6X6 TACTICAL TRUCK CHASSIS. TOR-M2E EXPORT VERSION HAS A NEW JAM-RESISTANT TA

RADAR. MAX ENGAGEMENT ALTITUDE IS 10,000 M.

NOTES

SA-15B IS DESIGNED TO BE A COMPLETELY AUTONOMOUS AIR DEFENSE SYSTEM (AT DIVISION LEVEL), CAPABLE OF SURVEILLANCE, COMMAND AND CONTROL, MISSILE LAUNCH AND GUIDANCE FUNCTIONS FROM A SINGLE VEHICLE. THE BASIC COMBAT FORMATION IS THE FIRING BATTERY CONSISTING OF FOUR TLARS AND THE RANGIR BATTERY COMMAND POST. THE TLAR CARRIES EIGHT READY MISSILES STORED IN TWO CONTAINERS HOLDING FOUR MISSILES EACH. THE SA-15B HAS THE CAPABILITY TO AUTOMATICALLY TRACK AND DESTROY 2 TARGETS SIMULTANEOUSLY IN ANY WEATHER AND AT ANY TIME OF THE DAY.

Recent Developments in Medium-Range Air Defense (MRAD) Systems

In the past, the US and Russia dominated military markets in medium-range SAM systems. Most well-fielded MRAD systems are Russian systems, or license-produced copies or variants of those systems. Most still have some effectiveness for AD, especially with upgrade programs. But new systems and new producers are expanding options for their MRAD choices.

For military forces in most countries, with substantial portions of their territory lacking strategic targets or vulnerabilities, MRAD SAMs (aka: MSAMs) are more practical AD systems than the more expensive and restricted mobility long-range SAM systems. Requirements for these systems include ranges from <1 km to 20-50 km, and altitudes of 5 m to 6-50 km. Many MRAD SAMs operate within these range limits, which are less than LRAD SAMs, but offer high-altitude protection against flight profiles of most fixed-wing aircraft and many missiles.

The most proliferated MRAD SAMs are former Warsaw Pact, e.g., SA-2, SA-3, SA-3b, SA-6/SA-6b, SA-11, Buk-M1-2, or US HAWK and I-HAWK. These include towed semi-mobile and vehicle-mounted mobile systems. Most legacy systems have seen many upgrades. In recent years the pace of upgrades increased with availability of digital data systems, computer integration, imaging fire control systems, and radar improvements. Improved supporting target acquisition and fire control radars are adding improvements in overall systems capabilities. Several towed systems are now mounted on vehicle chassis. Missile improvements include missile motor/range upgrades, new warhead designs, and improved missile guidance modes. Many MRAD systems are upgraded to meet recent AD challenges (e.g., stealth, SEAD, cruise missiles, low-flying helicopters, air-launched munitions, UAVs, and ballistic missiles).

The widely fielded Russian SA-6/Kub system has seen many upgrades, including improved missiles (Kub-M1 and Kub-M3), and unit changes. In the Soviet era, it was being obsolesced by SA-11/Buk systems. Most SA-6 units were upgraded and converted to SA-6b (pg 6-76). SA-11 units also saw upgrades. Dissolution of the USSR left Russia with fewer modern units and many older Kub units. New upgrade packages were fielded, and offered to export customers. Meanwhile, delays and cost issues with the forecasted SA-17 led them to a deep modernization to Buks with Buk-M1-2 (see pg 6-78). An economical and clever change was to add launcher-loaders to batteries, to supplement TELAR missile loads and increase launch rates. New Russian MRAD designs, such as Vityaz, are in development. The trend for increased missile loads on Russians LRADs will further delay domestic fielding of MRAD systems. The SA-17 is not well fielded; but Russia is upgrading and exporting AD, and modernization options.

Other countries have entered the development arena for indigenous MRAD systems. A number of air-to-air missiles have been adapted for ground mounts as medium-range SAMs. Others are indigenous developments, which offer export capabilities and flexible adaptation to meet specific customer needs. See some of the many variant examples with the Aspide 2000 missile (next page). Other systems have been developed by Sweden (RBS 23/BAMSE) Israel (Spyder-MR), and South Korea. Israel is also developing Arrow as an anti-theater ballistic missile (ATBM) system. India and several other countries have foreign system acquisition/upgrade programs, as well as indigenous development programs underway. European countries (SAMP-T), Norway (NASAMS, with the AMRAAM missile), and Turkey are currently in MRADS development programs. China is offering its KS-1A system. European firms are adapting the IRIS-T AAM for ground launchers.

Italian Aspide 2000 Medium-Range SAM System (in Skyguard Battery)

Aspide 4-canister configuration

Weapons & Ammunition Types	Typical Combat Load
Launch canisters	4/6 (depending on configuration)
Total missiles	12

SYSTEM	SPECIFICATIONS	FIRE CONTROL	SPECIFICATIONS
Alternative Designation	Missile formerly called Aspide Mk II. System is also called Skyguard gun/missile air defense system	Onboard Fire Control	Remote controlled K-band tracking radar and RC illuminator radars, I/J-band on launcher.
Date of Introduction	1986 for Mk I	Off-carriage	
Proliferation	At Least 18 Countries	Name	Skyguard II radar and CP unit
Target	FW, heli, CM, UAV, ASM, bombs	Platform	Towed compartment
Launcher		EO Sights	SEC-Vidicon TV system
Name	INA	EO Auto-Tracker	TV tracking system
Description	Towed 4/6 canister MEL	Range	25 km day only
Reaction Time (sec)	11	Laser Rangefinder	Yes
Time Between Launches (sec)	INA	Radars	
Fire on Move	No	Name	Skyguard Mk II (SW)
Number of Fire Channels	2	Function	Dual (TA and FC)
Emplacement Time (min)	15	Detection Range (km)	45
ARMAMENTS	**SPECIFICATIONS**	Tracking Range (km)	25
Missile		FC Radar Frequency	-20 GHz
Name	Aspide 2000 (aka: Aspide Mk II)	Frequency Band	I/J doppler MTI
Range (km)	Max. Range: 45 Min. Range: 0.75	Rotation Rate/min	60
Altitude (m)	Max. Altitude: 6,000+ Min. Altitude: 10 0 with degraded Ph	Mean Power (W)	200

Dimensions	Length (m): 3.65 Diameter (mm): 203	Link	Digital data invulnerable to ECM, including frequency jumps
Weight (kg)	230	Other Assets	Skyguard links to the IADS to get digital AD warnings and. Data. Associated radar for EW and TA data is radar at Bde and Div Tier 1 and 2. System can also pass data to the net.
Missile Speed (m/s)	1,288		
Velocity (mach)	4.0		
Maneuver Capability (Gs)	35-40		
Propulsion	Solid fuel booster		
Guidance	J-band semi-active radar homing, active or passive homing, and home-on-jam		
Warhead Type	Frag-HE		
Warhead Weight (kg)	33		
Fuze Type	Proximity and Contact		
Probability of Hit (Ph%):	80 FW and heli		
Simultaneous missiles	2 per Target		

VARIANTS

SKYGUARD ADA COMPLEXES CAN VARY WIDELY. ORGANIZED AROUND THE SKYGUARD RADAR AND CP UNIT, THEY MAY HAVE GUNS ONLY OR MISSILES ONLY. THE MOST EFFECTIVE CONFIGURATION IS A GUN/MISSILE SYSTEM.

ASPIDE 2000: THE SYSTEM CAN BE MISSILES ONLY OR GUN/MISSILE, WITH THE SKYGUARD II RADAR, TRUCKS AND GENERATORS.

SKYGUARD II/ASPIDE 2000: OPFOR TIER 2 GUN/MISSILE SYSTEM, WITH RADAR/CP AND MISSILE AND GDF-005 GUN. IT ALSO LINKS TO GIRAFFE OR OTHER RADARS. A BATTERY HAS 2 GUNS AND 2 MISSILES.

SKYGUARD III: GDF-005 GUN, SKYGUARD III I-BAND RADAR AND SKYGUARD RETROFIT KIT.

SKYGUARD III/ASPIDE 2000: OPFOR TIER 1 GUN/MISSILE SYSTEM WITH ABOVE CHANGES..

SKYGUARD IS COMPATIBLE WITH OTHER DIGITAL ADA FCS FORMATS. GDF-003 GUN AND ALLENIA ASPIDE MISSILE ARE ALSO EMPLOYED WITH RADAR AND CP UNITS OTHER THAN SKYGUARD.

SKYGUARD MK I RADAR RANGE WAS 20 KM.

SKYGUARD RETROFIT KIT: GUN UPGRADE FCS, RADAR, AND FITTED FOR AHEAD AMMUNITION.

OTHER GUNS AND MISSILES CAN BE USED WITH THE SKYGUARD RADAR AND CP UNIT.

AMOUN: EGYPTIAN ASPIDE/SPARROW SYSTEM

ARAMIS: BRIGADE SAM SYSTEM WITH 6-CANISTER LAUNCHER.

LY-60: CHINESE NAVAL VARIANT

PL-11: CHINESE VARIANT WITH UPGRADES. RANGE FOR PL-11C IS 75 KM.

SPADA: ITALIAN AIR FORCE LAUNCHER VERSION.

SPADA 2000: KUWAITI SYSTEM USES ABOVE

LAUNCHER AND ASPIDE 2000 MISSILE.

SPARROW: SYSTEM FROM WHICH ASPIDE WAS DERIVED - INTERCHANGEABLE IN THE LAUNCHER.

OTHER COMPATIBLE MISSILES INCLUDE: ADATS, ASRAD, AIM-7E/SPARROW, SAHV-IR, AND LY-60.

NOTES

GPS IS USED FOR SURVEYING SYSTEMS IN POSITION. SKYGUARD CONNECTION LINK IS 1,000-M CABLE LINK OR 5000-M RADIO LINK.

TO COUNTER SEAD JAMMING OPERATIONS, THE FIRE CONTROL SYSTEM TRACKER IS K-BAND. THE ASPIDE MISSILE SEEKER CAN USE HOME-ON-JAM MODE. SKYGUARD FIRE CONTROL SYSTEM INTEGRATES ACQUISITION RADAR WITH REMOTE CONTROLLED ILLUMINATION (GUIDANCE) RADARS.

Russian SAM System SA-2/GUIDELINE Russian SAM System

		Weapons & Ammunition Types	Typical Combat Load
		Single rail ground mounted	1 Six launchers per battery
SYSTEM	**SPECIFICATIONS**	**FIRE CONTROL**	**SPECIFICATIONS**
Alternative Designation	Volga-75SM, S-75 Dvina, V-75 Volkhov	Radar	
Date of Introduction	1959	Name	FAN SONG, A-F variants
Proliferation	At Least 41 Countries	Function	Fire control
Target	FW, heli, CM	Control Range (km)	60-120 A, B 70-145 for C, D, E INA for F
ARMAMENT	**SPECIFICATION**	Frequency Band	E/F for A-B, G for C-E, INA for F
Launcher		Location	Within battery formation
Description	Single-rail, ground- mounted, not mobile but transportable	Radar:	
Name	INA	Name	SPOON REST, P-12
Dimensions	INA	Function	Target acquisition, early warning
Weight (kg)	INA	Detection Range (km)	275
Reaction Time (sec)	8 lock-on 2-3 Volga-M	Frequency Band	A=A (VHF) B=VHF below A band
Time Between Launches (sec)	INA	Location	Outside battery formation
Reload Time (min)	10-12	Radar	
Fire on Move	No	Name	FLAT FACE, P-15
Emplacement Time (min)	< 4 hours	Function	Target acquisition, early warning,
Displacement Time (min)	< 4 hours	Detection Range (km)	250
Simultaneous Missiles	3 at 6-second intervals	Frequency Band	C
Missile	V750K/Volga Volga-2A	Location	At regimental HQ
Name	INA	Radar	
Range (m)	Max. Range: 35,000-50,000 60,000 Volga-2A Min. Range: 6,000-7,000	Name	SIDE NET, PRV-11
Altitude (m)	Max. Altitude: 30,000	Function	Height finding radar

	Min. Altitude: 100		
Dimensions	Length (m): 10.6 to 10.8 Diameter (m): 0.50	Detection Range (km)	180
Weight (kg)	2,300-2,450 at launch	Frequency Band	E
Missile Speed (mach)	4.5	Location	At regimental HQs in some cases
Propulsion	Solid fuel booster 5 sec duration Sustainer liquid <70 sec duration	Radar	
Guidance	Command RF	Name	KNIFE REST A
Warhead Types	HE, Nuc	Function	Early warning radar
Warhead Weight (kg)	195 HE	Detection Range (km)	370
Bursting Radius (m)	125-135	Frequency Band	A
Kill Radius (m)	65	Location	INA older system
CEP (m)	76.3		
Fuze Type	Proximity or Command		
Probability of Hit (Ph%)	50 FW, 40 heli Volga-2A: 75 FW, 60 heli		
Simultaneous Missiles	3 per Target		
Command Destruction at (sec)	115		

VARIANTS

SA-2A (MOD 0): FAN SONG A

SA-2B (MOD 1): FAN SONG B, LONGER MISSILE

SA-2C (MOD 2): FAN SONG C, LONGER RANGE, LOWER ALTITUDE ENGAGEMENT

SA-2D (MOD 3): FAN SONG E, EW ENHANCED

SA-2E (MOD 4): FAN SONG E NUC VARIANT

SA-2F (MOD 5): FAN SONG F, EW ENHANCED BACKUP OPTICAL, HOME-ON JAM MISSILE

SA-N-2: NAVAL TEST VERSION, UNSUCCESSFUL

HQ-2: CHINESE VARIANT (CSA-1), WITH A 30 KM RANGE.

HQ-2B: CHINESE UPGRADE, WITH GIN SLING FC RADAR AND IMPROVED MISSILE, DIGITAL ENCRYPTED C2, COMPUTER FCS, EO PASSIVE ALTERNATIVE FC, AND TRACKED LAUNCH VEHICLE. RANGE IS 40 KM.

IRAQI MOD: INFRARED TERMINAL GUIDANCE/MISSILE.

KS-1A/HQ-12: CHINESE HQ-2 UPGRADE TO 50 KM, ON A WHEELED LAUNCHER VEHICLE.

VOLGA-M: MID 90'S UPGRADE, WITH DIGITAL SUBSYSTEMS, 41 MILES RANGE, LESS MAINTENANCE. SYSTEM USES VOLGA-2A MISSILE.

UPGRADED RADARS MAY BE ASSOCIATED WITH THIS SYSTEM. FOR INSTANCE, P-12M AND SPOON-REST-B/P-12NP UPGRADES ARE FIELDED.

NOTES

THE SA-2/GUIDELINE IS A TWO-STAGE MEDIUM-TO-HIGH ALTITUDE, RADAR-TRACKING SAM. BECAUSE ITS RANGE IS GENERALLY IN THE 35-50-KM BAND, IT IS MORE MRAD SYSTEM THAN LRAD. THE WEAPON IS A NATIONAL-LEVEL ASSET USUALLY FOUND IN THE REAR AREA WITH THE MISSION OF SITE DEFENSE OF STATIC ASSETS SUCH AS SUPPLY AND COMMAND INSTALLATIONS. IT IS FIRED FROM A SINGLE-RAIL GROUND-MOUNTED LAUNCHER THAT CAN BE MOVED BY A TRUCK. THE MISSILES ARE CARRIED ON A SPECIAL TRANSLOADER-SEMI-TRAILER TOWED BY A ZIL TRUCK. AN SA-2 REGIMENT CONSISTS OF THREE BATTALIONS, EACH HAVING A SINGLE FIRING BATTERY. EACH BATTERY HAS SIX LAUNCHERS ARRANGED IN A STAR FORMATION, CENTRALLY POSITIONED FAN SONG FIRE CONTROL RADAR, AND A LOADING VEHICLE. THE TWO FORWARD BATTERIES USUALLY LOCATE 40 TO 50 KM BEHIND FRONT LINES; THE THIRD BATTERY LOCATES APPROX 80 KM BEHIND.

LIMITATIONS INCLUDE LIMITED EFFECTIVENESS AGAINST UPDATED ECM, RESTRICTED MOBILITY, AND LIMITED EFFECTIVENESS AGAINST LOW-ALTITUDE TARGETS.

Russian SAM System SA-3/GOA, Pechora-2M Launcher Vehicle

		Weapons & Ammunition Types	Typical Combat Load
		Launch Rails	2 or 4

SYSTEM	SPECIFICATIONS	FIRE CONTROL	SPECIFICATIONS
Alternative Designations	S-125 Neva, S-125 Pechora (export)	Radar	
Date of Introduction	Twin launcher 1961/ quadruple launcher 1973.	Name	LOW BLOW
Proliferation	At least 39 countries	Function	Fire control (tracking and command guidance)
Target	FW, heli, CM Also ASMs, UAVs Pechora-M	Control Range (km)	85
LAUNCHER		Detection Range (km)	110
Description	Towed twin or quad-rail launcher	Frequency Band	I
Name	INA	Tracking Capacity	1 target (1-2 missiles) 2 tgts UNV Model 1999 mod
Dimensions	INA	Radar	
Weight (kg)	INA	Name	FLAT FACE/P-15
Reaction Time (sec)	8 2-3 Pechora-M	Function	Target acquisition
Time Between Launches (sec)	INA	Detection Range (km)	250
Reload Time (min)	50 (quad launcher)	Frequency Band	C
Displacement Time (min)	100 30 Pechora-M	Radar	
ARMAMENT	SPECIFICATION	Name	SQUAT EYE/P-15M
Missile:		Function	Target acquisition (low altitude, instead of FLAT FACE)
Name	5V24, Pechora-2A, 5V27DE	Detection Range (km)	128
Range (m)		Frequency Band	C
Max Range (m)	25,000 28,000 Pechora-2A	Tracking Capability	6 targets

	35,000 5V27DE Coun		
Min Range (m)	2,400	Radar	
Altitude (m)		Name	Kasta-2E2for Pechora-M/-2/-2M
Max Altitude (m)	18,300	Function	Target acquisition and EW
Min Altitude (m)	20, 7.5 blast radius	Detection Range (km)	150 EW 95 TA FW 55 heli
Dimensions		Frequency Band	INA
Length (m)	6.1	Tracking capability	50 targets
Diameter (mm)	550	Countermeasure	Frequency agile, phase modulation
Weight (kg):	946		
Missile Speed (m/s):	650-1,150		
Velocity (mach):	3.5		
Propulsion	Solid fuel booster		
Guidance:	Command RF		
Warhead Type	Fragmenting Rod-HE		
Warhead Weight (kg):	73		
Kill Radius (m)	12.5		
Fuze Type	Proximity RF, 20 m detection		
Probability of Hit (Ph%)	70 FW, 70 heli 80 Pechora-M, -2M		
Simultaneous missiles	2 per target		

VARIANTS

SA-3A: TWO-RAIL LAUNCHER. MISSILES WITHOUT INTERSTAGE FINS.

SA-3B (GOA MOD 1): TWO-RAIL LAUNCHER. MISSILES HAVE INTER-STAGE FINS.

SA-3C: FOUR-RAIL LAUNCHER.

NEWA SC: POLISH MODERNIZED SYSTEM

PECHORA-M: UPGRADE FIELDED IN 1994 AND USED IN AT LEAST 3 COUNTRIES. IT HAS DIGITIZED FCS, AND LASER/EO/THERMAL AUTO-TRACKER FOR USE WITHOUT A RADAR. IT ADDED THE KASTA-2E2 TA EW RADAR.

PECHORA-2/UNV MODEL 1999: FURTHER UPGRADE WITH TRUCK-MOUNTED LOW BLOW FC RADAR), TRACKS 2 TARGETS. IT IS RESISTANT TO AIRCRAFT ECM.

PECHORA-2M: RUSSIAN MOBILE VARIANT OF -2, WITH LAUNCHER MOUNTED ON A TRUCK CHASSIS MODIFIED INTO A TRANSPORTER-ERECTOR-LAUNCHER (TEL). OTHER CHANGES: THE 2-RAIL LAUNCHER HAS A STORAGE COMPARTMENT UNDERNEATH FOR SUPPORT AND TEST EQUIPMENT. NAVIGATION AND AUTOMATED FIRE CONTROL TERMINAL ARE MOUNTED ONBOARD. THE CAB HAS ROOM FOR TWO OR THREE CREW MEMBERS. THE LATEST MISSILE IS 5V27DE. THE TRAILER-MOUNTED UNV MODEL 1999 FC RADAR (UP TO 300 M AWAY) CAN EMPLACE AND DISPLACE IN 5 MINUTES OR LESS. THIS SYSTEM HAS BEEN EXPORTED TO SEVERAL COUNTRIES.

WWW.MVDV.RU

NOTES

THE SA-3/GOA IS A TWO-STAGE, LOW- TO MEDIUM-ALTITUDE SAM. TWO READY MISSILES TRAVEL IN TANDEM ON A MODIFIED TRUCK OR TRACKED VEHICLE FROM WHICH THE CREW LOADS THE MISSILES ONTO A GROUND-MOUNTED, TRAINABLE LAUNCHER FOR FIRING. IT IS PRINCIPALLY A POINT/SMALL AREA DEFENSE WEAPON. SA-3 IS NOT MOBILE. IT IS MOVABLE, WITH CONSIDERABLE DISPLACEMENT TIME. PECHORA-2M (ABOVE) IS A HIGHLY MOBILE SYSTEM, IS PICKING UP SALES.

Russian SAM System SA-4b/GANEF Mod 1

SA-4a launcher with earlier miss

SA-4b launcher with 9M8M2 missile

Weapons & Ammunition Types	Typical Combat Load
Launch rails	2

SYSTEM	SPECIFICATIONS	Guidance	RF command guidance Semi-active radar-homing
Alternative Designations	Krug-M1. Complex is 2K11 or ZRD-SD (anti-aircraft missile system - medium range).	Missile Beacon	CW radar transponder
Date of Introduction	1974 for -M1 variant	Warhead Type	Frag-HE
Proliferation	At least 8 countries for SA-4	Warhead Weight	135
Target	FW, RW, CM	Fuze Type	RF command or prox
Description	System (battery) has 3 twin-launch TELs, up to 4 TZM transloaders, a missile guidance station (with radar), and technical support. Battalion has up to six batteries, 36-72 missiles, a command post van, radar vehicle, and support vehicles. At bde level, add LONG TRACK and THIN SKIN radars, 9S44 C2 complex and support assets.	Probability of Hit (Ph%)	70 FW and heli
Launcher Vehicle		Simultaneous missiles	2 per target
Name	2P24M1 or SA-4b	FIRE CONTROL	SPECIFICATIONS
Description	Transporter-Erector-Launcher	Launcher	
Chassis	GM 123, 7-roadwheel tracked chassis	Sights w/Magnification	Mounted on TEL, remotely controls msl cmd radar EO day sighting system IR night vision system
Crew	3-5	Missile Guidance Station :	
Combat Weight (mt)	28.2	Name:	1S32
Length (m)	7.5, 9.46 with missiles	Chassis	GANEF tracked variant
Height (m)	4.47	Function	Battery fire control vehicle
Width (m)	3.2	Radar	PAT HAND
Automotive Performance		Frequency Band	H

Engine Name, Type	520-hp diesel	Function	Fire control and guidance
Cruising Range (km)	450	Range (km):	
Speed (km/h)		Detection:	120-130
Max Road	35-45	Tracking/Guidance	80-90
Max Off Road	20-30	IFF	Yes
Fording Depth (m)	1.5	**ASSOCIATED SYSTEMS**	**SPECIFICATIONS**
Radio	R-123M, initial system	Radar	
Protection		Name:	LONG TRACK
Armor, Turret Front (m)	15	Function:	Battlefield surveillance, target acquisition, early warning
NBC Protective System	Collective	Chassis:	AT-T tracked P-40 variant
ARMAMENT	**SPECIFICATION**	Unit level	AD brigade
Launcher		Detection Range (km)	167
Name	2P24M1 (same as above vehicle)	Tracking Range (km)	150
Time Between Launches		Frequency:	2.6 GHz
Simultaneous Target Launcher	1	Frequency Band	E
Simultaneous Targets Battery	1, 3 if launchers are operating autonomously in the battery	Radar	
Simultaneous Missiles per Battery	1-6	Name	THIN SKIN on Prw-16 vehicle
Simultaneous Missiles Launcher	1or 2	Function	Height finding
Reaction Time (min)	1	Chassis	AT-T tracked variant
Reload Time (min)	10-15 per missile	Unit and Level	AD brigade
Emplace/Displace time (min)	5	Detection Range (km)	240
Fire on Move	No	Tracking Range (km)	INA
Missile		Frequency Band	H
Name	9M8M2/SA-4b	Transloader	
Range (m)		Name	TZM (generic)
Max Range	50,000	Chassis	URAL-375 truck
Min Range	6,000	Unit and Level	AD battery and above
Altitude (m)		Missiles per Vehicle	1
Max Altitude	24,500	Automated Fire Control Complex	
Min Altitude	150	Name	9S44, K-1 (Krab)
Dimensions		Chassis	Van
Length (m)	8.30	Unit and Level	Ad brigade

Diameter (mm)	800		
Weight (kg)	2,450		
Missile Speed (m/s)	800-1000		
Propulsion	Solid fuel		

VARIANTS

SA-4A: ORIGINAL 1967 SYSTEM WITH EARLIER LONG-NOSED MISSILE (9M8/-8M/-8M1) AND TERMINAL HOMING. BUT MIN RANGE (9 KM) AND ALTITUDE (3 KM) MEANS A LARGE DEAD SPACE.

SA-4B/KRUG-M1: USES 9M38M2 MISSILE, WHICH DECREASED MINIMUM RANGE AND ALTITUDE (SEE LEFT) TO REDUCE DEAD SPACE. THE MISSILE HAS A SHORTER NOSE SECTION THAN EARLIER VERSIONS. THE 2P24M1 IMPROVED TEL ADDED ELECTRO-OPTICAL FIRE CONTROL.

9M8M3: MODIFIED VERSION OF EARLIER SERIES(9M8 - 9M8M1) MISSILE WITH CHARACTERISTIC LONGER NOSE, BUT ADAPTED TO SA-4B LAUNCHER

NOTES

A VARIETY OF MORE MODERN AUTOMATED CONTROL COMPLEXES, SUCH AS POLYANA, CAN BE USED TO UPGRADE THE SYSTEM AND PROCESS DATA MORE RAPIDLY. BATTERIES MAY USE A MIX OF SA-4A AND SA-B MISSILES TO MAXIMIZE RANGE, ALTITUDE, AND GUIDANCE MODES AVAILABLE, WHILE REDUCING DEAD SPACE.

Russian SAM System SA-6/GAINFUL and SA-6b/GAINFUL Mod 1

SA-6/SA-6a TEL

		Weapons & Ammunition Types	Typical Combat Load
		Launch rails	3
SYSTEM	**SPECIFICATIONS**	Propulsion	2-stage, solid fuel
Alternative Designation	2K12 system, also SA-6a or Kub/Kvadrat (export) For SA-6b and Kub-M4 see VARIANTS	Guidance	Semi-active radar terminal-homing, 2-3 channels
Date of Introduction	1966, 1976 Kub-M3	Warhead Type	Frag HE
Proliferation	At least 22 countries	Warhead Weight (kg)	50
Target	Low to medium altitude FW and heli for SA-6a. FW, heli, CM for SA-6b FW, heli, TBM, CM, UAV, and ground targets for SA-6b/Kvadrat-M4.	Fuze Type	Proximity RF
Description	Battery has 4 triple-launcher TELs, battery control truck, STRAIGHT FLUSH, and two TZM reload vehicles (3 missiles each).	Probability of Hit (Ph%)	70, 80 heli SA-6b 80 FW/heli
Launcher Vehicle		Simultaneous Missiles	2-3/target
Name	SA-6/2P25M2 common upgrade. Launcher is called SA-6a.	**FIRE CONTROL**	**SPECIFICATIONS**
Description	Transporter-Erector-Launcher	Sights w/Magnification	
Chassis	Modified PT-76	EO sighting system	TV
Crew	3	Range (km):	30
Combat Weight (mt)	14	Commander and driver	IR
Length (m)	6.09	IFF	Pulse-doppler
Height (m)	4.45	Radar and fire control vehicle	
Width (m)	3.04	Name	1S91M2E/STRAIGHT FLUSH
Automotive Performance		Function	Dual (battery target acquisition and fire control)
Engine Name	V-6R, 6 cyl diesel	Frequency	G/H-med altitude acquisition H-illumination-med alt tracking

			I-low altitude tracking
Cruising Range (km)	250	Range (km)	60-90 detection 28 tracking
Speed (km/h)		Radar	
Max Road	45	Name	LONG TRACK
Max Swim	N/A	Function	Surveillance, target acq, early warning, on vehicle
Radio	INA	Detection Range (km)	4-167
Protection	NBC Protection System: Collective	Altitude (m)	25-14,000
ARMAMENT	**SPECIFICATION**	Tracking Range (km)	150
Launcher		Frequency	2.6 GHz, Band E
Name	2P25M2 (same as vehicle)	Radar	
Reaction Time (min)	22-24	Name	THIN SKIN
Time Between Launches (sec):	INA	Function	Height Finding
Reload Time (min):	10	Detection Range (km)	240
Fire on Move	No	Tracking Range (km)	INA
Simultaneous targets launcher	1	Frequency Band	H
Simultaneous targets battery	1	Other Radars	Links to IADS for EW and TA data from radars: Links to EW/TA radars at echelons above division, and radars in SAM units.
Simultaneous missiles battery	1-4		
Emplacement Time (min)	5 or less		
Displacement Time (min)	15 for a battery		
Missile			
Name	Kub-M3/3M9M3		
Range (m)	4,000-25,000		
Altitude (m)	30-14,000		
Dimensions	6.20 m length, 335 mm diameter		
Weight (kg)	630		
Missile Speed (m/s)	700		

VARIANTS

MANY EARLY SA-6/KUB/KVARDRAT SYSTEMS WERE PRODUCED AND EXPORTED. BY THE 1970S, THE SOVIETS FIELDED KUB-M1, M2, AND M3 VERSIONS, AND UPGRADED OLDER SYSTEMS WITH IMPROVED FC (1S91M1), REDUCED SIGNATURE, AND IMPROVED 2P25M2 LAUNCHER WITH AND TV/EO SIGHTS. UPGRADE MISSILE RANGES ROSE FROM 20 KM UP TO 25.

KVADRAT-M: FIELDING FOR THE LATEST SERIES OF UPGRADES BEGAN IN 1996, STARTING WITH SA-6A/ KUB-M3 UNITS. THEY ADDED MODERN SUPPORT SYSTEMS, E.G., AN IMPROVED BATTALION CP VEHICLE, ORION IW SYSTEM (6-21), AND MODERN RADARS. BATTALION STRUCTURE CHANGED TO 6 BATTERIES, EACH WITH 3 TELS AND A 1S91M1/2 FC VEHICLE. AN UPGRADE PACKAGE WAS LATER OFFERED FOR EXPORTS.

SA-6B/ KUB/KVADRAT-M4: THE SA-11/BUK SYSTEM WAS DEVELOPED TO REPLACE SA-6 MRAD UNITS. DUE TO DELAYS, A 1980 INTERIM FIX WAS TO REPLACE ONE SA-6 TEL PER BATTERY WITH A BUK/ 9A38 TELAR, FORMING HYBRID SA-6B BATTERIES AND BATTALIONS. THE 9A38 WAS BATTERY CP/FC VEHICLE (REPLACING/SUPPLEMENTING 1S91M2E) WITH THE FIRE DOME DUAL MODE TA/FC RADAR AND COULD DIGITALLY LINK TO THE IADS. THE 9A38 LAUNCHER COULD BE FIT TO LAUNCH SA-6 MISSILES TO 25

KM, OR SA-11 TO 3.

LATER, SA-6B/KUB-M4 UNITS USED IMPROVED 9A310 SERIES TELARS (E.G., BUK-M1-2 TELARS WITH 42 KM RANGE), AND THEIR UPGRADE SUPPORT ASSETS. A NEW HYBRID SA-6B STRUCTURE WAS DEVELOPED WITH ADDITION OF LAUNCHER-LOADERS (LLS. NOW BATTALIONS HAVE 3 KUB-M3 BATTERIES (ABOVE), AND 3KUB-M4 BATTERIES. KUB-M4 BATTERIES HAVE 1 BUK-M1/M1-A2 TELAR AND 1 LL, FOR 12 MISSILES AND 42 KM RANGE. THE -M4 BATTERIES CAN HANDLE A WIDER RANGE OF TARGETS AND DISTANCES, FOR LESS COST.

2K12 KUB CZ: RECENT CZ CONVERSION OF SA-6 WITH 3 ASPIDE 2000 SAMS. IT DIGITALLY LINKS TO SA-6 FCS, WITH UPDATED DISPLAY. SAM ENGAGEMENT IS TO 23 KM RANGE AND 25-12,000 M ALTITUDE. SPEED AND PH/PK ARE HIGHER. FIELDING IS IMMINENT.

NOTES

THE ASSOCIATED STRAIGHT FLUSH FIRE CONTROL/TARGET ACQUISITION RADAR VEHICLE USES THE SAME CHASSIS AS THE SA-6A TEL.

Russian SAM System SA-11/GADFLY

	Weapons & Ammunition Types	Typical Combat Load
	Self-Propelled launcher TELAR	4
	Launcher-loaders On launch rails On transport rails	8 4 4

SYSTEM	SPECIFICATIONS	FIRE CONTROL	SPECIFICATIONS
Alternative Designations	Buk-M1, Gang For OPFOR Buk-M1 is a Tier 2 system.	Sights	TV optical auto-tracker
Date of Introduction	1979/ 83 for -M1	Acquisition range (km):	20
Proliferation	At least 5 countries	Navigation systems	Available on all
Target	FW, heli, CM, UAV, guided bomb, artillery rocket, ground targets, ships	Onboard Radar	
Description	Brigade assets include bde/btry CPs and radars, TELARs, launcher-loaders, TM-9T229 missile transporter, maintenance and test units. The 6 batteries have 1 TELAR and 1 LL each.	Name	FIRE DOME
Launcher Vehicle		Function	Dual (acquisition and fire control)
Name	9A310M1 or BUK-M1	Detection Range (km)	80 (2 m2), 100 (3m2)
Description	TELAR	Targets Tracked	1 per SPL vehicle
Crew	4	Frequency	6-10 GHz (H/I band)
Combat Weight (mt)	32.34 for TELAR	Guidance Range	42 km
Chassis	GM-569 armored tracked for CP, radar, TELAR, launcher-loader	Other Assets	SA-11 digitally links to the IADS (e.g., aircraft, intel , and other SAM units. SA-10/20/11 FO radars share data with other units in the IADS net. Other assets are FOs and ELINT, e.g., Orion.
Description	TELAR	Radar	
Length (m)	9.3	Name	9S18M1/SNOW DRIFT
Height (m)	3.8 travel/7.72 deployed	Function	Battery target acquisition radar

Width (m)	3.25	Description	Armored tracked chassis w/ phased array radar and dipole antenna
Automotive Performance		Detection Range (km)	100-150
Engine Name, Type	700-hp diesel	Range Precision	400
Cruising Range (km)	500	Detection Altitude (km)	25
Max Road Speed (km/h)	65, 30 TELARs up	Targets Tracked	75
Fording Depth (m)	1	Frequency	Centimetric 3-D phased array
APU	Yes for TELARs, LL, radars, CP	Azimuth Coverage (°):	360 with rotation
Radio	INA	Emplace/Displace (min):	5
Protection		Other Radars	Regiment/Bde will have EW/TA radars, such as SPOON REST or Kasta-2E2.
Armor Protection	Small arms (est)	Launcher-loader (LL)	
NBC Protection System	INA	Name	9A39M1
ARMAMENTS	**SPECIFICATIONS**	Function	Battery resupply and TEL
Launcher		Fire Control	None, TELARs guide
Missiles per Launcher	4	Missile Load	8
Reaction Time (min)	0.25-0.5 0.1 for low-flyers	Reload Time (min)	15
Time Between Launches (sec)	3	Emplacement Time (min)	5
Reload Time (min)	12	Use of LLs is transforming SA-11 units. They appear to be updated and modified SA-6 TELs (2P25s), economically converted to expand the unit missile load, yet requiring SA-11 batteries to have only one expensive TELAR. Thus the force can expand with existing stocks of SA-6.	
Fire on Move	No	C2 Vehicle	
Emplacement Time from March (min)	5	Name	9S470M1
Displacement time (min)	5	Function	Battery Command Post
Emplace Time for Reposition (sec)	20 for a 100-200 m survivability move.	Data Links	Wire and radio AD net, to IADS net, and to SA-10/Osnova
Simultaneous targets per launcher	1	Targets Tracked	15 (with 6 at TELs)
Simultaneous Missiles per Launcher	2		
Missile			
Name	9M38M1		
Range (m):			
Max. Range	36,000		
Min. Range	3,000		
Altitude (m):			
Max. Altitude	22,000		

Min. Altitude	15, 0 with degraded Ph		
Dimensions			
Length (m)	5.55		
Diameter (mm)	400		
Weight (kg)	690		
Max target speed (m/s)	830		
Max missile Speed (m/s)	1,200		
Propulsion:	Solid fuel		
Guidance:	RF command, inertial correction, semi-active radar homing		
Warhead Type	Frag HE		
Warhead Weight (kg)	70		
Warhead lethal radius (m)	17		
Fuze Type	Proximity RF		
Probability of Hit (Ph%)	80 FW and heli		
Simultaneous missiles	2 per target		
PROTECTION/COUNTERMEASURES			
Jam ECCM:	Noise jam 240-330 w/MHz		
Passive Jam ECCM	3 Packets/100m		
Measures	One launcher operates radar, while others are passive. Other guidance modes reduce radar illumination time.		
IFF	Pulse-doppler		

VARIANTS

SA-6B/ KUB-M4/KVADRAT-M4: HYBRID UNIT WITH SA-6, AND OR BUK-M1/SA-11 TYPE TELARS.

BUK-M: SYSTEM WITH SA-11 MISSILE. IT HAD THE INADEQUATE TUBE ARM, REPLACED BY SNOW DRIFT. FEW BNS WERE FIELDED. MOST SA-11 UNITS USE BUK-M1.

BUK-M1-2 AND SA-17/GRIZZLY: UPGRADE SYSTEMS.

NOTES

TELARS CAN OPERATE AUTONOMOUSLY. LAUNCHER-LOADERS CAN LAUNCH WITH TELAR COMMAND. SA-11 CAN LAUNCH SAMS AGAINST GROUND TARGETS.

Russian SAM System Buk-M1-2 (SA-11 FO) and Buk-M2E (SA-17)

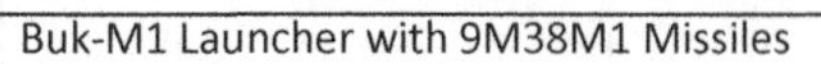

Buk-M1 Launcher with 9M38M1 Missiles

New 9M317 Missiles

Weapons & Ammunition Types	Typical Combat Load
Self-Propelled Launcher	4
TELAR	
Launcher-loader	8
On launch rails	4
On transport rails	4

SYSTEM	SPECIFICATIONS		
Alternative Designations	9K37M1-2 In OPFOR this is a Tier 1 system.	Jam ECCM	Noise jam 240-330 w/MHz
		Passive Jam ECCM	3 Packets/100m
Date of Introduction	1997	Measures:	One launcher operates radar, while others are passive. Other guidance modes reduce radar illumination time
Proliferation	At least 3 countries, export	IFF:	Pulse-doppler
Target	FW, heli, TBM, CM, ASM, UAV, artillery rocket, ships, ground targets	**fire control**	**SPECIFICATIONS**
Primary Component	System is a modernized version of the SA-11/Buk-M1 system. It adds elements of the SA-17/GRIZZLY system (missile, LRF fire control) to the system. Battalion/Complex: CP vehicle, radar, 6 transport, maintenance, mobile test vehs. Chassis: GM-569 armored tracked for CP, radar, TELAR, launcher-loader	Laser Range-finder	New addition to FCS. This permits system to engage ground targets to 15 km, waterborne targets 25 km.
Launcher Vehicle		Sights:	TV optical auto-tracker
Name	9A310M1-2	Acquisition range (km):	20, permits passive missile guidance, day and night
Description	TELAR	Navigation systems	Available on all
Crew	4	Onboard Radar:	
Combat Weight (mt)	32.34	Name	FIRE DOME
Description	TELAR	Radar:	

Dimensions (m)	9.3 length x 3.25 width 3.8 travel/7.72 deployed height	Name	9S18M1-1/SNOW DRIFT
Automotive Performance	See SA-11	Note	It is similar to 9S1M1
Radio	INA	Other Radars	Brigade will have EW/TA radars, such as Kasta-2E2, or one similar to Giraffe AMB. Upgrade options include radars and support vehicles from the SA-17 System.
Protection		Other Assets	The SA-11 digitally links to the IADS (e.g., aircraft, intel , and other SAM units. SA-10/20/11 FO radars share data with other units in IADS net. Assets include FOs and IW, e.g., Orion.
Armor	Small arms (est)	Launcher-loader (LL):	9A39M1-1, see 9A38M1
NBC Protection System	Collective	C2 Vehicle	9S470M1-2, see 9S470M1
ARMAMENT	**SPECIFICATIONS**		
Launcher			
Missiles per Launcher	4		
Reaction Time (min)	0.25-0.5 0.1 for low-flyers		
Time Between Launches (sec)	2		
Reload Time (min)	12		
Fire on Move	No		
Emplace/Displace Time (min)	5		
Emplace Time, Reposition (sec)	20 for a 100-200 m survivability move.		
Missile			
Name	9M317		
Range (km)	3-42, 15 with TV sights		
Altitude (m)			
Max Altitude	25,000		
Min Altitude	0 with degraded Ph		
Dimensions	5.5 m length, 400 mm diam		
Weight (kg)	715		
Max Target and Missile Speed (m/s)	1,200		
Propulsion	Solid fuel		
Guidance:	RF command, inertial correction, Semi-active radar homing (9M317A)		
Warhead Type	Frag HE		
Warhead Weight (kg):	70		

Warhead lethal radius (m):	17		
Fuze Type:	Proximity RF or contact		
Probability of Hit (Ph%):	70 TBM, 80 other		
Simultaneous missiles	2 per target		
Other Missile	9M317A is an anti-radiation homing missile/attack missile interceptor		

VARIANTS

BUK AND BUK-M1: PREDECESSORS,

HQ-16: CHINESE BUK-M1-2 UPGRADE IN R&D.

SA-N-12: NAVAL VERSION WITH 12 X 9M17M/ SHTIL-1 MISSILES IN A VERTICAL-LAUNCH CANISTER.

SHTIL-1: NEW SHIP CANISTERED VERTICAL LAUNCHER VERSION WITH 9M317ME MISSILE TO 32 KM.

SA-17/GRIZZLY/BUK-M2E/URAL: RUSSIAN REDESIGN/FOLLOW-ON OF SA-11. IT USES 9M317 MISSILES AND 2 NEW RADARS. THE SYSTEM HAS 2

GIRAFFE VEHICLES (WITH DUAL MODE RADARS ON TELESCOPE ARMS), 4 TELARS, 8 LLS, ORION RF INTEL SYSTEM, AND A SUPPORT COORDINATION VEHICLE. ALL BATTERY RADARS ARE CHAIRBACK PHASED ARRAY WITH 160 KM DETECTION, 120 FOR LOW FLYERS. SYSTEM SIMULTANEOUSLY TRACKS 10 TARGETS AND ENGAGES 4 (OR 24/BN). EFFECTIVE RANGE IS 45 KM WITH PH OF 90% FOR FW/HELI, 80 TBMS. MINIMUM ALTITUDE IS 0 M WITH 80% P-HIT. IT NOW AS LIMITED FIELDING IN 1 COUNTRY. IT IS LIKELY THAT INITIAL RUSSIAN UNITS WILL HAVE BATTERIES WITH A TELAR AND 1-4 LLS, SIMILAR TO BUKS IN KUB-M4 BATTERIES.

A WHEELED VERSION OF SA-17 IS BUK-M2EK ON A 6X6 BELORUSSIAN CROSS-COUNTRY CHASSIS.

BUK-M3: AN UPGRADE IN TESTING FOR USE IN ALL PREVIOUS BUK-M UNITS, WITH A NEW RADAR, AND TBM INTERCEPT CAPABILITY TO MACH 4.

NOTES

THE BUK-M1-2 IS A MULTI-ROLE SYSTEM FOR SAM AND SURFACE-TO-SURFACE MISSILE (SSM) GROUND/SEA TARGET ATTACK MISSIONS.

Recent Developments in Long-Range and High-Altitude Air Defense Systems

The worldwide trend in modernization of long-range AD (LRAD) and high-altitude continues, even in times of shrinking military budgets. The trend is driven by expanding strategic threats of aerodynamic systems (e.g., ballistic and cruise missiles, UAVs, and stealth aircraft), and deadly munitions (e.g., weapons of mass destruction and effects, and precision weapons).

Challenges of rising costs and constricted budgets affect the modernization patterns. Most countries continue to focus most of their air defense modernization programs on upgrading and reconfiguring existing systems. There are upgrade missiles, C^2 and fire control assets for Russian SA-5 and other older SAM systems (pgs 6-73, 6-80, etc.). A few other countries are developing new systems, including anti-ballistic missile (ABM) systems.

The non-US strategic systems which have received most world attention in recent years are the Russian SA-10, SA-12, and SA-20 series missile systems. S-300P (SA-10/-20) series have seen a lot of changes, and a confusing mix of names and designators. To clarify those nomenclatures, the table appears below. Export and Russian forces systems in each series may have same capabilities; but in some cases, due to the lengthy export contract negotiation process, export systems may be upgraded by time of shipment. Upgrades succeed only when radars and fire control match missiles with range and altitude coverage to use their capabilities.

In recent years we have heard much about the new Russian 4th-generation missile system, **S-400/Triumf**. Due to developmental delays and budget issues, the program was delayed. The delays expedited Russian efforts to modernize SA-10 and SA-12 systems, and to incorporate comparable missile/support capabilities into them. Thus **SA-10d** and **SA-10e** upgrades were further modernized and re-designated **SA-20a** and **SA-20b**. An upgraded SA-12 is fielded and designated **SA-23**. The S-400/Triumf is now fielded, and designated **SA-21a**. SA-20b and S-400 systems can launch two different sizes of missiles (see SA-20b at pg 6-86, and SA-21 at pg 6-87). The large missiles offer superior performance for ballistic missile defense (BMD), and for long-range defeat of AWACS, RISTA, stealth, and SEAD targets. With the changes in SA-20 and SA-21 programs, many sources have confused those systems and their details.

Changes in strategic systems may impact fielding of medium-range air defense systems (MRADs). As the 9M96-series small missiles improve, they will form the majority of missiles on S-300/400 launchers, to service most aerial targets. Some countries may choose not to acquire MRADs (e.g., Buk-M1-2), rather upgrade strategic systems like SA-10/20a to SA-20b capability. But ground forces also want long-range AD. Most MRADSs range only to 50 km, yet lack surge capacity of the SA-20b and later long-range systems (up to 16 SAMs, pg 6-81). SA-12/23 units currently have limited surge capacity. Users can now add canisters of small missiles to existing LRADS TELs for increased surge capacity, without the need to add new expensive MRADs.

Russia intends to upgrade strategic SAM systems and upgrade all S-300 and S-400 systems into an integrated network. Priorities are for every battery to be able to counter ballistic missiles, surges, and high-value systems (stealth, AWACS, and SEAD). China is upgrading its SA-10/SA-20 systems and to compete with Russian systems. Many forces are adding new long-range EW and TA radars and other sensors, and upgrading older systems to extend range and digitally integrate them into IADS. These include ELINT, other passive sensors, and responsive, jam-resistant, secure C^3 networks to destroy UCAVs and stealth aircraft.

Russian Long Range SAM System SA-5b/GAMMON

	Weapons & Ammunition Types	Typical Combat Load
	Single-rail ground mounted	1 Six launchers per Battalion

SYSTEM	SPECIFICATIONS	FIRE CONTROL	SPECIFICATIONS
Alternative Designations	S-200V, S-200M, or Vega	Radar	
Date of Introduction	1963	Name	SQUARE PAIR
Proliferation	At least 15 countries	Function	Dual mode - target acquisition and fire control
Target	FW, CM	Effective Range (km)	350
ARMAMENT	**SPECIFICATIONS**	Frequency (GHz)	6.62-6.94
Launcher		Frequency Band	H
Description	Single-rail ground-mounted not mobile but transportable	Located	With firing units
Dimensions	INA	Associated Radars	
Weight (kg)	INA	Name	BAR LOCK B (P-50) follow-on (BACK NET initially)
Reaction Time (sec)	INA	Function	Target acquisition/early warning
Time Between Launches (sec)	INA	Range (km)	250/ 390
Reload Time (min)	INA	Frequency Band	E/F-band (2-2.5 GHz),
Fire on Move	No	Location	Generally with separate EW or signals recon bns
Emplacement Time (min)	Days	Name	BIG BACK
Displacement Time (min)	Days	Function	Very long-range early warning
Missile		Effective Range (km)	600
Name	5V28M/S-200M	Frequency Band	3-d L-band
Range (km)		Location	Brigade Level
Max Slant Range	300	Name	TALL KING
Effective Range	250	Function	Very long-range early warning
Min Range	17	Effective Range (km)	500-600
Altitude (m)		Frequency Band	A-band (150-180 MHz)
Max Altitude	29,000	Location:	Generally with separate early warning or Signals Recon battalions
Effective Ceiling	30,000	Name	BACK TRAP

Min Altitude	300	Function	Very long-range early earning
Dimensions		Effective Range (km)	410
Length (m)	10.7	Frequency Band	A-band (172 MHz)
Diameter (mm)	750	Location	Brigade Level
Weight (kg)	7,100	Name	ODD PAIR, E-band follow-on (SIDE NET/PRV-11 initially)
Wrap Around Boosters		Function	Height finding radar
Length (m)	4.9	Range (km)	INA
Diameter (mm)	500	Frequency Band	E-band
Missile Speed (m/s)	1,100	Location	Generally with separate early warning or Signals Recon bns
Propulsion	2-stage liquid fuel, four wrap-around solid fuel rockets	Other Radars	The SA-5 can also link to the IADS or to other AD units to get analog AD data. Newer radars, such as the Nebo-SVU mobile radar, are marketed, and can be used with SA-5 series systems.
Guidance	Semi-active homing, active radar homing terminal phase, home on jam		
Warhead Type	Conventional (HE) or nuclear		
Warhead Weight (kg)	60 HE		
Fuze Type	INA		
Probability of Hit (Ph%):	75 FW/85 large		
Simultaneous missile	INA		
Self-Destruct (sec):	INA		
Booster separation at (km):	2		
Reload Time (min):	5		
Other Missiles			
S-200A	Original missile, 160 km		
S-200 Vega/SA-5b	Improved to 300 km, 40 km ceiling		
S-200VE	Export, range 250 km, 29 ceiling		
S-200M/5V28M	Improved to 300 km, 29 ceiling. It can replace S-200VE as upgrade.		
S-200D/SA-5c	Upgrade 400 km, 40 ceiling		

VARIANTS

RUSSIAN ARTICLES HAVE PREDICTED MODERNIZATION PROGRAMS, IN ADDITION TO MISSILE UPGRADES.

THERE ARE REPORTS THAT THE SQUARE PAIR CAN BE LINKED WITH AND (PERHAPS) SLAVED TO S-300P SERIES TARGET ACQUISITION RADAR, TO ENGAGE TARGETS TRACKED BY THAT RADAR. THUS AN SA-10 OR SA-20 UNIT COULD INTEGRATE LAUNCHES WITH THE SA-5B TO ENGAGE TARGETS BEYOND THEIR OWN 200 KM RANGE (WITH LIMITED THREAT FROM ATMOSPHERIC SYSTEMS), AND COULD PROTECT THE SA-5 LAUNCHERS WITH THEIR BALLISTIC MISSILE CAPABILITIES. SA-10/20 PHASED ARRAY RADARS GREATLY REDUCE DETECTABLE RF SIGNAL.

IRAN CLAIMS TO HAVE UPGRADED ITS SYSTEMS WITH BETTER RADARS AND DIGITAL C^2.

NOTES

THE SA-5/GAMMON IS A LONG-RANGE, STRATEGIC SEMI-ACTIVE GUIDED MISSILE SYSTEM FOR TARGETING MEDIUM-TO-HIGH ALTITUDE HIGH-SPEED AIRCRAFT.

THE MISSILE HAS A LONG CYLINDRICAL BODY WITH A CONICAL NOSE, FOUR LONG CHORD CRUCIFORM DELTA WINGS, FOUR SMALL CRUCIFORM RECTANGULAR CONTROL SURFACES AT THE EXTREME REAR, AND FOUR JETTISONABLE, WRAPAROUND SOLID-FUEL BOOSTERS WITH CANTED NOZZLES. IT USES A LIQUID PROPELLANT, DUAL THRUST ROCKET ENGINE, AND THE MISSILE TRAVELS ABOUT 2 KM BEFORE BOOSTER SEPARATION. THE SUSTAINER HAS FOUR CROPPED DELTA WINGS AND STEERABLE REAR FINS. CONTROL IS ASSISTED BY AILERONS.

S-300P Series Strategic Air Defense Systems Comparison*

NATO DESIGNATOR	SA-10b GRUMBLE	SA-10c GRUMBLE	SA-20a (SA-10d) GARGOYLE	SA-20b (SA-10e) GARGOYLE
LAUNCHERS	5P85SU cmd TEL** 5P85DU slaveTEL** 5P85 trailer lchr w/KrAZ-260V	5P85SU cmd TEL** 5P85DU slaveTEL** 5P85T trlr lchr w/KrAZ-260V	5P85SE cmd TEL** 5P85TE trlr lchr w/KrAZ-260V	5P85SE2 cmd TEL** 5P85TE2 Trailer w/KrAZ-260V
MISSILES Range (km) Altitude (km)	5V55R 7-75 0-25 blast radius Also 5V55V (nuc option) 5V55KD (upgrade variant of 5V55K)	5V55RUD 5-90 0-27 blast radius Also 5V55V nuc 5V55PM anti- radiation (ARM) 48N6E (upgrade option)	48N6/ 48N6E export 5-150 0-27 blast radius Also 5V55V nuc 5V55PM anti-radiation (ARM) 48N6E2 (upgrade option)	48N6M /48N6E2 export 5-200 0-27 blast radius ***"Small missile" (4 per canister) 9M96 /9M96E 9M96M /*9M96E2 5-40 5-40 5-120 5-120 0-35 0-35 0-35 0-35 Near term small missiles will range 200 km (upgrade option).
RADARS	64N6/BIG BIRD Bd* bde TA radar vehicle 30N6/FLAP LID-B Battery FC rdr veh 76N6/CLAM SHELL TA on tower trailer (36D6/TIN SHIELD TA trlr in older units)	64N6/BIG BIRD D* (in 83M6 Bd C^2 sys) 30N6/FLAP LID-B 76N6/CLAM SHELL TA on tower trailer (Optional 96L6E Bn TA radar vehicle)	64N6E/ BIG BIRD E** (in 83M6E1 Bd C^2 sys) 30N6E1/TOMBSTONE Battery FC rdr veh 96L6E Bn TA rdr veh (76N6/CLAM SHELL Optional supplement) Option: NEBO-SVU	64N6E2/ BIG BIRD E** bde TA radar vehicle 30N6E2/TOMBSTONE Battery FC rdr vehicle 96L6E2 Bn TA radar vehicle 76N6/CLAM SHELL bn option sup NEBO-SVU target track radar (Bn)
OTHER SUPPORT	54K6 CP veh (in the 83M6 Bde C^2 system) 1T12 survey trk 22T6 loading trk Baikal-1 Bde Intel Ctr 5157 power station MAZ-537 for rdr twr	54K6/Baikal-1 Bde Intel Ctr (in 83M6 Bde C^2 system) 1T12-2M, 22T6 5157 power station MAZ-537 for rdr twr 48III6y MRepair Base	54K6E CP veh (in the 83M6E Bde C^2 system) 1T12-2M survey trk 22T6 loading trk Baikal-1 Bde Intel Ctr 5157 power station MAZ-537 tows rdr twr 48III6y M Repair Base	54K6E2 CP veh (battle management center in 83M6E2 Bde C^2 system) 1T12-2M survey trk 22T6 loading trk Baikal-1 Bde Intel Ctr 5157 power station trailer MAZ-537 tows the radar tower 48III6y Mobile Repair Base

TA radar = Target Acquisition (surveillance, detection, target tracking, IFF)

FC radar = Fire Control (illumination and guidance, missile tracking, IFF).

Many modern FC radars are dual-mode (capable of TA and FC functions). The 30N6 series radars are dual-mode.

System radars and most others are phased-array. They offer SEAD rejection, low detection, and high jam resistance.

* Fielded systems may adopt radars or missiles of earlier or later versions. Supporting vehicles carry forward, or are upgraded/replaced with new versions. Thus 30N6 on SA-10b and SA-10c is replaced by 30N6E1 on SA-10d. For SA-10b, a76N6 TA radar replaced the36D6 TA radar. An exception to upgradability is the obsolete SA-10a, missile which used radio command guidance, incompatible with later systems. SA-10a units were converted to SA-10b. Missiles with E designators are for use in exported systems, but could be used in domestic Russian launchers.

Mobile AD radars with counter-stealth ability, e.g., Nebo-SVU, and older EW radars, can be used with SA-10/20.

Substantial numbers of air observers will be used. SHORAD systems (including 2 MANPADS/TEL are co-located).

** The TELs are variants of MAZ-543M. Radar and C^2 vehicles are on MAZ-543M or MAZ-7910 chassis. Various other trucks and vans are used for support. Radar tower trailers have supporting units for erection and disassembly.

*** Some strategic anti-ballistic missile (ABM) SA-20b units only have 48N6-type "big missiles" and ARMs. In other units, one or more canisters of 4 small missiles will be used. As the smaller (9M96 series) missiles improve in range closer to the big missiles, more launch pods will convert from big missiles to small missiles. Thus the firing units will be able to disperse more widely, with up to four times the target-handling capacity of current firing units.

**** In SA-20a and 20b systems, there are no slave versions of the TELs, only command. Many have the trailer launchers operating out of battalion as primarily transport vehicles for resupplying firing units. They can, however, be used as launchers during air surge activities. Firing units which lose trailer-launchers may then add more TELs.

Russian SAM System SA-10b/GRUMBLE

		Missiles	Typical Combat Load
		In canisters onboard TEL	4
		SA-16 MANPADS	2
SYSTEM	**SPECIFICATIONS**	**VARIANTS**	**SPECIFICATIONS**
Alternative Designations	S-300PM	SA-10A/S-300P	First system, semi-fixed on trailers, with 5V55K (50 km) missile. Early SA-10b units used the 36D6/TIN SHIELD TA radar, later supplemented or replaced by 76D6/CLAM SHELL
Date of Introduction	1980	SA-10b	Added TELs, 5V55R (75 km) missiles, and FLAP LID B improved radar
Target	FW, heli, TBM, CM, ASM, UAV	HQ-2	Chinese copy, indigenous launchers
Proliferation	At least 8 countries	HQ-9	Chinese variant and upgrade
Primary Components per Battery	1 5P85S cmd TEL 1 5P85D slaveTEL 1 5P85 trailer launcher 1 30N6 radar/fire control vehicle A 5P85SU launcher has a command shelter behind the cab. A 5P58DU TEL does not.	SA-10c	Russian export upgrade system (aka: S-300PMU) with improved missile
ARMAMENTS	**SPECIFICATIONS**	SA-10f/SA-N-6	Russian naval version. For other variants.
Transporter-Erector Launcher (TEL):		Forces may mix earlier and later assets. Thus a system may start as SA-10b, and upgrade to SA-10c or SA-20b.	
Name	5P85S or 5P58D (see NOTES)	**ARMAMENTS**	**SPECIFICATIONS**

Time Between Launches (sec)	3	Name	64N6
Reaction Time (sec)	8-10 (vertical-launch missiles for no slew time)	NATO Designation	BIG BIRD B
Reload Time (min)	INA	Function	Early warning, target acquisition
Crew	6	Unit	Grouping (brigade) level, supports 3-6 90Zh6E complexes (bns), and 12-36 launchers
Fire on Move	No	Mobility	MAZ-7910 van
Emplace/Displace Time (min)	5/30 TEL 30/30 trailer launcher	Detection Range (km)	300 FW/heli, 127 TBM
Automotive Performance, 5P85S TEL		Number of Targets Detected	up to 200
Chassis	MAZ-7910 (8x8)	Targets for Simultaneous Lock and Track	100
Engine	D12A-525 525-hp diesel	Frequency Band	F, 3-D phased array
Cruising Range (km)	650	Azimuth Coverage (°)	180, 360 with rotation
Max Road Speed (km/h)	63	Name:	30N6
Weight (kg)	42.15 with missiles	NATO Designation	FLAP LID-B
Missile		Function:	Dual (tgt acquisition/fire control)
Name	5V55R	Mobility:	MAZ-7910 8x8 van
Range (km)	7-75aircraft, 5-35 TBMs	Dimensions (m):	14.5 L x 3.2 W x 3.8 H
Altitude (m)		Unit Associated With	Firing battery
Max Altitude	25,000	Interception Altitude (m):	25 and higher
Min Altitude	25, 0 with blast radius	Targets Engaged Simultaneously	6
Speed (m/sec)		Missiles Guided Simultaneously	12
Target	50-1,200	Frequency Band	I/J phased Array
Max SAM	2,000	Linked to Integrated Air Defense	Yes
Dimensions		Detection range (km):	200
Length (m)	7.25	Guidance Range (km):	90+, auto-track
Diameter (mm)	508	Azimuth Coverage (°):	120, 360 with rotation

Weight (kg)	2,340 in canister	Many SA-10B units were fielded with 36D6/ TIN SHIELD TA radars. Most were later replaced with 76N6/CLAM SHELL.	
Guidance	Track-Via-Missile (TVM) and missile radar-homing	Other Assets	
Warhead Type	Frag-HE		
Warhead Weight (kg)	130		
Fuze Type	Radio Command		
Probability of Hit (Ph%):	80 FW and heli		
Simultaneous missiles	2/target (2 x P-hit)		

PRIMARY COMPONENTS

GROUP (EQUALS A BRIGADE) HAS 83M6 C2/ BATTLE MANAGEMENT COMPLEX (WITH BAYKAL-1/54K6 CP VEHICLE AND 64N6 SURVEILLANCE RADAR VEHICLE). THE C2 CAN CONTROL 6X 90ZH6E COMPLEXES (BNS). A GROUP ALSO HAS TECHNICAL SUPPORT FACILITIES. STATIONARY GROUP FOR AREA DEFENSE HAS UP TO 72 LAUNCHERS. OPFOR BDE IS 18. A 90ZH6 MISSILE COMPLEX TOTALS 6-12 LAUNCHERS WITH BN CP, 76N6 OR 36D6 BN TA RADAR AND 2-4 FIRE UNITS. IT ALSO HAS VEHICLES (TRUCKS,) UAZ-452T2 SURVEY VEHICLE, ETC.), AND EQUIPMENT. TACTICAL AD ASSETS (E.G., MANPADS), ARE INCLUDED.

OTHER ASSETS

THE SA-10B LINKS TO THE IADS TO GET DIGITAL AD DATA FROM EW ASSETS, AD AIRCRAFT, AD INTEL, AND OTHER SAM UNITS. SA- 10 RADARS SHARE DATA WITH OTHER AD UNITS. FORWARD OBSERVERS ARE DISTRIBUTED THROUGHOUT THE COVERAGE AREA. OTHER EW AND TA RADARS CAN USED IN SA-10 GROUPS AND COMPLEXES.

NOTES

ALTHOUGH MANY SA-10B UNITS WERE FIELDED WITH 36D6/TIN SHIELD TA RADARS, MOST WERE LATER REPLACED WITH 76N6/CLAM SHELL. THE PHASED-ARRAY RADARS FEATURE LOW DETECTION, AND HIGH JAM RESISTANCE.

Russian SAM System SA-10c/GRUMBLE (export)

		Missiles	Typical Combat Load
		TEL and trailer launcher	4
		5V55RUD	4
		5V55PM/HQ-2 ARM	4/battery
		SA-18 MANPADS	2
SYSTEM	**SPECIFICATIONS**	**ASSOCIATED RADARS**	**SPECIFICATIONS**
Alternative Designation	S-300PMU Original fielding was Russian only. This was a commonly exported version of the S-300PM system, including upgrades.	Name	64N6
Date of Introduction	1984	NATO Designator	BIG BIRD D
Proliferation	At least 6 countries	Function	Early warning, target acquisition
Target	FW, heli, TBM, CM, ASM, UAV	Unit	Grouping (brigade) level, supports 3-6 90Zh6E complexes (bns), and 12-36 launchers
Primary Components	SeeSA-10b. TELs are designated 5P85SU and 5P85DU. A new semi-trailer permits faster employment from the move.	Mobility	MAZ-7910 8x8 van
ARMAMENTS	**SPECIFICATIONS**	Detection Range (km)	300 aircraft, 127 TBMs
TEL and New Semi-trailer Launcher		Number of Targets Detected	up to 200
Name	5P85T (road-mobile only)	Targets for Simultaneous Lock and Track	100
Missiles per Launcher	4	Frequency Band	F, 3-D phased array
Reaction Time (sec)	8-10, vertical- launch missiles for no slew time	Azimuth Coverage (°)	180, 360 with rotation
Time Between Launches (sec)	3	Name:	30N6 (FLAP LID-B) See SA-10b.
Reload Time (min)	INA	Name	76N6
Crew	4-6	NATO Designation	CLAM SHELL
Fire on Move	No	Function	Low altitude target acquisition

Emplace/ Displace Time (min)	5/30	Unit Associated With	Battalion and bde
Automotive Performance	For TEL see SA-10b, except	Mobility	Mounted atop 40V6 trailer tower. Antenna station is on a 5T58 truck
Cruising Range (km)	800	Operation	Station can operate 500m from radar
Road/ Dirt Road Speed (km/h)	60/30 The 5P85TE trailer-launcher is normally towed by a KRAZ-260B 6x6 truck.	Emplacement Time (hrs)	1-2
		Detection Range (km)	
Missile		@ 500 m altitude	93
Name	5V55RUD	@ 1,000 m altitude	120
Range (km)	5-90	Targets Tracked Simultaneously	Up to 180
Max Altitude (m)	27,000	Target Generation Time/Target (sec)	3
Min Altitude (m)	25, 0 with blast radius	Resolution of Target RCS	.02 m2 @ 1400 kts
Speed (m/sec)		Frequency Band	I, 3-D radar
Max Target	1,200	Azimuth Coverage (°):	120, 360 with rotation
Max SAM	2,100	Recent upgrade 96L6E all-altitude target acquisition radar vehicle can replace the CLAM SHELL towed (stationary) site radar.	
Length (m)	7		
Diameter (mm)	513		
Weight (kg)	2,300 in canister		
Guidance	Track-Via-Missile, missile radar homing, home on jam		
Warhead Type	Frag-HE		
Warhead Weight	133		
Fuze Type	Radio Command		
Probability of Hit (Ph%):	80 FW and heli		
Simultaneous missiles	2 per target, doubles the probability of hit		
Other Missiles:	5V55R, in early units 5V55PM anti-radiation missile (radar homing missile), 6Zh48 nuclear warhead missile. An optional upgrade is		

	48N6. HQ-2 Chinese ARM for FT-2000.		

VARIANTS

RADARS, MISSILES, AND C^2 ARE COMPATIBLE AMONG SYSTEM VARIANTS. FORCES MAY USE A MIX OF EARLIER AND LATER ASSETS. LATER C^2, MISSILES, AND RADARS ARE COMPATIBLE, AND OTHER UPGRADE ASSETS ARE ALSO COMPATIBLE. FOR OTHER SA-10/20 VARIANTS.

HQ-9: CHINESE UPGRADE SA-10B SYSTEM TO NEAR SA-10C, WITH INDIGENOUS TELS AND MISSILES (100-KM) AND HQ-2 75-KM ARM.

FT-2000: ADDS A 100-KM PASSIVE ARM.

NATIONAL WAR COLLEGE PHOTO

COMMAND AND CONTROL

THE 83M6 BDE AUTOMATED C^2 SYSTEM INCLUDES THE 54K6/BAIKAL CP VAN AND THE 64N6 RADAR. THE BAIKAL CONTAINS THE BDE BATTLE MANAGEMENT CENTER AND DIGITAL DATA TRANSMISSION SYSTEM.

WITH THIS C^2 AND OTHER COMPATIBLE NETS, THE SA-10 COMPLEX CAN BE USED AS THE BASE FOR AN AREA INTEGRATED AIR DEFENSE SYSTEM. THE SA-10C DIGITALLY LINKS TO EW ASSETS, AD AIRCRAFT, AD INTEL. SA-10 RADARS SHARE DATA WITH OTHER UNITS IN THE IADS NET. THE SYSTEM CAN BE LINKED DIRECTLY OR THRU IADS WITH OTHER AD MISSILE SYSTEM COMPLEXES, SUCH AS SA-5, EARLIER SA-10, AND SA-11. THE 83M6E CAN PASS DETECTIONS (OF UP TO 60 TARGETS) DIRECTLY TO THE RUBEZH-2M AIR INTERCEPT CONTROL NET.

THE OSNOVA-1E INTEGRATED AIR DEFENSE SYSTEM C^2 VEHICLE CAN PROCESS 120 TARGETS AT A TIME. IT CAN SIMULTANEOUSLY SORT OUT AIRCRAFT ECM (WITH THE AKUP-22 SYSTEM) AND PASS UP TO 80 TARGETS TO BAIKAL-1E OR OTHER AD MISSILE SYSTEMS, AS WELL AS TO RUBEZH-2M.

EVEN IF IADS AND BRIGADE NETS ARE TAKEN OUT OF OPERATION, DUAL-MODE RADARS ON 30N6 PERMIT A FIRE UNIT (BATTERY) TO OPERATE AUTONOMOUSLY.

OTHER ASSETS

FORWARD OBSERVERS ARE DISTRIBUTED THROUGHOUT THE COVERAGE AREA. SA-10C GROUP INCLUDES 85V6E/ORION ELINT. THE NEBO-SV MOBILE COUNTER-STEALTH RADAR SYSTEM OR NEWER NEBO-SVU CAN LINK TO SA-10C, WITH COUNTER-STEALTH DETECTION TO 350 KM.

NOTES

MOST UNITS USE TELS ONLY, NOT SEMI-TRAILER MELS (MOBILE ERECTOR LAUNCHERS). CHINESE UPGRADES SIMILAR TO SA-10C ARE CALLED HQ-10 AND HQ-15.

THE PHASED-ARRAY RADARS FEATURE LOW DETECTION, AND HIGH JAM RESISTANCE.

Russian SAM System SA-12a and SA-12b, SA-23, and S-300V4

SA-12a/GLADIATOR on 9A83 TELAR	SA-12b/GIANT on 9A82 TELAR	Missiles SA-12a canisters on TELAR SA-12b canisters on TELAR Launcher-loader Vehicle SA-24 MANPADS	Typical Combat Load 4 2 2 or 4 2
SYSTEM	**SPECIFICATIONS**	**COMMAND AND CONTROL**	**SPECIFICATIONS**
System Designation	Antey S-300V	Name	9S457-1
Date of Introduction	1982	Function	Command Post tracked vehicle
Proliferation	At least 6 countries	Unit	Brigade, links to up to 4 9S15
Targete	FW, heli*, TBM, CM, ASM, UAV	Targets Detected	200
Primary Components	System (brigade) has 9S457-1 CP vehicle, 12-24 TELARs or (heavy or light) launcher-loaders, and radars Brigade has 2-4 batteries. A battery has 2-4 SA-12a TELARS, 1-2 SA-12b TELARSl.	Targets Tracked	70, 24 assigned at a time All vehicles have link for response/set-up. System can use SA-10c C2/radar assets, including Osnova-1 automated complex.
Launcher Vehicle	9A83, GLADIATOR, SA-12a 9A82, GIANT, SA-12b	**ASSOCIATED RADARS**	**SPECIFICATIONS**
Name	TELAR	Name	9S15MTS
Description	4	NATO Designator	BILL BOARD-A
Crew	MT-T heavy tracked chassis	Function	Early warning, target acquisition
Chassis		Unit Associated With	Brigade
Weight (mt)	48	Mobility	Tracked vehicle-mounted
Dimensions (m)		Detection Range (km)	10-250
Length	12.3 LLVs & 9A85, 14.5 9A82	Range Accuracy (m)	250
Width and Height	3.38 and 3.78	Azimuth Coverage/Sweep:	360°, 6-12 sec
Automotive Performance		Number of Targets tracked	up to70

Engine		Frequency Band	F (3-4GHz), phased array
Cruising Range (km)		ECCM:	Operates in jam 1-2kW/MHz
Max Road Speed (km/h)		Emplace/Displace (min):	5
ARMAMENT	SPECIFICATIONS	Name	
Transporter-Erector-Launcher		NATO Designator	HIGH SCREEN
Reaction Time (sec):	40 alert, 15 launch	Function	Sector TA for TBMs
Time Between Launches (sec):	1.5	Unit Associated With	Brigade
Brigade missile load	96-192 (4-8/TELAR)	Mobility	Tracked vehicle-mounted
Fire on Move	No	Detection Range (km)	200
Emplacement/displacement time (min):	5	Range Accuracy (m)	
Navigation equipment	FCS embedded	Number of Targets tracked	16-20 based on jamming
Onboard fire control	Illum/guidance radar	Frequency Band	INA 3-D phased array
	180, 360 per rotation	Azimuth Coverage (°):	90, 360 with rotation
Missiles		Name	9S32-1
Name	9M83 aka GLAD or GLADIATOR, also Zur-2, SA-12a	NATO Designation	GRILL PAN
Type	Two-Stage, solid-fuel	Function	In FC tracks missile and remote controls TELAR guidance radars
Primary Targets	Dual - aircraft/missiles	Unit	Battery, receives mission from CP
Launch Mode	Vertical launch	Mobilty	Tracked vehicle-mounted
Range (km)	6-80, 30 TBMs	Detection Range (km)	150, 140 automatic
Altitude (km)	0.025 - 25	Targets Tracked Simultaneously	up to 12
Max Speed (km/sec)	3.0 target, 1.7 SAM	Missiles Guided Simultaneously	up to 6
Dimensions	7.9 m x 915 mm diameter	Frequency Band	INA 3-D phased array
Weight (kg)	2,400	Azimuth Coverage (°):	42, 360 with rotation
Guidance	inertial/radar SAH Home on jam	LAUNCHER-LOADER VEHICLES (LLVS)	SPECIFICATIONS
Warhead Type	Focused Frag-HE	Name	9A84 - GIANT, 9A85 - GLADIATOR
Warhead Weight (kg)	150	Function	Primary role is to reload TELARs, but can launch

			with TELARs nearby. Vehicles use same chassis.
Fuze Type	radio cmd or proximity		
Probability of Hit (Ph%):	90 FW, 70 heli		
Simultaneous missiles	2 per target		
Name	9M82 aka GIANT, Zur-1, SA-12b		
Type	Two-Stage, solid-fuel		
Primary Target	TBMs-IRBMs		
Launch Mode	Vertical launch		
Range (km)	13-100 aircraft, 40 TBMs		
Altitude (km)			
Max Altitude	25 TBMs, 30 aircraft		
Min Altitude	2 TBMs, 1.0 aircraft		
Max Speed (km/s)	3.0 target, 2.4 SAM		
Dimensions	9.9 m x 1215 mm diameter		
Weight (kg)	4,600		
Guidance	Inertial, radar semi-active homing (SAH), home on jam		
Warhead Type	Focused Frag-HE		
Warhead Weight (kg)	150		
Fuze Type	radio command or proximity		
Probability of Hit (Ph%)	80 FW, 70 TBM		
Simultaneous missiles	2 per target		

OTHER ASSETS

THE SA-12 SYSTEM DIGITALLY LINKS TO THE IADS (E.G., EW ASSETS, AIRCRAFT, INTEL, AND OTHER SAM UNITS. RADARS SHARE DATA WITH OTHER UNITS IN THE IADS NET. OTHER ASSETS ARE FOS AND ELINT (ORION). THE NEBO-SV/BOX SPRING OR NEBO-SVU COUNTER-STEALTH RADAR CAN BE USED.

VARIANTS

SA-23/S-300VM: INTERIM UPGRADE OF 5 UNITS AROUND MOSCOW, WITH 9M82M /3M MISSILES. 9M82M RANGES 200 KM, IS IMMUNE TO ECM, AND CAN INTERCEPT BALLISTIC MISSILES AT 4,500 M/S (MRBMS FROM 2,500 KM). THE 9M83M RANGES TO 110 KM. EXPORT NAME IS ANTEY-2500.

S-300V4: A DEEPER MODERNIZATION BEYOND THE ABOVE INTERIM UPGRADE. NEW MISSILES ON THE 12A AND 12B LAUNCHERS RANGE 120+ AND 300+ KM, RESPECTIVELY. THE SYSTEM MAY LINK TO OTHER S-300 SYSTEMS FOR RANGE OF 300 ± KM AND BETTER INTEGRATION. FIELDING IS UNDERWAY.

SAMODERZHETS: SOME SOURCES SAY THIS IS THE PROGRAM FOR INTEGRATING/UPGRADING ALL S-300 SYSTEMS. IT MAY INCLUDE A MISSILE TO BRING OLDER TELS UP TO MODERN STANDARDS. RECENT UPGRADES MAY HAVE OVERTAKEN IT.

NOTES

THE SYSTEM GENERALLY DOES NOT TARGET HELICOPTERS, BUT WILL FOR SELF-DEFENSE. THE PHASED-ARRAY RADARS FEATURE LOW DETECTION, AND HIGH JAM RESISTANCE

Russian SAM System SA-20a/GARGOYLE

		Missiles	Typical Combat Load
		TEL and trailer launcher	4
		48N6E	4
		5V55PM/HQ-2 ARM	4/battery
		SA-18 MANPADS	2
SYSTEM	**SPECIFICATIONS**	**ASSOCIATED RADARS**	**SPECIFICATIONS**
Alternative Designation	S-300PMU1 Previously, system was called SA-10d.	Radar	
Date of Introduction	1990-93	Name	64N6E
Proliferation	At least 6 countries	NATO Designator	BIG BIRD E
Target	FW, heli, TBM, CM, ASM, UAV	Function	Early warning, target acquisition
Primary Components	See SA-10c, above. Note the updated equipment in the Comparison Table.	Unit	Grouping (brigade) level, supports 3-8 90Zh6E complexes (bns - 12-36 launchers)
ARMAMENT	**SPECIFICATIONS**	Mobility	MAZ-7910 8x8 van
TEL and Trailer Launcher		Detection Range (km)	300
Name	5P85SE TEL only, ground units 5P85TE trailer launcher for site defense See SA-10c, above. Note other updated equipment in the Comparison Table.	Number of Targets Detected	up to 200
Automotive Performance	For 5P85SE TEL on MAZ-5910, see SA-10b, except the following.	Targets Simultaneous Lock and Track	100
Cruising Range (km)		Frequency Band	F, 3-D phased array
Road/Dirt road Speed (km/h):		Azimuth Coverage (°):	360 with rotation
The 5P85TE trailer-launcher is normally towed by a KRAZ-260B 6x6 truck.		Emplace/Displace Time (m):	5
Missile		Name	30N6E1

Name	48N6/ 48N6E export	NATO Designator	TOMBSTONE
Type	Single-Stage, solid-fuel	Function	Dual (TA/FC) and battery CP
Launch Mode	Vertical launch	Unit	Battery (SAM system), for 3 launchers
Launch Range (km)	5-150	Mobility	MAZ-7930 8x8 van
Max Range	40	Detection Range (km)	300
Targets .5-1 km altitude	28-38	Guidance Range (km)	200 auto-track
Altitude (m)	6-27,000 0 with blast radius	Targets Engaged Simultaneously	up to 6
Speed (m/sec)		Missiles Guided Simultaneously	up to 12
Max Target	2,800	Frequency Band	I/J, 3-D phased array
Max SAM	2,100, 25g turn	Name	96L6E
Dimensions	7.5 m length 519 mm diameter	Function	All-altitude target acquisition and processing center - replaces CLAM SHELL
Weight in Canister (kg)	2,580	Unit Associated With	Battalion (2-6 btry)
Guidance	Track-Via-Missile, missile radar homing, home on jam	Mobility	MAZ-7930 8x8 van
Warhead Type	Frag-HE	Operation	Up to 5 remote workstations
Warhead Weight (kg)	145, twice the previous KE from warhead fragments	Emplacement/ Displacement Time (min)	5 for truck, 30 towed, 120 for mast mounted
Fuze Type	Radio command	Range (km)	300, more with 40V6M tower
Probability of Hit (Ph%):	90 FW/heli, 70 others.	Targets Tracked Simultaneously	up to 100
Simultaneous missiles	up to 2 per target	Frequency Band	Centimeter L-band, 3-D phased-array
Other Missiles	5V55R, original missile. First export missile was 5V55RUD. Optional export upgrade (see above) is 48N6E. 5V55PM anti-radiation missile, 6Zh48 nuclear warhead missile. HQ-2 Chinese ARM for FT-2000.	Azimuth Coverage (°):	120, 360 with rotation

		The antenna can be mounted on a 40V6M tower with same height as CLAM SHELL	
		96L6E TA radar/processing center	
		Nebo-SVU/1L119	VHF counter-stealth radar is in SA-20/SA-21 brigades and IADS. Its range is up to 350 km, with 100 km altitude. Deployment time is 25 minutes. Coordinating with 9L96E and TOMBSTONE, and using triangulation, the radar can digitally acquire stealth and other LPI aircraft, and cue the IADS. Azimuth Coverage (°): 120 est. Some forces (Tier 3) may still use CLAM SHELL with SA-20a. Others employ new, indigenous or legacy target acquisition radars.

VARIANTS

SA-10C: THIS IS A COMMONLY EXPORTED VERSION OF S-300. OPTIONAL UPGRADES OF C^2, MISSILES AND RADARS ARE AVAILABLE.

SA-20A/SA-10D/S-300PMU1: THIS SYSTEM UPGRADE WAS DESIGNED FOR 48N6/48N6E MISSILES. MOST EQUIPMENT IS COMPATIBLE WITH

SA-10C. CHINA HAS ACQUIRED SA-20A, AND IS UPGRADING EARLIER LAUNCHERS TO THIS CAPABILITY. IT IS ALSO TRYING TO UPGRADE TO SA-20B.

SA-20B/FAVORIT: RUSSIAN IMPROVED SYSTEM WITH UPGRADE TO 200-KM 48N6E2 MISSILE AS WELL AS 9M96 SERIES "SMALL MISSILE."

OTHER ASSETS

THE SA-20A DIGITALLY LINKS TO THE IADS, INCLUDING AD AIRCRAFT, AND OTHER SAM UNITS IN THE IADS NET. FORWARD OBSERVERS ARE DEPLOYED THROUGHOUT THE COVERAGE AREA. EACH BRIGADE ALSO HAS AN 85V6E/ORION ELINT. FOR OSNOVA-1E IADS C2 VEHICLE AND 83M6E AUTOMATED C2 SYSTEM.

NOTES

THE "BIG MISSILE" COULD BE REPLACED WITH 4 N6M/4 N6E2 (NEXT PAGE). THE 30N6E1 MAY NOT BE ABLE TO USE THE FULL 200 KM MISSILE RANGE AGAINST SOME SMALLER AERIAL TARGETS; BUT IT CAN AGAINST LARGER TARGETS. THE PHASED-ARRAY RADARS FEATURE LOW DETECTION, AND HIGH JAM RESISTANCE.

Russian Universal SAM System SA-20b/Favorit

Favorit with 3 x 48N6-type missiles and a 9M96-type canister set (4 x "small missiles")		Missiles	Typical Combat Load
		TEL and trailer launcher	7
			(4-16)
		48N6E/ 48N6E2/ARM	3
		("big missile")	(1-4)
			4
		9M96E2	
		("small missile")	
		Near Term	(16-0)
		SA-18S MANPADS	2
SYSTEM	**SPECIFICATIONS**	Name	9M96E2/9M96M "small missile"
Alternative Designation	S-300PMU2/ GARGOYLE or GARGOYLE B. The system has several stages of upgrade. Favorit is Russian forces and export (-E2).	Type	Single-Stage, solid-fuel "Hittile" – (agile "hyper-maneuver" with small HE)
Date of Introduction	1996	Launch Mode	Vertical launch
Proliferation	Fielded in 6 countries	Launch Range	1-120
Target	FW, MRBM, heli, CM, ASM, UAV, and artillery rockets	Altitude	30,800 5, 0 with blast radius
Primary Components	Group (equal to brigade) Nomenclature for system components ends with –E2 (e.g., 83M6E2 battle management complex, versus -E for SA-10C, -E1 for SA-20A). The 83M6E2 has improved ABM ability. A battery (firing unit) has 3 launchers.	Speed (m/sec)	
ARMAMENT	**SPECIFICATIONS**	Max Target	4,800
TEL and Trailer Launcher		Max SAM	1,800 and 20+ g turns with thrust vectoring
Name	5P85SE2 TEL and 5P85TE2 TL	Length (m)	5.65
Chassis	MAZ-5910 chassis for TEL KRAZ-260 tractor for TL (MEL)	Diameter (mm)	240

Missiles per Launcher	4 for 48N6E2 16 for 9M96E/E2, in 4 pods	Weight (kg)	420, 2,700 for container of 4
Automotive Performance	5P85SE2 TEL on MAZ-5910	Guidance	Track-via-missile, active radar homing, also ARM and home on jam
Missile		Warhead Type	Controlled frag-HE
Name	N6M/4 N6E2 "big missile"	Warhead Weight (kg)	24
Type	Single-Stage, solid-fuel	Fuze Type	"Smart" prox, frag shaping
Launch Mode	Vertical launch	Probability of Hit (Ph%):	90 FW, and heli 80 others
Launch Range (km)	5-200	Simultaneous missiles	up to 2 per target
Max Range TBMs	40	Other Missiles	Previous 48NE missile (150 km) can be used. 9M96E ranges 40 km. Domestic 9M96M ranges 120 km.
Targets .5-1 km high	28-38	ASSOCIATED RADARS	SPECIFICATIONS
Altitude (m):		Name	64N6E2
Max. Altitude	27,000	NATO Designator	BIG BIRD E See 64N6E at SA-20a
Min. Altitude	6, 0 with blast radius	Detection Range (km)	400
Speed (m/sec):		Name	30N6E2
Max Target	2,800	NATO Designation	TOMBSTONE See 30N6E1 at SA-20a
Max SAM:	2,100, 25g turn	Guidance Range (km)	200
Dimensions:		Name	96L6E Target acquisition radar and battle mgt center at battalion.
Length (m)	7.5	Nebo-SVU	VHF (counter-stealth) TA radar is located at brigade level.
Diameter (mm)	519	Other Assets	The SA-20b digitally links to the IADS, and shares data with other units in the net. Forward observers are deployed in the coverage area. Each brigade also has an 85V6E/Orion ELINT system. For discussion of the Osnova-1E IADS C2

			vehicle, Baikal-1E, Rubezh-2M, 83M6E2 automated C2 system and other assets.
Weight in Canister	2580		
Guidance:	Track-Via-Missile, missile radar homing, home on jam		
Warhead Type	Frag-HE		
Warhead Weight (kg):	180		
Fuze Type	radio command		
Probability of Hit (Ph%):	90 FW, 70 for high-speed missiles and TBMs, 80 others		
Simultaneous missiles	up to 2 per target		

VARIANTS

THIS SYSTEM (ORIGINALLY CALLED SA-10E) WAS DESIGNED AGAINST BALLISTIC MISSILES AND LOW MANEUVERABLE SYSTEMS SUCH AS UAVS, ARTILLERY ROCKETS (LIKE MLRS), AND AIR-LAUNCH MISSILES. IMPROVED FROM SA-10D, IT IS COMPATIBLE WITH MOST OF THE EQUIPMENT FOR SA-10B, C, AND D (SA-10A). STRATEGIC ABM UNITS HAVE ONLY BIG MISSILES.

CHINA IS ORDERING SA-20B AND UPGRADING OTHER LAUNCHERS TO SA-20B CAPABILITY.

S-400/SA-21A: THE SYSTEM WAS FIELDED IN 2007 WITH RUSSIAN VEHICLES. IT SHARES 9M96-SERIES MISSILES WITH SA-20B. THE SYSTEM WILL USE NEW, MORE POWERFUL RADARS, INCLUDING PROTIVNIK-GE AND NEBO-M, AND WILL INTERLINK WITH SA-20 LAUNCHERS.

SA-21B/SAMODERZHETS: NEAR-TERM UPGRADE FOR ALL S-300/S-400 SYSTEMS. THE PROGRAM HAS IMPROVED INTEGRATION AND MISSILE.

NOTES

ABOVE PHOTO SHOWS FAVORIT WITH 1 CANISTER OF 9M96E2 MISSILES. BY SHIFTING FROM 1 SMALL-MISSILE POD PER LAUNCHER TO 2-4, THE NUMBER OF MISSILES PER LAUNCHER CAN INCREASE FROM 7 TO 10, 13, OR 16. THE PHASED-ARRAY RADARS FEATURE LOW DETECTION, AND HIGH JAM RESISTANCE.

Russian SAM System SA-21a/GROWLER/S-400

		Missiles	Typical Combat Load
		TEL and trailer launcher	7
		40N6	
		(with 1x 9M96E2 canister)	3
			Near Term 2
		9M96M/E2	4
		("small missile")	Near Term 8
		SA-18S MANPADS	2

SYSTEM	SPECIFICATIONS	ARMAMENT	SPECIFICATIONS
Alternative Designations	Triumf, Triumph as a translation	Name	91N6E2 BIG BIRD E It is an improved SA-20B EW/TA radar, with an AD intel processing center on a MAZ-7930 towed van trailer, co-located with the brigade CP/battle management center. See 64N6E at SA-20a.
Date of Introduction	2007	Detection Range (km)	At least 400
Proliferation	Fielded in 1 country	Azimuth Coverage (°):	360
Target	FW, IRBMs to 3,500 km, heli, CM, ASM, UAV, and artillery rocket	Name	Nebo-SVU/1L119 This VHF target acquisition radar is at Brigade level. The first search priority is stealth aerial systems. Because of limited sector coverage, it is likely that up to 4 will be used.
Primary Components	Group/brigade 2-8 bns and 91N6 E2. Each bn has 6-12 trailer launchers (TLs, aka mobile erector-launchers or MELs) , 55K6E 8x8 van, 5T58-2 SAM transporter, 22T6-2 loading crane, and radars. Battery (firing unit) has 3 TLs.	Name	96L6E TA radar/battle mgt center is initially at bn until 59N6 replaces it.
ARMAMENT	**SPECIFICATIONS**	Name	59N6/Protivnik-GE

Trailer Launcher (TL) or MEL		Function	All-altitude target acquisition and Unit Associated With: Battalion (2-6 btry)
Name	5P85TE2	Unit Associated With	Trailer with KrAZ-260 tractor
Tractor	BAZ-64022 6x6 tractor	Mobility	Trailer with KrAZ-260 tractor
Missiles per Launcher	3 x 40N6 4 x 9M96E2 (current likely mix)	Operation	Digital links to battery, battalion, and brigade/IADS processing center
Automotive Performance	For 5P85TE2 TL	Emplacement/Displacement time (min):	15
Cruising Range (km)	800 (est)	Range (km):	400
Road/ Dirt Road Speed (km/h)	60/30 (est)	Targets Tracked Simultaneously	up to 150
Missile		Frequency Band	AESA Decimetric L-band, 3-D phased-array
Name	40N6 “big missile”	Azimuth Coverage (°):	120, 360 with rotation
Type	Solid-fuel	Name	92N2E
Launch Mode	Vertical launch	NATO Designation	GRAVESTONE
Launch Range (km)	5-400	Function	Dual (TA/FC) radar vehicle and CP
Max Range TBMs	40	Unit	Battery (SAM system), for 3 launchers
Targets	28-48	Mobility	MAZ-7930 8x8 van
Altitude (m)		Detection and Guidance (km)	400 auto-track
Max Altitude	0,000+	Targets Engaged Simultaneously	up to 6 (est)
Min Altitude	5, 0 with blast radius	Missiles Guided Simultaneously	up to 12 (est)
Speed (m/sec)		Frequency Band	I/J, 3-D phased array
Max Target	5,000	Azimuth Coverage (°):	120, 360 with rotation
Max SAM	4,800		
Dimensions			
Length (m)	7.5		
Diameter (mm)	519		

Weight (kg)	2,000, 2,800 in canister		
Guidance	Track-Via-Missile, missile active radar homing, home on jam		
Warhead Type	Frag-HE		
Warhead Weight (kg)	180+		
Fuze Type	Radio command		
Probability of Hit (Ph%):	90 FW. 80 heli		
Simultaneous missiles	up to 2 per target (doubles probability of hit)		
Name	9M96E2/9M96M "small missile."A canister of 4 can fit on the SA-21 launcher in place of a big missile. It is possible that most launchers in most batteries (by the Near Term) will have 2 canisters of small missiles (8 total).		
Other Missiles	The system can also launch older missiles for SA-10 and SA-20 systems. There are reports of a 48N6DM missile, which offers longer range than the 48N6. This may have been an interim missile for use until 40N6 was fielded.		

OTHER ASSETS

THE SA-21A DIGITALLY LINKS TO THE IADS, AND SHARES DATA WITH OTHER UNITS IN THE NET. FOR DISCUSSION OF OSNOVA-1E IADS C2 VEHICLE, BAIKAL-1E, RUBEZH-2M, 83M6E2 AUTOMATED C2 SYSTEM , AND OTHER ASSETS. FORWARD OBSERVERS ARE DEPLOYED THROUGHOUT THE COVERAGE AREA. EACH BRIGADE ALSO HAS AN 85V6E/ORION ELINT.

AN IADS DIGITALLY INTERFACES THE NEBO-SVU COUNTER-STEALTH RADAR SYSTEM, PROTIVNIK, AND 96L6E, TO OVERLAY DETECTIONS.

A RECENT COUNTER-STEALTH RADAR SYSTEM IS THE NEBO-M MOBILE MULTI-BAND SYSTEM, WITH THREE VEHICLES. RLM-D HAS L-BAND RADAR. RLM-S HAS X-BAND. RLM-M HAS A VHF RADAR, SIMILAR TO NEBO-SVU. THE SYSTEM IS SPECIFICALLY DESIGNED AGAINST STEALTH AIRCRAFT AND F-35. A RECENT RUSSIAN CONTRACT CALLS FOR 100 SYSTEMS TO REPLACE NEBO-SVU IN SA-20B/S-400 UNITS, AND WILL BE INCLUDED IN SA-23/S-300V4 UNITS.

VARIANTS

THE S-400 SERIES USES A NEW ARRAY OF TRUCKS, TRACTORS, AND TRAILERS. DUE TO S-400 PRODUCTION DELAYS, THE SA-20 SERIES WAS CONFUSED WITH IT. MANY S-400 UPGRADES CAN BE APPLIED TO SA-10, SA-12, AND SA-23. CHINA IS ORDERING SA-20B AND UPGRADING OTHER LAUNCHERS TO SA-20B CAPABILITY.

S-400/SA-21A: THE SYSTEM WAS FIELDED IN 2007 WITH RUSSIAN VEHICLES. EARLY UNITS ARE STRATEGIC AND USE ONLY 40N6 400-KM BIG MISSILES. MOST LAUNCHERS CAN ALSO MOUNT CANISTERS OF 9M96 SERIES SMALL MISSILES.

SA-21B/S-400M/SAMODERZHETS: UNLIKE THE OTHER. SAMS, SA-21A'S 40N6 WILL RANGE 400 KM.

NOTES

THERE ARE ALSO REPORTS OF A SYSTEM IN DEVELOPMENT CALLED S-500, WITH LONGER RANGE AND A DESIGN VELOCITY OF 10,000 M/S. NO DETAILS ARE AVAILABLE. THE PHASED-ARRAY RADARS FEATURE LOW DETECTION, AND HIGH JAM RESISTANCE.

Made in the USA
Monee, IL
15 October 2022